Late Palaeozoic and Early Mesozoic
Circum-Pacific Events and
Their Global Correlation

WORLD AND REGIONAL GEOLOGY SERIES

Series Editors: M. R. A. Thomson and J. A. Reinemund

This series comprises monographs and reference books on studies of world and regional geology. Topics will include comprehensive studies of key geological regions and the results of recent IGCP projects.

1. *Geological Evolution of Antarctica* – M. R. A. Thomson, J. A. Crame & J. W. Thomson (eds.)

2. *Permo-Triassic Events in the Eastern Tethys* – W. C. Sweet, Yang Zunyi, J. M. Dickins & Yin Hongfu (eds.)

3. *The Jurassic of the Circum-Pacific* – G. E. G. Westermann (ed.)

4. *Global Geological Record of Lake Basins, Vol. 1* – E. Gierlowski-Kordesch & K. Kelts (eds.)

5. *Earth's Glacial Record* – M. Deynoux, J. M. G. Miller, E. W. Domack, N. Eyles, I. Fairchild & G. M. Young (eds.)

6. *Tertiary Basins of Spain: The Stratigraphic Record of Crustal Kinematics* – P. F. Friend & C. J. Dabrio (eds.)

7. *The Quaternary History of Scandinavia* – J. Donner

8. *The Tectonic Evolution of Asia* – A. Yin and M. Harrison (eds.)

9. *The Pleistocene Boundary and the Beginning of the Quaternary* – J. A. Van Couvering (ed.)

10. *Late Palaeozoic and Early Mesozoic Circum-Pacific Events and Their Global Correlation* – J. M. Dickins, Yang Zunyi, Yin Hongfu, S. G. Lucas & S. K. Acharyya (eds.)

11. *Paleocommunities: A case study from the Silurian and Lower Devonian* – A. J. Boucot & J. D. Lawson (eds.)

Late Palaeozoic and
Early Mesozoic
Circum-Pacific Events
and Their
Global Correlation

Edited by

J. M. DICKINS
Australian Geological Survey Organization

Editorial Board:

Yang Zunyi
 China University of Geosciences
Yin Hongfu
 China University of Geosciences
S. G. Lucas
 New Mexico Museum of Natural History
S. K. Acharyya
 Geological Survey of India

CAMBRIDGE
UNIVERSITY PRESS

PUBLISHED BY THE PRESS SYNDICATE OF THE UNIVERSITY OF CAMBRIDGE
The Pitt Building, Trumpington Street, Cambridge CB2 1RP, United Kingdom

CAMBRIDGE UNIVERSITY PRESS
The Edinburgh Building, Cambridge CB2 2RU, United Kingdom
40 West 20th Street, New York NY 10011-4211, USA
10 Stamford Road, Oakleigh, Melbourne 3166, Australia

First published 1997

Printed in the United States of America

Typeset in Times Roman

Library of Congress Cataloging-in-Publication Data
Late Palaeozoic and early Mesozoic circum-Pacific events and their
  global correlation / edited by J. M. Dickins . . . [et al.].
      p.   cm. – (World and regional geology series : 10)
    Includes index.
    ISBN 0-521-47175-3
    1. Geology, Stratigraphic – Carboniferous.   2. Geology,
  Stratigraphic – Triassic.   3. Geology – Pacific Area.
  4. Stratigraphic correlation.   5. Paleontology – Carboniferous.
  6. Paleontology – Triassic.   7. Paleontology – Pacific Area.
  I. Dickins, J. M. (James MacGregor)   II. Series: World and
  regional geology : 10
  QE671.L28   1997
  551.7′5′09 – dc20   96-21449
                                                                       CIP

A catalogue record for this book is available from the British Library

ISBN 0 521 47175 3 hardback

Contents

Contributors

Christoph Breitkreuz
Institut für Geologie und Paläontologie, Technische Universität Berlin, Ernst-Reuter Platz 1, 10587 Berlin, Germany

Ivan V. Burij
Far East Geological Institute, Far Eastern Branch, Russian Academy of Sciences, Stoletiya Vladivostoka Prospekt, Vladivostok 690022, Russia

Galina I. Buryi
Far East Geological Institute, Far Eastern Branch, Russian Academy of Sciences, Stoletiya Vladivostoka Prospekt, Vladivostok 690022, Russia

J. D. Campbell
Geology Department, University of Otago, P.O. Box 56, Dunedin, New Zealand

G. Cassinis
Dipartimento di Scienze della Terra, Università degli Studi, Via Abbiategrasso 209, I-27100 Pavia, Italy

Maria Alessandra Conti
Dipartmento di Scienze della Terra, Università degli Studi "La Sapienza," Rome, Italy

J. M. Dickins
Australian Geological Survey Organization, P.O. Box 378, Canberra, A.C.T. 2601, Australia

Yoichi Ezaki
Department of Geosciences, Faculty of Science, Osaka City University, Sugimoto 3-3-138, Sumiyoshi-ku, Osaka 558, Japan

C. R. González
Institute de Paleontologia, Fundacion Miguel Lillo, Miguel Lillo 251, Tucaman, Argentina

Carlos González-León
ERNO, Instituto de Geologia, Universidad Nacional Autónoma de México, Apartado Postal 1039, Hermosillo, Sonora 83000, México

Vitaliy G. Gvozdev
Far East Geological Institute, Far Eastern Branch, Russian Academy of Sciences, Stoletiya Vladivostoka Prospekt, Vladivostok 690022, Russia

Vladimir V. Ivanov
Far East Geological Institute, Far Eastern Branch, Russian Academy of Sciences, Stoletiya Vladivostoka Prospekt, Vladivostok 690022, Russia

Lev N. Khetchikov
Far East Geological Institute, Far Eastern Branch, Russian Academy of Sciences, Stoletiya Vladivostoka Prospekt, Vladivostok 690022, Russia

Vadim G. Khomich
Far East Geological Institute, Far Eastern Branch, Russian Academy of Sciences, Stoletiya Vladivostoka Prospekt, Vladivostok 690022, Russia

Galina V. Kotlyar
All-Russian Geological Research Institute (VSEGEI), Sredny Prospekt 74, St. Petersburg 199026, Russia

Kiyoko Kuwahara
Department of Geosciences, Faculty of Science, Osaka City University, Sugimoto 3-3-138, Sumiyoshi-ku, Osaka 558, Japan

Oscar R. López-Gamundí
Frontier Exploration Department, Texaco Inc., P.O. Box 430, Bellaire, Texas 77402-2324, USA

Spencer G. Lucas
New Mexico Museum of Natural History, 1801 Mountain Road NW, Albuquerque, New Mexico, 87104-1375, USA

Christopher A. McRoberts
Department of Geological Sciences, State University of New York, Binghamton, New York, 13903-6006, USA.

Nino Mariotti
Dipartimento di Scienze della Terra, Università degli Studi "La Sapienza," Rome, Italy

Umberto Nicosia
Dipartimento di Scienze della Terra, Università degli Studi "La Sapienza," Rome, Italy

Alexander Oleinikov
Far East Geological Institute, Far Eastern Branch, Russian Academy of Sciences, Stoletiya Vladivostoka Prospekt, Vladivostok 690022, Russia

Vera A. Pakhomova
Far East Geological Institute, Far Eastern Branch, Russian Academy of Sciences, Stoletiya Vladivostoka Prospekt, Vladivostok 690022, Russia

Eugene S. Panasenko
Far East Geological Institute, Far Eastern Branch, Russian Academy of Sciences, Stoletiya Vladivostoka Prospekt, Vladivostok 690022, Russia

Rachel K. Paull
Geosciences Department, University of Wisconsin–Milwaukee, Milwaukee, Wisconsin, 53201, USA

Richard A. Paull
Geosciences Department, University of Wisconsin–Milwaukee, Milwaukee, Wisconsin, 53201, USA

C. R. Perotti
Dipartimento di Scienze della Terra, Università degli Studi, Via Abbiategrasso 209, I-27100 Pavia, Italy

Phan Cu Tien
Research Institute of Geology and Mineral Resources, Thanxuan, Dongda, Hanoi, Vietnam

Paola Pittau
Dipartimento di Scienze della Terra, Università degli Studi, Cagliari, Italy

Tatiana A. Punina
Far East Geological Institute, Far Eastern Branch, Russian Academy of Sciences, Stoletiya Vladivostoka Prospekt, Vladivostok 690022, Russia

Vladimir V. Ratkin
Far East Geological Institute, Far Eastern Branch, Russian Academy of Sciences, Stoletiya Vladivostoka Prospekt, Vladivostok 690022, Russia

Jaime Roldán-Quintana
ERNO, Instituto de Geologia, Universidad Nacional Autónoma de México, Apartado Postal 1039, Hermosillo, Sonora 83000, México

Valeria S. Rudenko
Far East Geological Institute, Far Eastern Branch, Russian Academy of Sciences, Stoletiya Vladivostoka Prospekt, Vladivostok 690022, Russia

Sergey V. Rybalka
Far East Geological Institute, Far Eastern Branch, Russian Academy of Sciences, Stoletiya Vladivostoka Prospekt, Vladivostok 690022, Russia

Vladimir G. Sakhno
Far East Geological Institute, Far Eastern Branch, Russian Academy of Sciences, Stoletiya Vladivostoka Prospekt, Vladivostok 690022, Russia

C. Venturini
Dipartimento di Scienze della Terra, Università degli Studi, Via S. Maria 53, I-56100 Pisa, Italy

Alexander A. Vrzhosek
Far East Geological Institute, Far Eastern Branch, Russian Academy of Sciences, Stoletiya Vladivostoka Prospekt, Vladivostok 690022, Russia

Wu Shunbao
China University of Geosciences, Beijing 100083, China

Yang Jiduan
Institute of Geology, Chinese Academy of Geological Sciences, Beijing 100037, China

Yang Zunyi
China University of Geosciences, Beijing 100083, China

Yin Hongfu
Department of Geology, China University of Geosciences, Wuhan 430074, China

Zhan Lipei
Institute of Geology, Chinese Academy of Geological Sciences, Beijing 100037, China

Yuri D. Zakharov
Far East Geological Institute, Far Eastern Branch, Russian Academy of Sciences, Stoletiya Vladivostoka Prospekt, Vladivostok 690022, Russia

Vera G. Zimina
Far East Geological Institute, Far Eastern Branch, Russian Academy of Sciences, Stoletiya Vladivostoka Prospekt, Vladivostok 690022, Russia

Preface

Project 272 of the International Geological Correlation Programme (IGCP-272), "Late Palaeozoic and Early Mesozoic Circum-Pacific Events and Their Global Correlation," is a successor project to IGCP-203, "Permo-Triassio Events of East Tethys and Their Intercontinental Correlation," and has been succeeded by IGCP-359, on Tethyan, circum-Pacific, and marginal Gondwanan late Palaeozoic and early Mesozoic correlations (biota, facies, formations, geochemistry, and events). The 25 reports that make up this volume represent the final report of IGCP-272. During the five years of the project (1988 to 1992) the participants met and carried out field work in Brazil (São Paulo), Australia and New Zealand (Newcastle, Hobart, and Dunedin), Argentina (Buenos Aires), the United States (Washington and Chicago), Russia (Perm), Japan (Kyoto), and the Russian Far East (Vladivostok).

The events reported here were diverse in nature and timing, and considerable attention has been devoted to improving the time scale. A special feature has been to emphasize the time relationships of different events that otherwise might seem to have been unrelated. The first chapter in this volume, by the co-leaders of IGCP-272, provides an overview that describes a framework for location of those events that reflects the recognized worldwide standard stratigraphical and time scales for systems, series, and stages and emphasizes their real, physical basis. The work of this project has shown the importance of a major geological and biological event reflected in Upper Permian strata between the Midian and the Dzhulfian stages that had already been recognized in the Vietnamese, Chinese, and Japanese stratigraphic scales and has elucidated its geologic and biologic character. This volume includes reports from the Russian Far East that contain material either not previously available or not readily available in English.

The authors herein have been encouraged not necessarily to confine their conclusions to comments on the existing theory but to bear in mind that some of that theory will be superseded in the future and to look toward possible new theoretical approaches. Thus, not all of these authors are in agreement regarding either their theoretical approaches or their conclusions, and, where pertinent, this is mentioned in the overview given in Chapter 1. We regard this as a healthy sign of a vigorous and progressive science.

The co-leaders of the project, J. M. Dickins, Yang Zunyi, and Yin Hongfu, together with S. G. Lucas and S. K. Acharyya, were appointed to be the editorial board for this volume at the Kyoto meeting of Project 272. All chapters have been reviewed by at least two members of the editorial board and, in cases where it seemed warranted, have been examined by additional reviewers. J. M. Dickins has acted as chief editor, as decided at the Kyoto meeting, and has coordinated the interactions among the authors, the editorial board, and the publisher.

The Australian Geological Survey Organization (previously the Bureau of Mineral Resources) has provided extensive support. Other participants in the Chinese working group for IGCP Project 272 are as follows: in the marine group, Yin Hongfu, Li Zishun, Yang Fengqing, Xu Guirong, Zhang Kexing, Ding Meihua, and Bi Xianmei, and in the continental group, Qu Lifan, Zhou Huiqin, Cheng Zhengwu, Zhou Tong Shun, Hou Jingpeng, Li Peixian, Sun Suying, and Wu Xiaozu.

J. M. DICKINS

1 Major global change: framework for the modern world

J. M. DICKINS, YANG ZUNYI, and YIN HONGFU

In so many features the Permian and Triassic periods differed from the present that if we could be transported back to that distant past, surely we would think that we were on another planet. The plants and animals would be largely unrecognizable – no mammals, and none of our modern trees or grasses. Although the nature and dispositions of the seas, lands, and mountains during that time are subject to different interpretations and conflicting views, who can doubt that the world's geography would be unrecognizable? The climate would have varied widely, from very cold at the beginning of the Permian (perhaps not unlike the glacial climate of the world a few thousand years ago) to a hot, dry climate during the Triassic, like nothing we see at present, followed by a warm climate continuing into the Jurassic.

Major changes in faunas were associated with the beginning of the Jurassic period, as reviewed by Boucot (1990), Dickins (1993a), and Hallam (1981). Boucot stated that "the modern marine . . . benthic fauna and communities really begin with the Jurassic." In the latter part of the Permian and during the Triassic the earth was wracked by vast earthquakes, and great mountain chains were formed. There was magmatic-volcanic activity on an enormous scale. Some of those events are described in this volume, and some of the most interesting problems for future investigation are considered.

Carboniferous and Lower Permian

Neither the Carboniferous events associated with the Hercynian Orogeny (or Hercynian orogenic phase) nor the major regressions and progressive continentality shown by the Upper Carboniferous sequences (the Hercynian Gap) are covered in this volume, except in the area of South America, where López-Gamundí and Breitkreuz, in Chapter 2, describe the widespread hiatus between the Upper Devonian and the Carboniferous strata. Their information supplements prior studies of the changes associated with the massive adaptive radiation of faunas at the beginning of the Carboniferous (Boucot, 1990).

The events marking the end of the Carboniferous and the beginning of the Permian have already been described (Dickins, 1988, 1989, 1991, 1992a,b). An extensional episode has been reported by Visscher (1993) for southern Africa and by Yang,

Liang, and Nie (1993) for Tibet and western China. Now, in Chapter 2, a significant episode of extension and accompanying basaltic-volcanic activity is described by López-Gamundí and Breitkreuz for southern South America. Parallel events, with tensional faulting and basaltic volcanism, seem to have occurred in the western and northern parts of North America (Beauchamp, Harrison, and Henderson, 1989; Miller and Harewood, 1990). Lower Permian rocks in Mexico, mostly of a platform type, are described in Chapter 3, by González-León, Lucas, and Roldán-Quintana, and are distinguished from Upper Permian and Upper Triassic deposits in an area where thrusting has been shown to have occurred from the mid-Permian to the middle Triassic. In this volume it is the mid-Permian events and those of later times that are especially well described.

Mid-Permian (Kungurian–Ufimian)

"Mid-Permian" is used herein to refer to the sequences associated with the boundary between the Kungurian and the Ufimian stages of the twofold Russian subdivisions of the Permian. Traditionally the Kungurian is placed in the Lower Permian, and the Ufimian in the Upper Permian. Some possible modifications of that placement are discussed later in relation to Kotlyar's Chapter 4 in this volume.

In Chapter 5, Cassinis, Perotti, and Venturini describe the tectonic and sedimentary events during the mid-Permian in the European Southern Alps. They present a mechanism for the development of extensional (pull-apart) basins that they ascribe to a transcurrent regime associated with the late Hercynian interval. However that interval may be named, certainly in the European Southern Alps it represents a remarkable tectonic episode (or perhaps episodes) in which the early Permian marine basins were entirely eliminated and massive uplifts resulted in widespread unconformities, with the subsequent sedimentation beginning with thick conglomerates lying on well-developed weathering profiles (Cassinis, 1988; Italian IGCP-203 group, 1988). They believe that that tectonic activity had long-lasting effects and subsequently influenced Alpine orogenic development. In Chapter 6, Conti and co-workers examine the tetrapod footprints and palynology of that interval and the overlying sequences up to the end of the Permian. They draw important conclusions about the correlations for those sequences based on combined studies of tetrapod footprints and the palynology in the European Southern Alps. Their work indicates that the sequences are quite complicated, reflecting large-scale movements not only in the mid-Permian but also in later Permian times, perhaps especially associated with the Dzhulfian and Zechstein stages. Their work raises further questions about many of the traditional terms and concepts associated with the nonmarine geology of western Europe.

Kotlyar's correlation in Chapter 4 offers the possibility of a better understanding of those events. She refers to the two parts of the Ufimian type area and associates the lower part (the Solikamsk) with the underlying Lower Permian, and the upper part (the Sheshminsk) with the overlying beds, and she correlates that with the Roadian Stage. This calls for some reconsideration of the boundary between the Lower and Upper Permian in the Russian type area, currently placed at the base of the Ufimian. Detailed field data will be needed to elucidate the character of the Ufimian and its biota in the type area. In Australia, *Daubichites* occurs with a Kungurian-type ammonoid and other pre–Upper Permian elements (Dickins et al., 1989), and therefore it cannot be taken as defining the base of the Upper Permian if that base is to be regarded as post-Kungurian.

Yang and co-workers (Chapter 7) outline those events in China, and Dickins and Phan (Chapter 8) discuss the same material in regard to Vietnam. Yang and co-workers redefine the Chihsia beds and the units on either side, and now further field work seems desirable to clarify those relationships and to describe the faunas, especially the various species of *Misellina* and their ranges. According to Leven (1993), there are unconformities in South China for that interval, and more information on those, together with data on the faunal relationships and the tectonic movements, will be illuminating. Leven's work indicates that there seems to be considerable confusion at present regarding recognition of species such as *Misellina claudiae* and regarding the definition of the base of the Chihsia (Yang, Chen, and Wang, 1986). The type areas for the Chihsian Stage and the Maokouan Stage are separated by a considerable distance, and that may be a hindrance to an understanding of their relationships. This work is particularly important for the implications it may have for the International Standard Stratigraphical Scale. As in Japan, important movements appear to have been likely at that time in Vietnam, and detailed studies of the fusulinid faunas should throw light on this matter. Thus far, good examples of conodont faunas have not been found in those sequences in Vietnam.

In the Russian Far East, important changes occurred in the mid-Permian, including major changes in the types and distributions of basins, and there were distinctive patterns of magmatic and volcanic activity. Those are dealt with in the chapters by Zimina (Chapter 9, on land plants) and Vrzhosek (Chapter 10, on volcanism). Those changes paralleled the changes in many other parts of the world, as, for example, in South America (López-Gamundí and Breitkreuz, Chapter 2) and in Vietnam

(Dickins and Phan, Chapter 8). Chapter 11, by Khetchikov and co-workers, is particularly interesting, indicating the mineralization patterns associated with various parts of the Permian and Triassic. Mineralization seems to have been periodic, but its relationships to the various events studied in this project are poorly known. Gold mineralization seems to have been very common during the mid-Permian and is known, for example, from eastern Australia (Olgers, Flood, and Robertson, 1974; Bottomer, 1986) and was studied in our field visit to western North America (Vallier, 1992). Epithermal gold, accompanied especially by silver, is also a feature of these sequences in Vietnam (J. M. Dickins, personal observation), in Primorye (Khetchikov et al., Chapter 11, this volume), and in southern South America (Delpino et al., 1993).

Chapter 12, by Dickins, reviews the mid-Permian interval for selected areas around the world, detailing the various names for the orogenic activities and the surprisingly coeval character of the movements. It describes the widespread regression reflected in the upper part of the Lower Permian and the tectonic movements and foldings associated with intrusive and explosive volcanic activity, transgressive unconformity, and basin reconstruction whose beginnings are reflected in the Upper Permian. Chapter 12 also discusses the faunal changes, ascribing them to the environmental changes. In eastern Australia, the name Hunter-Bowen has been applied to the orogenic activity beginning in the late Permian and extending into the Triassic. It has been argued elsewhere (Dickins, 1987) that such activity continued to the end of the Triassic and that the name can usefully be combined with the Indosinian, as Hunter-Bowen (Indosinian), to indicate an orogenic cycle between the Hercynian and the Alpine-Himalayan cycles. The name Indosinian has been used in southern and eastern Asia generally for movements that took place during the late Permian and Triassic or parts thereof. In Chapter 8, Dickins and Phan attempt a description of the Indosinian Orogeny in Indochina, from where the name is derived. Elsewhere in this volume, the names "late Hercynian" and "Hercynian" are used for the Permian orogenic movement, and this is a matter on which we might reach consensus through discussion.

Upper Permian

A major success of IGCP Project 272 has been to recognize the outstanding features associated with the Midian–Dzhulfian boundary. This important event has long been recognized in Japan as marking the boundary between the Middle and Upper Permian in the Japanese traditional scale. In China it marks the boundary between the Yanghsingian (Yangxinian) and Lopingian series, generally, in the past, referred to as the Lower and Upper Permian. That time was marked by an outpouring of a "flood basalt," the Emeishan Basalt. Elsewhere in the Tethyan region that corresponds to the Midian–Dzhulfian boundary (Kotlyar, Chapter 4, this volume), frequently considered as the upper boundary of the Middle Permian in a threefold Permian subdivision. The significance of that event was recognized by

Dickins (1991) from his work in Japan and was reported at the IGCP-272 meeting in Sendai. That strong tectonic activity is reflected in the separation of the Kanokura and Toyoma series (Middle and Upper Permian, as traditionally used in Japan). At that time, carbonate platforms, some of which had existed since the Carboniferous, were broken up and deluged with largely siliceous volcanic and volcano-lithic material (Sano and Kanmera, 1991), and deep crustal fracturing led to large-scale basic magmatic and volcanic activity, associated with periods of intermediate and acidic magmatism ("ophiolites"). At the Sendai meeting, Phan (1991a) also reported on an unconformity at that level in Vietnam.

Subsequently, coeval changes were recognized in China (Yang et al., 1993), and extensive faunal changes were tabulated by Jin (1993). Further information was provided by Phan (1993), from Vietnam, where the unconformity at that level was accompanied by bauxite formation, strong magmatic and volcanic activity featuring a wide range of compositions, and apparently important mineralization. Dickins and Phan (Chapter 8, this volume) describe that episode further in Vietnam and its relationship to the Indosinian. No doubt that phase of the earth's development will eventually be recognized in other parts of the world, and perhaps it may explain some of the puzzling features associated with the Zechstein Stage. Yang et al. (Chapter 7, this volume) suggest recognition of a stage in China at that level. They ascribe the tectonic activity to the Dongwu movement. Some problems remain concerning the ranges and definitions at that level, especially in regard to the *Lepidolina kumaensis* fauna. In Japan, *L. kumaensis* appears to have been confined to the Toyoma (Dickins, 1991), but elsewhere it has been reported from an earlier time (Kotlyar, Chapter 4, this volume). That needs confirmation, and possibly the missing section now being studied in China may help in the solution of this problem.

Permian–Triassic boundary sequences and events

The information presented in these chapters adds to the detailed account given in a previous volume reporting on IGCP-203 (Sweet et al., 1992) and the further accounts provided by Zishun Li et al. (1991) on South China and by Kamada (1991) on Japan. Erwin (1994) has recently reviewed "the Permo-Triassic extinction" and has drawn similar conclusions. Wignall and Hallam (1993) have emphasized the effects of anoxia, but their conclusions about the importance of a single factor remain to be substantiated.

Yang et al. (Chapter 7, this volume) emphasize the natural causes of the extinctions, mainly the widespread regression and the volcanic activity. Ezaki (Chapter 13, this volume) traces the extinction events in the corals and concludes that the changes in the biota were not confined to a single, short-lived crisis, but represented changes over a long period of time during the Permian, culminating at the boundary. Already before the end of the Permian, restriction of carbonate platforms had adversely affected the colonial waagenophyllids, and the last rugose corals did not survive the Changhsingian. Chapter 14, by Ezaki and

Kuwahara, details the changes in the cherts of Japan and the radiolaria they contain. Particularly interesting is the absence, thus far, of the Griesbachian and Dienerian in the lower part of the Lower Triassic. This is consistent with the absence of marine deposits elsewhere in Japan in the lower part of the Triassic (e.g., Kamada, 1991). Kamada (1991) reported the unconformable relations of the Permian and Lower Triassic in northeastern Honshu (Japan) and a source of pebbles in the Triassic, not only from the west but also from the east. The search for an explanation for that raises some interesting possibilities. For example, were there very large scale changes in sea-floor level?

The late Changhsingian (Changxingian) ammonoids, bivalves, and brachiopods from South Primorye (Russian Far East), discussed by Zakharov, Oleinikov, and Kotlyar in Chapter 15, record a very late part of the Permian in that region. However, an important structural change and general hiatus can be shown above the Permian, as in all of Japan, and extensive volcanic activity (as in the Chinese region) is reflected in the latest part of the Permian, associated with the boundary. In Chapter 16, Rudenko, Panasenko, and Rybalka report very late Permian radiolaria and conodonts from Primorye (Sikhote-Alin). Olenekian radiolaria are well known, but they also record probable Induan radiolaria, which confirms distinctive changes from the Permian radiolaria to the Triassic radiolaria.

It has recently been suggested that the eruption of the Siberian traps either caused or had a strong effect on the Permian–Triassic extinctions (e.g., Baski and Farrar, 1991; Renne and Basu, 1991; Campbell et al., 1992; Erwin, 1994). Some sweeping claims regarding the accuracy of the dating and the geologic context of the traps have not been helpful in clarifying this problem. The radiometric dating is not sufficiently accurate, and the geology of the traps not well enough known, to allow a firm conclusion. Even if those eruptions did not coincide with the boundary, it is clear enough that they were associated with a time of very strong volcanic and tectonic activities. Large-scale eruptions at the boundary would not explain the changes that were already going on during the Permian.

Lower Triassic

Although it is fairly easy to distinguish the Upper Triassic from the Middle Triassic on the basis of its geologic nature and geographic distribution, at present it is not so easy to distinguish between the Middle Triassic and the Lower Triassic. A distinctive regressive–transgressive phase has been described for the Arctic region by Embry (1988) and Moerk (1994), and we have observed the marked hiatus at the base of the Anisian in the Southern Alps of Italy. There is sufficient evidence of a major change associated with the Lower–Middle Triassic boundary for the two units to be considered separately in this review. The widespread, large-scale volcanic and magmatic activities of the Upper Permian, with a large acidic component (Dickins, 1992b), continued into the Triassic. In Chapter 17, Sakhno identifies two major phases of basic-trap formation: in the Lower Triassic and in the Upper Triassic–Lower Jurassic.

Khetchikov et al. (Chapter 11, this volume) report epigenetic mineralization associated with the volcanic and intrusive activities, not only during the Upper Permian but also during the Lower Triassic and throughout the rest of the Triassic. They report alluvial gold in the basal conglomerates of the Lower Triassic, which rest unconformably on older rocks, similar to occurrences in Vietnam. Dickins and Phan (Chapter 8, this volume) report continued igneous activity during the Lower Triassic. In the Songda Depression, basic activity was important, a feature of that structure also during the Permian and Jurassic. In the areas to the north and south, acidic igneous activity was predominant, although in some places basaltic activity was associated with deep fracturing. Folding was occurring intermittently through the Triassic, and flysch-like deposits are found in Lower Triassic strata. The Permian–Triassic boundary sequences can vary from geometrically conformable to unconformable, with formation of transgressive conglomerates. Similar trends have been reported in southern South America, but no marine deposition is known in the Lower Triassic strata.

In Chapter 18, Paull and Paull review the conodont faunas during the Lower Triassic. They recognize three periods of relative sea-level rise. They conclude that conodont faunas were virtually cosmopolitan during the Lower Triassic. That has important implications for correlations, and it also suggests widespread warm water temperatures and climates and the need for caution before identifying distinct provinces such as cold, temperate and tropical assuming water temperatures and climatic differences for such provinces, according to present world differences in climate.

In Chapter 19, Yin reports the distributions of various faunal and floral elements and their provincial distributions in eastern Asia. The sea was distributed north–south over that region, with no sea connection to the west through northern Asia, but with a relatively broad oceanic connection to the west through southern Asia, corresponding to the Tethys of Suess.

Middle Triassic

Yin, in Chapter 19, shows a similar distribution of sea and land in eastern Asia and the continuation of a western connection through southern Asia.

In Vietnam, the Middle Triassic was a time of tremendous, largely acidic, igneous activity, and there were considerable changes in the distribution of basins, although not as many as in the Upper Triassic, when nonmarine deposition predominated (Dickins and Phan, Chapter 8, this volume).

In New Zealand (J. M. Dickins, personal observation) and in southern South America (López-Gamundí and Breitkreuz, Chapter 2, this volume), large-scale tectonic upheavals are indicated by the changed dispositions of structures and sedimentations.

The adverse conditions for life that characterized the Permian–Triassic boundary continued in the Lower Triassic, as witnessed by a generally relatively low faunal diversity and an absence of reef development. Those conditions changed in the Middle Triassic, and Punina (Chapter 20, this volume) describes reef

developments in Sikhote-Alin during the Middle and Upper Triassic. For descriptions of that change in the nature of carbonate deposition during the Triassic, see Flügel and Flügel-Kahler (1992) and Talent (1988).

Upper Triassic

Nowhere in all the Triassic were the changes more drastic than those that separated the Upper from the Middle Triassic. In Chapter 19, Yin tabulates the marked changes in the distribution of sea and land in eastern Asia. The events in Vietnam are reported in Chapter 8, and in the late Triassic the previously widespread marine sedimentation became confined to the western part of the Songda Depression, where it was connected westward with the Tethys seaway into western Asia and Europe. Eastern Southeast Asia formed a land area marking the eastern boundary of Tethys. In Vietnam, the positions changed again distinctly in the Lower Jurassic, whose strata are largely unconformable on Upper Triassic and earlier formations and are mainly nonmarine. In the southern part of Vietnam, a marine transgression took place from the south and east, either not connected or perhaps indirectly connected with Tethys (Phan, 1991b).

In Mexico, marine sedimentation began once again (González-León, Lucas, and Roldán-Quintana, Chapter 3, this volume) after a break following the Permian. That paralleled widespread deposition in the western United States (Miller and Harewood, 1990). Also in the nonmarine basins of the western United States, the Upper Triassic strata represent a distinctive period of time that Lucas (Chapter 23, this volume) proposes as a nonmarine standard for the Upper Triassic.

In Chapter 22, McRoberts discusses *Halobia* and its distribution in the uppermost Middle Triassic strata and the Upper Triassic strata. This bivalve is especially important for understanding the correlation of the marine rocks of that age and for discriminating between Middle and Upper Triassic strata.

The upper part of the Upper Triassic strata, the Rhaetian, has a distinctive development (Hallam, 1981; Hallam and El Shaarawy, 1982). In Chapter 24, Campbell discusses the marine fauna of that interval in New Zealand and its special character, as also found in other parts of the world.

Triassic–Jurassic boundary

The character of the Triassic–Jurassic boundary has been treated by Hallam (1981, 1990) and elsewhere in reports on Project 272 (Dickins, 1993a). Its significance is again emphasized in the trap volcanism identified by Sakhno (Chapter 17, this volume) in the Russian Far East. The worldwide significance of that tectonic and magmatic (as well as biologic) event lies in the unique geochemistry of the basaltic volcanism that has been described by Puffer (1992). The basalts from eastern North America have the same predominant character as those from Morocco, southern Africa, southern South America, and Siberia. Puffer ascribed their character to rapid decompressive melting caused by an event "uniquely powerful enough to allow for rapid delivery of large quantities of unfractionated and uncontaminated magma directly from undepleted or enriched mantle sources" (Puffer, 1992, p. 95). There is also ample evidence for that event in the basinal, stratigraphic, sedimentary, and biologic changes at the Triassic–Jurassic boundary (Dickins, 1989, 1993a), no doubt reflecting major subcrustal changes in the earth.

Climate

Although great climatic changes occurred during the late Palaeozoic and early Mesozoic, the connections with the other events outlined herein are not clear, nor are the connections with other postulated changes, such as the formation and movement of Pangaea.

In the early Carboniferous, the widespread areas of warm and equable climate were reflected in cosmopolitan marine faunas. The late Carboniferous apparently was generally cooler, but there seems to have been at least one important warmer interval, and at the beginning of the Permian, glaciation was widespread. Warming took place during the Permian, so that by the beginning of the Triassic there was a universally warm and largely dry climate. A warm climate then continued into the Jurassic. Those changes have been reviewed by Dickins (1993b) as part of Project 272.

In Chapter 25, González gives a detailed account of the changes in South America from the early Carboniferous to the early Permian and the faunas that accompanied the various phases. For the early Carboniferous he identifies a warm, cosmopolitan fauna, replaced in the late part of the early Carboniferous and the early part of the late Carboniferous by the *Levipustula* fauna associated with glacials. For the uppermost Carboniferous strata he reports fauna and sedimentation patterns indicating warmer conditions, before a return to less hospitable conditions in the early Permian. Whether or not that warm interval can be identified in other parts of the world is of considerable interest. In Chapter 2, López-Gamundí and Breitkreuz describe the development of arid conditions later in the Permian, and in Chapter 18, Paull and Paull, on the basis of the cosmopolitan character of the conodont fauna, leave little doubt that there was a widespread warm climate for the early Triassic.

Yin (1991) has reviewed the palaeobiogeography of the Triassic in China, and in Chapter 19 he describes the biostratigraphy and the palaeogeography of eastern Asia during the early, middle, and late Triassic. He describes the distribution of sea and land – no sea through central Asia, but sea between the present Himalayas and the Kunlun (often called Neotethys, but the Tethys of Suess) – and the faunal and floral provinces and discusses all that in relation to commonly held plate-tectonic interpretations. Despite the widespread warm climate apparent throughout the Triassic, he is able to delineate provinces, and further examination of the meaning of those provinces could be of great significance.

Discussion and conclusions

To understand geologic events, a reliable time scale that can be made increasingly precise is a basic requirement. It has been natural, therefore, for the participants in Project 272 to devote considerable attention to biostratigraphic correlations. The way toward further refinement of that scale has been indicated. The radiometric data available from the Siberian traps have revealed the difficulties associated with that method, and there is a need for more radiometric work tied closely to sampling data from well-established sequences dated stratigraphically and palaeontologically.

The main purpose of these studies has been to look at those circum-Pacific events in time and to assess the nature of the events and the reliability and precision in their time assignments. In some cases, analysis in terms of tectonic explanation has been attempted, but the data are important not only for examining the validity of existing theory but also because of their potential for use in developing new approaches.

Some attempts have been made to examine the relationships of metallogenesis to the other circum-Pacific events, and the studies from Vietnam and the Russian Far East have been particularly useful in indicating the relationships between periodicities and other factors. Although, strictly speaking, they are perhaps outside the scope of this project, the large-scale polymetallic mineralizations associated with the magmatic and tectonic events of the Cretaceous–Tertiary boundary are commonly encountered in Triassic carbonates. Such mineralization is widespread in the circum-Pacific region close to the Cretaceous–Tertiary boundary, and it has been extensively described for the western United States (Emmons, Irving, and Loughlin, 1927; Behre, 1953; Pearson et al., 1962; Russell, 1991).

References

Baski, A. K., and Farrar, E. 1991. ^{40}Ar/^{39}Ar dating of the Siberian traps, USSR: evaluation of the ages of the two major extinction events relative to episodes of flood-basalt volcanism in USSR and the Deccan Traps, India. *Geology* 19:461–4.

Beauchamp, B., Harrison, J. C., and Henderson, C. M. 1989. Upper Paleozoic stratigraphy and basin analysis of the Sverdrup Basin, Canadian Arctic Archipelago. Part 1: Time frame and tectonic evolution. *Geological Survey of Canada, Current Research, Part G*, pp. 105–11.

Behre, C. H., Jr. 1953. *Geology and Ore Deposits of the West Slope of the Mosquito Range.* U.S. Geological Survey, professional paper 235. Washington, DC: U.S. Geological Survey.

Bottomer, L. R. 1986. Epithermal silver-gold mineralization in the Drake area, New South Wales. *Australian Journal of Earth Sciences* 33:457–73.

Boucot, A. J. 1990. Phanerozoic extinctions: How similar are they to each other? In *Extinction Events in Earth History*, ed. E. G. Kauffman and O. H. Walliser, pp. 5–30. Berlin: Springer-Verlag.

Campbell, I. H., Czamanske, G. K., Federenko, V. A., Hill, R. I., and Stepanov, V. 1992. Synchronism of the

Siberian Traps and the Permo–Triassic boundary. *Nature* 258:1760–3.

Cassinis, G. (ed.) 1988. Permian and Permian–Triassic boundary in the South-Alpine segment of the western Tethys, and additional regional reports. *Memorie della Società Geologica Italiana* 34:1–366.

Delpino, D., Pezzutti, N., Godeas, M., Donnari, E., Carullo, M., and Nunez, E. 1993. Un cobre porforico paleozoico superior en el centro volcanico San Pedro, disrito minero El Nevado, provincia de Mendoza, Argentina. In *Douzième Congres International de la Stratigraphie et Géologie du Carbonifère et Permien, Comptes Rendus*, vol. 1, ed. S. Archangelsky, pp. 477–90. Buenos Aires.

Dickins, J. M. 1987. Tethys – a geosyncline formed on continental crust? In *Shallow Tethys 2*, ed. K. G. McKenzie, pp. 149–58. Rotterdam: Balkema.

Dickins, J. M. 1988. The world significance of the Hunter/Bowen (Indosinian) mid-Permian to Triassic folding phase. *Memorie della Società Geologica Italiana* 34:345–52.

Dickins, J. M. 1989. Major sea level changes, tectonism and extinctions. In *Onzieme Congres International de Stratigraphie et de Géologie du Carbonifère*, vol. 4, ed. Jin Yugan and Li Chun, pp. 135–44. Beijing.

Dickins, J. M. 1991. Permian of Japan and its significance for world understanding. In *Shallow Tethys 3*, ed. T. Kotaka, J. M. Dickins, K. G. McKenzie, K. Mori, K. Ogasawara, and G. D. Stanley, Jr., pp. 343–51. Special publication 3. Sendai: Saito Ho-on Kai.

Dickins, J. M. 1992a. Permo-Triassic orogenic, paleoclimatic, and eustatic events and the implications for biotic alteration. In *Permo–Triassic Events in the Eastern Tethys: Stratigraphy, Classification, and Relations with Western Tethys*, ed. W. C. Sweet, Yang Zunyi, J. M. Dickins, and Yin Hongfu, pp. 169–74. Cambridge University Press.

Dickins, J. M. 1992b. Permian geology of Gondwana countries: an overview. *International Geology Review* 34:986–1000.

Dickins, J. M. 1993a. The Triassic–Jurassic boundary: sea-level, tectonic and magmatic change and the biological change. In *Douzième Congres International de la Stratigraphie et de Géologie du Carbonifère et Permien*, vol. 2, ed. S. Archangelsky, pp. 522–32. Buenos Aires.

Dickins, J. M. 1993b. Climate of the Late Devonian to Triassic. *Palaeogeography, Palaeoclimatology, Palaeoecology* 100:89–94.

Dickins, J. M., Archbold, N. W., Thomas, G. A., and Campbell, H. J. 1989. Mid-Permian correlation. In *Onzieme Congres International de Stratigraphie et de Géologie du Carbonifère*, vol. 2, ed. Jin Yugan and Li Chun, pp. 185–92. Beijing.

Embry, A. F. 1988. Triassic sea-level changes: evidence from the Canadian Arctic Archipelago. *Society of Economic Paleontologists and Mineralogists, Special Publication* 42:249–59.

Emmons, S. F., Irving, J. D., and Loughlin, G. F. 1927. *Geology and Ore Deposits of the Leadville Mining District, Colorado*. U.S. Geological Survey, professional paper 148.

Erwin, D. H. 1994. The Permo-Triassic extinction. *Nature* 367:231–6.

Flügel, E., and Flügel-Kahler, E. 1992. Phanerozoic reef evolution: basic questions and data base. *Facies* 26:167–278.

Hallam, A. 1981. The end Triassic extinction event. *Palaeogeography, Palaeoclimatology, Palaeoecology* 35:1–44.

Hallam, A. 1990. The end Triassic mass extinction event. In *Global Catastrophes in Earth History; An Interdisciplinary conference on impacts, volcanism and mass mortality*, ed. V. L. Sharpton and P. D. Ward, pp. 577–83. Geological

Society of America special paper 247. Boulder: Geological Society of America.

Hallam, A., and El Shaarawy, Z. 1982. Salinity reduction of the end-Triassic sea from the Alpine Region into northwestern Europe. *Lethaia* 15:169–78.

Italian IGCP-203 group. 1988. *Field Conference on Permian and Permian–Triassic Boundary in the South-Alpine Segment of the Western Tethys*. Excursion guidebook, Società Geologica Italiana.

Jin Yugan. 1993. Pre-Lopingian benthos crisis. In *Douzième Congres International de la Stratigraphie et Géologie du Carbonifère et Permien*, vol. 2, ed. S. Archangelsky, pp. 269–78. Buenos Aires.

Kamada, K. 1991. Upper Permian to Middle Triassic sedimentation and its tectonic implication in the southern Kitakami Belt, Japan: preliminary report. In *Shallow Tethys 3*, ed. T. Kotaka, J. M. Dickins, K. G. McKenzie, K. Mori, K. Ogasawara, and G. D. Stanley, Jr., pp. 423–34. Special publication 3. Sendai: Saito Ho-on Kai.

Leven, E. Y. 1993. Main events in Permian history of the Tethys and fusulinids. *Stratigraphy and Geological Correlation* 1:51–65.

Miller, M. M., and Harewood, D. S. 1990. Paleogeographic setting of Upper Paleozoic rocks in the northern Sierra and eastern Klamath terranes, northern California. In *Paleozoic and Early Mesozoic Paleogeographic Relations; Sierra Nevada, Klamath Mountains, and Related Terranes*, ed. D. S. Harewood and M. M. Miller, pp. 175–200. Geological Society of America special paper 255. Boulder: Geological Society of America.

Moerk, A. 1994. Triassic transgressive–regressive cycles of Svalbard and other Arctic areas: a mirror of stage subdivision. In *Recent Developments in Triassic Stratigraphy*, ed. J. Geux and A. Baud. *Memoires de Geologie (Lausanne)* 22:69–81.

Olgers, F., Flood, P. G., and Robertson, A. D. 1974. *Palaeozoic geology of Warwick and Goondiwindi, 1:250 000 Sheet Areas, Queensland and New South Wales*. Publication 162. Canberra: Bureau of Mineral Resources.

Pearson, R. C., Tweto, O., Stern, T. W., and Thomas, H. H. 1962. *Age of Laramide porphyries near Leadville, Colorado*. U.S. Geological Survey, professional paper 450-C, pp. 78–80. Washington, DC: U.S. Geological Survey.

Phan Cu Tien. 1991a. Stratigraphic correlation of Permian and Triassic in Vietnam. In *Shallow Tethys 3*, ed. T. Kotaka, J. M. Dickins, K. G. McKenzie, K. Mori, K. Ogasawara, and G. D. Stanley, Jr., pp. 359–70. Special publication 3. Sendai: Saito Ho-on Kai.

Phan Cu Tien. (ed.) 1991b. *Geology of Cambodia, Laos and Vietnam (Explanatory Note to the Geological Map of Cambodia, Laos and Vietnam at 1 : 1,000,000 Scale)*, 2nd ed. Hanoi: Geological Survey of Vietnam.

Phan Cu Tien. 1993. Upper Carboniferous–Permian volcano-sedimentary formations in Vietnam and adjacent territories. In *Gondwana Eight, Assembly, Evaluation and Dispersal*, ed. R. H. Findlay, R. Unrug, M. R. Banks, and J. J. Veevers, pp. 299–306. Rotterdam: Balkema.

Puffer, J. H. 1992. Eastern North American flood basalts in the context of the incipient breakup of Pangea. In *Eastern North American Mesozoic Magmatism*, ed. J. H. Puffer, J. H. Ragland, and P. C. Ragland, pp. 95–119. Geological Society of America, special publication 268.

Renne, P. R., and Basu, A. R. 1991. Rapid eruption of the Siberian Traps flood basalts at the Permo–Triassic boundary. *Science* 253:176–9.

Russell, C. W. 1991. Gold mineralization in the Little Rocky Mountains, Phillips County, Montana. In *Guidebook of the Central Montana Alkalic Province: Geology, Ore Deposits and Origin*, ed. D. W. Baker and R. B. Berg, special publication 100. Helena: Montana Bureau of Mines and Geology.

Sano, H., and Kanmera, K. 1991. Collapse of ancient reef complex. *Geological Society of Japan* 97:113–33.

Sweet, W. C., Yang Zunyi, Dickins, J. M., and Yin Hongfu. 1992. Permo–Triassic events in the Eastern Tethys: stratigraphy, classification, and relations with the Western Tethys. In *World and Regional Geology*, vol. 2, ed. W. C. Sweet, Yang Zunyi, J. M. Dickins, and Yin Hongfu. Cambridge University Press.

Talent, J. A. 1988. Organic reef building: episodes of extinction and symbiosis? *Senckenbergiana Lethaea* 69:315–68.

Vallier, T. L. 1992. *A Geologic Guidebook to Hell's Canyon*. NAPC/IGCP field trip, July 9 and 10, 1992.

Visscher, J. N. J. 1993. The tectonogeographic evolution of part of Southwestern Gondwana during the Carboniferous and Permian. In *Douzième Congres International de la Stratigraphie et Géologie du Carbonifère et Permien*, vol. 1, ed. S. Archangelsky, pp. 447–54. Buenos Aires.

Wignall, P. B., and Hallam, A. 1993. Griesbachian (earliest Triassic) palaeoenvironmental changes in the Salt Range, Pakistan and southeast China, and their bearing on the Permo-Triassic mass extinction. *Palaeogeography, Palaeoclimatology, Palaeoecology* 102:215–37.

Yang Zunyi, Chen Yuqi, and Wang Hongzhen. 1986. *The Geology of China. Oxford Monographs on Geology and Geophysics* 3. Oxford University Press.

Yang Zunyi, Liang Dinyi, and Nie Zetong. 1993. On two Permian submarine extension sedimentation events along the north margin of Gondwanaland the west margin of the Yangzte Massif. In *Douzième Congres International de la Stratigraphie et Géologie du Carbonifère et Permien*, vol. 1, ed. S. Archangelsky, pp. 467–74. Buenos Aires.

Yin Hongfu. 1991. Triassic paleobiogeography of China. In *Shallow Tethys 3*, ed. T. Kotaka, J. M. Dickins, K. G. McKenzie, K. Mori, K. Ogasawara, G. D. Stanley, Jr., pp. 403–21. Special publication 3. Sendai: Saito Ho-on Kai.

Zishun Li, Lipei Zhan, Jianxin Yao, and Yaoqi Zhou. 1991. On the Permo-Triassic events in South China – probe into the end Permian abrupt extinction and its possible causes. In *Shallow Tethys 3*, ed. T. Kotaka, J. M. Dickins, K. G. McKenzie, K. Mori, K. Ogasawara, and G. D. Stanley, Jr., pp. 371–85. Special publication 3. Sendai: Saito Ho-on Kai.

2 Carboniferous-to-Triassic evolution of the Panthalassan margin in southern South America

OSCAR R. LÓPEZ-GAMUNDÍ and CHRISTOPH BREITKREUZ

The continental margin of southern South America facing Panthalassa (or the proto-Pacific) has been characterized by subduction and transcurrent movements at least since mid-Palaeozoic times (Dalziel and Forsythe, 1985; Ramos, 1988a; Breitkreuz et al., 1989). Between the late Palaeozoic and the Triassic, that convergent margin subsided during synchronous stages of Pangaean extension punctuated by diachronous subduction that formed a series of foreland basins by cratonward thrusting of the foldbelt/magmatic arc (Veevers et al., 1994). Those basins were formed by extension (E-I), subsequent foreland shortening (FS), and final extension (E-II). That tectonomagmatic evolution of the continental margin had subtler effects in the interior basins of South America, where sedimentation, although influenced, was not interrupted by major discontinuities. A key element in understanding that evolution is the presence of a rich stratigraphic record, mainly in western Argentina and northern Chile, spanning the Carboniferous–Triassic interval that helps to identify the transition from a compressional to an "extensional" convergent margin. The objective of this chapter is to synthesize our current knowledge of the evolution of the Panthalassan margin of southern South America during the late Palaeozoic and Triassic. For detailed descriptions and discussions, the reader is referred to the numerous publications cited.

On the basis of distinct characteristics to be described later, the convergent margin of South America can be subdivided into three segments (Figure 2.1), namely, (A) a northern segment (north of 20°S), (B) a central segment (20–40°S), and (C) a southern segment (south of 40°S) (López-Gamundí et al., 1994).

Northern (A) segment. The Devonian-to-Permian evolution of the northern segment was mainly controlled by the interaction between the Arequipa Massif, a Precambrian basement block, and the South American continent, with periodic episodes of extension and compression/transpression (Forsythe et al., 1993). In western Bolivia and southwestern Perú, Devonian and younger deposits rest on a basement largely composed of Precambrian intrusives and high-grade metamorphic rocks of the Arequipa Massif and adjacent paraautochthonous terranes (e.g.,

Mejillones, Belén) in northern Chile. Palaeocurrent evidence from early Palaeozoic deposits of that ensialic seaway suggests that the Arequipa Massif constituted the main provenance terrane (Isaacson, 1975). During the late Devonian and early Carboniferous the Eo-Hercynian Orogeny affected the deposits presently exposed in the Eastern Cordillera of Perú and Bolivia (Laubacher and Megard, 1985). The Tarija foredeep basin (Figure 2.1), developed since the Carboniferous, was emplaced over the axis of the early Palaeozoic basin between the Arequipa Massif and the Brazilian Shield.

Central (B) segment. Formation of this segment was characterized by subduction and associated arc magmatism. In the northern part (20–27°S), magmatism did not start until the late Carboniferous. There, magmatic (intrusive and extrusive) activity lasted until the Triassic, a time span during which volcano-sedimentary successions up to 4 km thick (Peine Group) (Bahlburg and Breitkreuz, 1991) were formed. Earlier, during the Devonian and early Carboniferous, a marine marginal basin existed in northern Chile between the southernmost part of the Arequipa Massif on the west and the Puna Arch on the east (Niemeyer et al., 1985; Bahlburg, 1991). Its formation, locally accompanied by marine rift volcanism (Breitkreuz et al., 1989), and later closure were most likely controlled by north–south-trending dextral strike-slip movements (Bahlburg and Breitkreuz, 1991, 1993).

In the southern part of the central segment, between 27°S and 40°S, an Andean-type continental margin developed, probably as early as during the Devonian. Well-preserved accretionary prism sequences (Davidson et al., 1987; Hervé, 1988) and tectonic mélanges (Dalziel and Forsythe, 1985) indicate the presence of a subduction complex along that part of the continental margin. The inception of a subduction-zone-related magmatic arc occurred most probably in the late Devonian, and back-arc basins developed later in the late Palaeozoic. Although terrane collision has been proposed as an important factor in the development of that margin during the early and middle Palaeozoic (cf. Ramos et al., 1986), other views tend to minimize its influence (Charrier et al., 1991; González Bonorino and González Bonorino, 1991).

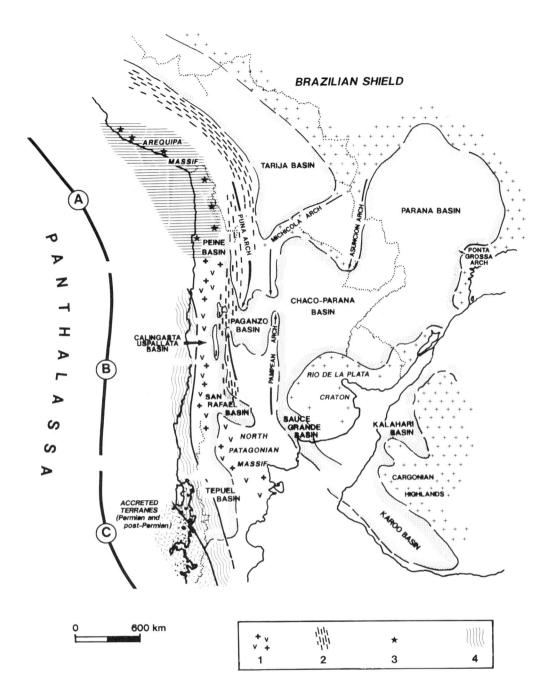

Figure 2.1. Late Palaeozoic geotectonic framework of the Panthalassan (proto-Pacific) margin of southern South America. The margin has been subdivided into three segments (A–C) according to their distinct geotectonic characteristics. Late Palaeozoic basins superimposed in stippled pattern: 1, late Palaeozoic magmatic arc; 2, Eo-Hercynian (late Devonian–early Carboniferous) deformation indicated by orientation of schistosity and fold axes (Dalmayrac et al., 1980); 3, Precambrian basement rocks of the Arequipa Massif; 4, Devonian–late Palaeozoic accretionary prism and subduction complexes (Dalziel and Forsythe, 1985; Hervé, 1988). (Adapted from López-Gamundí and Rossello, 1993.)

Southern (C) segment. South of 43°S, the Panthalassan continental margin mostly resulted from accretion of terranes, like the one represented by the Permian limestones of the Madre de Dios Archipelago of southern Chile (Mpodozis and Forsythe, 1983; Dalziel and Forsythe, 1985). The late Palaeozoic magmatic arc that developed along the middle segment of the continental margin extended southward, being deflected into the continent, as indicated by the presence of coeval, crust-contaminated magmatism in Patagonia. That southeastward swing of the magmatic tract occurred along a line coincident with the old

Gondwanan margin. Forsythe (1982) and Uliana and Biddle (1987) related that magmatic activity in the North Patagonian Massif to the presence of an Andean-type margin with a wide magmatic-arc–back-arc system across Patagonia (Figure 2.1). This wide system has been subdivided into an outer arc with metaaluminous granitoids and an inner arc with peraluminous and weakly metaaluminous granitoids and silicic volcanics (Cingolani et al., 1991). Associated sedimentation in fore-arc settings is exemplified by the thick (>5,000 m) sequence of the Tepuel Basin in western Patagonia (Figure 2.1).

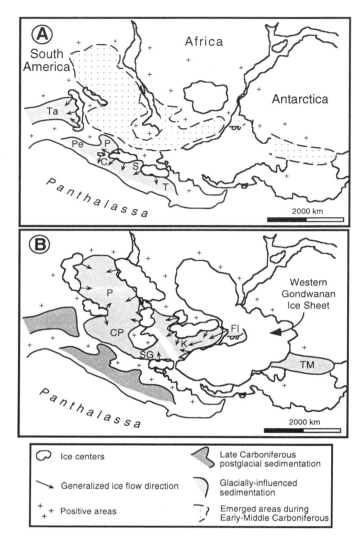

Figure 2.2. Glacial palaeogeography of western Gondwana during the late Palaeozoic. A: Middle Carboniferous (Namurian–early Westphalian) palaeogeography. Glacially influenced sedimentation was confined to basins along the Panthalassan margin of southern South America: Tarija (Ta), Calingasta-Uspallata (C), Paganzo (P), San Rafael (S), Tepuel (T), and, probably, Peine (Pe) basins. Isolated ice-spreading centers along the cratonic margins of those basins shed glacial detritus in mostly marine environments. Ice cover may have been present over elevated areas of central Africa. The rest of western Gondwana remained uplifted until generalized subsidence led to sedimentation by the end of the Carboniferous (Stephanian). B: Late Carboniferous (Stephanian) and earliest Permian palaeogeography. The growth and amalgamation of ice-spreading centers in Africa and Antarctica led to the formation of the western Gondwanan ice sheet (WGIS). The WGIS and satellite ice centers influenced sedimentation from the Transantarctic Mountains (TM) in Antarctica and the Karoo (K) basin (including the Falkland Islands, FI) of southern Africa to the Paraná, Chaco-Paraná (CP), and Sauce Grande (SG) basins of southeastern South America. Glacially influenced sedimentation started in those basins when glacial conditions ceased along the proto-Pacific basins.

Alternatively, Ramos (1984, 1986) related the late Palaeozoic magmatism in Patagonia to a collisional model. This model proposes a southwestward-dipping subduction zone under a separate Patagonian plate in the context of a continent–continent collision. However, the available evidence from radiometric dates (Cingolani et al., 1991), palaeomagnetic data (Rapalini et al., 1993), palaeogeographic aspects, and the continuity of the magmatic belt (López-Gamundí et al., 1994) favors autochthony of Patagonia since at least the Carboniferous.

The continental-margin evolution occurred under a drastic palaeoclimatic transition characterized by widespread glaciation during the mid-Carboniferous, followed by climatic amelioration evinced by the presence of extensive peats during the late Carboniferous, culminating with arid to semiarid conditions, illustrated by the presence of redbeds with eolian deposits in the middle to late Permian (López-Gamundí, Limarino, and Cesari, 1992). The combination of high palaeolatitude and high relief induced the onset and development of the mid-Carboniferous glaciation along the Andean basins of southwestern South America. That glacial episode preceded the generalized glaciation that affected eastern South America, Africa, and Antarctica during the latest Carboniferous and early Permian (Figure 2.2). After a moderately humid interlude during the latest Carboniferous and early Permian, extremely arid conditions induced the development of extensive sand seas behind the continental margin in western Argentina by the middle to late Permian (Limarino and Spalletti, 1986). A similar trend, although starting in the early Carboniferous with glacial deposits followed by mid-Carboniferous deposits and culminating with platform carbonates in the early Permian, has recently been documented from the Antiplano of Bolivia (Sempere, 1987; Díaz Martínez, Isaacson, and Sablock, 1993).

In the following sections we review the evolution of this margin between 20°S and 38°S during the Carboniferous–Triassic time interval, a sector roughly coincident with segment B as defined earlier. To this end, we review the existing stratigraphic and structural information on the Argentinian and northern Chilean outcrops and document the evolution of that portion of the Panthalassan margin.

Initial extension (early and middle Carboniferous to early Permian)

Early–middle Carboniferous: orogeny and inception of the magmatic arc

Late Devonian–early Carboniferous unconformity. A widespread unconformity that separates Lower and Middle (Devonian) Palaeozoic marine deposits from marine to nonmarine Carboniferous deposits is thoroughly documented in southern South America. In southeastern Perú and Bolivia, Devonian marine sediments tightly folded during the Eo-Hercynian orogenic phase (Dalmayrac et al., 1980) are overlain by early Carboniferous coarse deposits (Laubacher and Megard, 1985). This unconformity has been correlated to the equivalent Chañic Orogeny (Turner and Méndez, 1975; Coira et al., 1982) identified in northwestern Argentina and later extrapolated to adjacent areas farther south. That phase seems to have been associated with pervasive deformation, documented by schistosity and folding (Figure 2.1). The persistent north-northwest–south-southeast (NNW–SSE) orientation of the schistosity and fold axes gives a distinctive fabric to the pre-Carboniferous basement (Dalmayrac et al., 1980).

The hiatus involved ranged from, at the minimum, the late Devonian to the early Carboniferous, or, at the maximum, from the late Devonian through the late Carboniferous (López-Gamundí and Rossello, 1993). The initial stages of those basins, especially in the west, are documented by infrequent early Carboniferous syntectonic sedimentation. The scarcity of deposits of that age suggests that the early Carboniferous was dominated by uplift and erosion, followed by widespread subsidence during the latest part of the early Carboniferous and during the late Carboniferous. Interestingly, the Devonian–Carboniferous shelf succession in northern Chile shows no evidence of a late Devonian–early Carboniferous hiatus (Bahlburg and Breitkreuz, 1993). The Puna Arch, on the western flank of which those sediments accumulated, probably protected the area from deformation. The stress from the convergence of the Arequipa Massif and the South American Craton could have been absorbed by the intervening Puna Arch, an early Palaeozoic mobile belt located between two rigid nuclei (Ramos, 1988b).

Some interpretations suggest a collisional cause for this orogeny, as proposed by Ramos et al. (1986). Those authors inferred a collision between the South American continent and a hypothetical terrane named Chilenia at around 30°S as the cause of the sharp unconformity between the tightly folded Devonian deep-marine (meta)sediments and the less tectonically disturbed Carboniferous deposits present at several localities in western Argentina.

Birth of the late Palaeozoic magmatic arc. Late Devonian–early Carboniferous arc magmatism was present along the present-day Andes, probably representing the early stage of the arc magmatism that developed along the same margin after the late Carboniferous. That early arc magmatism is marginally represented inward of the margin in western Argentina (31–33°S) by igneous bodies and dikes of early Carboniferous age (337 ± 10 Ma, 330 ± 6 Ma) (Caminos, Cordani, and Linares, 1979; Sessarego et al., 1990) that intruded Devonian deposits.

Sediments. The mid-Carboniferous sag phase (E-I) led to the formation of several basins located behind the magmatic arc that later were extended, during the late Carboniferous, into the interiors of some parts of eastern South America (Chaco-Paraná and Paraná basins, Figure 2.1). The middle (Namurian–Westphalian) Carboniferous was characterized by increased subsidence and consequent encroachment of the depositional areas.

The basal fill in some of those basins contains the oldest (early to middle Carboniferous) well-documented record of Gondwanan glaciation. Palaeogeographic and palaeomagnetic studies suggest that the mid-Carboniferous glacial episode in southwestern South America was triggered by a combination of topography and pole positioning (López-Gamundí, 1987; Veevers and Powell, 1987). Those basins were fed by ice centers that preceded the extensive growth of the western Gondwanan ice sheet (WGIS) after the late Carboniferous (Figure 2.2). That ice sheet probably was centered in southern Africa and Antarctica during most of its existence (Visser, 1989).

Glacial-marine deposits derived from local ice centers fed the Tarija, Calingasta-Uspallata, Paganzo, and San Rafael basins (Figure 2.3). Glacial sedimentation prevailed in the Tarija, Calingasta-Uspallata, San Rafael, and western Paganzo basins, whereas continental (fluvial and lacustrine) sedimentation was confined to small palaeovalleys formed along old zones of weakness in the Pampean basement of the eastern Paganzo Basin (Andreis, Leguizamón, and Archangelsky, 1986; Fernández-Seveso, Pérez, and Alvarez, 1990). The glacial-marine facies consist of massive to crudely bedded diamictites, thin-bedded diamictites, and dropstone shales, with occasional evidence of grounded ice activity (Frakes and Crowell, 1969; López-Gamundí, 1987). That early glacial period ended by the late part of the middle Carboniferous (Westphalian) and was followed by a sea-level rise, of probable glacio-eustatic origin, eventually augmented by foreland-type subsidence and consequent expansion of the marine domain in the Andean basins (López-Gamundí, 1989).

According to Charrier (1986) and Breitkreuz (1986), northern Chile lacks late Palaeozoic glacial deposits. However, we (C. Breitkreuz, unpublished data) have recently observed varves, dropstones, and slumps in lacustrine sediments of the Peine Formation (Ramírez and Gardeweg, 1982) exposed along the eastern margin of the Peine Basin in the Salar de Atacama area. Those strata might represent a western expression of the mid-Carboniferous glacial facies prominent in Bolivia and Argentina. However, no biostratigraphic data are available so far.

Late Carboniferous–early Permian: generalized subsidence and extension of the magmatic arc along the entire central segment

Basin development. Foreland basins were subjected to generalized subsidence during that time interval. In Argentina, late Carboniferous (Stephanian) sedimentation expanded fully into the eastern Paganzo Basin. Both the eastern and western sectors of that basin were integrated in a major fluvio-deltaic drainage system. Evidence from palaeocurrent patterns and facies distributions suggests a general direction of flow from east to west and transition to a shallow-marine domain in the Calingasta-Uspallata Basin (Figure 2.3). That palaeogeographic expansion was coupled with a drastic climatic amelioration that led to the

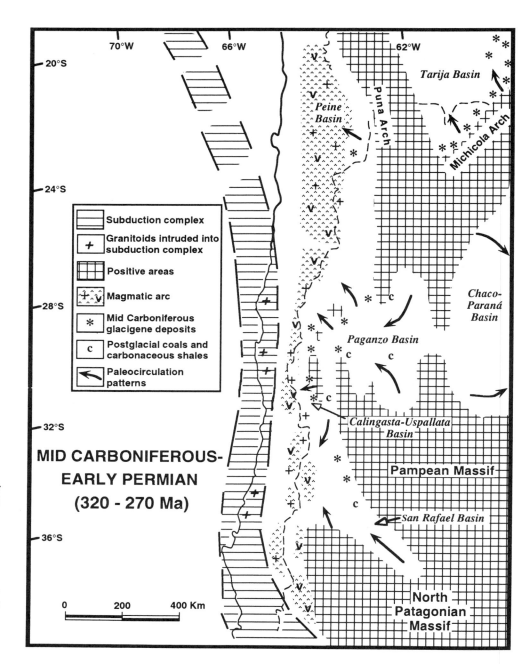

Figure 2.3. Palaeogeographic sketch of the proto-Pacific margin between 20°S and 40°S for the mid-Carboniferous–early Permian interval. The subduction complex includes the accretionary complex. The magmatic arc became progressively emergent toward the end of the early Permian. Generalized sedimentation in the western Paganzo and Chaco-Paraná basins commenced during the late Carboniferous. (Adapted from López-Gamundí et al., 1994.)

formation of peats in the Paganzo Basin (López-Gamundí, Cesari, and Limarino, 1993).

Volcanicity coeval with sedimentation is evinced by tuffs and subordinate volcanic agglomerates interbedded with fluvial deposits and volcanogenic detritus in the eolian facies (Limarino et al., 1986) of the Paganzo Basin and in the shallow-marine deposits of the Calingasta-Uspallata and Arizaro basins.

Shallow-marine conditions have been inferred for the late Carboniferous intra-arc Las Placetas Formation (29–30°S) and the fore-arc Huentelauquen Formation (31–32°S) west of the Calingasta-Uspallata Basin in Chile (Nasi and Sepúlveda, 1986). In its upper section, the Huentelauquen Formation contains a mixed siliciclastic-calcareous shallow-marine facies, with a Foramol fauna of cold-to-temperate waters (Rivano and Sepúlveda, in Nasi and Sepúlveda, 1986).

North of 24°S, outcrops of early Permian shallow-marine limestones and volcanic rocks are known from Chile ("Cerro 1584," Quebrada Blanca, Juan de Morales) (Niemeyer et al., 1985; Breitkreuz, 1986) and Argentina (Salar Arizaro) (Benedetto, 1976; Donato and Vergani, 1985). The limestones represent the southernmost extension of the Copacabana transgression, extensively documented in western Bolivia and southern Perú (Helwig, 1972; Megard, 1978). However, the main magmatic arc in northern Chile was never flooded by the

Copacabana Sea, because terrestrial depositional facies prevailed from the late Carboniferous until the late Triassic, at least for the area around Salar de Atacama (Breitkreuz, 1986; Breitkreuz, Bahlburg, and Zeil, 1988). Magmatic activity started in the late Carboniferous (Davidson et al., 1985). An intra-arc basin setting, which existed during the latest Carboniferous and Permian, accommodated volcanic rocks and alluvial/lacustrine sediments (Breitkreuz, 1991; Breitkreuz et al., 1992; Flint et al., 1993).

Origin of arc magmatism. The late Carboniferous–earliest Permian bimodal magmatism located in Chile between 27°S and 31°S has been subdivided into two subcycles (Nasi et al., 1985; Nasi and Sepúlveda, 1986; Mpodozis and Kay, 1992). The older subcycle represents arc magmatism formed along an active margin and contaminated with metasedimentary crustal components, probably derived from the accretionary prism. The younger subcycle, mostly composed of hypersilicic granites and consanguineous volcanics, shows evidence of increasing crustal melting (Llambías and Sato, 1990; Mpodozis and Kay, 1992). Likewise, magmatic rocks that formed during the late Carboniferous–early Permian in the Salar de Atacama region (north of 24°S) show isotope compositions indicative of melting of the early Proterozoic crust (Mpodozis et al., 1993; Breitkreuz and Van Schmus, 1996).

Foreland shortening (late Permian–early Triassic)

Argentina. The diachronous foreland shortening (FS) that followed the initial sag phase can be traced along the Panthalassan margin. The deformation was followed by a foreland-basin phase established in western Argentina by the mid-Permian (ca. 270 Ma), slightly later in southern Africa, and in eastern Australia by the late Permian (260–255 Ma) (Powell et al., 1991). In western Argentina, the change from extensional sag basins to compressional foreland basins was related to the San Rafael Orogeny of early Permian age (Figure 2.4), most likely correlatable to the Tardi-Hercynian Orogeny of Dalmayrac et al. (1980). The associated Choiyoi magmatic activity, situated principally in the Andean and northern Patagonian regions and traceable along the Panthalassan margin (Kay et al., 1989), peaked during the late part of the early Permian and during the late Permian.

The Choiyoi volcanic episode seems to have been partially contemporaneous with the sedimentation of early–late Permian continental redbeds in the Paganzo Basin and shallow-marine sediments in the Calingasta-Uspallata Basin, as documented by interbedded tuffs (Alvarez and Fernández-Seveso, 1987; López-Gamundí et al., 1989). The redbeds are associated with extensive Lower-to-Upper Permian eolian deposits identified in the Paganzo, Calingasta-Uspallata, and San Rafael basins (Limarino and Spalletti, 1986). Palaeocurrent data indicative of a general wind pattern from west to east suggest the presence of a long volcanic arc developed along the western margin of the continent

that contributed to the deflection of winds toward the east and to the desiccation of westerlies because of orographic effects (Figure 2.4) (Limarino and Spalletti, 1986).

The advance of the Choiyoi volcanic front into the foreland basins caused drastic change in the configurations of the depositional areas. That volcanic activity was coupled with cratonward thrusting, general uplift and erosion, and consequent withdrawal of the sea. Thus, early Permian volcanics and associated nonmarine deposits (foreland-basin phase, FS) rest unconformably on late Carboniferous or early Permian shallow-marine deposits along the Calingasta-Uspallata Basin (Cortés, 1985) and on early Permian continental deposits farther south in the San Rafael Basin (Espejo, 1990). A K-Ar radiometric date of 275 ± 10 Ma (Vilas and Valencio, 1982) suggests an early Permian age for the volcanic complex on the western margin of the Calingasta-Uspallata Basin. Similar radiometric dates (between 270 and 280 Ma) were obtained farther south, at the latitude of the San Rafael Basin, from volcanics associated with alluvial-fan deposits that rest unconformably on early Permian sediments (Dessanti and Caminos, 1967; Toubes and Spikermann, 1976; Linares, Manavella, and Piñeiro, 1979). In contrast, farther from the orogenic front, continental sedimentation continued without major interruptions throughout most of the Permian in the Paganzo Basin (Aceñolaza and Vergel, 1987). The San Rafael Orogeny is thus seen as a deformation phase genetically linked to the tectonomagmatic activity along the Panthalassan margin, with waning influence toward the interior of the continent.

Although the magmatic activity along the Panthalassan/proto-Pacific-plate margin was intermittent during the late Palaeozoic, the Choiyoi silicic volcanism along the Andean Cordillera and its equivalent in Patagonia peaked between the late part of the early Permian and the late Permian. Extensive rhyolitic ignimbrites and consanguineous airborne tuffaceous material erupted in the North Patagonian Massif (Figure 2.1) during that period. In northern Patagonia, two magmatic episodes have been identified – in the early Permian or the early part of the late Permian (280–260 Ma) and in the early Triassic (240–230 Ma) – both correlative to the Choiyoi acid-magmatic province of western Argentina (Kay et al., 1989). All the rocks are calc-alkaline, but those of the younger episode become peraluminous, with a tendency to a peralkaline composition (Caminos et al., 1988). According to Mpodozis and Kay (1992), the Choiyoi volcanism and consanguineous granitoids were emplaced in a thickened crust, possibly in the early stages of extension that followed the San Rafael compression. Although it seems reasonable to infer extensional conditions for the Choiyoi province, the limited areal extent of the subsequent Triassic extension suggests that that stage was related to an extensional stage within the evolution of an active margin, rather than being associated with cessation of subduction.

Recent studies on granitoid intrusions and consanguineous extrusions of the North Patagonian Massif indicate that several intrusives previously considered to be of early and middle Carboniferous age (Caminos et al., 1988) now have been assigned late Permian ages (258 ± 15 Ma and 259 ± 16 Ma),

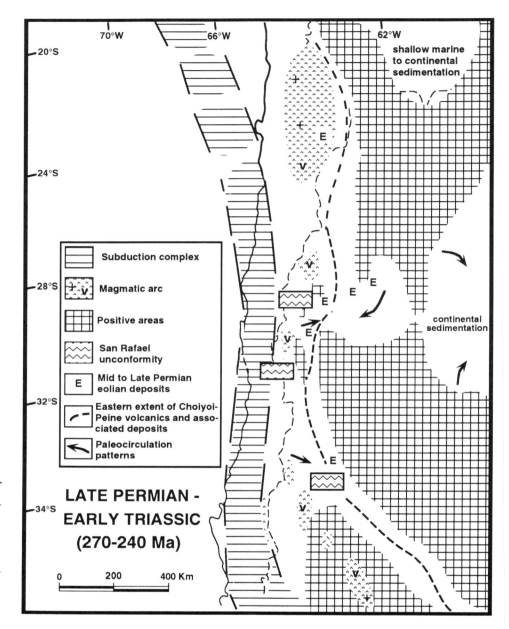

Figure 2.4. Palaeogeographic sketch of the proto-Pacific margin between 20°S and 40°S during the late part of the early Permian and the early Triassic. The Choiyoi-Peine volcanics have been identified along the entire margin and farther south in the North Patagonian Massif. Note the incipient uplift in the eastern Paganzo basin and the widespread presence of eolian deposits. In the Tarija Basin (Cangapi-Vitiacua basin of Gohrbandt, 1992), early-to-middle Permian fluvial and eolian deposits pass upward to late Permian shallow-marine limestones and evaporites (Sempere et al., 1992). (Adapted from López-Gamundí et al., 1994.)

with associated volcanics extending into the early Triassic (Pankhurst et al., 1992). Further evidence of volcanic activity contemporaneous with the sedimentation of the Tunas Formation in the Sauce Grande Basin (Figure 2.1), also reflected farther from the plate margin in equivalent units in the Paraná and Karoo basins (López-Gamundí, in press), is provided by radiometric dates ranging from 270 to 230 Ma from rhyolitic ignimbrites in northern Patagonia (Sruoga and Llambías, 1992). The age of the Tunas Formation is constrained between the early Permian (*Glossopteris* flora found in the lower half of the unit) and the latest (pre-Tatarian) Permian, on the basis of palaeomagnetic data. The Palaeozoic strata of the Sauce Grande Basin were deformed during the late Permian–early Triassic by right-lateral wrenching (Cobbold et al., 1986). That

deformation, also seen in other localities in southern South America (Figure 2.5), has been interpreted as structural inversion of previous normal faults.

Northern Chile. Strong late Permian–early Triassic magmatic activity is documented by numerous radiometric ages from northern Chilean plutons, many of which belong to caldera complexes (Breitkreuz and Zeil, 1994). Although the stratigraphic controls regarding depositional and tectonic developments in that segment of the continental margin are poor, the lack of intercalations of lacustrine/floodplain deposits in the latest Permian ignimbrites and lavas points to a cessation of intra-arc extension, which had been significant during the preceding period (E-I).

Extension II (middle–late Triassic)

The second stage of Pangaean extension (E-II) was expressed as rifting and graben formation along the Panthalassan margin. After the climax of Choiyoi volcanism during the early Permian–early Triassic, western Argentina underwent uplift/erosion, in some cases expressed as structural inversion (Figure 2.5). Subsequent extensional relaxation established a sedimentary regime in fault-bounded depressions or grabens because of collapse of portions of the late Palaeozoic and early Triassic fore-arc, arc, and back-arc terranes (Uliana and Biddle, 1988). Most of those fault-bounded basins were floored by (meta)sedimentary rocks of Palaeozoic age and Choiyoi volcanics, as extensively documented by surface and subsurface studies. That extensional phase, probably the transtensional by-product of oblique subduction and sinistral strike-slip motion during the Triassic (Pichowiak, Buchelt, and Damm, 1990), appears to have been confined to a relatively narrow belt parallel to, and inboard of, the NNW–SSE-oriented Gondwanan margin (Figure 2.5). The areally restricted Triassic rifting seems to have been intimately related to oblique plate convergence, rather than to extension associated with cessation of subduction. That limited Triassic rifting, however, predated the generalized extension and associated Jurassic magmatism that characterized the fragmentation of Gondwanaland (Dalziel et al., 1987). The middle–late Triassic sediments and associated volcanics of Chile and western Argentina were deposited within rapidly subsiding fault-bounded troughs that the sea entered from the west (Charrier, 1979; Stipanicic, 1983).

In Argentina, that mid-Triassic extensional phase resulted in restricted, nonmarine basins. The Cuyo Basin was the largest of those fault-bounded troughs located in west-central Argentina. It was characterized by local volcanism and initial rift subsidence, followed by thermal subsidence (Kokogian and Mancilla, 1989). Sedimentation in that rift basin was initiated around the mid-Triassic (Anisian–Ladinian) (Zavattieri, 1990) and was characterized by basal alluvial-fan conglomerates and braid-plain cross-bedded coarse sandstones, followed by lacustrine bituminous shales and tuffaceous mudstones. The section culminates with fluvial tuffaceous sandstones, carbonaceous shales, and subordinate alluvial-fan conglomerates. Basalts with intraplate affinities are intercalated in the sediments (Ramos and Kay, 1991).

Most Triassic syn-rift deposits in northern Chile consist largely of conglomerates, sandstones, subordinate mudstones, and basaltic-to-silicic volcanic rocks deposited in alluvial fans, braided and ephemeral rivers, lakes, and mud and evaporitic (sabkha) flats (Suárez and Bell, 1992). A transition to shallow-marine conditions during the late Triassic marked a transgression that covered large areas of northern Chile (Chong and Hillebrandt, 1985; Gröschke et al., 1988).

Concluding remarks

Convergence along the entire Panthalassan margin of southern South America was active throughout the late Palaeozoic and

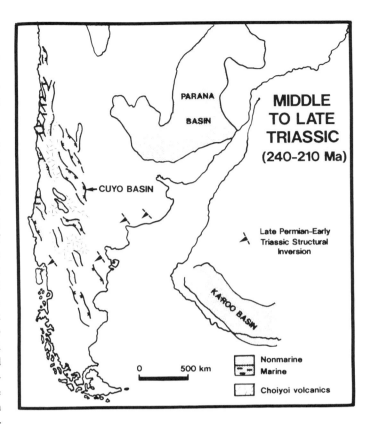

Figure 2.5. Framework for Triassic rift basins in southern America. Note NNW–SSE orientation of basins. Marine sedimentation was confined to the westernmost parts of the troughs. Data from Uliana and Biddle (1988), Ramos and Kay (1991), and Cobbold et al. (1992). Continental sedimentation prevailed in the Paraná and Karoo basins.

Triassic. The margin can be subdivided into three segments: (A) a northern segment (north of 20°S) whose evolution was dominated by pulses of extension and compression-convergence between the Arequipa Massif and the South American Craton, (B) a central part (20–40°S) characterized, from west to east in present coordinates, by a subduction complex and accretionary prism, a magmatic arc, and back-arc basins, and (C) a southern segment (south of 40°S) characterized by terrane accretion during Permian and post-Permian times and by a wide magmatic arc.

The Carboniferous-to-Triassic evolution of the central segment was characterized by (1) subduction and arc magmatism associated with an initial extensional sag phase (E-I), (2) subsequent shortening due to cratonward thrusting of the magmatic-arc foldbelt and associated subsidence (FS), and (3) limited extension associated with rifting (E-II), a precursor of the Jurassic–Cretaceous extension that led to the fragmentation of Gondwana.

The early Chañic compressional episode was followed by generalized subsidence during the middle Carboniferous through the early Permian (E-I) that in turn was terminated by the San Rafael Orogeny in the mid-Permian. The late Devonian–early

Carboniferous Eo-Hercynian (Chañic) Orogeny was followed by the first extension (E-I), which led in turn to the initial filling (mostly glacigenic) of the Calingasta-Uspallata, Paganzo, and San Rafael back-arc basins. In contrast, sedimentation was continuous through the Devonian and the lower part of the Carboniferous in the Peine Basin of northern Chile. At that latitude, the Chañic deformation probably was concentrated farther east between the rigid southeastern flank of the Arequipa Massif and the South American Craton.

The more subtle effects of both the Eo-Hercynian (Chañic) and Tardi-Hercynian (San Rafael) orogenies in basins located farther from the active continental margin reinforce the notion that tectonomagmatic activity at the plate margin triggered both episodes of compressive deformation. Whereas both episodes were characterized by angular unconformities along the margin, more tenuous effects have been identified in more distant regions. For instance, Zalán et al. (1987, 1990) stated that the Eo- and Tardi-Hercynian orogenies triggered, respectively, (1) the onset of sedimentation in the Paraná Basin and (2) the progradation of a sand-rich wedge from the eastern and western basin margins by the mid-Permian.

The Carboniferous–Permian arc magmatism and basin development of that segment of the Panthalassan margin were strongly controlled by dextral transtensional movements parallel to the margin (Rapalini and Vilas, 1991; Mpodozis and Kay, 1992). That tectonic regime may have been caused by oblique convergence of Panthalassan plates and/or by a Baja California–type plate configuration (Bahlburg and Breitkreuz, 1991). It also matches with a postulated northward drift of Laurentia away from Gondwanaland (Pichowiak et al., 1990; Dalla Salda et al., 1992; Bahlburg, 1993). The Permian–Triassic transition marked the fundamental change from an accretionary to an erosive plate margin (Kay et al., 1989; Breitkreuz, 1990). The prelude to the latter represented the sinistral transtensional setting established during the late Triassic.

In western Argentina, the change from the extensional sag phase (E-I) to compressional formation of foreland basins (FS) was related to the San Rafael Orogeny (ca. 270 Ma). Cratonward, associated foreland subsidence generated accommodation space for the deposition of thick synorogenic sections in the Sauce Grande and Karoo basins (López-Gamundí and Rossello, 1992; Cole, 1992) during the late part of the early Permian and the early part of the late Permian. The associated Choiyoi magmatic episode preceded and partially overlapped the Triassic rifting (E-II). The Triassic basins were confined to a narrow belt parallel and inboard of the continental margin, following the preexisting northwest-trending basement structural grain (Uliana and Biddle, 1988; Light et al., 1993). Those narrow troughs were filled with nonmarine sediments in Argentina and continental-to-marine deposits in Chile. Their limited areal extent indicates that the Triassic extension can be seen as a transtensional by-product of oblique convergence.

The tectonomagmatic evolution of that part of the Gondwanan margin occurred under a drastic palaeoclimatic shift characterized by widespread glaciation during the mid-Carboniferous, climatic amelioration during the late Carboniferous, and arid to semiarid conditions by the middle–late Permian.

References

Aceñolaza, F., and Vergel, M. 1987. Hallazgo del Pérmico superior fosilífero en el Sistema de Famatina. In *10th Congreso Geológico Argentino, Actas,* vol. 3, pp. 125–9.

Alvarez, L., and Fernández-Seveso, F. 1987. Estratigrafía del Cerro Horcobola (La Rioja). Su importancia como elemento de correlacíon. Nueva localidad fosilífera del Pérmico Inferior en la Argentina. In *10th Congreso Geológico Argentino, Actas,* vol. 3, pp. 121–4.

Andreis, R. R., Leguizamón, R. R., and Archangelsky, S. 1986. *El paleovalle de Malanzán: nuevos criterios para la estratigrafía del Neopaleozoico de la Sierra de los Llanos, La Rioja, República Argentina.* Boletín de la Academia Nacional de Ciencias de Córdoba, no. 57.

Bahlburg, H. 1991. *The Ordovician Back-Arc to Foreland Successor Basin in the Argentinian–Chilean Puna: Tectono-sedimentary and Sea-Level Changes.* International Association of Sedimentologists special publication 12, pp. 465–84.

Bahlburg, H. 1993. Hypothetical southeast Pacific continent revisited: new evidence from the middle Paleozoic basins of northern Chile. *Geology* 21:909–12.

Bahlburg, H., and Breitkreuz, C. 1991. Paleozoic evolution of active margin basins in the southern Central Andes (northwestern Argentina and northern Chile). *Journal of South American Earth Sciences* 4:171–88.

Bahlburg, H., and Breitkreuz, C. 1993. Differential response of a Devonian–Carboniferous platform–deeper basin system to sea-level change and tectonics, North Chilean Andes. *Basin Research* 5:21–40.

Benedetto, J. L. 1976. Foraminíferos pérmicos de la Formacion Arizaro (Provincia de Salta, Argentina). In *2nd Congreso Latinoamericano de Geología, Memorias,* vol. 2, pp. 1009–24.

Breitkreuz, C. 1986. Das Paläozoikum in den Kordilleren Nordchiles (21°–55°S). *Geotektonische Forschungen* 70:1–88.

Breitkreuz, C. 1990. Late Carboniferous to Triassic magmatism in the Central and Southern Andes: The change from an accretionary to an erosive plate margin mirrors the Pangea history. In *Symposium International "Geodynamique Andine,"* pp. 359–62. Paris: Colloques et Seminaires, ORSTOM (Office de Recherche Scientifique et Technique de Outre Mer).

Breitkreuz, C. 1991. Fluvio-lacustrine sedimentation and volcanism in a Late Carboniferous tensional intra-arc basin, northern Chile. *Sedimentary Geology* 74:173–87.

Breitkreuz, C. 1993. Epiclastic sedimentation events in a Late Paleozoic arc basin in Chile, triggered by eruption-related torrential rainfalls. *Geological Society of America, Abstracts with Programs* 25(6):A-465.

Breitkreuz, C., Bahlburg, H., Delakowitz, B., and Pichowiak, S. 1989. Paleozoic volcanic events in the Central Andes. *Journal of South American Earth Sciences* 2:171–89.

Breitkreuz, C., Bahlburg, H., and Zeil, W. 1988. The Paleozoic evolution of northern Chile: geotectonic implications. In *Lectures and Notes on Earth Sciences,* vol. 17, pp. 87–102. Berlin: Springer-Verlag.

Breitkreuz, C., Helmdach, F. F., Kohring, R., and Mosbrugger, V. 1992. Late Carboniferous intra-arc sediments in the North Chilean Andes: stratigraphy, paleogeography and paleoclimate. *Facies* 26:67–80.

Breitkreuz, C., and Van Schmus, R. 1996. U/Pb systematics of Permian ignimbrites in northern Chile. *Journal of South American Earth Sciences.*

Breitkreuz, C., and Zeil, W. 1994. The Late Carboniferous to Triassic volcanic belt in Northern Chile. In *Tectonics of the Southern Central Andes,* ed. K. J. Reutter, E. Scheuber, and P. K. Wigger, pp. 227–92. Berlin: Springer-Verlag.

Caminos, R., Cordani, U. G., and Linares, E. 1979. Geología y geocronología de las rocas metamórficas y eruptivas de la Precordillera y Cordillera Frontal de Mendoza, República Argentina. In *2nd Congreso Geológico Chileno, Actas,* vol. 1, pp. F43–61.

Caminos, R., Llambías, E., Rapela, C., and Parica, C. 1988. Late Paleozoic–early Triassic magmatic activity of Argentina and the significance of new Rb-Sr ages from northern Patagonia. *Journal of South American Earth Sciences* 1:137–45.

Charrier, R. 1979. El Triásico en Chile y regiones adyacentes de Argentina – Una reconstrucción paleogeográfica y paleoclimática. *Comunicaciones, Santiago de Chile* 26:1–47.

Charrier, R. 1986. The Gondwana glaciation in Chile: description of alleged glacial deposits and paleogeographic conditions bearing on the extension of the ice cover in southern South America. *Palaeogeography, Palaeoclimatology, Palaeoecology* 56:151–75.

Charrier, R., Godoy, E., Rebolledo, S., and Hervé, F. 1991. Is Early Paleozoic accretion of the Choapa metamorphic complex in central Chile compatible with the Precordillera terrane? *Comunicaciones, 5th International Circum Pacific Terrane Conference (Santiago, Chile)* 42:44–8.

Chong, G., and Hillebrandt, A. 1985. El Triásico preandino de Chile entre los 23° 30' y 26° 00' de latitud Sur. In *4th Congreso Geológico Chileno, Actas,* vol. 1, pp. I162–209.

Cingolani, C., Dalla Salda, L., Hervé, F., Munizaga, F., Pankhurst, R. J., Parada, M. A., and Rapela, C. W. 1991. The magmatic evolution of northern Patagonia: new impressions of pre-Andean and Andean tectonics. In *Andean Magmatism and Its Tectonic Setting,* ed. R. S. Harmon and C. W. Rapela, pp. 29–43. Geological Society of America special paper 265.

Cobbold, P. R., Gapais, D., Rossello, E. R., Milani, E. J., and Sztmari, P. 1992. Permo-Triassic deformation in SW Gondwana. In *Inversion Tectonics of the Cape Fold Belt, Karoo and Cretaceous Basins of Southern Africa,* ed. M. J. de Wit and I. G. D. Ransome, pp. 23–6. Rotterdam: Balkema.

Cobbold, P., Massabie, A. C., and Rossello, E. A. 1986. Hercynian wrenching and thrusting in the Sierras Australes foldbelt, Argentina. *Hercynica II* 2:135–48.

Coira, B. L., Davidson, J. D., Mpodozis, C., and Ramos, V. A. 1982. Tectonic and magmatic evolution of the Andes of northern Argentina and Chile. *Earth Science Review* 18:451–70.

Cole, D. I. 1992. Evolution and development of the Karro Basin. In *Inversion Tectonics of the Cape Fold Belt, Karroo and Cretaceous Basins of Southern Africa,* ed. M. J. de Wit and I. G. D. Ransome, pp. 87–100. Rotterdam: Balkema.

Cortés, J. M. 1985. Vulcanitas y sedimentitas lacustres en la base del Grupo Choiyoi al sur de la Estancia Tambillos, Mendoza, Argentina. In *4th Congreso Geológico Chileno, Actas,* vol. 1, pp. 89–108.

Dalla Salda, L. H., Dalziel, I. W. D., Cingolani, C. A., and Varela, R. 1992. Did the Taconic Appalachians continue into southern South America? *Geology* 20:1059–62.

Dalmayrac, B., Laubacher, G., Marocco, R., Martinez, R., and Romasi, P. 1980. La Chaine hercynienne d'Amerique du Sud, structure et evolution d'un orogene intracratonique. *Geologische Rundschau* 69:1–21.

Dalziel, I. W. D., and Forsythe, R. D. 1985. Andean evolution and the terrane concept. In *Tectonostratigraphic Terranes of the Circum-Pacific Region,* ed. D. G. Howell, pp. 565–81. Earth sciences series no. 1. Houston, TX: Circum-Pacific Council for Energy and Mineral Resources.

Dalziel, I. W. D., Storey, B. C., Garrett, S. W., Grunow, A. M., Herrod, L. D. B., and Pankhurst, R. J. 1987. Extensional tectonics and the fragmentation of Gondwanaland. In *Continental Extensional Tectonics,* ed. M. P. Coward, J. F. Dewey, and P. L. Hancock, pp. 433–41. Geological Society of London special publication no. 28.

Davidson, J., Mpodozis, C., Godoy, E., Hervé, F., Pankhurst, R., and Brook, M. 1987. Late Paleozoic accretionary complexes on the Gondwana margin of southern Chile; Evidence from the Chonos Archipelago. In *Gondwana Six: Structure, Tectonics and Geophysics,* ed. G. D. McKenzie, pp. 221–7. Geophysical monograph series no. 40. Washington, DC: American Geophysical Union.

Davidson, J., Ramírez, C. F., Gardeweg, M., Hervé, M., Brook, M., and Pankhurst, R. 1985. Calderas del Paleozoico Superior–Triasico Superior y mineralización asociada en la Cordillera de Domeyko, Norte de Chile. *Communicaciones* 35:53–7.

Dessanti, N. R., and Caminos, R. 1967. Edades Potasio–Argón y posición estratigráfica de algunas rocas ígneas y metamórficas de la Precordillera, Cordillera Frontal y Sierras de San Rafael, provincia de Mendoza. *Revista Asociación Geológica Argentina* 22:135–62.

Díaz Martínez, E., Isaacson, P. E., and Sablock, P. E. 1993. Late Paleozoic latitudinal shift of Gondwana: stratigraphic, sedimentologic and biogeographic evidence from Bolivia. *Documents des Laboratoires de Géologie de Lyon* 125:119–38.

Donato, E. O., and Vergani, G. 1985. Geología del Devónico y Neopaleozoico de la Zona del Cerro Rincón, Provincia de Salta, Argentina. In *4th Congreso Geológico Chileno, Actas,* pp. 262–83.

Espejo, I. S. 1990. Análisis estratigráfico, paleoambiental y de proveniencia de la Formación El Imperial en los alrededores de los ríos Diamante y Atuel (Provincia de Mendoza). Ph.D. thesis, University of Buenos Aires.

Fernández-Seveso, F., Pérez, M. A., and Alvarez, L. 1990. Análisis estratigráfico del ámbito occidental de la cuenca de Paganzo, en el rango de grandes ciclos deposicionales. In *11th Congreso Geológico Argentino, Actas,* vol. 2, pp. 77–80.

Flint, S., Turner, P., Jolley, E. J., and Hartley, A. J. 1993. Extensional tectonics in convergent margin basins: an example from the Salar de Atacama, Chilean Andes. *Geological Society of America Bulletin* 105:603–17.

Forsythe, R. D. 1982. The late Paleozoic to early Mesozoic evolution of southern South America: a plate tectonic interpretation. *Journal of the Geological Society of London* 139:671–82.

Forsythe, R. D., Davidson, J., Mpodozis, C., and Jesinkey, C. 1993. Lower Paleozoic relative motion of the Arequipa block and Gondwana: paleomagnetic evidence from Sierra de Almeida of northern Chile. *Tectonics* 12:219–36.

Frakes, L. A., and Crowell, J. C. 1969. Late Paleozoic Glaciation: I. South America. *Geological Society of America Bulletin* 80:1007–42.

Gohrbandt, K. H. A. 1992. Western Gondwana margin as an environment for Paleozoic hydrocarbon accumulations. *Journal of South American Earth Sciences* 6:267–87.

González Bonorino, G., and González Bonorino, F. 1991. Precordillera de Cuyo y Cordillera Frontal en al Paleozoico Temprano: terrenos 'bajo sospecha' de ser autóctonos. *Revista Geológica de Chile* 18:97–107.

Gröschke, M., von Hillebrandt, A., Prinz, P., Quinzio, L. A., and Wilke, H. G. 1988. Marine Mesozoic paleogeography in northern Chile between 21°–26°S. In *Lectures and Notes in Earth Sciences*, vol. 17, pp. 105–17. Berlin: Springer-Verlag.

Helwig, J. 1972. Stratigraphy, sedimentation, paleogeography, and paleoclimates of Carboniferous ("Gondwana") and Permian of Bolivia. *American Association of Petroleum Geologists Bulletin* 56:1008–33.

Hervé, F. 1988. Late Paleozoic subduction and accretion in southern Chile. *Episodes* 11:183–8.

Isaacson, P. E. 1975. Evidence for a western extracontinental land source during the Devonian period in the central Andes. *Bulletin of the Geological Society of America* 86:39–46.

Kay, S. M., Ramos, V. A., Mpodozis, C., and Sruoga, P. 1989. Late Paleozoic to Jurassic silicic magmatism at the Gondwana margin: analogy to the middle Proterozoic in North America? *Geology* 17:324–8.

Kokogian, D. A., and Mancilla, O. H. 1989. Análisis estratigráfico secuencial de la Cuenca Cuyana: secuencias deposicionales continentales. In *Cuencas Sedimentarias Argentinas*, ed. G. Chebli and L. Spalletti, pp. 169–202. Serie Correlación Geológica no. 6, Universidad Nacional de Tucumán.

Laubacher, G., and Megard, F. 1985. The Hercynian basement: a review. In *Magmatism at a Plate Edge – The Peruvian Andes*, ed. W. S. Pitcher, M. P. Atherton, E. J. Cobbing, and R. D. Beckinsale, pp. 29–35. London: Wiley.

Light, M. P. R., Keeley, M. L., Maslanyj, M. P., and Urien, C. M. 1993. The tectonostratigraphic development of Patagonia, and its relevance to hydrocarbon exploration. *Journal of Petroleum Geology* 16:465–81.

Limarino, C. O., Sessarego, H., Cesari, S. N., and López-Gamundí, O. R. 1986. El perfil de la Cuesta de Huaco, estratotipo de referencia (hipoestratotipo) del Grupo Paganzo en la Precordillera Central. *Academia Nacional de Ciencias Exactas, Físicas y Naturales, Buenos Aires, Anales* 38:81–109.

Limarino, C. O., and Spalletti, L. 1986. Eolian Permian deposits in west and northwest Argentina. *Sedimentary Geology* 49:109–27.

Linares, E., Manavella, M. A., and Piñeiro, A. 1979. Geocronología de las rocas efusivas de las zonas de los yacimientos "Dr. Baulíes" y "Los Reyunos," Sierra Pintada de San Rafael, Mendoza. República Argentina. In *7th Congreso Geológico Argentino, Actas*, vol. 2, pp. 13–21.

Llambías, E. J., and Sato, A. M. 1990. El batolito de Colangüil (29°–31°S): Estructura y marco tectónico. *Revista Geológica de Chile* 17:89–108.

López-Gamundí, O. R. 1987. Depositional models for the glaciomarine sequences of Andean late Paleozoic basins of Argentina. *Sedimentary Geology* 52:109–26.

López-Gamundí, O. R. 1989. Postglacial transgressions in late Paleozoic basins of western Argentina: a record of glacioeustatic sea level rise. *Palaeogeography, Palaeoclimatology, Palaeoecology* 71:257–70.

López-Gamundí, O. R. In press. Permian volcanic activity along the proto-Pacific plate margin reflected in the basins of southern South America. In *Proceedings, 9th Gondwana Symposium, Hyderabad (India)*.

López-Gamundí, O. R., Alvarez, L., Andreis, R. R., Bossi, G. E., Espejo, I. S., Fernández-Seveso, Kokogian, D., Legarreta, L., Limarino, C. O., and Sessarego, H. 1989. Cuencas intermontanas. In *Cuencas Sedimentarias Argentinas*, ed. G. A. Chebli and L. A. Spalletti, pp. 123–68. Serie Correlación Geológica no. 6, Universidad Nacional de Tucumán.

López-Gamundí, O. R., Cesari, S. N., and Limarino, C. O. 1993. Paleoclimatic significance and age constraints of the Carboniferous coals of Paganzo basin. In *Gondwana Eight*, ed. R. H. Findlay, R. Unrug, M. R. Banks, and J. J. Veevers, pp. 291–8. Rotterdam: Balkema.

López-Gamundí, O. R., Espejo, I. S., Conaghan, P. J., and Powell, C. M., 1994. Southern South America. In *Permian–Triassic Basins and Foldbelts Along the Panthalassan Margin of Gondwanaland*, ed. J. J. Veevers and C. M. Powell, pp. 281–330. Geological Society of America memoir no. 184.

López-Gamundí, O. R., Limarino, C. O., and Cesari, S. N. 1992. Late Paleozoic paleoclimatology of central western Argentina. *Palaeogeography, Palaeoclimatology, Palaeoecology* 91:305–29.

López-Gamundí, O. R., and Rossello, E. A. 1992. La cuenca interserrana de Claromecó, Argentina: un ejemplo de cuenca de antepaís hercínica. In *8th Congreso Geológico Latinoamericano (Salamanca, Spain), Actas*, pp. 55–9.

López-Gamundí, O. R., and Rossello, E. A. 1993. Devonian–Carboniferous unconformity in Argentina and its relation to the Eo-Hercynian orogeny in southern South America. *Geologische Rundschau* 82:136–47.

Megard, F. 1978. *Étude geologique des Andes de Perou central: contribution à l'étude des Andes*, vol. 1. ORSTROM memoir no. 83. Paris: Office de Recherche Scientifique et Technique de Outre Mer.

Mpodozis, C., and Forsythe, R. D. 1983. Stratigraphy and geochemistry of accreted fragments of the ancestral Pacific Ocean floor in southern South America. *Palaeogeography, Palaeoclimatology, Palaeoecology* 41:103–24.

Mpodozis, C., and Kay, S. M. 1992. Late Paleozoic to Triassic evolution of the Gondwana margin: evidence from Chilean Frontal Cordilleran batholiths (28° to 31°S). *Geological Society of America Bulletin* 104:999–1014.

Mpodozis, C., Marinovic, C., Smoje, I., and Cuitiño, L. 1993. Estudio geológico-estructural de la Cordillera de Domeyko entre Sierra Limón Verde y Sierra Mariposas, Región de Antofagasta. Unpublished report. Santiago de Chile: SERNAGEOMIN-CODELCO.

Nasi, C., Mpodozis, C., Cornejo, P., Moscoso, R., and Maksaev, V. 1985. El Batolito Elqui-Limari (Paleozoico superior-Triásico): Características petrográficas, geoquímicas y significado tectónico. *Revista Geológica de Chile* 24/25:77–111.

Nasi, C., and Sepúlveda, P. 1986. Avances en el conocimiento del Carbonífero en el norte de Chile. In *Annual Meeting of the Working Group, IUGS-IGCP Project 211, Late Paleozoic of South America, Abstracts*, pp. 27–43.

Niemeyer, H., Urzúa, F., Aceñolaza, F. G., and González, C. 1985. Progresos recientes en el conocimiento del Paleozoico de la región de Antofagasta. In *4th Congreso Geológico Chileno, Actas*, vol. 1, pp. 410–38.

Pankhurst, R. J., Rapela, C. W., Caminos, R., Llambías, E. J., and Parica, C. 1992. A revised age for the granites of the central Somuncura batholith, North Patagonian Massif. *Journal of South American Earth Sciences* 5:321–6.

Pichowiak, S., Buchelt, M., and Damm, K.-W. 1990. *Magmatic Activity and Tectonic Setting of the Early Stages of the Andean Cycle in Northern Chile.* Geological Society of America special paper 241, pp. 127–41.

Powell, C. M., Veevers, J. J., Collinson, J., Conaghan, P. J., and López-Gamundí, O. R. 1991. Late Carboniferous to Early Triassic foreland basin behind the Panthalassa margin of Gondwanaland. In *8th International Symposium on Gondwana (Hobart, Australia), Abstracts,* p. 67.

Ramírez, C. F., and Gardeweg, M. 1982. *Hoja Toconao – Carta Geológica de Chile,* vol. 54. Santiago de Chile: Servicio Nacional de Geología y Minería.

Ramos, V. A. 1984. Patagonia: un continente a la deriva? In *9th Congreso Geológico Argentino, Actas,* vol. 2, pp. 311–28.

Ramos, V. A. 1986. Discussion on "Tectonostratigraphy as applied to analysis of South African Phanerozoic basins," by H. de la R. Winter. *Transactions of the Geological Society of South Africa* 89:427–9.

Ramos, V. A. 1988a. The tectonics of the Central Andes; 30° to 33°S latitude. In *Processes in Continental Lithospheric Deformation,* ed. S. P. Clark, Jr., B. C. Burchfiel, and J. Suppe, pp. 31–54. Geological Society of America memoir 218.

Ramos, V. A. 1988b. Late Proterozoic–early Paleozoic of South America – a collisional history. *Episodes* 11:168–74.

Ramos, V. A., Jordan, T., Allmendinger, R., Mpodozis, C., Kay, S. M., Cortés, J., and Palma, M. 1986. Paleozoic terranes of the central Argentine–Chilean Andes. *Tectonics* 5:855–80.

Ramos, V. A., and Kay, S. M. 1991. Triassic rifting and associated basalts in the Cuyo basin, central Argentina. In *Andean Magmatism and Its Tectonic Setting,* ed. R. S. Harmon and C. W. Rapela, pp. 79–91. Geological Society of America special paper 265.

Rapalini, A. E., Turner, P., Flint, S., and Vilas, J. F. 1993. Paleomagnetism of the Carboniferous Tepuel Group, central Patagonia, and the late Paleozoic Gondwana path. *Annales Geophysicae (Suppl. I)* 10:25.

Rapalini, A. E., and Vilas, J. F. 1991. Tectonic rotations in the late Paleozoic continental margin of southern South America determined and dated by paleomagnetism. *Geophysical Journal International* 107:333–51.

Sempere, T. 1987. Caracteres geodinámicos generales del Paleozoico superior de Bolivia. In *Annual Meeting IGCP Project 211, Late Paleozoic of South America, Santa Cruz (Bolivia), Abstracts,* pp. 9–19.

Sempere, T., Aguilera, E., Doubinger, J., Janvier, P., Lobo, J., Oller, J., and Wenz, S. 1992. La Formation de Viticua (Permien moyen a supérieur – Trias ?inférieur), Bolivie du Sud: stratigraphie palynologie et paléontologie. *Neues Jahrbuch für Geologie und Paläontologie, Abhandlungen* 185:239–53.

Sessarego, H. L., Amos, A. J., Texeira, W., Kawashita, K., and Remesal, M. 1990. Diques eocarbónicos en la Precordillera Occidental, margen oeste de las Sierras del Tigre, provincia de San Juan. *Revista Asociación Geológica Argentina* 45:98–106.

Sruoga, P., and Llambías, E. J. 1992. Permo–Triassic leu-

corhyolitic ignimbrites at Sierra de Lihue-Calel, La Pampa province, Argentina. *Journal of South American Earth Sciences* 5:141–52.

Stipanicic, P. 1983. The Triassic of Argentina and Chile. In *The Phanerozoic Geology of the World. II. The Mesozoic,* ed. M. Moullade and E. Nairn, pp. 181–99. Amsterdam: Elsevier.

Suárez, M., and Bell, C. M. 1992. Triassic rift-related sedimentary basins in northern Chile (24°–29°S). *Journal of South American Earth Sciences* 6:109–21.

Toubes, R. O., and Spikermann, J. P. 1976. Algunas edades K-Ar para la Sierra Pintada, provincia de Mendoza. *Revista Asociación Geológica Argentina* 31:118–26.

Turner, J. C. M., and Méndez, V. 1975. Geología del sector oriental de los Departamentos de Santa Victoria e Iruya, provincia de Salta, República Argentina. *Boletín Academia Nacional de Ciencias de Córdoba* 51:11–24.

Uliana, M., and Biddle, K. T. 1987. Permian to Late Cenozoic evolution of Northern Patagonia: main tectonic events, magmatic activity and depositional trends. In *Gondwana Six: Structure, Tectonics and Geophysics,* ed. G. D. McKenzie, pp. 271–86. Geophysical monograph series 40. Washington, DC: American Geophysical Union.

Uliana, M. A., and Biddle, K. T. 1988. Mesozoic–Cenozoic paleogeographic and geodynamic evolution of southern South America. *Revista Brasileira de Geociencias* 18:172–90.

Veevers, J. J., and Powell, C. M. 1987. Late Paleozoic glacial episodes in Gondwanaland reflected in transgressive-regressive depositional sequences in Euramerica. *Bulletin of the Geological Society of America* 98:475–87.

Veevers, J. J., Powell, C. M., Collinson, J. W., and López-Gamundí, O. R. 1994. Synthesis. In *Permian–Triassic Basins and Foldbelts Along the Panthalassan Margin of Gondwanaland,* ed. J. J. Veevers and C. M. Powell, pp. 331–54. Geological Society of America memoir no. 184.

Vilas, J. F., and Valencio, D. A. 1982. Implicancias geodinámicas de los resultados paleomagnéticos de formaciones asignadas al Paleozoico Tardío–Mesozoico del centro-oeste argentino. In *5th Congreso Latinoamericano de Geología, Actas,* vol. 3, pp. 743–58.

Visser, J. N. J. 1989. The Permo–Carboniferous Dwyka Formation of southern Africa: deposition by a predominantly subpolar marine ice sheet. *Palaeogeography, Palaeoclimatology, Palaeoecology* 70:377–91.

Zalán, P. V., Wolff, S., Astolfi, M. A., Vieira, I. S., Conceiçao, J. C., Appi, V. T., Neto, E. V., Cerqueira, J. R., and Marques, A. 1990. The Paraná basin, Brazil. In *Interior Cratonic Basins,* ed. M. W. Leighton, D. R. Kolata, D. R. D. F. Oltz, and J. J. Eidel, pp. 681–708. American Association of Petroleum Geologists memoir 51.

Zalán, P. V., Wolff, S., Conceiçao, J. C., Astolfi, I., Vieira, S., Appi, V., and Zanotto, O. 1987. Tectonica e sedimentaçao da Bacia do Paraná. In *3rd Simposio Sul-Brasileiro de Geologia, Atas,* vol. 1, pp. 441–77.

Zavattieri, A. M. 1990. Stratigraphic and paleoecologic evaluation of the palynofloras of the Triassic Las Cabras Formation at the type locality Cerro Las Cabras, Mendoza, Argentina. *Neues Jahrbuch für Geologie und Paläontologie Abhandlungen* 181:117–42.

3   Permian and Triassic geologic events in Sonora, northwestern Mexico

CARLOS GONZÁLEZ-LEÓN, SPENCER G. LUCAS, and JAIME ROLDÁN-QUINTANA

The Palaeozoic and Mesozoic strata of northwestern Mexico are exposed in small, scattered areas in Sonora and Baja California because of the Tertiary tectonic events in normal faulting, the pervasive effects of Tertiary intrusives, and the widespread volcanic cover of the Sierra Madre Occidental (Figure 3.1). However, the early investigations, followed by more recent geologic work in that region, allow us to unravel the late Palaeozoic and early Mesozoic sedimentologic and tectonic history of that part of Mexico.

In this chapter we examine the sedimentologic and tectonic history of the Permian and Triassic geologic events in northwestern Mexico, with special reference to the state of Sonora. The younger Palaeozoic rocks of Sonora are of early and middle Permian age. These rocks are parts of thicker Palaeozoic successions that are of cratonic-platform character in the northeastern part of the state, of shallow-water miogeoclinal character in the northwestern and central parts of Sonora, and of deep-water eugeoclinal character in the central part of the state (Figure 3.2). Permian rocks of eugeoclinal character have been reported from one locality in Baja California. The Triassic rocks of Sonora, which are of Carnian and Norian age, can be grouped into two lithotectonic assemblages. In the central part of the state, these rocks are recognized as part of the Barranca Group (Alencaster, 1961a), whereas in northwestern Sonora they are included in the Antimonio Formation (González-León, 1980). Upper Triassic rocks are also reported from a few localities in the west-central part of the Baja California Peninsula, where they are parts of ophiolitic sequences included in accreted terranes.

The late Palaeozoic–early Mesozoic geologic history of northwestern Mexico was marked by an important event of tectonic thrusting of the Palaeozoic eugeoclinal strata against the Palaeozoic miogeoclinal strata. Highly deformed eugeoclinal rocks of the Sonoran Orogen (Poole, Madrid, and Oliva-Becerril, 1991) in the central part of the state record that tectonic event. It has been suggested that this zone of deformation extends eastward into Chihuahua to join the Marathon Belt, whereas westward it continues northwestward along the eastern coast of Baja California to join the Antler Orogen (Poole et al., 1991). The miogeoclinal strata of Sonora were originally part of the North American Terrane, but during the middle Jurassic they

were offset and translated southeastward to form the Caborca Terrane of Coney and Campa (1987). On the other hand, strata of the eugeoclinal zone have been recognized as a different terrane, named the Cortes Terrane by Coney and Campa (1987).

Permian rocks

Outcrops of Permian strata in Sonora representing mainly marine deposits are known from the northwestern, northeastern, and central regions of the state (Figure 3.3–3.5), although in most of those areas they have been poorly described stratigraphically.

Palaeozoic rocks in northeastern Sonora crop out in several ranges (Figure 3.1). The most nearly complete Palaeozoic section (1,200 m thick) has been measured in Sierra del Tule (González-León, 1986), where Permian strata of the Upper Pennsylvanian–Lower Wolfcampian Earp Formation, the Wolfcampian Colina Formation, and the Leonardian Epitaph, Scherrer, and Concha formations (Figure 3.3) have thicknesses ranging between 200 and 650 m. This Permian section is well correlated with the same formations in southeastern Arizona (Schreiber et al., 1990) and is mainly composed of fossiliferous carbonates and interbedded siliciclastic rocks representing shallow- and marginal-marine deposits of the northern Pedregosa Basin.

Permian rocks in northwestern Sonora belong to the Monos Formation (Cooper and Arellano, 1946), whose only outcrop is in the northern part of the Sierra del Alamo (Figures 3.2 and 3.6). The Monos Formation is a 500-m-thick fossiliferous, shallow-water, terrigenous carbonate unit whose base is not exposed. It is of Guadalupian age and is the youngest Permian formation in Sonora.

In central Sonora, the Permian rocks consist of both miogeoclinal and eugeoclinal strata (Figure 3.2). Outcrops of these strata occur in several ranges located between Hermosillo in the west and the Sahuaripa Valley in the east.

Miogeoclinal strata of central Sonora

The miogeoclinal Palaeozoic section of central Sonora is a very thick succession formed mainly by carbonates. The lower part

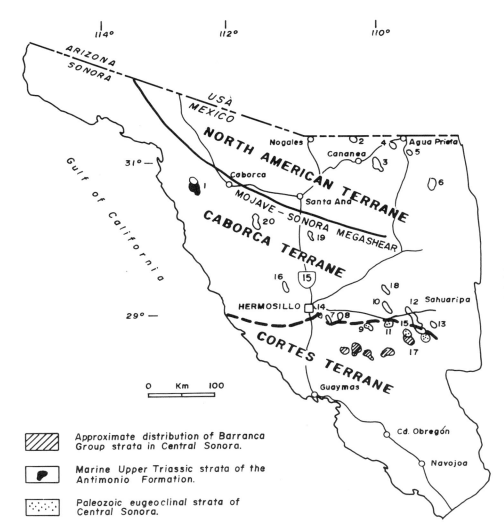

Figure 3.1. Distribution of Permian and Triassic localities mentioned in this report: (1) Sierra del Alamo, (2) Sierra del Tule, (3) Sierra los Ajos, (4) Cerro Morita, (5) Cerro Cabullona, (6) Sierra del Tigre, (7) Cerros Willard, (8) Sierra Santa Teresa, (9) Cerro Cobachi, (10) Sierra Martinez–Cerro los Chinos, (11) Barita de Sonora mine, (12) Sierra El Encinal–Sierra los Pinitos, (13) Arivechi, (14) Sierra la Flojera, (15) Sierra El Aliso, (16) Sierra de López, (17) Sierra San Javier, (18) Sierra Agua Verde, (19) Sierra Caracahui, (20) Sierra de Santa Rosa. Also indicated are the boundaries of the North American, Caborca, and Cortes terranes and the approximate trace of the Mojave–Sonora megashear.

AGE		NE SONORA	C SONORA (miogeoclinal)	C SONORA (eugeoclinal)	NW SONORA
TRIASSIC J	E			Barranca Group	Antimonio Fm.
	L				
	M				
	E				
	O				
PERMIAN G	G				Monos Fm.
	L	Concha Scherrer Epitaph	Picacho Colorado Limestone & El Tigre Formation	Sierra Santa Teresa section; Mina Mexico Fm.; La Vuelta Colorada Fm.	
	W	Colina Earp			

Figure 3.2. Correlation chart indicating the most important lithostratigraphic units of Permian and Triassic age in Sonora.

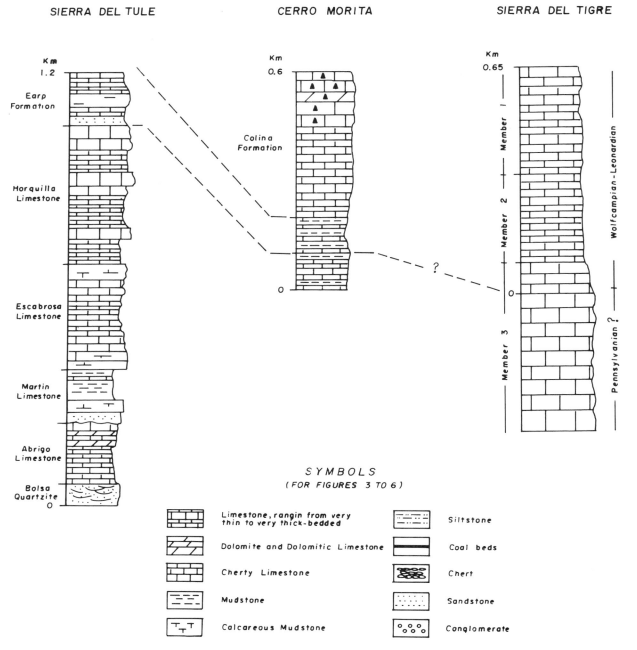

Figure 3.3. Typical Palaeozoic stratigraphic succession in northeastern Sonora, in the Sierra del Tule (according to González-León, 1986), and correlative Permian rocks of Cerro Morita (according to Peiffer-Rangin, 1982) and the Sierra del Tigre (according to Imlay, 1939).

of this section (Cambrian-to-Devonian strata) is best exposed in the Sierra de López (Stewart et al., 1990) and in the nearby area of Rancho las Norias, whereas its upper part is well represented in Sierra Santa Teresa (Stewart and Amaya-Martinez, 1993) (Figure 3.4). The Permian miogeoclinal section of central Sonora is best exposed in the Cerro Cobachi area (Figure 3.4), as reported by Noll (1981). The Upper Palaeozoic section of Cerro Cobachi consists of more than 1,450 m of (?)Mississippian-to-Permian rocks. According to Noll (1981),

the Permian section at that locality, which forms the upper part of the Picacho Colorado Limestone, should be designated the type section for the Lower Permian rocks of central Sonora. The Picacho Colorado Limestone consists of thinly bedded to very thickly bedded cherty limestones and marbles, with minor beds containing bryozoans, gastropods, crinoids, and fusulinids of Leonardian age.

Other well-studied outcrops of Lower Permian rocks in central Sonora include the 1,019-m-thick El Tigre Formation (Schmidt,

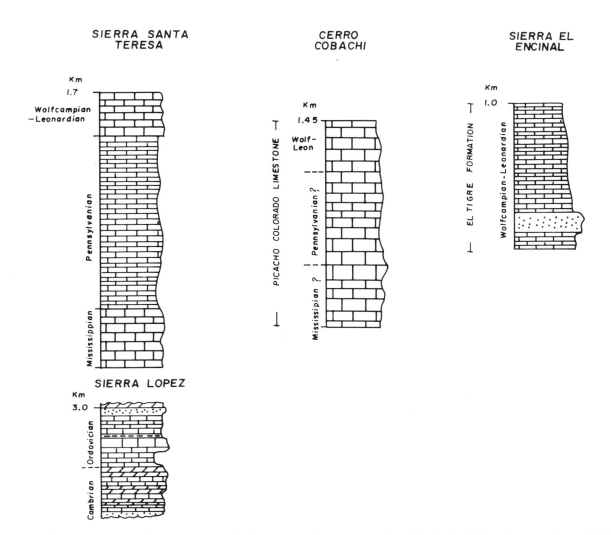

Figure 3.4. Generalized sections. Left: Composite miogeoclinal Palaeozoic section for central Sonora (Sierra de López section from Stewart et al., 1990; Sierra Santa Teresa section drawn from information taken from Stewart and Amaya-Martinez, 1993); poorly studied

Devonian rocks from Rancho las Norias (Vega and Araux, 1987) are not indicated. Middle: Miogeoclinal Permian formation of Cerro Cobachi (drawn according to information taken from Noll, 1981). Right: Section of the Sierra El Encinal (according to Hewett, 1978).

1978) from Sierra El Encinal (Figure 3.4), which consists of a lower member of dolomitic mudstone, a middle member of quartz-arenite, and an upper member of limestone and subordinate sandstone, with fusulinids of late Wolfcampian–early Leonardian age. From the Arivechi region, Perez-Ramos (1992) described a 170-m-thick section of fossiliferous massive limestone with interbedded red sandstone of Wolfcampian–Leonardian age. From the Sierra Santa Teresa, near Hermosillo, Stewart and Amaya-Martinez (1993) reported a carbonate section several hundred metres thick, with fusulinids of Wolfcampian–Leonardian age. Other poorly studied outcrops of Lower Permian strata, also consisting of fossiliferous, shallow-water carbonate sequences of Wolfcampian-to-Leonardian age, occur in several areas, including the Sierra Martinez–Cerro los Chinos, Rancho las Norias, Cerros Willard, and Sierra los Pinitos (Peiffer-Rangin, 1982).

Eugeoclinal strata of central Sonora

Lower Permian deep-water facies in central Sonora (Figure 3.1) cap a thick Ordovician, Devonian, Mississippian, and Pennsylvanian eugeoclinal assemblage that has been reported from the Barita de Sonora area (Poole et al., 1991) (Figure 3.5).

Lower Permian eugeoclinal rocks are known from the Sierra la Flojera area (Stewart et al., 1990), where a 290-m-thick section of siltstone, fine-grained sandstone, detrital limestone, and conglomerate is in close, concealed contact with miogeoclinal Lower Permian rocks of the Cerros Willard (Figure 3.5). The rocks of this section have been interpreted by Stewart et al. (1990) as turbiditic and debris-flow deposits of Wolfcampian age, based on fusulinids transported into the detrital limestone. A sequence lithologically similar to that of the Sierra la Flojera, with fusulinids of Leonardian age, has been reported to be in

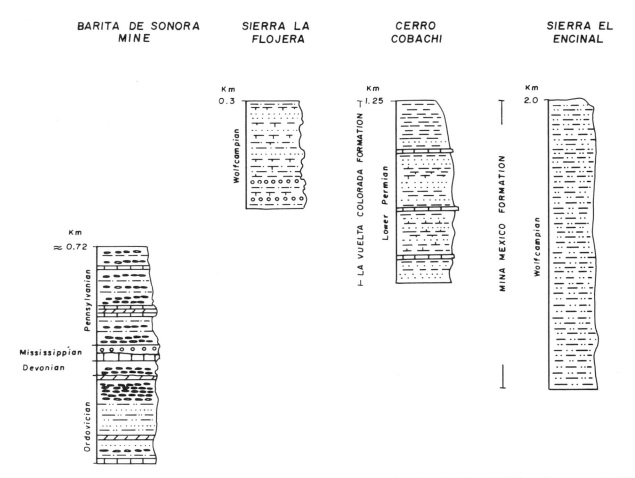

Figure 3.5. Generalized Palaeozoic eugeoclinal section for the Barita de Sonora mine area (according to Poole et al., 1991) and Permian eugeoclinal sections from the Sierra la Flojera (Stewart et al., 1990), Cerro Cobachi (Noll, 1981), and Sierra El Encinal (Hewett, 1978).

unclear contact with miogeoclinal strata in the Sierra Santa Teresa (Stewart and Amaya-Martinez, 1993).

In the Cerro Cobachi area, the Vuelta Colorada Formation (Noll, 1981), with a minimum thickness of 1,225 m, consists of calcareous siltstone, argillite, lenses of carbonate debris, sandy limestone, bedded chert, and nodular and bedded barite (Figure 3.5). The limestone contains fragments of crinoids and fusulinids. Ketner and Noll (1987) interpreted the Vuelta Colorada Formation as a Permian and Pennsylvanian succession forming part of the siliceous, deep-water assemblage of Cerro Cobachi.

The approximately 2,000-m-thick Mina Mexico Formation (Hewett, 1978; Schmidt, 1978) of the Sierra El Encinal, which is a lithologically homogeneous unit consisting of argillite and siltstone, has been considered by Stewart et al. (1990) as a largely turbiditic deposit, with abundant *Nereites*-association trace fossils and conodonts of Wolfcampian age (Figure 3.5). From the Sierra El Aliso, Bartolini (1988) reported a turbiditic unit of conglomeratic limestone with reworked fusulinids of Wolfcampian age. A turbiditic succession of Leonardian age, lithologically very similar to that from Sonora, has been reported from the eastern coast of Baja California (Delattre, 1984).

Triassic rocks

Triassic rocks have been recognized in the northwestern and central parts of Sonora (Figures 3.1 and 3.2). In northwestern Sonora, the Upper Triassic sedimentary succession that forms the lower member of the Antimonio Formation in the Sierra del Alamo is overlain, with a probable disconformity (Lucas, 1993), by the upper member of that formation, which is dated as early Jurassic (Figures 3.6 and 3.7G–H). Lower Jurassic strata similar to the upper member of the Antimonio Formation are found at various localities in north-central Sonora, including the Sierra de Santa Rosa (Hardy, 1981) and the Sierra Caracahui (Palafox, Mendoza, and Minjarez, 1992) (Figure 3.7E–F).

The Upper Triassic succession in the Sierra del Alamo is 1,350 m thick and consists in its lower part of shallow-marine siliciclastic and carbonate sediments, with minor interbeds of lenticular conglomerates, that grade upward to a deep-marine interval of mudstone, siltstone, and bioclastic limestone, with an abundant fauna of ammonites and belemnites of late Carnian age. The uppermost part of the Upper Triassic succession is composed of an interval of shallow-water fossiliferous limestone and minor

A

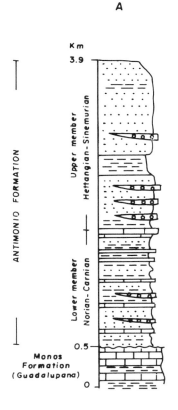

B

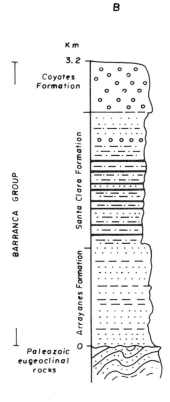

Figure 3.6. Stratigraphic successions of Triassic rocks of Sonora: (A) Upper Triassic–Lower Jurassic Antimonio Formation of the Sierra del Alamo (data from González-León, 1980) and (B) Upper Triassic Barranca Group of the Sierra San Javier (data from Stewart and Roldán-Quintana, 1991).

interbeds of siliciclastic rocks; the fossils in that interval are varied and abundant, including 30 taxa of colonial corals, sponges, spongiomorphs, brachiopods, gastropods, bivalves, and ammonites of Norian age (Stanley et al., 1994) (Figure 3.5). Other rocks correlative with the Carnian and Norian intervals of the Antimonio Formation are fossiliferous strata exposed in the Sierra la Flojera (Lucas and González-León, 1994) and in the eastern foothills of the Sierra Santa Teresa (Stanley and González-León, 1995).

Significantly different rocks of Carnian age in Sonora are those of the approximately 3,000-m-thick Barranca Group, whose outcrops are widely distributed in the central parts of the state (Figure 3.1). The type locality for the Barranca Group, however, is located in the Sierra San Javier (Figure 3.6). This group consists, from the base upward, of the Arrayanes, Santa Clara, and Coyotes formations (Alencaster, 1961a). The Arrayanes consists of sandstone, siltstone, and minor conglomerate beds of probable fluvial origin (Stewart and Roldán-Quintana, 1991) (Figure 3.7A). The Santa Clara Formation consists of interstratified siltstone, sandstone, conglomerate, carbonaceous shale, and coal beds (Figure 3.7B–C). Abundant plant fossils (134 genera) have been reported from the finer-grained parts of that formation (Silva-Pineda, 1961; Weber, 1980; Weber, Zambrano, and Amozurrutia, 1980; Weber et al., 1980a,b), whereas localized occurrences of shallow-marine fossils in restricted horizons of shale have been reported by Alencaster (1961b) and Stewart and Roldán-Quintana (1991). The Santa

Clara Formation has been interpreted as a series of deposits of deltaic origin and is the only fossiliferous unit of the Barranca Group. Indeed, cyclothemic deltaic sedimentation can be easily documented in the Santa Clara Formation at some sections where brackish-marine *Lingula* shales grade upward into shallow-marine mollusk-bearing shales and are disconformably overlain by delta-front sandstones (Figure 3.8). That repetitive pattern fits well the deltaic sedimentation model proposed for the Santa Clara Formation by Cojan and Potter (1991).

The flora and invertebrate fossils of the Santa Clara Formation indicate that it is of late Carnian age, although other outcrops in southern Sonora that possibly may correspond to that formation have yielded the ammonite *Acanthinites*, of Norian age (Martinez, 1984). The Coyotes Formation is a predominantly conglomeratic unit (Figure 3.7D) interpreted as an alluvial-fan- related deposit.

Tectonic relationships of Permian and Triassic strata

According to the interpretation of Coney and Campa (1987), as modified by González-León (1989), three main tectonostratigraphic terranes are recognized in Sonora. The different Palaeozoic successions discussed here correspond to those terranes. The cratonic-platform succession of northeastern Sonora belongs to the North American Terrane, the miogeoclinal succession of northwestern and central Sonora occurs within the Caborca Terrane, and the eugeoclinal succession of central Sonora forms the Cortes Terrane.

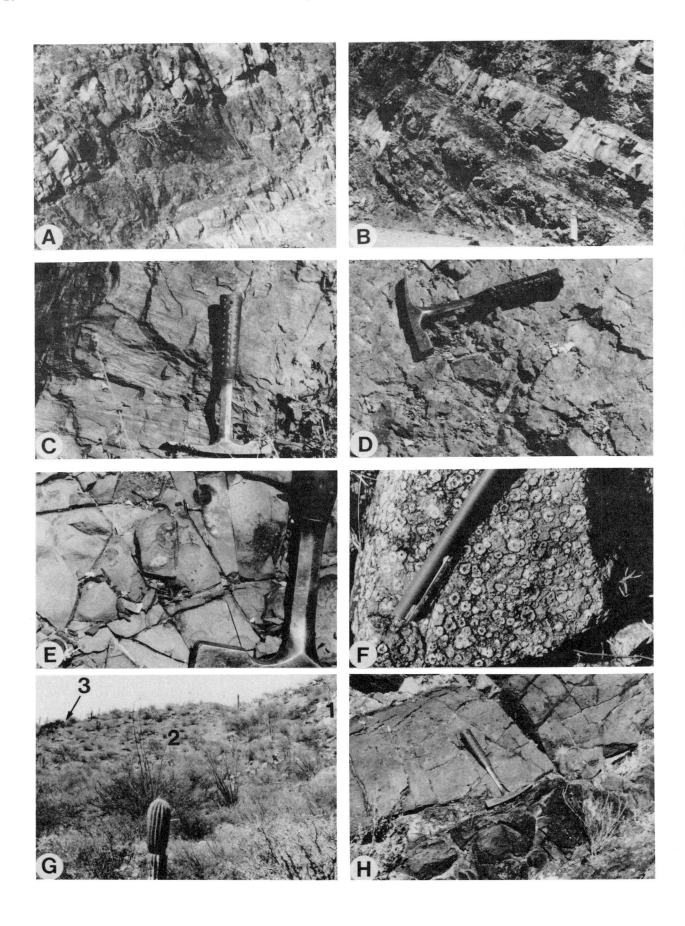

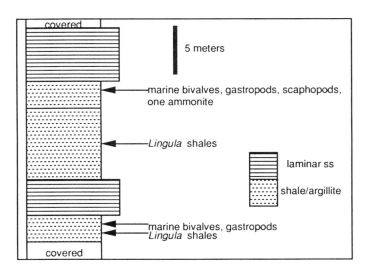

5 meters

marine bivalves, gastropods, scaphopods, one ammonite

Lingula shales

laminar ss

shale/argillite

marine bivalves, gastropods
Lingula shales

Figure 3.8. Measured stratigraphic section for part of the Santa Clara Formation along Arroyo Tarahumara in the Sierra San Javier, encompassing units 15–18 of Wilson and Rocha's section (1949, p. 18). The uppermost marine-fossil horizon is the principal fossil locality of Alencaster (1961b).

It has been suggested that the Caborca Terrane is a fragment of the North American Terrane that was displaced, during Jurassic time, along the left-lateral Mojave–Sonora megashear (Anderson and Silver, 1979). The Proterozoic and Palaeozoic succession of the Caborca Terrane is considered to be an offset fragment of the Cordilleran miogeocline of the southwestern United States. The Cortes Terrane of eugeoclinal strata, of Ordovician–early Permian age, is considered to be thrust against the Caborca Terrane; the zone of structural deformation that records that tectonic event is located in central Sonora along the boundary between the two terranes, forming the Sonoran Orogen of Poole et al. (1991). As assumed by Stewart et al. (1990), the time of juxtaposition of the strongly deformed eugeoclinal assemblage against the miogeoclinal rocks is presumed to have been late Permian to middle Triassic. The late Triassic Barranca Group of central Sonora unconformably overlies the eugeoclinal strata and also contains clasts derived from the miogeoclinal strata (Stewart and Roldán-Quintana, 1991), thus constraining the estimate for the earliest possible time for that deformation.

On the other hand, the Guadalupian Monos Formation of northwestern Sonora is considered to be an allochthonous unit

forming the lower part of the Antimonio Terrane (González-León, 1989). That formation is disconformably overlain by the Upper Triassic Antimonio Formation. The two sedimentary units form the Antimonio Terrane, which was thrust over the Caborca Terrane during the middle Jurassic.

The Barranca Group extends in an east–west-trending belt in central Sonora and has been interpreted by Stewart and Roldán-Quintana (1991) as a series of deposits formed in a rift basin. The basin developed during late Triassic time over deformed eugeoclinal Palaeozoic rocks after they had been juxtaposed to the miogeoclinal strata. The Barranca sediments also have clasts derived from both miogeoclinal and eugeoclinal rocks, and thus the unit is considered to be an overlapping, autochthonous deposit. An alternative interpretation for the deposition of the Barranca Group, favored by one of us (S. G. Lucas, unpublished data), views the Arrayanes(?) and Santa Clara formations as coastal-plain deposits (Cojan and Potter, 1991) that were *not* deposited in a rift basin; instead, those late Carnian strata are seen as having been deposited along the Pangaean western shoreline, possibly far to the northwest of their present location, before translation along the Mojave–Sonora megashear. The Coyotes Formation conglomerates (Figure 3.7D) resemble Lower Jurassic siliceous conglomerates in the Sierra del Alamo (Lucas, 1993) and the Sierra de Santa Rosa (Hardy, 1981) and thus may be much younger than the underlying Santa Clara Formation strata, to which they may have little or no genetic relationship.

Although Alencaster (1961a) suggested that the Barranca Group and Antimonio Formation were deposited in a single continuous basin that extended along western Sonora, the contrasts in lithologic succession, palaeontologic content, petrographic composition, and basement type between these two units indicate that they represent different lithotectonic assemblages (Stanley and González-León, 1995). Viewed as a whole, the Antimonio Formation comprises a 3,400-m-thick section of Upper Triassic–Lower Jurassic rocks. Correlative Lower Jurassic strata of that formation show widespread outcrops in western Sonora southwest of the Mojave–Sonora megashear and northwest of the Barranca Group outcrops. The original depositional setting for that formation is not clear, but it may have corresponded to a fore-arc setting, as its position is sandwiched between a subduction complex as old as the late Triassic (Sedlock and Isozaki, 1990) and volcanic-arc sequences of middle–late Jurassic age to the west, in Baja California (Kimbrough, 1985), and a late Triassic–middle Jurassic continen-

Figure 3.7 (opposite). A: Redbed mudstones, siltstones, and sandstones of the Arrayanes Formation near the abandoned Luz del Cobre mine in the eastern part of the Sierra San Javier. B: Cyclothemic deposits of shale, carbonaceous shale, and sandstone in the Santa Clara Formation along the highway just south of the Sierra San Javier. C: Laminar and flasser-bedded delta-front sandstone of the Santa Clara Formation along Arroyo Tarahumara, eastern part of the Sierra San Javier (this is the uppermost sandstone in Figure 3.8). D: Quartzite-boulder conglomerate of the Coyotes Formation in the eastern part of the Sierra San Javier along the highway to the town of San Javier. E: Lower Jurassic

ammonite beds in the Sierra Santa Rosa about 3 km east of Rancho San Carlos. F: Lower Jurassic *Pentacrinus* encrinite in the Sierra Caracahui, about 2 km southeast of Rancho San Fernando. G: Triassic–Jurassic boundary section in the Antimonio Formation of the Sierra del Alamo (see Lucas, 1993, for precise location); late Triassic carbonates (1) and clastics (2) are disconformably overlain (3) by Lower Jurassic (late Liassic) strata. H: Close-up of point 3 shown in photograph G. The hammer head rests on the topmost Triassic carbonate bed, disconformably overlain by Lower Jurassic limestone.

tal volcanic arc (Saleeby and Busby-Spera, 1992) to the north and northeast. Mainly on the basis of palaeontologic and lithostratigraphic data, González-León (1980) and Stanley and González-León (1995) considered that the Antimonio Formation and the Monos Formation were allochthonous over the Caborca Terrane. Also on the basis of tectonic relationships, those authors suggested that the time of thrusting of those formations was middle Jurassic.

Summary

The Permian and Triassic rocks of northwestern Mexico record important sedimentologic and tectonic events. The Permian rocks of that region, which are parts of thick Palaeozoic successions, consist of the following: (1) cratonic-platform carbonate and siliciclastic marine strata in the northeastern part of Sonora, where several formations of Wolfcampian–Leonardian age have been recognized; (2) miogeoclinal shallow-water successions of Wolfcampian–Leonardian age extensively distributed in central Sonora; (3) deep-water eugeoclinal successions of early Permian age that occur in central Sonora; and (4) the shallow-marine Monos Formation of Guadalupian age in the northwestern part of the state. Two lithotectonic assemblages of late Triassic (Carnian–Norian) age have been recognized: (1) In northwestern Sonora, the lower member of the Antimonio Formation is of shallow-marine origin, with a fossiliferous middle interval of deep-marine facies, whereas (2) in central Sonora the mainly terrigenous Barranca Group comprises either rift-basin or delta-plain (in part) deposits. Those rocks also form parts of several different terranes. The Permian cratonic-platform succession of northeastern Sonora is part of the North American Terrane; the miogeoclinal rocks of central Sonora belong to the Caborca Terrane, whereas the eugeoclinal succession is part of the Cortes Terrane. The Monos and Antimonio formations are parts of the Antimonio Terrane.

The tectonic events recorded by those strata include thrusting of the Cortes Terrane against the Caborca Terrane during middle Permian and middle Triassic times. The Barranca Group may be an overlapping assemblage, thus constraining the youngest age for that event. The Antimonio Terrane is considered to represent thrusting of the Caborca Terrane during middle Jurassic time.

Acknowledgments

We thank David Sivils and Pete Reser for valuable assistance and companionship during field work. The National Geographic Society and CONACYT project 3934-T supported the research reported here. Helpful reviews by Drs. Yang Zunyi and S. K. Acharyya greatly improved the manuscript.

References

Alencaster, G. 1961a. Estratigrafía del Triásico Superior de la parte central del Estado de Sonora. *Universidad Nacional Autónoma de México, Instituto de Geología, Paleontología Mexicana* 11:1–18.

Alencaster, G. 1961b. Fauna fósil de la Formación Santa Clara (Carnico) del Estado de Sonora. *Universidad Nacional Autónoma de México, Instituto de Geología, Paleontología Mexicana* 11:19–45.

Anderson, T. H., and Silver, L. T. 1979. The role of the Mojave-Sonora megashear in the tectonic evolution of northern Sonora. In *Geology of Northern Sonora*, ed. T. H. Anderson and J. Roldán-Quintana, pp. 59–68. San Diego, CA: Geological Society of America.

Bartolini, C. 1988. Regional structure and stratigraphy of Sierra Aliso, central Sonora, Mexico. Unpublished M.Sc. thesis, University of Arizona.

Cojan, I., and Potter, P. E. 1991. Depositional environment, petrology, and provenance of the Santa Clara Formation, Upper Triassic Barranca Group, eastern Sonora, Mexico. In *Studies of Sonoran Geology*, ed. E. Pérez-Segura and C. Jacques-Ayala, pp. 37–50. Special paper 254. Boulder, CO: Geological Society of America.

Coney, P. J., and Campa, M. F. 1987. *Lithotectonic Terrane Map of Mexico (West of the 91st Meridian)*. U.S. Geological Survey, miscellaneous field studies map MF-1874-D.

Cooper, G. A., and Arellano, A. R. 1946. Stratigraphy near Caborca, northwestern Sonora, Mexico. *American Association of Petroleum Geologists Bulletin* 30:606–19.

Delattre, M. 1984. Permian miogeoclinal strata at El Volcan, Baja California, Mexico. In *Geology of the Baja California Peninsula*, vol. 39, ed. V. A. Frizzell, Jr., pp. 23–9. Los Angeles, CA: Society of Economic Paleontologists and Mineralogists, Pacific Section.

González-León, C. 1980. La Formación Antimonio (Triásico Superior–Jurásico Inferior) en la Sierra del Alamo, Estado de Sonora. *Universidad Nacional Autónoma de México, Instituto de Geología Revista* 4:13–18.

González-León, C. 1986. Estratigrafía del Paleozoico de la Sierra del Tule, noreste de Sonora. *Universidad Nacional Autónoma de México, Instituto de Geología Revista* 6:117–35.

González-León, C. 1989. Evolución de terrenos mesozoicos en el noroeste de México. *Universidad de Sonora, Departamento Geología Boletin* 6:39–54.

Hardy, L. R. 1981. Geology of the central Sierra de Santa Rosa, Sonora, Mexico. In *Geology of Northwestern Mexico and Southern Arizona*, ed. L. Ortlieb and J. Roldán-Quintana, pp. 73–98. Hermosillo, México: Geological Society of America.

Hewett, R. L. 1978. Geology of the Cerro la Zacatera area, Sonora, Mexico. Unpublished M.Sc. thesis, Northern Arizona University.

Imlay, R. W. 1939. Paleogeographic studies in northeastern Sonora. *Geological Society of America Bulletin* 50:1723–44.

Ketner, K. B., and Noll, J. H. 1987. *Preliminary Geologic Map of the Cerro Cobachi Area, Sonora, Mexico*. U.S. Geological Survey, miscellaneous field studies map MF-1980.

Kimbrough, D. L. 1985. Tectonostratigraphic terranes of the Vizcaino Peninsula and Cedros and San Benito Islands, Baja California, Mexico. In *Tectonostratigraphic Terranes of the Circum-Pacific Region*, ed. D. G. Howell, pp. 285–98. Houston, TX: Circum-Pacific Council for Energy and Mineral Resources.

Lucas, S. G. 1993. The Triassic–Jurassic boundary section in the Sierra del Alamo, northwestern Sonora. In *Libro de Resumenes, III Simposio Geología de Sonora*, ed. C.

González-León and E. L. Vega-Granillo, pp. 66–8. Hermosillo, México.

Lucas, S. G., and González-León, C. 1994. Marine Upper Triassic strata at Sierra la Flojera, Sonora, Mexico. *Neues Jahrbuch für Geologie und Paláontologie, Monatshefte* 1:34–40.

Martinez, R. 1984. Prospección geológica del grafito microcristalino del Triásico del Municipio de Alamos, Sonora. Unpublished thesis, Instituto Politecnico Nacional.

Noll, J. H., Jr. 1981. Geology of the Picacho Colorado area, northern Sierra de Cobachi, central Sonora, Mexico. Unpublished M.Sc. thesis, Northern Arizona University.

Palafox, J. J., Mendoza, A., and Minjarez, V. A. 1992. Geología de la región de la Sierra Caracahui, Sonora, México. *Universidad de Sonora, Departamento Geología Boletin* 9:19–34.

Peiffer-Rangin, F. 1982. *Biostratigraphic Study of Paleozoic Rocks of Northeastern and Central Sonora.* Internal report, Universidad Nacional Autónoma de México, Instituto de Geología.

Perez-Ramos, O. 1992. Permian biostratigraphy and correlation between southeast Arizona and Sonora. *Universidad de Sonora, Departamento de Geología Boletin* 9:1–74.

Poole, F. G., Madrid, R. J., and Oliva-Becerril, J. F. 1991. Geological setting and origin of stratiform barite in central Sonora, Mexico. In *Geology and Ore Deposits of the Great Basin,* vol. 1, ed. G. L. Raines, R. E. Lisle, R. W. Schafer, and W. H. Wilkinson, pp. 517–22. Reno, NV: Geological Society of Nevada.

Saleeby, J. B., and Busby-Spera, C. 1992. Early Mesozoic tectonic evolution of the western U.S. Cordillera. In *The Geology of North America, Vol. G-3: The Cordilleran Orogen: Conterminous U.S.,* ed. B. C. Burchfield, P. W. Lipman, and M. L. Zoback, pp. 107–68. Boulder, CO: Geological Society of America.

Schmidt, G. T. 1978. Geology of the northern Sierra el Encinal, Sonora, Mexico. Unpublished M.Sc. thesis, Northern Arizona University.

Schreiber, J. F., Jr., Armin, R. A., Connolly, W. M., Stanton, R. J., Jr., Armstrong, A. K., Lyons, T. W., and Wrucke, C. T. 1990. Upper Paleozoic stratigraphy of the Whetstone Mountains, Cochise and Pima Counties, Arizona. In *Geologic Excursions Through the Sonoran Desert Region, Arizona and Sonora,* ed. G. E. Gehrels and J. E. Spencer, pp. 104–13. Special paper 7. Tucson, AZ: Arizona Geological Survey.

Sedlock, R. L., and Isozaki, Y. 1990. Lithology and biostratigraphy of Franciscan-like chert and associated rocks in west-central Baja California, Mexico. *Geological Society of America Bulletin* 102:852–64.

Silva-Pineda, A. 1961. Flora fósil de la Formación Santa Clara (Cárnico) del Estado de Sonora. *Universidad Nacional Autónoma de México, Instituto de Geología, Paleontología Mexicana* 11:1–37.

Stanley, G. D., Jr., and González-León, C. 1995. Upper Triassic fossils from the Antimonio Formation, northwestern Mexico, and their paleogeographic and tectonic implications. In *The Mesozoic of Northwestern Mexico,* special paper 301, ed. C. J. Ayala, C. González-Léon, and J. Roldán-Quintana, pp. 1–16. Boulder, CO: Geological Society of America.

Stanley, G. D., Jr., González-León, C., Sandy, M. R., Senowbari-Daryan, B., Doyle, P., Tamura, M., and Erwin, D. H. 1994. *Upper Triassic Invertebrates from the Antimonio Formation, Sonora, Mexico.* Paleontological Society memoir 36. *Journal of Paleontology (Suppl.)* 68(4):1–33.

Stewart, J. H., and Amaya-Martinez, R. 1993. Stratigraphy and structure of Sierra Santa Teresa near Hermosillo, Sonora, Mexico: a preliminary appraisal. In *Libro de Resumenes, III Simposio Geología de Sonora,* ed. C. González-León and E. L. Vega-Granillo, pp. 118–19. Hermosillo, México.

Stewart, J. H., Poole, F. G., Ketner, K. B., Madrid, R. J., Roldán-Quintana, J., and Amaya-Martinez, R. 1990. Tectonic and stratigraphy of the Paleozoic and Triassic southern margin of North America, Sonora, Mexico. In *Geologic Excursions Through the Sonoran Desert Region, Arizona and Sonora,* ed. G. E. Gehrels and J. E. Spencer, pp. 183–202. Special paper 7. Tucson, AZ: Arizona Geological Survey.

Stewart, J. H., and Roldán-Quintana, J. 1991. Upper Triassic Barranca Group; nonmarine and shallow-marine rift-basin deposits of northwestern Mexico. In *Studies of Sonoran Geology,* ed. E. Pérez-Segura and C. Jacques-Ayala, pp. 37–50. Special paper 254. Boulder, CO: Geological Society of America.

Vega, R., and Araux, E. 1987. Estratigrafía del Paleozoico en el área del Rancho las Norias, Sonora central. *Universidad de Sonora, Departamento de Geología Boletin* 4:41–50.

Weber, R. 1980. Megafósiles de Conifera del Triásico Tardío y del Cretácico Tardío de México y consideraciones generales sobre las coníferas Mesozoicas de México. *Universidad Nacional Autónoma de México, Instituto de Geología Revista* 4:111–24.

Weber, R., Trejo-Cruz, R., Torres-Romo, A., and Garcia-Padilla, A. 1980a. Hipotesis de trabajo acerca de la paleoecología de comunidades de la tafoflora Santa Clara del Triásico de Sonora. *Universidad Nacional Autónoma de México, Instituto de Geología Revista* 4:138–54.

Weber, R., Zambrano, A., and Amozurrutia, E. 1980b. Nuevas consideraciones al conocimineto de la tafoflora de la Formación Santa Clara (Triásico tardío) de Sonora. *Universidad Nacional Autónoma de México, Instituto de Geología Revista* 4:125–37.

Wilson, I. F., and Rocha, V. S. 1949. *Coal Deposits of the Santa Clara District near Tonichi, Sonora, Mexico.* U.S. Geological Survey bulletin 962-A.

4 Advances in the correlation of the Permian

GALINA V. KOTLYAR

The geocratic character of the Permian stage in the development of the earth's crust is shown in the destruction of the marine basins and in the endemism of faunas. Those circumstances, along with the progressively strengthening climatic gradient, make interregional correlations extremely difficult and in some cases even impossible. However, combined studies of the detailed biostratigraphic zonations and geologic analyses of basins and sequence stratigraphy have resulted in resolution of some of the problems of the Permian correlation.

Correlations of Permian deposits in the Tethys have traditionally been based on fusulinids, which were widespread and rapidly evolving in the Permian Tethys, as well as on ammonoids and conodonts. Most of the biostratigraphic zones have been defined on a phylogenetic basis and can be recognized throughout the Tethyan realm (Figures 4.1 and 4.2). Correlations within the Boreal realm have been based on the continuous and well-dated sections of the Taymir-Kolyma region (northeastern Russia) using ammonoid and brachiopod zonations. Those zones can be recognized from Novaya Zemlya to Mongolia (Figures 4.3 and 4.4). Recently the uppermost zone of the Dzhigdali horizon (*Kolymaella ogonerensis*) was defined in the upper part of the Solikamsk horizon of the Ufimian stratotype. Several bivalve and small-foraminifera zones established in northeastern Russia (Kashik et al., 1990) can also be recognized in the Boreal realm. These data allow reliable correlation of parts of the section and recognition of the significant biotic events.

Some of the most informative boundaries for correlations have been defined in the Tethyan, Boreal, and North American realms. The first boundary (base of the Upper Artinskian) can be recognized globally and corresponds to a major transgression in Tethys (Leven, 1993) and the largest palaeogeographic restructuring in the Boreal realm. Faunas of Asselian–early Artinskian age are represented by associations of fusulinids, ammonoids, conodonts, and brachiopods, which closely inherited Carboniferous-type assemblages allow us not only to correlate the deposits of the Tethyan and Boreal realms but also to define common stratigraphic stages. The biostratigraphic boundary within the Middle Artinskian is defined by the change from the Aktastinian ammonoid community to the Baigendzhinian community and can be easily recognized in many regions. The base of the *Neostreptognathodus pequopensis* Zone approximately corresponds to the aforementioned boundary and can be recognized throughout the Tethyan and Boreal realms and also approximately corresponds to the Wolfcampian–Leonardian boundary (middle part of the Skinner Ranch Formation) of North America (Figure 4.4). The boundary between the Lower and Upper Artinskian is relatively well defined in the Tethyan and Boreal realms. The Baigendzhinian ammonoid assemblage in the Tethys occurs in the *Chalaroschwagerina vulgaris* Zone of the Yakhtashian (Leonova and Dmitriev, 1989). The *Chalaroschwagerina solita* Zone of the lower part of the Yakhtashian lacks ammonoids, but the fusulinid assemblages of the two zones are very similar. The boundary between the Lower and Upper Artinskian in the Boreal realm is defined by Aktastinian ammonoids in the *Jakutoproductus rugosus* Zone of the upper part of the Munugudzhak horizon and by Baigendzhinian ammonoids in the lower part of the Dzhigdali horizon (Ganelin, 1984) (Figure 4.3). These data serve to correlate the Yakhtashian of the Tethys with the lower zone of the Dzhigdali horizon of the Boreal realm. In the Tethys, the Baigendzhinian assemblage is changed to a Bolorian assemblage, and in the Boreal realm it is changed to the upper Dzhigdalian (*Epijuresanites musalitini–Tumaroceras yakutorum* Zone) assemblage. Those associations have the same stratigraphic position between beds with Baigendzhinian and Roadian assemblages. M. Bogoslovskaya and T. Leonova (private communication) believe those assemblages to be of the same age because of their degree of evolutionary development.

The second correlative boundary that can be traced globally is defined by the appearance of the Roadian ammonoid assemblage (Figures 4.2 and 4.4). This boundary is defined within the evolutionary conodont cline from *Mesogondolella idahoensis* to *M. serrata* (= *M. nankingensis*). In the Tethyan realm the Roadian boundary coincides with the base of the Kubergandinian Stage and is marked by the appearance of the Ceratitida order and the *Paraceltites, Epiglyphioceras, Stacheoceras,* and *Daubichites* genera among ammonoids. The first *Armenina* and highly evolved *Misellina* (Leven, 1993) appear there among fusulinids. The conodonts had not appeared at that boundary in the Tethys. The significant changes in the faunal

Figure 4.1. Correlation of the Permian biostratigraphic zones of the Tethyan and Boreal realms.

The chart correlates the General scale with the Boreal Realm and the Tethyan Realm.

General scale (left columns):

System	Series	Stage	Substage
Permian	Upper (P u)	Tatarian	Upper
			Lower
		Kazanian	Upper
			Lower
	Middle (P m)	Ufimian	
	Lower (P l)	Kungurian	
		Artinskian	Upper
			Lower

Boreal Realm — East-European Subrealm Horizon:

Vyatka; Severnaya Dvina; Urzhum; Volga; Sok; Sheshma; Solikamsk; Iren; Filippov; Saran; Sarga; Irga; Burtsev

Boreal Realm — Taimyr–Kolyma Subrealm, Regional scheme:

Horizon	Brachiopods	Bivalves	Small foraminifers	Ammonoids
Continental deposits				
Khivach	Stepanoviella paracurvata ?	Intomodesma costata ?	Frondicularia maxima	Paramexicoceras Neogeoceras
		Maitaia tectensis		
Gizhiga	Concrinelloides curvatus	Maitaia bella	Frondicularia composita	Timorites
	Concrinelloides obrutshewi			
Omolon	Magadania bajarica	Kolymia multiformis	Frondicularia planiata	
	Terrarea korkodonensis			
	Terrarea borealis	Kolymia plicata	Frondicularia elongata	
	Omolonia sniatkovi	Kolymia inoceramiformis	Frondicularia ganelinae	Svendrdrupites Daubicites Anuites
	Mongolosia russiensis	Aphanaia ditatata		
		Aphanaia andrianovi–Aditateta		
	Kolymaella ogonerensis	Aphanaia andrianovi	Frondicularia prima	Epijuresanites musalitini
	Megousia kulixi			Tumaroceras yakutorum
	Anidanthus aagardi	Aphanaia lima	Frondicularia zavodovski	Neoschumardites triceps–Uraloceras ex gr simense
	Jakutoproductus burgaliensis	Edmondia nebrascensis		
	Jakutoproductus rugosus			

Tethyan Realm — Tethyan scale:

Stage	Fusulinids	Ammonoids	Conodonts
Dorashamian	Palaeofusulina sinensis	Pleuronodoceras / Paratirolites	Cl. latidentatus
	Shindella simplicata	Shevyrevites / Iranites / Phisonites	Clarkina subcarinata
Dzhulfian	Codonofusiella	Vedioceras	Clarkina orientalis
	Reichelina media	Arazoceras	Clarkina teveni–Cl. liangshanensis
		Eoarazoceras Anderssonoceras	
Midian — Upper (Abadehian)	Lepidolina kumaensis (Yabeina)	Stacheoceras	Clarkina ro-senranzi / Mesogondolella babcocki
Midian — Lower	L. multiseptata / Neoschwagerina "margaritae" (Yabeina)	Timorites Waagenoceras	Mesogondolella postserrata
Murgabian	Neoschwagerina "craticulifera" / Afganella schencki	Waagenoceras, Altudoceras, Sosiocrinites	Mesogondolella serrata–M. siciliensis
	Neoschwagerina simplex		
Kubergandian	Cancellina cutalensis / Armenina, Misellina ovalis	Medlicottia, Parapronorites, Paraceltites	Mesogondolella serrata
Bolorian	Misellina parvicostata	?	Nettaperrinites, Miklukhoceras
	Misellina dyhrenfurti		
Yakhtashian	Chalaroschwagerina vulgaris	Daraelites, Propinacoceras, Eothinites, Neorimites	Mesogondolella idahoensis
	Chalaroschwagerina solita		

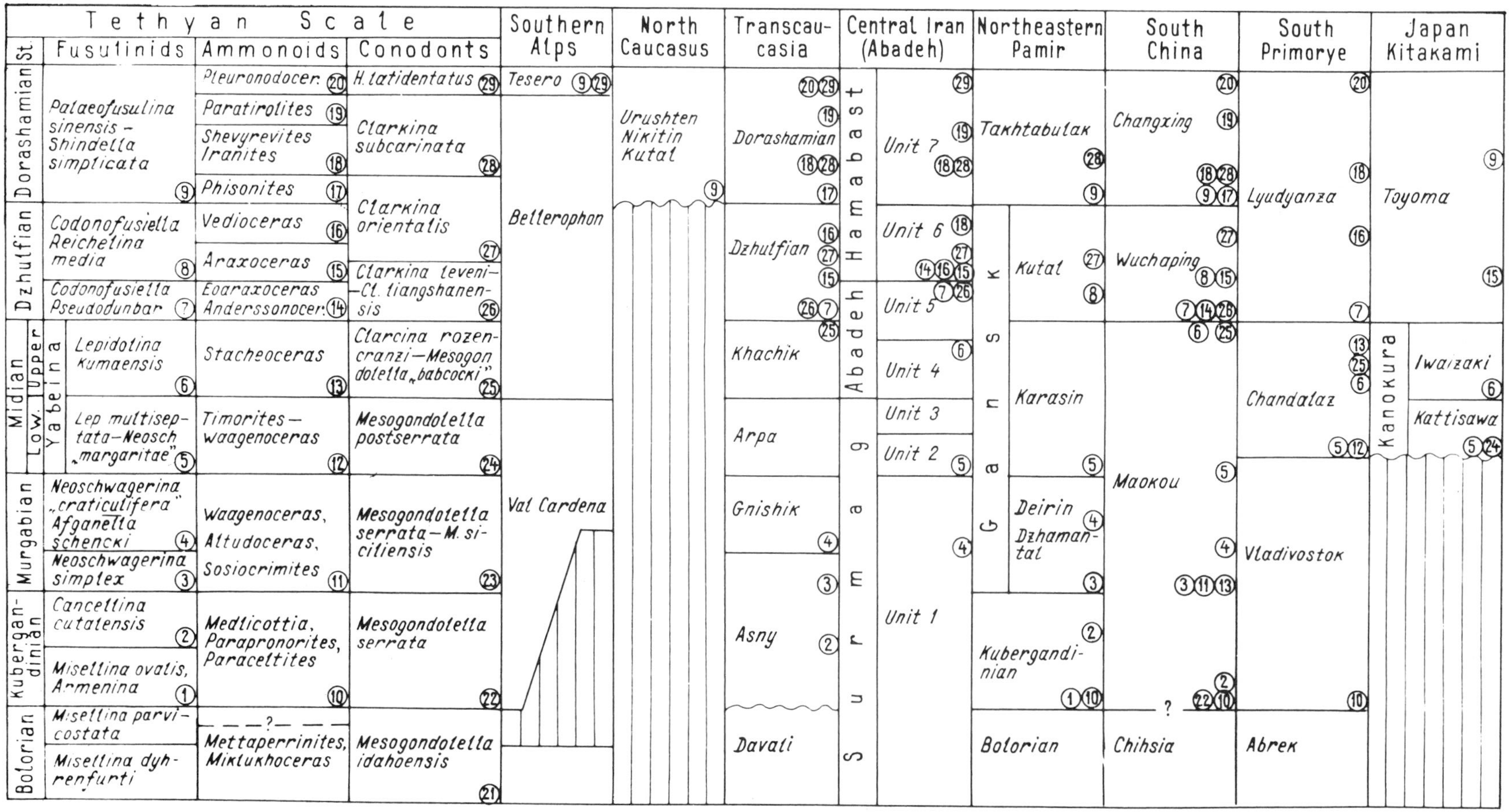

Figure 4.2. Correlation chart for the Upper Permian in the Tethyan realm. The numbers in the circles show the spreading of the guide faunas for the biostratigraphic zones.

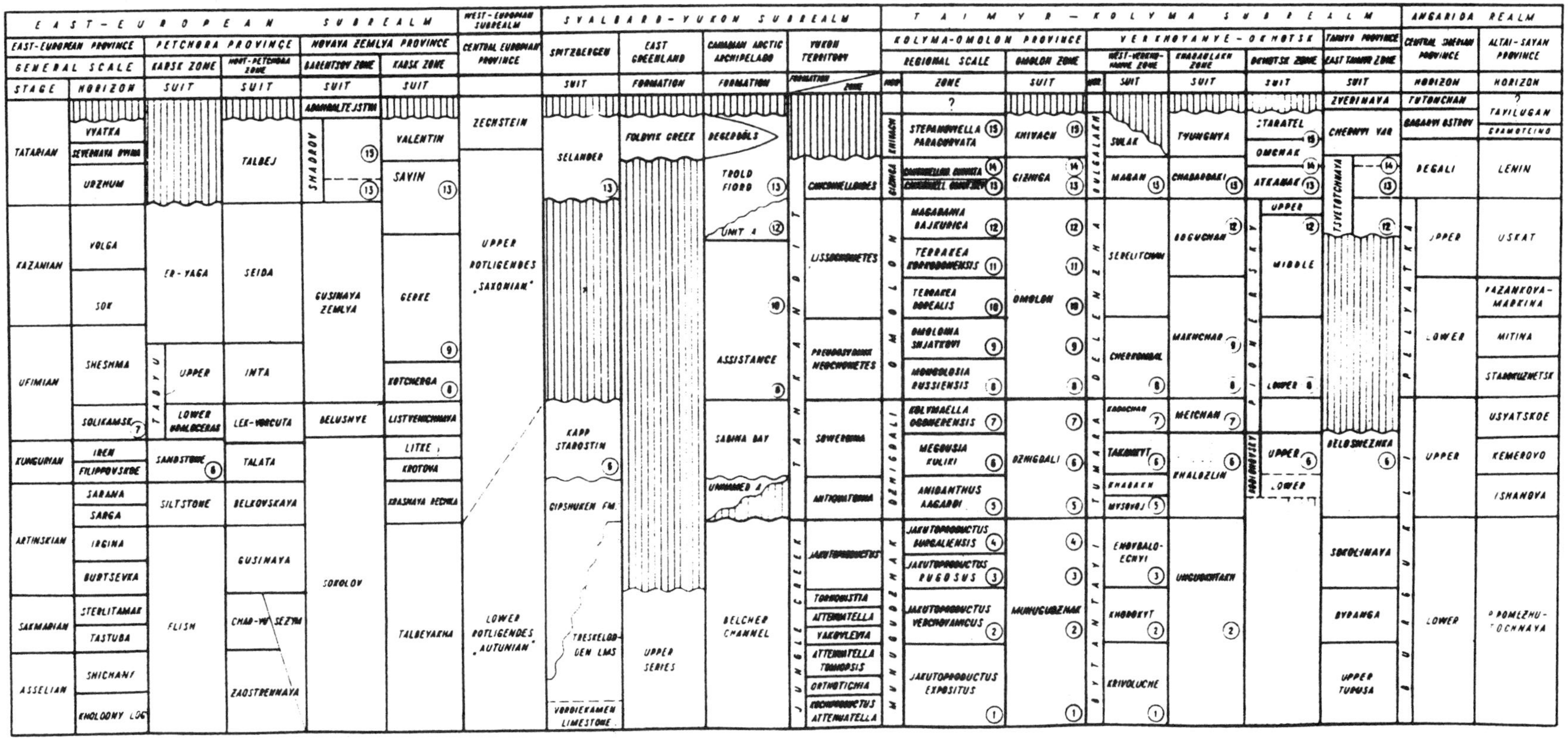

Figure 4.3. Correlation chart for the Permian in the Boreal realm. The numbers in the circles show the spreading of brachiopod and ammonoid zones.

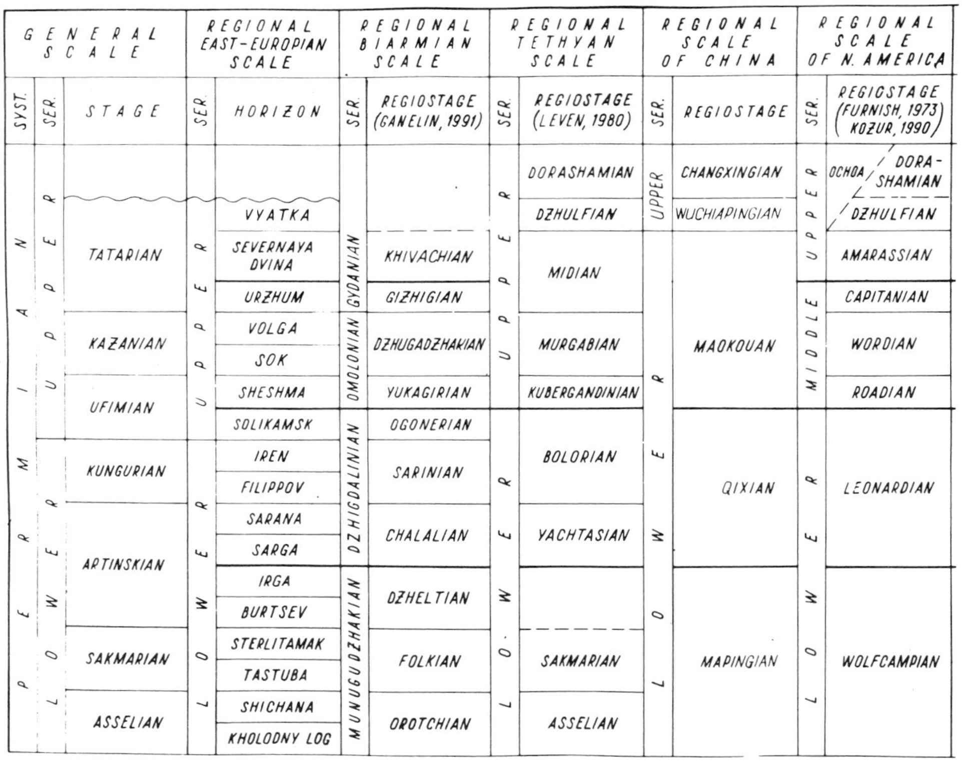

GENERAL SCALE			REGIONAL EAST-EUROPIAN SCALE		REGIONAL BIARMIAN SCALE		REGIONAL TETHYAN SCALE		REGIONAL SCALE OF CHINA		REGIONAL SCALE OF N. AMERICA	
SYST.	SER.	STAGE	SER.	HORIZON	SER.	REGIOSTAGE (GANELIN, 1991)	SER.	REGIOSTAGE (LEVEN, 1980)	SER.	REGIOSTAGE	SER.	REGIOSTAGE (FURNISH, 1973; KOZUR, 1990)
P E R M I A N	U P P E R						U P P E R	DORASHAMIAN	UPPER	CHANGXINGIAN	U P P E R	OCHOA DORASHAMIAN
		TATARIAN	U P P E R	VYATKA	GYDANIAN	KHIVACHIAN		DZHULFIAN		WUCHIAPINGIAN		DZHULFIAN
				SEVERNAYA DVINA				MIDIAN				AMARASSIAN
				URZHUM		GIZHIGIAN				MAOKOUAN	MIDDLE	CAPITANIAN
		KAZANIAN		VOLGA	OMOLONIAN	DZHUGADZHAKIAN		MURGABIAN				WORDIAN
				SOK								ROADIAN
		UFIMIAN		SHESHMA		YUKAGIRIAN		KUBERGANDINIAN				
	M I D D L E			SOLIKAMSK	DZHIGDALINIAN	OGONERIAN	L O W E R	BOLORIAN	L O W E R	QIXIAN	L O W E R	LEONARDIAN
		KUNGURIAN	L O W E R	IREN		SARINIAN						
				FILIPPOV								
				SARANA		CHALALIAN		YACHTASIAN				
		ARTINSKIAN		SARGA								
	L O W E R			IRGA	MUNUGUDZHAKIAN	DZHELTIAN						
				BURTSEV								
		SAKMARIAN		STERLITAMAK		FOLKIAN		SAKMARIAN		MAPINGIAN		WOLFCAMPIAN
				TASTUBA								
		ASSELIAN		SHICHANA		OROTCHIAN		ASSELIAN				
				KHOLODNY LOG								

Figure 4.4. Correlation of the main Permian scale for the Tethyan, Boreal, and North American realms.

assemblages at the Bolorian–Kubergandinian boundary allow it to be traced all over the Tethyan realm (Leven, 1993) (Figure 4.2). Besides *Daubichites,* some other genera (*Sverdrupites, Altudoceras, Alnuites*) appear at that level in the Boreal realm. The presence of the Roadian ammonoid assemblage is correlated to the *Mongolosia russiensis* and *Omolonia snjatkovi* zones at the base of the Omolon horizon in northeastern Russia (Kashik et al., 1990), to the Gerke and Kocherga formations of Novaya Zemlya (Bogoslovskaya, Ustritsky, and Chernyak, 1982), and to the lower part of the Assistance Formation in the Canadian Arctic Archipelago (Nassichuk, Furnish, and Glenister, 1965; Nassichuk, 1970) and can be traced within the whole Boreal realm (Kotlyar and Ganelin, 1989). The Kungurian ammonoid assemblages in the type region are not limited to the Kungurian Stage only. The discovery of *Uraloceras* in the marine analogues of the Solikamsk horizon in the middle part of the Tabyu Formation in the Pai-Khoi (Guskov, Puchanto, and Yatsuk, 1980) indicates the presence of Kungurian ammonoids in the lower part of the Ufimian Stage. That conclusion is supported by further finds of the fauna: *Gobioceras elenae* Bogoslovskaya has been found together with *Kolymaella ogonerensis* in Mongolia (Pavlova et al., 1991). The analogous brachiopod succession has been reported in that region as well as in the Kolyma-Omolon region. *Kolymaella ogonerensis* and *Gobioceras elenae* have been found immediately above beds that contain the brachiopod assemblage of the *Anidanthus aagardi* and *Megousia kulikii* zones (the Khovsgolian brachiopod assemblage). *Neogeoceras orientalis* Bogoslovskaya, which is characteristic of the upper

zone of the Dzhigdali horizon in the Kolyma-Omolon region, has been noted together with the Khovsgolian brachiopods. Bogoslovskaya believes that the *Gobioceras* genus undoubtedly is of Kungurian age and supposes it to be a transitional form from the Artinskian *Gobioceras lobulatum* Armstrong, Dear, and Runnegar to the Roadian *Spirolegoceras fischeri* Miller, Furnish, and Clark (Bogoslovskaya and Pavlova, 1988). In the Kolyma-Omolon region, the *Kolymaella ogonerensis* Zone is overlain by the *Mongolosia russiensis* Zone, which contains *Sverdrupites amundseni* Nassichuk. Thus the ammonoid assemblages of Kungurian and Roadian ages have been found in two frontier brachiopod zones (Figure 4.3). Taking into account that the Kungurian community is followed by the Roadian community and that the last Kungurian representative is present in the analogous Solikamsk horizon, the appearance of the Roadian ammonoids should be assigned to the Upper Ufimian. The Roadian ammonoid association has been reported from the base of the Guadalupian Series. Thus the lower boundaries of the Kubergandinian Stage of the Tethys, the Omolon horizon and its analogues in the Boreal realm, the Sheshma horizon of the type region, and the Roadian Stage of North America can be considered to be synchronous.

There are insufficient data for a direct correlation of the Murgabian Stage. However, its stratigraphic position above the Kubergandinian Stage reveals its correspondence to the upper part of the Omolon horizon (the *Terrakea borealis, T. korkodonensis,* and, probably, *Magadania bajcurica* zones). Those zones, as well as the Murgabian Stage, overlie beds with the

Roadian ammonoids. Some Omolonian bryozoan genera and species (*Maychella, Wjatkella, Streblascopora fasciculata*) have been found in the Vladivostok horizon of Murgabian age in South Primorye. Also, *Permopora kapitza* Romanchuk and *Girtyporina immemorata* Kiseleva from the Vladivostok horizon are found together with the Murgabian *Neoschwagerina craticulifera* (Schwager) in the Amur region of the Russian Far East (Kiseleva, 1986). According to Kiseleva and Morozova, the Vladivostokian bryozoans have some similarities to the Omolonian bryozoans in the Boreal realm (Morozova, 1981). The abundant *Orulgania* and *Tumarinia* of the Omolon horizon are very close to the Kazanian *Licharewia,* and they were identified earlier as *Licharewia* (Zavadovsky et al., 1970). In addition, there are some similarities in the small foraminifers (*Nodosaria noinskyi, N. incelebrata, Frondicularia tsaregradskyii, F. planilata, Rectoglandulina borealis, R. pygmaeformis*) of the Omolon horizon and the Kazanian Stage (Kashik et al., 1990). The Omolon horizon and the Wordian Stage of North America seem to be equivalent, because they contain practically identical *Terrakea belokhini* Ganelin and *Grandaurispina crassa* Grant. That conclusion is confirmed by the fact that the overlying Gizhiga horizon of the Boreal realm and the Capitanian Stage of North America have been correlated reliably on the basis of the presence of *Timorites* in both subdivisions. Those subdivisions can be assumed to be of the same age on the basis of the presence of *Pseudogastrioceras* at the base of the Kazanian Stage in the northern Russian Platform. That goniatite was defined by Pavlov (Kulikov, Pavlov, and Rostovcev, 1973) and is supposed to be very close to the early Wordian *Altudoceras.*

The Midian Stage can be correlated more definitely and reliably. The Midian can be divided into two parts. The Lower Midian contains *Timorites markevichi* Zakharov in South Primorye (Kotlyar et al., 1989) and *Timorites sigillarius* Ruzhencev and *Waagenoceras obliquum* Ruzhencev in the Amur region of the Russian Far East (Ruzhencev, 1976), as well as *Maitaia bella* Biakov in both regions. The latter is of great importance for correlation because it is widely distributed and its occurrence is limited to the *Cancrinelloides* Zone only. *Timorites, Cancrinelloides,* and *Maitaia bella* have been found together in Transbaikal (Kotlyar et al., 1990b; Okuneva and Zakharov, 1992). *Cancrinelloides obrutshewi* and *Maitaia bella* have been shown in the Gizhiga horizon by Ganelin and Biakov (Kashik et al., 1990) (Figure 4.1). Joint findings of *Cancrinelloides obrutshewi* and an early Midian fusulinid assemblage that contains *Yabeina globosa* (Yabe), *Neoschwagerina margaritae* Deprat, and *Lepidolina multiseptata* (Deprat) have been reported from the Midian of Koryakia (Epshtein, Terekhova, and Solovieva, 1985). Those data provide reliable correlations among the Midian of the Tethys, the Gizhiga horizon of the Boreal realm, and the Capitanian Stage (perhaps without its uppermost part) of North America. It can also be assumed that all the aforementioned subdivisions correspond to the Lower Tatarian Substage of the Russian Platform, on the basis of stratigraphic positions as well as magnetostratigraphic data.

Previous correlations of the Kazanian and Capitanian stages were based on the presence of *Stepanovites meyeni* Kozur and Movschovitsch in both subdivisions (Movschovitsch, 1986). However, the Kazanian Stage belongs to the Kiama hyperzone, but a paleomagnetic zone of normal polarity is known in the Upper Capitanian (Peterson and Nairn, 1971). Moreover, the discovery of *S. meyeni* in the Upper Midian Abadeh Formation of Iran (Iranian-Japanese Research Group, 1981) indicates the wide distribution of that species.

Many authors have discussed the boundary within the Midian or Capitanian Stage, and the Amarassian and Abadehian have been used as independent stages (Taraz, 1971; Furnish, 1973; Kozur, 1991). The large number of analyses of events at that boundary indicates its importance. The Upper Midian of the Tethys corresponds to the *Lepidolina kumaensis* Zone. The boundary at the base of the *L. kumaensis* Zone can be defined throughout the whole Tethyan realm on the basis of fusulinids, small foraminifers, ammonoids, conodonts, and brachiopods. *Mesogondolella* "*babcocki,*" *Clarkina wilcoxi,* and *C. rosenkrantzi* appear at that level. The Hemigordiopsidae, and particularly the *Baisalina pulchra* Zone assemblage among the small foraminifers and the *Spirigerella* and *Septospirigerella* genera among brachiopods, are widely distributed in the upper part of the Midian Stage (Kotlyar et al., 1989).

The *Lepidolina kumaensis* Zone has been attributed to the Lower Dzhulfian by some specialists (Ozawa, 1975), corresponding to the *Codonofusiella* Zone. However, the *L. kumaensis* fauna is restricted within the upper part of the Maokou Formation in China (Rui, 1983) and the upper part of the Chandalaz horizon in South Primorye (Kotlyar et al., 1990a). In both regions the *L. kumaensis* fauna is represented by abundant examples of the highest fusulinids such as the neoschwagerinids and verbeekinids. The conodonts *Mesogondolella* "*babcocki,*" *Clarkina wilcoxi, C. rosenkrantzi,* and *C. bitteri* are present in the *L. kumaensis* Zone in South Primorye and in the uppermost part of the Maokou Formation in China (Clark and Wang, 1988). Several similar species of the small foraminifers and corals (Sokolov, 1960) are known from the Chandalaz horizon in South Primorye and from the Khivach horizon of the Boreal realm. The Khachik horizon of Transcaucasia in the West Tethys contains some common species of small foraminifers with the Zechstein Limestone of central Europe. Among them are *Spandelina delicatula* (Yurkevich), *Cornuspira kinkeli* Spandel, and *Dentalina lineamargaritarum* Sherp (G. Pronina, personal communication). There are many similar common brachiopods and conodonts in the *L. kumaensis* Zone in South Primorye and in the Foldvik Creek Formation of eastern Greenland (Dunbar, 1955; Kotlyar, 1978). Among brachiopods, such taxa as *Cancrinella, Waagenoconcha, Kochiproductus, Muirwoodia, Pterospirifer, Cancrinella germanica* Frebold, *Muirwoodia greenlandica* Dunbar, *Rhynchopora kochi* Dunbar, *Pterospirifer alatus* Schlotheim, and *Kaninospirifer striatiparadoxus* Toula are known. *Clarkina rosenkrantzi* Bender and Stoppel and *Merrilina divergens* Bender and Stoppel have been reported in both subdivisions. *Cyclolobus* is known from the Foldvik Creek Formation (Nassichuk et al., 1965) and from the late Midian

Abadeh Formation in central Iran (Taraz, 1971). W. Nassichuk (personal communication) has recently found *Paramexicoceras* near the Permian–Triassic boundary in Greenland. Earlier, that species had been described from the uppermost Permian in the Verchoyanye (Popov, 1970).

The foregoing data allow us to correlate the Upper Midian in the Tethys, the Khivach horizon of the Boreal realm, the Foldvik Creek Formation of Greenland, and the Zechstein Limestone of central Europe.

The *Lepidolina kumaensis* Zone probably corresponds to the Lamar Limestone of North America. The latter, like the *L. kumaensis* Zone, contains *Yabeina, Paradoxiella, Codonofusiella, Lantschichites,* and *Mesogondolella "babcocki."* The occurrences of *Ishigaconus scholasticus* (Ormiston and Babcock) and *Follicucullus ventricosus* Ormiston and Babcock in the two comparable subdivisions seem to confirm their simultaneity. Only the uppermost part of the Altuda Formation (Kozur, 1992) can be considered as Dzhulfian.

The "Lower La Colorada beds" of Mexico, which contain *Timorites, Neocrimites, Stacheoceras, Kingoceras, Rauserella,* and *Reichelina* (Wilde, 1955; Spinosa, Furnish, and Glenister, 1970), are assumed to be of the same age as the Upper Midian.

The assumed correlation is supported by the palaeomagnetic data (Figure 4.5) collected from the Tethys (Kotlyar et al., 1984; Novikov, Suprychev, and Chramov, 1991), the Boreal realm (Gurevich and Slautsitajis, 1988; Kashik et al., 1990), Europe (Molostovsky, 1983; Menning, 1988), and North America (Peterson and Nairn, 1971). Although the boundary between the Illawarra and Kiama hyperzones in most sections has not yet been established, the linkage of the Upper Midian and the Khivach horizon and its equivalents to the Illawarra hyperzone is not in doubt. Moreover, we can say with confidence that those subdivisions correlate with the Upper Tatarian Substage of the stratotype.

The lower boundary of the Dzhulfian Stage in the type region was long considered to be inadequately demonstrated (the transitional Midian–Dzhulfian beds are not characterized by ammonoids), and its position was continually being debated. Recently, on the basis of new data on conodonts (Grigoryan, 1990), the lower boundary of the Dzhulfian Stage has been defined by the first appearance of *Clarkina leveni* Kozur, Mostler, and Pjatakova at the base of the *Codonofusiella kwangsiana–Pseudodunbarula arpaensis* Zone, not at the base of the *Araxilevis* beds, as was earlier believed (Kozur et al., 1978). According to Kozur, the top of the Capitanian Stage in North America can now be defined by the transition from *Mesogondolella "babcocki"* to *Clarkina altudaensis* within a phylomorphogenetic continuum discovered in the uppermost Altuda Formation in the western Glass Mountains (Kozur, 1992). That level corresponds to the Midian–Dzhulfian boundary (at the base of the *C. kwangsiana–P. arpaensis* Zone) in Transcaucasia. In China, the *Clarkina liangshanensis–C. bitteri* Assemblage Zone was established in the lowermost part of the Wujiaping Formation (Clark and Wang, 1988). That zone has been correlated with the uppermost part of the Abadeh Formation in Iran (= the *C.*

kwangsiana–P. arpaensis Zone in Transcaucasia) by Wang and Clark. Those data allow direct correlation of the base of the Wujiapingian in South China and the Dzhulfian in Transcaucasia (*C. kwangsiana–P. arpaensis* Zone) with the uppermost Capitanian. *Anderssonoceras* is present in the lower part of the Wujiaping Formation, and its appearance coincides closely with the base of the Dzhulfian Stage (*Codonofusiella* beds in the stratotype) (Figure 4.1). The highly evolved Araxoceratinae (*Araxoceras, Prototoceras, Vescotoceras*) appear immediately above the *C. kwangsiana–P. arpaensis* Zone. The presence of *Eoaraxoceras ruzhencevi, Kingoceras, Xenodiscus,* and *Timorites* in the *Araxoceras tectum* Subzone of the Abadeh Formation in central Iran and in the "Upper La Colorada beds" of Mexico (Spinosa et al., 1970) allowed Glenister (Glenister and Spinosa, 1991) to correlate those beds. Kozur considered that "the important discovery of the Mexican *Eoaraxoceras* fauna in the type Dzhulfian indicates at the moment only that [the] Middle Dzhulfian is younger than the Early Capitanian" (Kozur, 1992, p. 109). In spite of the very close similarities between the Mexican *Eoaraxoceras* fauna and the *A. tectum* assemblages in Iran, it should be taken into account that the *Eoaraxoceras* ammonoid fauna in Iran occurs together with the highly evolved Araxoceratinae genera *Araxoceras, Prototoceras,* and *Vescotoceras.* In Mexico, the *Eoaraxoceras* fauna is accompanied by the more ancient assemblage of *Stacheoceras toumanskayae* Miller and Furnish, *Neocrimites* sp., *Difuntites hidium* (Ruzhencev), and *Nodosageceras nodosum* Wanner, and highly evolved Araxoceratinae are completely absent there. Accepting that the *Eoaraxoceras* fauna of the "Upper La Colorada beds" is undoubtedly of Dzhulfian age, and taking into account that it immediately follows the "Lower La Colorada beds" assemblage (King et al., 1944) of latest Midian age, the *Eoaraxoceras* fauna should be placed only at the base of Dzhulfian (*C. kwangsiana–P. arpaensis* Zone).

The lower boundary of the Dzhulfian can be defined reliably on the basis of the Toyoma series in Japan, the Wujiaping Formation in China, the Lyudyanza horizon in South Primorye, and the Kutal member in northeastern Pamir and their collections of ammonoids, conodonts, fusulinids, and other faunas (Kotlyar, 1993).

Correlation of the Dzhulfian Stage on the basis of deposits in other biogeograhic realms is very difficult. In the Boreal realm, the lower parts of the volcano-terrigenous sequences of central Siberia, Novaya Zemlya, and Taymir may correspond to the Dzhulfian. Also, a Dzhulfian age for the Zechstein Limestone of central Europe, the Foldvik Creek Formation of Greenland, and the uppermost part of the Khivach horizon in northeastern Russia cannot be excluded. The possible correlation of the entire Upper Tatarian Substage, or only its lower part (Severnaya Dvina horizon), with the Upper Midian of the Tethys is yet to be resolved. The Vyatka horizon probably corresponds to the Lower Dzhulfian. The data obtained by Novozhilov on Conchostraca, by Foster on spores and pollen (Foster, 1979), and by Chudinov (1987) and Anders on vertebrates, as cited by Gomankov (1988), probably indicate correspondence between the Vyatka horizon and the Lower Dzhulfian.

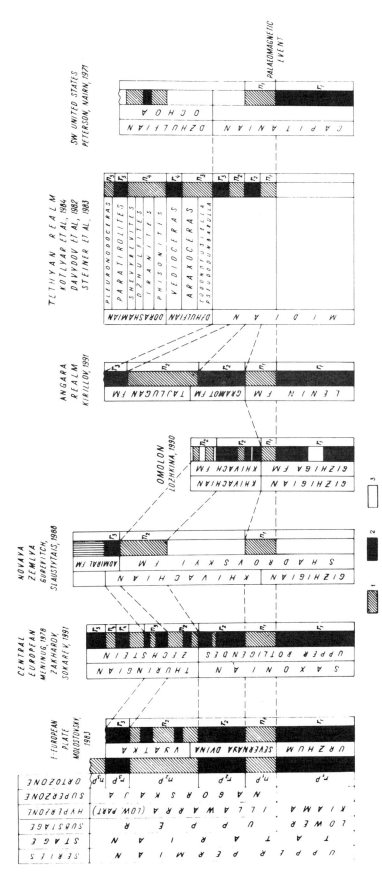

Figure 4.5. Correlation of the Upper Permian magnetostratigraphic schemes: 1, normal polarity; 2, reversed polarity; 3, uninvestigated interval.

Until recently it had been supposed that the Dorashamian marine deposits were restricted in distribution. It is now known that complete sections of the Dorashamian or the Changxingian Stage occur in Iran, Transcaucasia, and China. The Dorashamian age of the Nikitin and Urushten formations of the northern Caucasus has now been demonstrated by abundant *Palaeofusulina* and *Parananlingella,* which are characteristic only for the Changxingian (Kotlyar et al., 1989). The same age for the Tachtabulak Formation of northeastern Pamir is demonstrated by the Dorashamian conodonts *Clarkina subcarinata* and *C. planata* (Davydov et al., 1990).

The top of the Tatarian Stage is easily located within continental sequences, but its correlation in marine and continental facies is difficult. The definition of the base of the Triassic is still in dispute. Traditionally the Triassic has been defined by the base of the *Otoceras woodwardi* Zone (Tozer, 1988). The *Otoceras* species are characteristic only for the Boreal realm and for Gondwanan Tethys. The stratotype of the Permian–Triassic boundary has not yet been established, because there are no known sections in which clearly identifiable *Otoceras* are overlying the succession of Araxoceratinae.

The section at Meishan in China was once proposed as a candidate for the type of the Permian–Triassic boundary. It was considered that the lowermost Triassic beds (mixed bed 1) contained *Otoceras* in that section. However, according to Dagys and Dagys (1987), the ammonoid association cited by Wang (1984) includes Permian, Induan, and even Olenekian genera. In addition, the ammonoids are poorly preserved and may be close to the Permian genera. That possibility seems very likely, because "mixed bed 2" contains only the Permian brachiopods and the conodont *Hindeodus parvus.*

In the Boreal realm, *Otoceras woodwardi* is absent. Most specialists correlate the *O. woodwardi* Zone of Gondwanan Tethys with the *Otoceras boreale* Zone of the Boreal realm. If that correlation is correct, then the lower *Otoceras concavum* Zone should correspond to the Dorashamian Stage.

New discoveries of Permian conodonts in *Otoceras* beds in Greenland (H. Kozur, personal communication) and in the Canadian Arctic Archipelago (Henderson, 1993) seem to confirm a Permian age for the *Otoceras* beds, but they do not clarify the problem of the Permian–Triassic boundary.

The Permian–Triassic boundary in continental deposits is also in dispute. The Lower Buntsandstein, which immediately overlies the Zechstein, was long considered to be of Triassic age. Now it has been proposed to raise the lower boundary of the Triassic to the base of the Middle Buntsandstein, on the basis of palynologic data (Ecke, 1987; Fuglewicz, 1987; Rebello and Doubinger, 1988). It has been reported that the Upper Zechstein and Lower Buntsandstein are characterized by the same types of megaspores and miospores. Kozur (1989) correlated the Lower Buntsandstein with the lower part of the Werfen Formation (Tesero horizon). The latter contains Permian brachiopods, conodonts, fusulinids, and small foraminifers (Broglio Loriga et al., 1988). Also, Lozovsky has correlated the Lower Buntsandstein with the lower part of the Vetluga

highland of the Russian Platform and with the *Otoceras* beds of Greenland (Lozovsky, 1992). There is an omission at the end of the Permian in the type region. Most probably the Dzhulfian and Dorashamian stages of the Tethyan scale, as well as the lower part of the volcano-terrigenous sequences of central Siberia, correspond to that gap. They contain similar Permian palynocomplexes (Meyen, 1984; Gomankov, 1988; Yaroshenko, 1990; Sadovnikov, 1991) The Chhidruan species in the palynofloras of Dzhulfian age are present in the palynocomplexes of the Korvunchanian Series in central Siberia (Balme, 1970; Foster, 1979, 1982).

Thus the problem of the Permian–Triassic boundary in marine and continental facies remains open for further study.

Acknowledgments

My participation in the international symposium on Shallow Tethys 3 in Japan in 1990 was sponsored by the chairman of Project 272, Dr. J. M. Dickins, and by the organizing committee for Shallow Tethys 3. My work for the international field conference on Permian–Triassic biostratigraphy and the tectonics of the Far East was partly sponsored by Prof. Yu. D. Zakharov. Important conodont data have been provided by Prof. H. Kozur.

I wish to thank the institutes that provided generous financial support and many colleagues who provided valuable data and comments. I thank especially Dr. J. M. Dickins and Prof. Yang Zunyi for critically reading the manuscript and for many important comments.

References

Balme, B. E. 1970. Palynology of Permian and Triassic strata in the Salt Range and Surghar Range, West Pakistan. In *Stratigraphic Boundary Problems: Permian and Triassic of West Pakistan,* ed. B. Kummel and C. Teichert, pp. 306–453. University Press of Kansas special publication no. 4.

Bogoslovskaya, M. F., and Pavlova, E. E. 1988. O rasvitii ammonoidei semeistva Spirolegoceratidae. *Paleontological Journal* 2:111–14.

Bogoslovskaya, M. F., Ustritsky, V. I., and Chernyak, G. E. 1982. Permskie ammonoidei Novoyi Zemly. *Paleontological Journal* 4:58–67.

Broglio Loriga, G., Neri, G., Pasini, M., and Posenato, R. 1988. Marine fossil assemblage from Upper Permian to lowermost Triassic in the western Dolomites (Itali). *Memorie della Societa Geologica Italiana* 34:5–44.

Chudinov, P. K. 1987. Nazemnie posvonochnye i yarusnoye deleniye verkhei permi. In *Materialy po geologii vostoka Russkoi platformy,* pp. 54–73. Kazan.

Clark, D. L., and Wang Cheng-Yuan. 1988. Permian neogondolellids from south China: significance for evolution of the serrata and carinata groups in North America. *Journal of Palaeontology* 62:132–8.

Dagys, A. S., and Dagys, A. A. 1987. Biostratigrafia drevneishikh otlozhenyi Triasa i granitsa paleozoya i Mezozoya. *Geologia i Geofizika* 1:19–28.

Davydov, V. I., Kotlyar, G. V., Ivanova, Z. B., and Timofeeva, O. B. 1990. Analogi Dzhulfinskogo i gorashamskogo yarusov na Yugo-Vostochnom Pamire. In *Permskaya sistema Zemnogo Shara, Abstracts*, pp. 34–5. Sverdlovsk.

Dunbar, C. O. 1955. Permian brachiopod faunas of central East Greenland. *Meddelelser om Gronland* 110:1–169.

Ecke, H. H. 1987. Palynology of the Zechstein sediments in the Germanic basin. In *International Symposium, Zechstein, Abstracts*, p. 28. Bochum.

Epshtein, O. G., Terekhova, G. P., and Solovieva, M. N. 1985. Paleozoi Koryakskogo nagorya (fauna foraminifer, biostratigrafia). *Voprosy Micropaleontologii* 27:47–76.

Foster, C. B. 1979. *Permian Plant Microfossils of the Blair Athol Coal Measures, Baralaba Coal Measures, and Basal Reman Formation of Queensland*. Geological Survey of Queensland, publication 372, paleontological paper 45.

Foster, C. B. 1982. Spore-pollen assemblages of the Bowen Basin, Queensland (Australia), their relationship to the Permian–Triassic boundary. *Review of Paleobotany and Palynology* 36:165–83.

Fuglewicz, R. 1987. Permo–Triassic boundary in Poland. In *The Final Conference on Permo–Triassic Events of the East Tethys Region and Their Intercontinental Correlations*, vol. 6. Beijing.

Furnish, W. M. 1973. Permian stage names. In *The Permian and Triassic Systems and Their Mutual Boundary*, ed. A. Logan and L. V. Hills, pp. 522–48. Canadian Society of Petroleum Geologists, memoir 2.

Ganelin, V. G. 1984. Kolyma-Omolon province (in Russian). In *Main Features of the Stratigraphy of the Permian System in the USSR*, ed. G. V. Kotlyar and D. L. Stepanov, pp. 137–41. Trudy Vsesoyusnogo Nauchno-Issledovatelskogo Geologicheskogo Instituta, new ser., 268.

Glenister, B. F., and Spinosa, C. 1991. Korrelatsiya po ammonoideyam gvadelupsko-dzhulfinskoy granitsy Zakavkazya i Severnoyi Ameriki. In *Permskaya Sistema Zemnogo Shara*, pp. 126–7. Sverdlovsk.

Gomankov, A. V. 1988. Comment on correlation chart of the Upper Permian. *Permophiles, A, Newsletters of SCPS* 13:17–20.

Grigoryan, A. G. 1990. Konodonty pogranichnych otlozhenyi permi i triasa Armyanskoi SSR. Avtoreferat dissertatsii, Moskva.

Gurevich, E. L., and Slautsitajis, T. P. 1988. Paleomagnetnyi razrez verkhnepermskikh i nizhetriasovykh otlozhenyi poluostrova Admiralteistva (Ostrov Novaya Zemlya). *Izvestiya Akademii Nauk SSSR, Ser. Geol.* 1:102–10.

Guskov, V. A., Puchanto, S. K., and Yatsuk, N. E. 1980. Verkhnepermskie otlozhenia Severo-Vostochnogo Pai-Choya. *Sovetskaya Geologiya* 2:68–75.

Henderson, C. M. 1993. Are Permian–Triassic boundary events diachronous? Evidence from the Canadian Arctic. In *Carboniferous to Jurassic Pangea, Abstracts*, p. 136. Calgary.

Iranian-Japanese Research Group. 1981. The Permian and Lower Triassic system in Abadeh region, central Iran. *Memoires of the Faculty of Science, Kyoto University, Ser. Geol. Miner.* 47:1–133.

Kashik, D. S., Ganelin, V. G., Karavaeva, N. I., Byakov, A. S., Miclukho-Maclay, O. A., Stukalina, G. A., Lozhkina, N. V., Dorofeeva, L. A., Burkov, A. S., Guteneva, E. A., and Smirnova, L. N. (eds.) 1990. *Permian Key Section of the Omolon Massiv.* Leningrad: Nauka.

King, R. E., Dunbar, C. O., Cloud, P. E., and Miller, A. K. 1944. Geology and paleontology of the Permian area northwest of Las Delicias, South-Western Coahuila, Mexico. *Geological Society of America Special Papers* 52:1–172.

Kiseleva, A. V. 1986. Pozdnepermskie mshanki vladivostokskogo gorizonta Primorya. In *Permo–Triasovye sobytiya v razvitii organicheskogo mira Severo-Vostoka Azii*, pp. 48–56. Vladivostok.

Kotlyar, G. V. 1978. Sopostavlenie verkhnepermskikh otlozhenyi Yuzhnogo Primorya s odnovozrastnymi obrasovaniyami Tethycheskoyi i Borealnoyi oblasteyi. In *Verkhnyi Paleozoi Severo-Vostochnoi Azii*, ed. L. I. Krasny, pp. 5–23. Vladivostok.

Kotlyar, G. V. 1993. Subdivision and correlation of the Upper Permian of Tethys and Circum-Pacific. *Tikhookeanskaya Geologiya (Novosibirsk)* 3:5–24.

Kotlyar, G. V., and Ganelin, V. G. 1989. Permian. In *Interregional correlation of stratified formations and development of sedimentations in geological history of the USSR territory*, (in Russian), vol. 10, ed. A. I. Zhamoida, pp. 180–5. *Geologicheskoe stroenie SSSR i zakonomernosty rasmescheniya poleznykh iskopaemykh* 10(1):180–5.

Kotlyar, G. V., Komissarova, R. A., Chramov, A. N., and Chedia, I. O. 1984. Paleomagnitnaya kharakteristika verkhnepermskikh otlozhenyi Zakavkazya. *Doklady Akademii Nauk SSSR* 286:669–74.

Kotlyar, G. V., Kropatcheva, G. S., Sosnina, M. I., Pronina, G. P., and Chedia, I. O. 1990a. Zonalnoe rastchleneniye morskikh verkhnepermskikh otlozhenyi Yuzhnogo Primorya. In *Novye dannye po biostratigrafii Paleozoya i Mezozoya Yuga Dalnego Vostoka*, pp. 104–16. Vladivostok.

Kotlyar, G. V., Popeko, L. I., Oleksiv, B. I., and Afanasov, M. N. 1990b. Novye dannye po biostratigrafii verkhnepermskikh otlozhenyi Borzinskogo progiba Vostochnogo Zabaikalya. Khabarovsk.

Kotlyar, G. V., Zakharov, Y. D., Kropatcheva, G. S., Pronina, G. P., Chedia, I. O., and Burago, V. I. 1989. Evolution of the latest Permian biota (in Russian). In *Midian Regional Stage in the USSR*, pp. 1–84. Leningrad: Nauka.

Kozur, H. 1989. Biostratigraphic zonations in the Rotligendes and their correlations. *Acta Musei Reginaehradecensis S. A., Scientiae Naturales* 22:15–30.

Kozur, H. 1991. Late Permian Tethyan conodonts from West Texas and their significance for world-wide correlation of the Guadalupian–Dzulfian boundary. *Geologisch-paläontologische Mitteilungen (Innsbruck)* 18:179–86.

Kozur, H. 1992. Dzhulfian and Early Changxingian (late Permian) Tethyan conodonts from the Glass Mountains, West Texas. *Neues Jahrbuch für Geologie und Paläontologie, Abhandlungen* 187:99–114.

Kozur, H., Leven, E. Y., Lazovsky, V. P., and Pyatakova, M. V. 1978. Division of the Permian–Triassic boundary beds of Transcaucasus on the basis of conodonts (in Russian). *Bulletin Moscovskogo Obshchestva Ispytatelei Prirody, Otd. Geol.* 53:18–24.

Kulikov, M. V., Pavlov, A. M., and Rostovcev, V. N. 1973. O nakhodke goniatitov v nizhnekazanskikh otlozheniyakh severnoi chasti Russkoi platformy. *Doklady Academii Nauk SSSR* 211:1412–14.

Leonova, T. B., and Dmitriev, V. Y. 1989. Rannepermskie ammonoidei Yugo-Vostochnogo Pamira. *Trudy Paleontologicheskogo Instituta, Akademii Nauk SSSR* 235:1–197.

Leven, E. Y. 1993. Glavnye sobytia permskoyi istorii oblasti Tethys i fusulinidy. *Stratigraphiya i Correlatsia* 1:59–75.

Lozovsky, V. R. 1992. Rannetriasovyi etap razvitiya Zapadnoi Lavrasii. Avtoreferat dissertacii, Moskva.

Menning, M. 1988. On the Illawarra reversal. *Permophiles, A, Newsletters of SCPS* 13:12–16.

Meyen, S. V. 1984. Correlation of the Permian deposits of Angarida Realm (in Russian). In *Main features of the Stratigraphy of the Permian System in the USSR*, ed. G. V. Kotlyar and D. L. Stepanov, pp. 226–9. Trudy Vsesoyusnogo Nauchno-Issledovatelskogo Geologicheskogo Instituta, new ser., 286.

Molostovsky, E. A. 1983. *Paleomagnitnaya stratigrafiya verkhneyi permi i triasa vostoka evropeiskoyi chasti SSSR*. Saratov: Saratov University.

Morozova, I. P. 1981. *Pozdnepaleozoiskie mshanki Severo-Vostoka SSSR*. Trudy Paleontologicheskogo Instituta, 188. Moskva: Nauka.

Movschovitsch, E. V. 1986. Permskie konodonty SSSR i problemy korrelacii permi Lavrasii. In *Korrelacia permo–triasovikh otlozhenyi Vostoka SSSR*, pp. 33–49. Vladivostok.

Nassichuk, W. W. 1970. Permian ammonoids from Devon and Melville Islands, Canadian Arctic Archipelago. *Journal of Paleontology* 44:77–97.

Nassichuk, W. W., Furnish, W. M., and Glenister, B. F. 1965. *The Permian Ammonoids of Arctic Canada*. Geological Survey of Canada, bulletin 131.

Novikov, V. P., Suprychev, V. V., and Chramov, A. N. 1991. Korrelatsia po paleomagnitnym i litologicheskim dannym. In *Permskaya systema Zemnogo Shara*, p. 62. Sverdlovsk.

Okuneva, T. M., and Zakharov, Y. D. 1992. Pervye nakhodki permskikh ammonoideyi v basseine r. Borzya (Zabaikalye). *Izvestiya Akademii Nauk, Ser. Geol.* 4:142–4.

Ozawa, T. 1975. Evolution of Lepidolina multiseptata (Permian foraminifera) in East Asia. *Memoires of the Faculty Science, Kyushu University, ser. D*, 23:117–64.

Pavlova, E. E., Manankov, I. M., Morozova, I. P., Solovieva, M. N., Suetenko, O. D., and Bogoslovskaya, M. F. 1991. *Permskie bespozvonochnuye Yuzhnoi Mongolii*. Sovmestnaya Sovetsko-Mongolskaya paleontologicheskaya expeditsiya, vol. 40 Moskva: Trudy.

Peterson, D. N., and Nairn, E. M. 1971. Paleomagnetism of Permian redbeds from the southwestern United States. *Geophysical Journal* 23:191–205.

Popov, Y. N. 1970. Ammonoidei. In *Stratigrafia kamen-nougolnykh i permskikh otlozhenyi Severnogo Verkhoyanya*, pp. 113–40. Trudy Nauchno-Issledovatelskogo Instituta Geologii Arctiki, 154.

Rebello, M., and Doubinger, I. 1988. Stude palynologique dans le bassin evaporitique du Zechstein (Permian supereur): Aspects stratigraphiques paleoecologiques et paleoclimatologiques. *Cahiers Demicropaleontologie* 3:5–19.

Rui, Lin. 1983. On the *Lepidolina kumaensis* fusulinacean fauna. *Bulletin of the Nanjing Institute of Geology and Paleontology* 6:249–70.

Ruzhencev, V. E. 1976. Pozdnepermskie ammonoidei na Dalnem Vostoke. *Paleontologicheskyi Zhurnal* 3:36–50.

Sadovnikov, G. N. 1991. Stratigrafia verkhikh gorizontov permi i granica permi i triasa v kontinentalnykh otlozheniyakh Sibiri. In *Permskaya systema Zemnogo Shara, Abstracts*, p. 74. Sverdlovsk.

Sokolov, B. S. 1960. Permskie korally yugovostochnoi chasti Omolonslogo massiva. *Trudy Vsesoyusnogo Nauchno-Issledovatelskogo Neftyanogo Instituta* 154:38–76.

Spinosa, C., Furnish, W. M., and Glenister, B. F. 1970. Araxoceratidae, Upper Permian ammonoids from the Western Hemisphere. *Journal of Paleontology* 44:730–6.

Taraz, H. 1971. Uppermost Permian and Permo–Triassic transitional beds in central Iran. *Bulletin of the American Association of Petroleum Geologists* 55:1280–94.

Tozer, E. T. 1988. Towards a definition of the Permian–Triassic boundary. *Episodes* 11:251–5.

Wang, Y. 1984. Earliest Triassic ammonoid faunas from Jiangsu and Zhejiang and their bearing on the definition of Permo–Triassic boundary. *Acta Paleontologica Sinica* 23:257–70.

Wilde, G. L. 1955. *Permian Fusulinids of the Guadalupe Mountains*, pp. 59–62. Guidebook for the Permian field conference sponsored by the Permian Basin Section of the Society of Economic Paleontologists and Mineralogists.

Yaroshenko, O. P. 1990. Komplexy miospor i vozrast tufogenno-osadochnykh otlozhenyi Tungusskogo basseina. In *Paleofloristica i stratigrafia Fanerozoa*, pp. 44–84.

Zavadovsky, V. M., Stepanov, D. L., et al. 1970. *Polevoi atlas permskoi fauny i flory Severo-Vostoka SSSR*. Magadan.

In accordance with new data, all *Timorites* occurences are confined only to the upper half of the Midian Stage. Thus only the Upper Midian can be correlated with the Capitanian Stage and the Gizhiga horizon.

5 Examples of late Hercynian transtensional tectonics in the Southern Alps (Italy)

G. CASSINIS, C. R. PEROTTI, and C. VENTURINI

In the past decade the presence of a generally transcurrent tectonic regime in Europe at the end of the Hercynian collision has been emphasized. The Permo-Carboniferous rocks of the continental domains, including a number of sedimentary basins infilled with alluvial-to-lacustrine and volcanic deposits, were especially affected by that dynamic regime. For Spain, France, Germany, Italy, and other European regions the presence of transtensional and transpressional conditions is almost unanimously accepted, as such a regime can explain the larger part of the structural features generated during the late Hercynian interval. That activity lasted from the late Carboniferous until the Permian and perhaps more recent times.

In the Southern Alps, however, examples of such transcurrent tectonics have not been well documented. This chapter, which deals with late Palaeozoic transtensional activities in eastern Lombardy, western Trentino, and the Carnic Alps, is an early contribution to those topics (Figure 5.1). The former area was investigated by G. Cassinis and C. R. Perotti, and the latter was examined by C. Venturini.

Central Southern Alps

The Permian period in the central Southern Alps can be divided into two main tectonosedimentary cycles. Those two cycles were separated by a marked hiatus that has recently been dated to the early part of the late Permian (Italian I.G.C.P. 203 group, 1986; Cassinis and Doubinger 1991, 1992) (Figure 5.2).

The upper cycle, or cycle 2, is characterized by the Upper Permian Verrucano Lombardo–Val Gardena Sandstone, red fluvial deposits at the base of which lithologically different sandy-ruditic bodies crop out locally. Those sediments, with a maximum thickness of about 600 m, are widespread in the area examined and lie unconformably on all the older rocks, stepping down onto the Hercynian metamorphic basement.

The deposits of the lower cycle, or cycle 1, generally ascribed to the early Permian and slightly later Permian times, have a less extensive distribution. They infill some elongated troughs, of which the Collio (or Val Trompia), Tione, Tregiovo, and Orobic basins (the last not considered in this chapter) represent the best-known examples.

The stratigraphic succession of cycle 1 normally consists of both acidic-to-intermediate calc-alkaline volcanics and alluvial-to-lacustrine products, with a cumulative thickness of up to 2,000 m. This succession is affected by lithological and thickness variations, as well as by diachronous relationships of equivalent rock units from one basin to another. Thus, correlation among these basins becomes very problematic.

From a stratigraphic and sedimentologic point of view, the Collio, Tione, and Tregiovo basins display some common characteristics, which according to Cassinis (1966, 1985), Cassinis, Origoni Giobbi, and Peyronel Pagliani (1975), Ori, Dalla, and Cassinis (1988), Cassinis and Neri (1992), and Cassinis and Perotti (1994) can be described as follows:

1. All the aforementioned basins show marked longitudinal and lateral asymmetries. The Collio Basin, especially, corresponds to a pronounced asymmetric graben, with a considerable increase in volcanics toward the east, near the southern Giudicarie line.
2. All the basins are characterized by more or less abundant detrital deposition, indicating rapid and irregular subsidences of tectonic origin.
3. Abrupt facies changes and marked lateral variations, accompanied by sudden vertical developments of sedimentary successions over short distances, are clearly present in all the basins. Also, the volcanic deposits (lava flows, ignimbrites, and tuffs) show consistent changes in thickness, from zero to about 2,000 m. The most conspicuous lateral variations mainly affect the coarse fluvial conglomerates. Stratigraphic differences are particularly evident along the margins of these troughs. Consequently, they provide reliable palaeogeographic criteria for recognition of basin boundaries.
4. The numerous discontinuities observed in the basins clearly testify to syntectonic deposition strongly controlled by local structures.
5. Everywhere, the onlapping sedimentary successions generally are characterized by rather abundant deposition within the basins, as compared with their limited extensions along the borders. For example, the Dosso

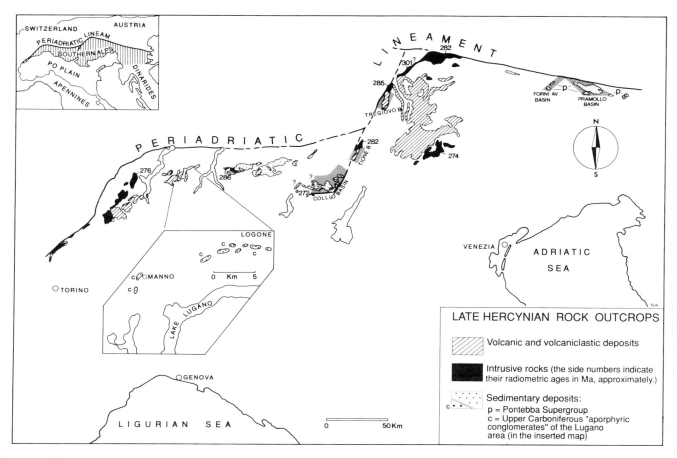

Figure 5.1. Schematic drawing showing the generalized distribution of the late Hercynian rocks in the Southern Alps and the locations of the basins examined (data from Castellarin, 1981, and Lehner, 1952, the latter only for the small area of the inset map). The radiometric dates (in Ma) for some intrusive bodies have been kindly furnished by A. Del Moro, and those for the Bergamasc and Brescian Alps by G. Liborio.

dei Galli Conglomerate, which in the internal areas of the Collio Basin has thicknesses of 500 m and more, unconformably oversteps the underlying Collio Formation and older Permian rocks for short distances along the southwestern margin of the same basin.

6. Alluvial-fan bodies linked to erosion of uplifted areas are widespread. These deposits are principally concentrated along the boundary faults of the basins and show interfingering with alluvial-to-lacustrine fine-grained distal sediments.

All these characteristics of the Collio, Tione, and Tregiovo basins represent typical elements in the sedimentary evolution of strike-slip basins (Ballance and Reading, 1980; Reading, 1980; Mann et al., 1983; Biddle and Christie-Blick, 1985; Ingersoll, 1988; Silvester, 1988). The initiation and development of the Permian basins in the Southern Alps appear to have been associated with the activities of some main tectonic lines, partially forming their margins (Figure 5.3). In fact, even though the later tectonics (in particular, during the Alpine Orogeny) caused tectonic inversion and transport of entire crustal masses southward, vast portions of those basins remained practically

undisturbed, thus permitting study and identification of their original margins, as well as their principal characteristics. From this point of view, the Giudicarie line is certainly the most important. According to some authors (e.g., Vecchia, 1957; Cassinis et al., 1976; Wopfner, 1984; Cassinis, 1985; Cassinis and Castellarin, 1988), that line became active during the Permian and perhaps during earlier times. The main findings supporting that interpretation are the following:

1. The Giudicarie line forms one of the margins of the basins examined. In particular, it more or less coincides with the western margins of the Tione and Tregiovo basins, which trend north-northeast–south-southwest (NNE–SSW), and demarcates the eastern boundary of the Collio Basin. In nearby areas the line changes sharply from a NNE–SSW trend to an approximately east–west trend.

2. During the Permian, the Giudicarie line coincided with a series of tectonic scarps situated on the borders of those basins, the presence of which strongly influenced the sedimentation in adjacent localities.

3. An intense magmatic resurgence took place along the

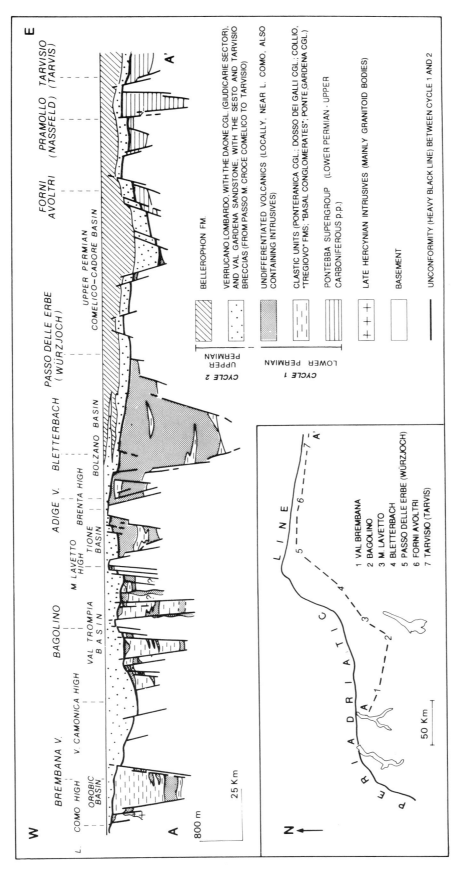

Figure 5.2. Diagrammatic, simplified nonpalinspastic cross section (trace on the inset map) through the Permian and Upper Carboniferous sedimentary and volcanic cover of the central-eastern Southern Alps. Recent investigations (Cassinis and Doubinger, 1992) have shown that both clastic and volcanic units reach, at least locally (*e.g.*, in the Tregiovo area), into the early part of late Permian times. Datum line: base of the Werfen Formation (= Servino Formation in Lombardy). (Adapted from Cassinis et al., 1988.)

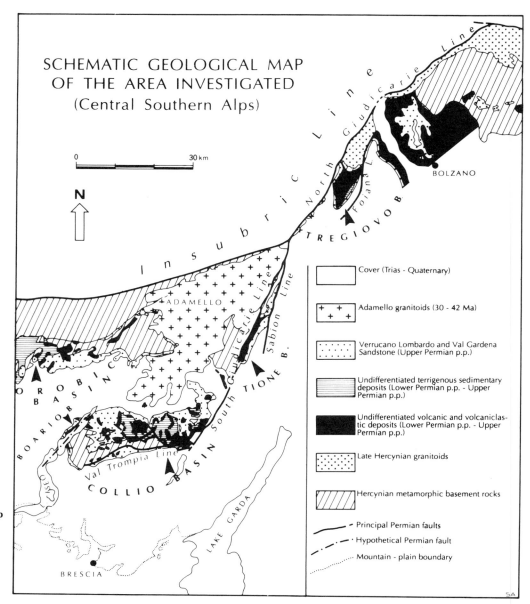

Figure 5.3. Schematic geological map of the investigated central sector of the Southern Alps including the Permian continental basins and neighbouring areas. (Adapted from Consiglio Nazionale delle Ricerche, 1992.)

line in question and in the surrounding areas; the eastern sector of the Collio Basin provides a clear example.

The development of those basins appears to have been influenced by other important tectonic lines running along their margins. In particular, the eastern margins of the Tione and Tregiovo basins coincide, respectively, with the Permian features of the Sabion and Foiana lines, which are subvertical faults almost parallel (at low angle) to the Giudicarie line. The Collio Basin is delimited to the south by the Permian Val Trompia palaeoline and to the west by a series of other Permian faults bounding the coeval M. Muffetto High.

The tectonic nature of the margins described is further

corroborated by the presence of marked, consistent slumping structures with compatible geometry. The rim positions and the contours of the alluvial fans within the basins, as well as the evolution of the sedimentary succession, were strongly influenced by the activity of uplifted areas and are locally characterized by stratigraphic discontinuities. The tectonic structures (faults and folds), observed primarily in the Collio Basin, also indicate intense Permian synsedimentary tectonics. That activity was extensional in nature and caused rapid and irregular subsidences, especially at the sides of subvertical faults.

The features described suggest the presence of a widespread tectonic regime, characterized by extensive transtensional processes during Permian times. The Giudicarie segment, especially, can be interpreted as a dextral transcurrent line, because only

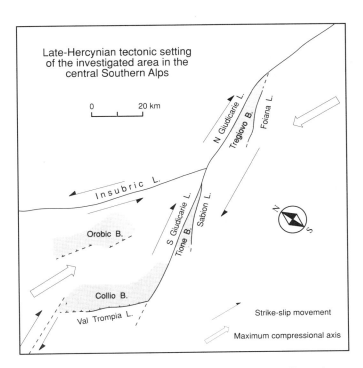

Figure 5.4. Interpretive schematic model showing the late Hercynian tectonic setting of the region examined, in the central Southern Alps, before deposition of the Upper Permian Verrucano–Val Gardena redbeds. The north–south orientation of that tectonic framework during Permian times is also shown.

such a movement would have permitted the opening of the Tione and Tregiovo basins, which are typical strike-slip basins, delimited to the east by other dextral faults like the Sabion and Foiana lines (Figure 5.4).

In all probability the Giudicarie line does not continue to the south of Lake Idro, but shifts in correspondence with the Val Trompia line and reappears, with the same NNE–SSW orientation, farther to the west, near or along the Camonica Valley. In that area, the existence of one or more Permian tectonic lines with a Giudicarie trend is suggested by the presence of the small Boario Basin (Cassinis, 1985) in the lower Camonica Valley, which was partly coeval with and probably was connected with the nearby Collio Basin, displaying a general NNE–SSW direction. In this regard, the assumed orientation (almost north–south) of the late Triassic and Jurassic rifting regime in the area between Lake Iseo and Gardone Val Trompia (Assereto and Casati, 1965; Castellarin and Picotti, 1990) seems to provide further indirect evidence.

As a consequence, the Collio Basin could be interpreted as a pull-apart basin, with the typical lazy-Z form (Mann et al., 1983), which is, however, clearly evident only in the eastern sector, where the basin suddenly changes direction and becomes parallel to the Giudicarie line. In this frame the Val Trompia palaeoline corresponded to a predominantly extensional fault determining the southern margin of the basin.

Within the same regional tectonic regime, the section of the

Insubric line to the west of the Giudicarie line (not considered in this chapter) could be interpreted as a sinistral transcurrent fault conjugated to the Giudicarie line at an angle of about 50–60° (Figure 5.4). Consequently, the Orobic Basin would also represent an extensional basin subparallel to, or at low angle with, the local direction of maximum regional compression, the birth and development of which could be conditioned by sinistral transcurrent movements along the Insubric line (Cadel, 1986). On the other hand, the parallelism between the Permian basins and the principal late Hercynian tectonic lines is, in general, confirmed all over Europe (Arthaud and Matte, 1977; Ziegler, 1986, 1988).

During the early part of late Permian time, the tectonic regime of the Southern Alps was subjected to marked changes because of reorganization of the regional stress field, giving rise to a new sedimentary cycle. Below the reddish detrital deposits of the Verrucano Lombardo–Val Gardena Sandstone, which generally represent the onset of that cycle, a clear angular unconformity with the underlying Permian succession can be observed locally.

The palinspastic reconstructions deduced from numerous sets of palaeomagnetic data (Van Hilten, 1960; Zijderveld and De Jong, 1969; Vandenberg and Wonders, 1976; Westphal, 1976; Heiniger, 1979; Heller, Lowrie, and Hirt, 1989) indicate that the central areas of the Southern Alps underwent a counterclockwise rotation of 50–60° with respect to stable Europe, probably around the end of Mesozoic time. If the present tectonic framework is rotated to its position during Permian times, the Giudicarie line and the associated lines assume an approximately east–west orientation, which appears compatible with the tectonic setting of late Hercynian Europe, crossed by an extended dextral-megashear zone, with about an east–west direction (Arthaud and Matte, 1977; Blès et al., 1989).

Eastern Southern Alps

In the Carnic Alps the Hercynian basement is unconformably covered by a thick, discontinuous sedimentary succession named the "Permo-Carbonifero Pontebbano," "Nassfeld Schichten," or "Pontebba Supergroup" (Selli, 1963; Kahler and Prey, 1963; Venturini, 1990, 1991; Krainer, 1992), whose time of deposition ranged from Westphalian (?)C-D (Moskovian) to Artinskian and perhaps slightly more recent times. The Pontebba Supergroup is followed upward by the Val Gardena Sandstone, of late Permian age, at the base of which siliciclastic and calcareous ruditic bodies are known.

As in the central Southern Alps, the aforementioned late-Hercynian–post-Hercynian succession comprises two depositional cycles. The Val Gardena Sandstone belongs to the upper cycle, whereas the underlying Pontebba unit, separated by a disconformity or a low-angle unconformity, represents the older cycle. That lower cycle began locally in earlier times, as compared with the central Southern Alps (Figure 5.2).

Fluvio-deltaic sediments and mainly marginal-marine terrigenous and carbonate deposits (Venturini, 1991; Massari, Pesavento, and Venturini, 1991) filled two narrow tectonic troughs: the N120°E-trending Pramollo Basin and the N50°E-trending Forni Avoltri Basin. Because of strong Alpine tectonics and scarce exposures of undeformed remnants in the latter basin, no detailed studies have been carried out, except for palinspastic restoration of its original shape and orientation (Venturini and Delzotto, 1992).

In contrast, the Pramollo Basin succession has been studied in great detail in respect of sedimentology, stratigraphy, petrography, and palaeomagnetic aspects; detailed field mappings were carried out for about 10 years. Moreover, the high content of fossils, mainly fusulinids and flora (Kahler, 1983; Fritz, Boersma, and Krainer, 1990), has been helpful in reconstructing the sedimentary-basin evolution. Northward, the Pramollo Basin ends against the N100°E-trending Gailtal line, the easternmost segment of the Insubric line, whereas southward the basin deposits are covered under the Alpine tectonic slices.

The Permo-Carboniferous sedimentary succession of the Pramollo Basin consists of basal, immature, often coarse conglomerates (Bombaso Formation). Those are followed by siliciclastic deposits interbedded with limestone strata (Auernig Group) and then by thick, shallow-water carbonates occasionally interfingered with siliciclastic sediments (Rattendorf and Trogkofel groups). On the whole, the Pontebba Supergroup can be considered a megasequence, terminated at about the Lower–Upper Permian boundary by a widespread regression related to tectonic activity (which has generally been linked by some authors to the so-called Saalian phase of middle Europe).

Consistent thickness changes are evident inside the Pramollo Basin. They range from about 2,000 m (maximum value) down to 120 m or even zero. Several data sets confirm that early tectonic uplifts were active inside the basin itself, during Kasimovian times (Venturini, 1983, 1990, 1991).

The debris-flow conglomerate facies of the Bombaso Formation represent alluvial-fan and fan-delta bodies bordering the Hercynian uplands. The thickness and age of the lithosome vary from one zone to another. The thickness changes from a few metres up to more than 230 m. Lacking any palaeontologic remnant, only the transition to the fossiliferous Auernig sediments can be dated, ranging from Westphalian D (latest Moskovian) to earliest Kasimovian times, or, in some cases, to early Kasimovian or even to late Kasimovian.

The overlying lowermost Auernig beds were deposited in marginal-marine environments, with an upper transition to slope settings with slumps and turbiditic processes. Palaeoflow directions were close to N120°E (toward ESE), which corresponds to the basin axis. The transgressive trend was abruptly interrupted by a sequence break documented over the entire basin and represented by fluvial facies overlying the open-sea deposits, with a marked paraconformity. A short-lived new subaerial drainage developed on a narrow, N120°E-trending tectonic block formed in the middle of the basin. The fluvial-transport direction was from south to north. Both deepening and sudden shallowing of the basin can be linked to strong tectonic control.

In the late Kasimovian the same tectonic block suffered a sudden uplift that caused prolonged emersions (up to the

Asselian, at least), rock slides with megabreccia deposits, and subaerial reworking. The erosion was stronger and deeper toward the southeast; it is attested by progressive exhumation of the Hercynian basement. The uplifted Devono-Dinantian Hercynian limestones gave rise to a 50-m-thick fan-delta body close to the northern fault scarp. All things considered, the late Kasimovian regressive trend must have been related to two close tectonic pulses.

From the Kasimovian to latest Gzelian times, the depositional rate in areas sideways to the uplifted and emerged block increased with time, from about 5 cm per 1,000 years to about 80 cm per 1,000 years, showing exponential growth. During the late Gzelian, more than 800 m of sediments piled up in the basin. Tectonic and partly isostatic subsidence was always balanced by sediment supply, represented by large amounts of siliciclastics and scattered interbedded carbonate deposits.

As a consequence of fast subsidence, the coastline orientation was modified, and the fluvial pattern was deflected toward the N120°E-trending Tröpolach–Camporosso line, which corresponds to the northeastern margin of the Pramollo Basin.

The strongest siliciclastic supply was related to erosion of a phyllitic basement (Fontana and Venturini, 1982). According to palaeocurrent analyses, the source rocks must have been located northwest of the Pramollo Basin, as the latter rests on nonmetamorphic units. The highest terrigenous supply (quartz sand and gravel) may have been connected with far-away tectonic uplifts active at the same time as the late Gzelian extensions of the Pramollo Basin.

According to the evidence presented, the evolution of the Pramollo Basin must be considered to have been strongly controlled and influenced by a transtensive evolution interrupted by a short transpressive pulse.

An interpretive tectonic scheme can be proposed to explain the geometric relationship between the two Permo-Carboniferous basins of the Carnic Alps. It must be taken into account that the Hercynian basement, west of the Forni Avoltri Basin, was affected by weak metamorphism. Field data show that that substratum was uplifted after the main Hercynian (Westphalian) compressive phase and before the sedimentation of the Upper Permian Val Gardena redbeds. The scheme depicted in Figure 5.5 is a speculative solution based on the geometric relationships identified by field mapping, with input from the tectonic models suggested by Rodgers (1980), Harding, Vierbuchen, and Christie-Blick (1985), and Woodcock and Fischer (1986) for strike-slip systems. The southern tectonic closure of the Permo-Carboniferous rock outcrops prevents any further investigation.

The Gailtal line is interpreted as a Permo-Carboniferous strike-slip fault in conjunction (releasing overstep) with a prominent fault farther to the south. At present, the aforementioned line may be buried under the thick Alpine tectonic slices of the region. A dextral strike-slip system was also considered. As a consequence, the following structures were suggested:

1. The Pramollo Basin opening as a pull-apart basin related to dextral strike-slip (Venturini, 1983; Massari, 1986).

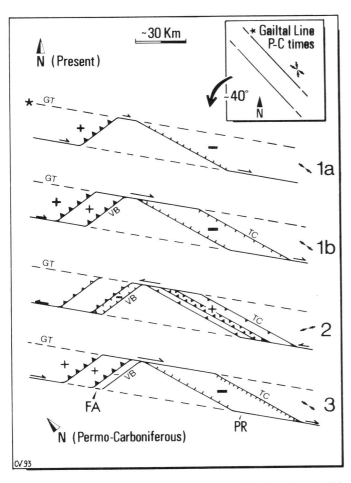

Figure 5.5. The Permo-Carboniferous tectonics of the Carnic area. This speculative model shows the evolution caused by transtensional events (1 and 3) interrupted by a short-lived transpressional pulse (2). In this context, extensional and contractional tectonics could operate in the sphere of influence of a strike-slip system, alternately acting in a dextral or sinistral sense, according to local shifts in the direction of maximum compression: 1, Westphalian (?)C–D, early Kasimovian; 2, late Kasimovian; 3, late Gzelian–early Permian. Black arrows show the inferred maximum compressional axes. FA, Forni Avoltri Basin; PR, Pramollo Basin; GT, Gailtal line; VB, Val Bordaglia line(s); TC, Tröpolach–Camporosso line.

2. The almost contemporaneous uplifting of the weakly metamorphic Hercynian basement, located just westward of the Forni Avoltri Basin. The raised block might be interpreted as a contractional duplex developed in the same dextral strike-slip system. Accordingly, the pervasive N50°E orientation of the foliation (Menegazzi, Pili, and Venturini, 1991) developed in that area during that deformation phase and was later emphasized by the Alpine tectonics.

3. The Forni Avoltri Basin (definitely smaller than the Pramollo Basin), born as isostatic flexure at the thrust front.

Moreover, it must be mentioned that strong tectonic activity (mainly strike-slip) of the palaeo-Gailtal line has been suggested by some authors (Vai, 1976, 1992; Spalletta and Venturini, 1988; Krainer, 1992).

The suggested tectonic scheme is, however, still debatable. According to the palaeomagnetic data (Manzoni, Venturini, and Vigliotti, 1989), the Pramollo Basin has belonged to the Southern Alpine domain since the late Carboniferous and as a whole has been rotated counterclockwise by about 40° with respect to stable Europe.

The scheme of Figure 5.6 depicts a close relationship between the Gailtal line and the induced fault systems, both extensional and contractional; the assumption is based on the present space relationships. It implies that the Gailtal line was also rotated with the basin itself. In that case, during late Carboniferous times the line would have been close to the N140°E direction.

The dextral-megashear zone developed in the circum-Mediterranean area during late Hercynian times (Arthaud and Matte, 1977) supported a NW–SE-oriented main stress. Such a direction would fit with sinistral strike-slip activation of N140°E faults.

A solution might be to assume an unsteady local stress, as suggested by Blès et al. (1989) for the Permo-Carboniferous basins of central France. In that area, the direction of the maximum compression changed from north–south to east–west. Such variance might explain how the Gailtal line could act alternately as a dextral and a sinistral strike-slip fault. That could fit well with the transtensive–transpressive–transtensive evolution of the Pramollo Basin. Moreover, following that concept, the opening of the Forni Avoltri Basin might be considered as a transtensive release, synchronous with the late Kasimovian transpressive event of the Pramollo Basin.

Conclusions

It is suggested that during the late Palaeozoic, transtensional tectonic movements characterized the Southern Alps. Those processes were active over widespread areas and contemporaneously affected the western continental regions as well as the eastern marine regions of that domain. The resultant stress regime led to the opening and development of the basins.

The interpretation of the Collio, Tione, Tregiovo, and Pramollo troughs as pull-apart or strike-slip basins fits well with the late Hercynian scenario for Europe that can explain the extensive distribution of such basins.

Other geological marks of that transcurrent tectonic regime appear significant in the regions examined. In fact, lineaments and structures derived from that activity would, during later times, influence several phases of Alpine evolution. In particular, the Mesozoic rifting arrangement of the central Southern Alps seems to have been substantially inherited from the late Hercynian structural framework that is reconstructed here. Also, many tectonic features that arose from the Alpine Orogeny display connections to the same ancestral architecture. In that regard, the work reported here clearly has brought to light the

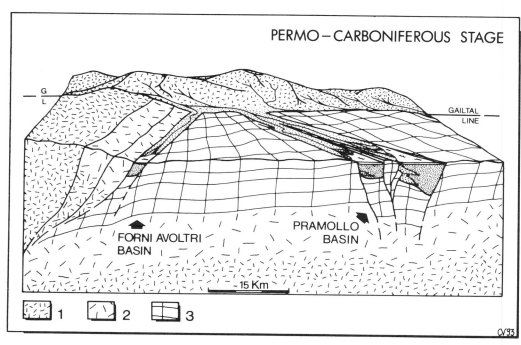

Figure 5.6. Interpretation of the tectonic setting of the Carnic area during the Permo-Carboniferous stage: 1, low-grade metamorphic basement; 2, very low grade metamorphic basement; 3, nonmetamorphic folded basement.

Permo-Carboniferous early activation of the Insubric line along some of its segments, undoubtedly the most prominent example of such connections.

Acknowledgments

We are grateful to M. Vanossi and H. Wopfner for critical readings of the manuscript, to J. M. Dickins for helpful comments, and to S. K. Acharyya for a careful review of the manuscript. We thank H. Wopfner for kindly helping to improve the English text.

This work was supported by grants from C.N.R. and M.U.R.S.T. (40%).

References

Arthaud, F., and Matte, P. 1977. Late Paleozoic strike-slip faulting in southern Europe and northern Africa: result of a right-lateral shear zone between the Appalachians and the Urals. *Geological Society of America Bulletin* **88:** 1305–20.

Assereto, R., and Casati, P. 1965. Revisione della stratigrafia permo–triassica della Val Camonica meridionale (Lombardia). *Rivista Italiana di Paleontologia e Stratigrafia* **71:**999–1097.

Ballance, P. F., and Reading, H. G. (eds.) 1980. *Sedimentation in Oblique-Slip Mobile Zones.* International Association of Sedimentologists, special publication 4.

Biddle, K. T., and Christie-Blick, N. 1985. Glossary – Strike-slip deformation, basin formation, and sedimentation. In *Strike-Slip Deformation, Basin Formation, and Sedimentation,* ed. K. T. Biddle and N. Christie-Blick, pp. 375–85. Society for Economic Paleontology and Mineralogy, special publication 37.

Blès, J. L., Bonijoly, D., Castaing, C., and Gros, Y. 1989. Successive post-Variscan stress fields in the French Massif Central and its borders (Western European plate): comparison with geodynamic data. *Tectonophysics* **169:**79–111.

Cadel, G. 1986. Geology and uranium mineralization of the Collio basin (central Southern Alps, Italy). *Uranium* **2:**215–40.

Cassinis, G. 1966. La Formazione di Collio nell'area-tipo dell'alta Val Trompia (Permiano inferiore bresciano). *Rivista Italiana di Paleontologia e Stratigrafia* **72:**507–88.

Cassinis, G. 1985. Il Permiano nel Gruppo dell'Adamello, alla luce delle ricerche sui coevi terreni delle aree contermini. *Memorie della Società Geologica Italiana* **26:**119–32.

Cassinis, G., and Castellarin, A. 1988. Il significato delle linee della Gallinera e delle Giudicarie sud nella geologia dell'Adamello e zone circostanti. *Atti Ticinensi di Scienze della Terra (Pavia)* **31:**446–62.

Cassinis, G., and Doubinger, J. 1991. On the geological time of the typical Collio and Tregiovo continental beds in the Southalpine Permian (Italy), and some additional observations. *Atti Ticinensi di Scienze della Terra (Pavia)* **34:**1–20.

Cassinis, G., and Doubinger, J. 1992. Artinskian to Ufimian palynomorph assemblages from the central Southern Alps, Italy, and their regional stratigraphic implications. In *Contribution to Eurasian Geology,* ed. A. E. M. Nairn and V. Koroteev, pp. 9–18. Columbia: University of South Carolina Press.

Cassinis, G., Massari, F., Neri, C., and Venturini, C. 1988. The continental Permian in the Southern Alps (Italy). A review. *Zeitschrift für Geologische Wissenschaften (Berlin)* **16:**1117–26.

Cassinis, G., Montrasio, A., Potenza, R., von Raumer, J. F., Sacchi, R., and Zanferrari, A. 1976. Tettonica ercinica nelle Alpi. *Memorie della Società Geologica Italiana (Suppl. 1)* **13:**289–318.

Cassinis, G., and Neri, C. 1992. Sedimentary and palaeotectonic evolution of some Permian continental basins in the central Southern Alps, Italy. *Cuadernos de Geologia Iberica* **16**:59–89.

Cassinis, G., Origoni Giobbi, E., and Peyronel Pagliani, G. 1975. Osservazioni geologiche e petrografiche sul Permiano della bassa Val Caffaro (Lombardia orientale). *Atti dell'Istituto Geologico della Università di Pavia* **25**:17–71.

Cassinis, G., and Perotti, C. R. 1994. Interazione strutturale permiana tra la Linea delle Giudicarie ed i Bacini di Collio, di Tione e Tregiovo (Sudalpino centrale, N Italia). *Bollettino della Società Geologica Italiana* **112**:1021–36.

Castellarin, A. (ed.) 1981. *Carta tettonica delle Alpi Meridionali alla scala 1:200.000. C.N.R., Progetto Finalizzato Geodinamica, Sottoprogetto 5 "Modello Strutturale," Pubbl. 441.* Bologna: Tecnoprint.

Castellarin, A., and Picotti, V. 1990. Jurassic tectonic framework of the eastern border of the Lombardian basin. *Eclogae Geologicae Helvetiae* **83**:683–700.

Consiglio Nazionale delle Ricerche. 1992. *Structural Model of Italy and Gravity Map. Progetto Finalizzato Geodinamica, Quaderni de "La Ricerca Scientifica," n. 114*, **3**. Florence: S.E.L.C.A.

Fontana, D., and Venturini, C. 1982. Evoluzione delle mode detritiche nelle arenarie permo–carbonifere del Bacino tardo-ercinico di Pramollo (Alpi Carniche). *Memorie della Società Geologica Italiana* **24**:43–9.

Fritz, A., Boersma, M., and Krainer, K. 1990. Steinkohlenzeitliche Pflanzenfossilien aus Kärtnen. *Carinthia II* **49**:1–189.

Harding, T. P., Vierbuchen, R. C., and Christie-Blick, N. 1985. Structural styles, plate tectonic settings and hydrocarbon traps of divergent (transtensional) wrench faults. In *Strike-Slip Deformation, Basin Formation, and Sedimentation*, ed. K. T. Biddle and N. Christie-Blick, pp. 51–77. Society for Economic Paleontology and Mineralogy, special publication 37.

Heiniger, C. 1979. Palaeomagnetic and rockmagnetic properties of the Permian volcanics in the Western Southern Alps. *Journal of Geophysics* **46**:397–411.

Heller, F., Lowrie, W., and Hirt, A. M. 1989. A review of palaeomagnetic and magnetic anisotropy results from the Alps. In *Alpine Tectonics*, ed. M. P. Coward, D. Dietrich, and R. G. Park, pp. 399–420. Geological Society special publication 45.

Ingersoll, R. V. 1988. Tectonics of sedimentary basins. *Geological Society of America Bulletin* **100**:1704–19.

Italian, I.G.C.P. 203 group. 1986. *Permian and Permian–Triassic Boundary in the South-Alpine Segment of the Western Tethys.* Field guide-book, S.G.I. and I.G.C.P. Project 203. Pavia: Tipolit. Comm. Pavese.

Kahler, F. 1983. Fusuliniden aus Karbon und Perm der Karnischen Alpen und der Karawanken. *Carinthia II* **41**:1–107.

Kahler, F., and Prey, S. 1963. *Erläuterungen zur Geologischen Karte des Nassfeld-Gartnerkofel Gebietes in den Karnischen Alpen.* Wien: Geologische Bundesanstalt.

Krainer, K. 1992. Fazies, Sedimentationsprozesse und Palaeogeographie im Karbon der Öst- und Südalpen. *Jahrbuch der Geologischen Bundesanstalt* **135**:99–193.

Lehner, P. 1952. Zur Geologie des Gebietes der Denti della Vecchia, des M. Broglia, des M. Brè und des M. San Salvatore bei Lugano. *Eclogae Geologicae Helvetiae* **45**:85–159.

Mann, P., Hempton, M. R., Bradley, D. C., and Burke, K. 1983. Development of pull-apart basins. *Journal of Geology* **91**:529–54.

Manzoni, M., Venturini, C., and Vigliotti, L. 1989. Paleomagnetism of Upper Carboniferous limestones from the Carnic Alps. *Tectonophysics* **165**:73–80.

Massari, F. 1986. Hypothesis on the role of tectonics during Permian times. In *Permian and Permian–Triassic Boundary in the South-Alpine Segment of the Western Tethys*, ed. Italian I.G.C.P. 203 group, p. 9. S.G.I. and I.G.C.P. Project 203. Pavia: Tipolit. Comm. Pavese.

Massari, F., Pesavento, M., and Venturini, C. 1991. The Permian–Carboniferous cyclothems of the Pramollo Basin sequence (Carnic Alps). In *Workshop Proceedings on Tectonics and Stratigraphy of the Pramollo Basin (Carnic Alps)*, ed. C. Venturini. *Giornale di Geologia, Ser. 3a* **53**:171–85.

Menegazzi, R., Pili, M., and Venturini, C. 1991. Preliminary data and hypothesis about the very-low metamorphic Hercynian sequence of the western Palaeocarnic Chain. In *Workshop Proceedings on Tectonics and Stratigraphy of the Pramollo Basin (Carnic Alps)*, ed. C. Venturini. *Giornale di Geologia, Ser. 3a* **53**:139–50.

Ori, G. G., Dalla, S., and Cassinis, G. 1988. Depositional history of the Permian continental sequence in the Val Trompia–Passo Croce Domini area (Brescian Alps, Italy). *Memorie della Società Geologica Italiana* **34**:141–54.

Reading, H. G. 1980. Characteristics and recognition of strike-slip fault system. In *Sedimentation in Oblique-Slip Mobile Zones*, ed. P. F. Ballance and H. G. Reading, pp. 7–26. International Association of Sedimentologists, special publication 4.

Rodgers, D. A. 1980. Analysis of pull-apart basin development produced by en echelon strike-slip faults. In *Sedimentation in Oblique-Slip Mobile Zones*, ed. P. F. Ballance and H. G. Reading, pp. 27–41. International Association of Sedimentologists, special publication 4.

Selli, R. 1963. Schema geologico delle Alpi Carniche e Giulie occidentali. *Giornale di Geologia, Ser. 2a* **30**:1–136.

Silvester, A. G. 1988. Strike-slip faults. *Geological Society of America Bulletin* **100**:1666–703.

Spalletta, C., and Venturini, C. 1988. Conglomeratic sequences in the Hochwipfel Fm.: a new palaeogeographic hypothesis on the Hercynian flysch stage of the Carnic Alps. *Jahrbuch der Geologischen Bundesanstalt* **131**:637–47.

Vai, G. B. 1976. Stratigrafia e paleogeografia ercinica delle Alpi. *Memorie della Società Geologica Italiana* **13**:7–37.

Vai, G. B. 1992. Exhuming age and structural setting of lower crust wedges outcropping in Italy. *International Geological Correlation Program 276, Newsletter* **5**:421–39.

Vandenberg, J., and Wonders, A. 1976. Paleomagnetic evidence of large fault displacement around the Po Basin. *Tectonophysics* **33**:301–20.

Van Hilten, D. 1960. Geology and Permian paleomagnetism of the Val-di-Non area. W Dolomites, N Italy. *Geologica Ultraiectina* **5**:1–95.

Vecchia, O. 1957. Significato del fascio tettonico giudicario-atesino. Dal Benaco a Merano: un problema geologico. *Bolletino della Società Geologica Italiana* **76**:81–135.

Venturini, C. 1983. Il bacino tardo-ercinico di Pramollo: un'evoluzione regolata dalla tettonica sinsedimentaria. *Memorie della Società Geologica Italiana* **24**:23–42.

Venturini, C. 1990. *Geologia delle Alpi Carniche centro orientali.* Museo Friulano di Storia Naturale, Pubbl. 36. Udine: Grafiche Fulvio s.r.l.

Venturini, C. 1991. Introduction to the geology of the Pramollo Basin and its surroundings. In *Workshop Proceedings on Tectonics and Stratigraphy of the Pramollo Basin (Carnic*

Alps), ed. C. Venturini. *Giornale di Geologia, Ser. 3a* **53**:13–47.

Venturini, C., and Delzotto, S. 1992. Evoluzione deformativa delle Alpi Carniche centro occidentali: paleotettonica e tettonica neoalpina. In *Studi Geologici Camerti, Sottoprogetto CROP 1-1A,* ed. R. Capozzi and A. Castellarin, pp. 261–70.

Westphal, M. 1976. Contribution du paléomagnétisme à l'ètude des dérives continentales autour de la Méditerranée occidentale. Thèse de l'Université de Strasbourg.

Woodcock, N. H., and Fischer, M. 1986. Strike-slip duplexes. *Journal of Structural Geology* **8**:725–35.

Wopfner, H. 1984. Permian deposits of the Southern Alps as product of initial alpidic taphrogenesis. *Geologische Rundschau* **73**:259–77.

Ziegler, P. A. 1986. Geodynamic model for the Palaeozoic crustal consolidation of Western and Central Europe. *Tectonophysics* **126**:303–28.

Ziegler, P. A. 1988. *Evolution of the Arctic–North Atlantic and the Western Tethys.* American Association of Petroleum Geologists Memoir **43.**

Zijderveld, J. D. A., and De Jong, K. A. 1969. Paleomagnetism of some late Paleozoic and Triassic rocks from the eastern Lombardic Alps, Italy. *Geologie en Mijnbouw* **48**:559–64.

6 Succession of selected bioevents in the continental Permian of the Southern Alps (Italy): improvements in intrabasinal and interregional correlations

MARIA ALESSANDRA CONTI, NINO MARIOTTI, UMBERTO NICOSIA, and PAOLA PITTAU

It is well known that determinations of stratigraphic subdivisions and, especially, correlations of continental sediments often are hampered, even on the regional scale, by discontinuities in sedimentary strata, whose rates of sedimentation often were highly variable and discontinuous, and by the fact that there had been episodic palaeogeographic and environmental control of pull-apart basins, where floras and faunas could vary because of differing ecological situations. Virtually all of those obstacles can be bypassed by using "unconformity-bounded stratigraphic units" (NASC, 1983) in an intrabasinal lithostratigraphic district. Unfortunately, when considering interregional correlations, chronostratigraphic and geochronologic units must be used. These "high-level" units, in turn, need to have clearly defined stratigraphic limits (boundaries), and their fossil contents should already have been well studied and illustrated. These are recommendations from the *International Stratigraphic Guide* (ISG) (Hedberg, 1976). However, we rarely find ourselves working under conditions that allow full compliance with the ISG recommendations.

At present, our difficulties in chronostratigraphic correlations of continental Permian sediments have resulted from several causes: (1) The primary cause is their discontinuous depositional settings; (2) The second cause is that their biologic contents are not always completely known, having been poorly studied or poorly illustrated; (3) A third problem arises because of confused and unclear chronostratigraphical utilization over hundreds of years; (4) The stratotypes of the boundaries often have not been exactly defined, probably because the boundaries do not crop out at the type localities. In conclusion, the use of any chronostratigraphic unit in the continental Permian is still confusing.

In using biozones, we are considering isochronal surfaces, and this is surely the best method for correlations in a time-space framework. But even biozones, when used for continental successions, can involve problems, the largest of which originates from the absence of evolutionary lineages within directly superimposed sequences.

Thus the chronostratigraphic correlations of Permian continental sequences are very uncertain. That is due in part to the fact that they were made in a traditional way. That is, they were based on comparisons of biofacies more than on selection of meaningful taxa. The same uncertainty is seen in biostratigraphy, because fossil-bearing levels are scattered and only rarely are in continuous vertical successions.

Needing a general correlation framework, we tried a less traditional method that, even while using basic palaeontological principles, allows less fine but more certain conclusions. That was achieved by selection of taxa and bioevents, and by characterization of groups of taxa, on the basis of their evolutionary levels (faunal units).

Bioevents and faunal units

Selection of bioevents and taxa

Several different types of bioevents were selected. According to their relative importance, they had different degrees of utility for correlation. In order of importance, we considered the following:

1. First-appearance datum (FAD) and last-appearance datum (LAD). These concern a taxon, at either the genus or species level, and are very important at the regional level of correlation.
2. Disappearance event (DE). This concerns a faunal complex and thus has very high correlation power.
3. First occurrences (FO) and last occurrences (LO).

Bioevent selection was followed by an attempt to establish a calibration of bioevents with respect to the standard scale by direct correlation in the study areas or by indirect calibration of deposits in areas where marine fossils are present, even if only scattered. Even if calibration is impossible, selected bioevents represent time lines and thus do not lose their correlation power and sequentiality. Intervals between them are biochrons, time units that allow us to use scattered findings to single out a virtual events sequence.

Obviously basic to these concepts is correct taxonomy and at least homogeneous identification of taxa. Unfortunately, broad syntheses frequently have been based simply on lists of names. They have grouped under the same binomial different taxa, lacking taxonomic revision, and they have denied correlations based on the taxon more than on the formal names. Thus, before accepting or rejecting the distribution of a taxon, we need at least a subjective synonymic selection and a set of controls based on illustrations (or direct observations) of the taxa.

Faunal units

Such units have their own scientific integrity, on the theoretical basis of an assumed correspondence between the principle of the irreversibility of evolution and the principle of superposition. The faunal units (FU) are recognized by bioevents as limits, and by asssemblages as time spans, and are designated by the names of the most representative fauna. Indeed, they seem somewhat more similar to anthropological "cultures" than to classic biozones.

Whenever possible, the units are hierarchically ordered as faunal ages (FA) and faunal subages (FsA) or as faunal units and subunits (FU, FsU). The relative values of "ages" or "units" depend on their global, regional, or local distribution and on the degree of diversity.

In practice, they describe the time needed for evolution of the organisms, and with respect to the classic biozones they have the advantage of being useful simply on the basis of bioevents recognized in scattered fossil-bearing outcrops, even if those outcrops are not in direct vertical sequence. After a global check, they can be considered somewhat equivalent to chronozones, being somewhat time-equivalent units. Formalized faunal units are more and more frequently being used by vertebrate palaeontologists for defining biochronological successions of Cenozoic and Quaternary sediments; in such intervals, continental deposits are timed by mammal "ages."

This chapter offers a first application of faunal-unit concepts to Permian vertebrate associations interpreted on the basis of their abundant footprints.

Data analysis

We shall explain separately the selection of taxa and events relative to tetrapod footprints and to certain microfloras singled out from among those present in our study areas. Then we shall move on to the joint results, to the recognized interregional events, and to the definition of a new correlation framework. Obviously, the events examined here are derived mostly from footprints and sporomorphs; other available data are sketched out only briefly.

The succession of bioevents considered here was sorted out from the sedimentary sequences laid down in the Permian basins of the Central and Southern Alps. Such sequences are generally considered to reflect two major cycles – the intracratonic and subsequently the marginal – in the geological framework of northern Italy, as was well illustrated by the Italian IGCP-203 group (1988) (Figure 6.1). The continental Permian units from which we collected the faunas were the Collio Formation, the Dosso dei Galli Conglomerate, and the Tregiovo Formation in the Lombardy Basin. In the Dolomitic Basin, the fossiliferous

units were the Val Gardena Sandstone and the Bellerophon Formation (Figure 6.2).

Tetrapod footprints

Strata bearing tetrapod footprints have been intensively studied in our region during the past 20 years (Leonardi and Nicosia, 1973; Conti et al., 1977, 1991; Mietto, 1981; Ceoloni et al., 1987, 1988a, b). Following those studies, large amounts of data began to show the presence of two faunas, well differentiated at the evolutionary level. On the basis of those differences we began the study reported here.

Concerning the tetrapod footprints, by using a twofold subdivision of the Alpine continental Permian strata into two major cycles (Cassinis et al., 1988), we based our study on two well-known sections, one for each cycle, in which the ichnofossils were recognized in some vertically succeeding levels. The lower succession (cycle I) was sampled within the Collio Formation type area (Conti et al., 1991). The most representative section for the second major cycle, in its turn subdivided into three minor cycles, is the famous Bletterbach section, where the Val Gardena Sandstone and the Bellerophon Formation crop out (Massari et al., 1988). Data from all the other Permian tetrapod-bearing outcrops known in the area of the Southern Alps were used to check the findings based on the main sections. The support sections were Bagolino, Faserno, Tregiovo, Pizzo del Diavolo, Scioc Valley, Sette Crocette Pass, Pulpito, and Malga Stabul Maggiore for the first cycle (Conti et al., 1991), and Seceda, Nova Ponente, San Pellegrino Pass, Ligosullo, and Recoaro for the second cycle (Conti et al., 1989).

Selected taxa

In the past few years we have spent most of our time on a careful revision of our faunas, augmented by direct observations of tetrapod footprints from other European basins. After reexamination of most of the published European material, we reached the conclusion that too many names had been given to the too few footprints species. The footprints often differed from one another only in terms of the kind of impression (extramorphologies). Thus synonyms were frequent, and the need for a general, systematic revision was evident. Those opinions were shared by other footprint specialists.

Even though our first general overview indicated a daunting task ahead, after our early studies (Ceoloni et al., 1988b) we became convinced that the use of inclusive footprint data can allow the same confidence in naming the major taxa as do skeletal remains. On that basis we decided to increase the degree of confidence by a drastic reduction in the number of names, and in that first search for meaningful bioevents we used only a few selected taxa higher than the species level. Such taxa were well defined and easy to recognize.

Starting from more or less the same criticisms, and using a similar selection procedure, Gand and Haubold (1988) had begun to try to decipher, from the confused footprint system-

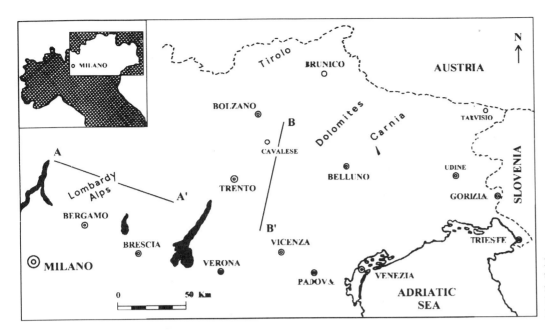

Figure 6.1. Study area; the profile traces refer to Figure 6.2: AA′, profile across Orobic and Val Trompia basins; BB′, profile across Dolomitic Basin.

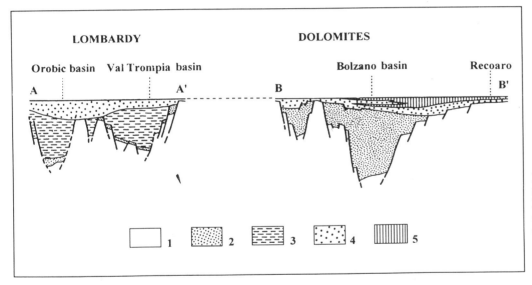

Figure 6.2. Schematic sections across the main Alpine basins: 1, pre-Permian basement; 2, volcanics; 3, clastic units of the first sedimentary cycle; 4, Verrucano Lombardo–Val Gardena Sandstone; 5, Bellerophon Formation. (Adapted from Italian ICGP-203 group, 1988.)

atics, the stratigraphic significance of the European tetrapod species studied through their footprints.

At the moment, while waiting for completion of the formal revision of systematics, which is in progress, we do not wish to address here the problems of nomenclature and synonymies. Thus, the selected taxa are indicated here simply by the best examples we were able to find:

> *Amphisauropus latus* Haubold 1970 – plate 59, fig. 7, of Haubold (1984)
>
> *Dromopus lacertoides* (Geinitz 1861) – plate 56 of Haubold (1984)
>
> *Dromopus didactylus* (Moodie 1930) – fig. 60 of Gand (1988)

> *Ichniotherium cottae* (Pohlig 1885) – plate 57 of Haubold (1984)
>
> *Ichniotherium* gen.
>
> *Rhynchosauroides* gen. (*R. pallinii* of Conti et al., 1977) – plate IV of Conti et al. (1977)
>
> *Pachypes* gen. (*P. dolomiticus* of Leonardi et al., 1975) – plate III of Conti et al. (1977)
>
> *Dicynodontipus* gen. [*D. geinitzi* (Hornstein 1876)] – plate 91, fig. 2, of Haubold (1984)

Simply on the basis of these few taxa it is possible to recognize useful bioevents; further changes or better definitions of other taxa could refine such efforts.

Findings were considered as valid bioevents only when

sediment types (local palaeoenvironmental conditions) could permit an older or younger occurrence of the selected taxa, but in reality they were not found. Other intervals were characterized by the peculiar state of the faunas (low diversity–high diversity).

Lombardy Basin. In the type area the volcano-sedimentary Collio Formation crops out, with a thickness of nearly 700 m, and it is informally subdivided into members A to G (Cassinis, 1966). The youngest member is directly overlain by the first member of the Dosso dei Galli Conglomerate, the Pietra Simona member of cavers (Ori, Dalla, and Cassinis, 1988).

The bioevents have been recognized at the chosen section on the basis of tetrapod assemblages found at 11 different levels, ranging from the lower boundary of the C member (nearly 180 m from the base) up to the first metres of the Pietra Simona member (700 m from the base and 300 m from the top of the Auccia Volcanite, which closes the first cycle); footprints have been found in members C, E, F, and G, that is, in each sedimentary member of the formation (the remaining members – A, B, and D – are volcanics).

The recognized bioevents are the local first and last occurrences and a peculiar faunal condition, namely: *Amphisauropus latus* FO–LO, *Ichniotherium cottae* FO–LO, *Dromopus lacertoides* FO – LO, *Dromopus didactylus* FO–LO, and an interval of low-diversity fauna. Two of them have been singled out as regionally important (Figure 6.3).

Dolomitic Basin. At the chosen section, the Val Gardena Sandstone and the overlying marine Bellerophon Formation crop out, for a total thickness of 252 m (Massari et al., 1988). Ten levels have yielded footprints, from 72 m up to 135 m from the base, 10 m below the boundary with the Bellerophon Formation, but footprints have also been found within continental levels interfingering with the marine sediments (162 and 174 m from the base, and 78 m below the Tesero horizon, the lowermost unit of the Werfen Formation and the assumed Permian–Triassic boundary). The recognized bioevents are the *Rhynchosauroides* FO, *Pachypes dolomiticus* FAD, *Ichniotherium* LO, *Dicynodontipus* FAD, and an interval of high-diversity fauna. Also in this case we have been able to recognize two regionally meaningful events (Figure 6.4).

Intrabasinal correlations

The previously listed bioevents locally characterize at least three intervals with different faunal assemblages. The first two intervals are characterized, from the evolutionary point of view, by a single evolving lineage (*Dromopus lacertoides–D. didactylus*), whereas the third features an absence of evolutionary changes. Additional data include the presence of a characteristic impoverished fauna at the top of the first-cycle sediments, whereas the faunal assemblage of the second cycle is the most diversified of all Permian tetrapod faunal assemblages.

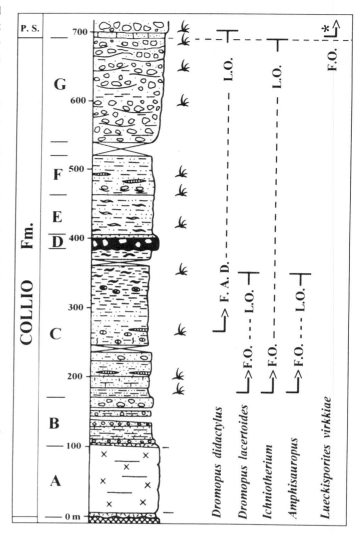

Figure 6.3. Schematic stratigraphic column for the Collio Formation outcropping at the Trompia Valley. The scheme shows the major bioevents related to the first Alpine Permian sedimentary cycle. *As explained in the text, the *virkkiae* FO actually occurs in the Tregiovo Formation (Tregiovo Basin), which is equivalent to the Pietra Simona member.

A shared character, an illusory absence of evolution in both assemblages, is clearly anomalous, for that was a time of rapid evolutionary changes among reptiles. Such data seem to be better explained as the results of a very short period of deposition for the footprint-bearing sediments of both cycles.

In this framework, some events seem to assume absolute meanings. For instance, the *Ichniotherium* LO may be considered as the genus LAD, because elsewhere the genus is present only in older sediments. The *Rhynchosauroides* FO must be considered as the genus FAD because of the absence of this well-known form in preceding deposits in different regions. In later deposits this genus characterizes the tetrapod faunas of continental Triassic sediments over a very large area. The first occurrence

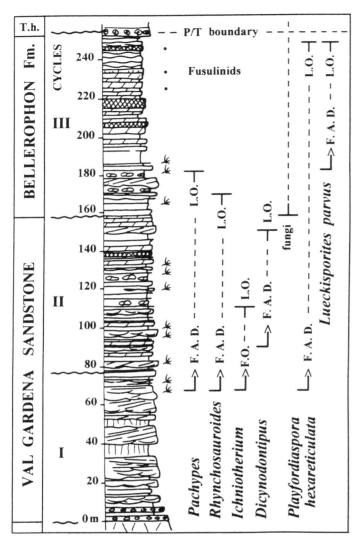

Figure 6.4. Schematic stratigraphic column for the Val Gardena Sandstone outcropping at the Bletterbach section. The bioevents reported are referable to the second Alpine Permian sedimentary cycle.

of the genus *Rhynchosauroides* is clearly diachronous in the sequences of the Southern Alps, as well as in the European sequences, but this time difference can be easily explained by a variety of causes involving localized situations such as depositional gaps or environments unsuitable for the establishment of such reptiles. If we hypothesize our area as a radiation starting point, the difference could even be explained by the time needed for migration, even if the time interval may seem too long. Anyway, its meaning is the first appearance in the Southern Alps area and the first occurrences in France and Germany. Such events cannot be directly correlated, but the FAD level retains its significance as a time line. From this standpoint, the boundary bioevents can be inferred as valid for large areas, thus retaining their status as datum levels, and on that basis we can distinguish in our sequences at least two major faunal units.

First faunal unit. This faunal unit is bounded by the *Amphisauropus latus* FO and the *Ichniotherium cottae* FO at the base and by the *Dromopus didactylus* LO at the top. The evolutionary level reached by the assemblage, on the whole, seems very similar to the level of the characters for the faunas typical of Permian outcrops in various European countries, which frequently share surprisingly uniform species compositions. It is named here the Collio FU, and it can in turn be subdivided into two subunits.

The lower subunit is characterized by the presence of *Dromopus lacertoides, Amphisauropus latus,* and an *Ichniotherium* sp. and by the appearance of *D. didactylus.* It crops out at various localities: Pulpito, Lago Dasdana, Bagolino, Sette Crocette, Scioc Valley, and Pizzo di Trona and is named the Pulpito FsU.

The upper subunit is bounded by the disappearance event (DE) for all but one of the preceding forms, yielding a drastic reduction in faunal diversity (Lower Permian tetrapod DE); indeed, the subunit is monotypic, being represented by very frequent specimens of the only relict species, *D. didactylus.* By chance, that disappearance event has also been recognized elsewhere: within the same Collio Basin, both in another locality (Faserno) and in a different facies (Pietra Simona member), and in the pull-apart Tregiovo Basin (Cassinis and Neri, 1990; Cassinis and Doubinger, 1991a,b). Within the Tregiovo Basin, tetrapod footprints are widespread across a measured section (see fig. 4 of Cassinis and Doubinger, 1991a), but the very frequent trackways present are only from *D. didactylus;* we ascribe that interval to a time after the Lower Permian tetrapod DE. Thus the disappearance event can be assumed not to have been controlled by local environmental conditions, so it keeps its meaning as a larger-scale bioevent. Here the second subunit is named the Tregiovo FsU, from the most characteristic locality (Tregiovo).

Second faunal unit. This complex is characterized by a modern and diversified fauna almost completely new as compared with other Permian tetrapod faunas. Bounded in its lower part by the *Rhynchosauroides* FAD, it records also the *Dicynodontipus* FAD and the *Ichniotherium* LAD; its upper boundary is hidden by the spreading of the Bellerophon Formation marine sediments. It has been recognized in some outcrops in the Southern Alps, allowing us to correlate the Bletterbach section with Seceda, San Pellegrino Pass, Pietralba, Redagno, Ligosullo and Recoaro. In other countries, only the very poor "terrestrischer Zechstein bei Gera" (Müller, 1959) seems to have reached a quite similar evolutionary level. It is named here the Bletterbach FU.

Interregional correlations

There had been some earlier correlation attempts based on tetrapod footprints (Haubold and Katzung, 1975; Kozur, 1980; Holub and Kozur, 1981; Boy and Fichter, 1988a; Gand, 1988). Some simply reported the tendencies of changes, but others erected formal associations, and some others biozones. We were not satisfied with those studies, perhaps with the exception of Gand's associations, though even those seem to us to be split into too many categories, still showing traces of the use of form names that need revision. We shall return to this subject in a later analysis.

Looking only at the studies conducted in more recent years, starting from the symposium on the Rotliegendes held in Erfurt in May 1987, steadily increasing concern has been expressed by researchers on the stratigraphy of the continental Permian. Diachronous events and difficulties in correlation have been emphasized, and explanations have been sought in terms of many different causes, such as migration delay (Schneider, 1989), environmental control on faunas and floras (Schneider, 1989), and confusion in basic systematics (Gand and Haubold, 1988). At the same time, Boy and Fichter (1988b) have denied that the distribution of tetrapods is strongly facies-controlled. We have tried to change such viewpoints, checking the distribution of tetrapod footprints with respect to major taxa (Ceoloni et al., 1988b) or evaluating faunal diversity as a stratigraphic key (Ceoloni et al., 1988a). Gand and Haubold (1988) achieved quite good results, suggesting the first major reduction in the number of taxa, even if not reaching the nearly complete rejection of some taxa that we favor in our current draft proposal.

All of those earlier attempts to subdivide the Permian continental sediments stratigraphically have been shown to be unsuitable for efficient and worthwhile correlations. Here we shall offer a further attempt to compare the bioevents selected in our regions and the assumed European correspondences. In fact, taking into consideration the literature data, as well as our direct observations of the materials stored at Autun, Dijon, Paris, Gotha, Halle, Kassel, Mainz, and Nierstein, the majority of the faunal assemblages seem to be made up of a quite restricted number of ichnospecies; in fact, we have been able to place many of the forms previously ascribed to different taxa into one or another of our selected taxa. The main drawback was that sometimes the wrong names were assumed to have been peculiar to associations or to have been indices of certain biozonal subdivisions based on footprints. Without entering into the taxonomic problems, we shall briefly examine the main faunal complexes of the European Permian.

Thuringian Forest. In the Thuringian Forest (Goldlauter, Oberhof, Rotterod, Tambach, and Eisenach formations), the faunal complex, the most thoroughly studied complex in the past (Haubold, 1971; Lützner, 1987), seems to correspond well to the fauna with *Amphisauropus,* preceding the *Dromopus didactylus* FAD. We cannot examine here the very low diversity faunas of the Gehren and Manebach formations, which might represent a further faunal unit bounded in its uppermost part by the *Ichniotherium cottae* FAD.

Saar-Nahe Group. Footprints reported by Fichter (1976, 1982, 1983a,b, 1984) and by Fichter and Kowalczyk (1983), basic for the zonation of Boy and Fichter (1988a), still are to be referred to the same faunal complex and to the same species typical of the Thuringian Forest.

Cornberg Formation. The footprints we examined, which were described and named by Schmidt (1959), seem to be, at least in large part, extramorphological impressions made by reptiles known elsewhere under different names. The morphological differences probably are due to the completely different environment (eolian sand dunes). Most of the names used are incorrect, rendering dubious the "Assoziation mit *Chelichnus, Laoporus* und *Phalangichnus*" of Kozur (1980) and both the *Anhomoichnium* Zone and the *Harpagichnus* Zone of Boy and Fichter (1988a), and correction of those errors will change the stratigraphic position of the formation.

France. The tetrapod associations in France numbered four, according to the general framework prepared by Gand (1988). At the moment, we recognize an association lower than the Rabejac, including associations I and II of Gand (1988) and mainly corresponding to the Thuringian Forest and to the Saar-Nahe associations, as well as the Rabejac association itself (Gand's III). The latter seems to correspond directly to the lower portion of the Collio FU, that is, to the Pulpito FsU, based on the presence of *D. didactylus* still associated with the other forms. Unfortunately, we have been unable to observe directly the La Lieude material (Gand's association IV); most of the names given by Ellenberger in 1983 (cited in Gand, 1988) seem to be related only to extramorphological impressions. On the basis of the few available illustrations, it seems to include *D. didactylus, Amphisauropus latus,* and *Ichniotherium cottae.*

Great Britain. Unfortunately, we could not directly examine the British material. On the basis of the illustrations and the paper by Haubold and Sarjeant (1974) on the Enville Group material, it seems that the forms present are the same as in the Thuringian Forest sequences. Concerning the footprints of Dumfriesshire, nothing can be said, because of lack of recent information and satisfactorily illustrated material.

In conclusion, only a few bioevents seem to be truly significant for a detailed correlation of the whole European Permian. For instance, *D. didactylus,* the characterizing form of the Tregiovo FsU, seems to occur in France for the first time at the Rabejac levels (Gand, 1988); the remaining clearly defined forms are present in the known footprint-bearing outcrops, with a few exceptions: Val Gardena Sandstone, Bellerophon Formation, "Zechstein bei Gera"(?), Dumfriesshire.

Spore-pollen floras

Palynological studies of the Permian Southern Alps have been carried out in recent decades by several authors (Klaus, 1963; Jansonius, cited in De Boer, 1963; Visscher, 1971; Clement Westerhof, 1974; Pittau, cited in Massari et al., 1988), and more recently by Doubinger (Cassinis and Doubinger, 1991a,b).

The successions investigated have been those of the Lombardy and Dolomitic basins, namely, the Collio Formation and the Tregiovo Formation in the former basin, and the Val Gardena Sandstone and the Bellerophon Formation in the latter. All of the four lithostratigraphic units have furnished rich microfloras, but of different states of preservation (Plate 6.1).

Lombardy Basin

Sporomorphs from the Collio Formation have been severely affected by thermal alteration. Doubinger (Cassinis and Doubinger, 1991a,b) has listed several taxa at the species level from the lower part of the Collio Formation (C member), where two levels have been productive. From a qualitative point of view, the two associations reported are similar; they differ only in the presence of *Lueckisporites microgranulatus* Klaus in one, and *L. globosus* Klaus in the other (Plate 6.1).

Those findings have revealed the first local appearance of *Lueckisporites* in the Permian deposits of the Italian Southern Alps. As shown in Figure 6.3, those two microfloras have been found interlayered with abundant tetrapod footprints, and they correspond to the *Dromopus didactylus* FAD. Worldwide, the first appearance of *Lueckisporites* is still in doubt. Formally, it dates back to the Artinskian in Pakistan, where Balme (1970) identified the species *Lueckisporites singhii*. It is the opinion of many palynologists, and of Balme himself, that that genus name is not proper for that taxon, but for 22 years it has remained, with doubtful assignment. Moreover, the specimen depicted by Doubinger (1974) from the A3 association of Autun very likely is not *Lueckisporites,* because of the laterally undivided proximal sexine; that is the same diagnostic character missing from *L. singhii.* Thus, given the current state of our knowledge, it seems that the first appearance of *Lueckisporites* is with Klaus's species *L. microgranulatus* and *L. globosus,* first at the LO3 level in the Lodève Basin, and likely later in the Collio Formation. For that reason, the recording of *Lueckisporites* in the Lombardy Basin is classified as a first-occurrence event. The end of the first cycle within the Lombardy Basin is marked by the levels ascribed to the Tregiovo Formation, where, in a rich microflora, *Lueckisporites virkkiae* makes its first appearance. On the basis of sequential stratigraphy and correlations based on ichnofaunas, the Tregiovo Formation corresponds to the lower member of the Dosso dei Galli Conglomerate (Pietra Simona), which comformably overlies the Collio Formation in the type area. Thus, the appearance of *L. virkkiae* in the fossiliferous sequence is of great importance, and we classify it as a first-appearance bioevent that allows direct correlations with the following:

1. Upper Leonardian to Lower Guadalupian (Kazanian) of the North America midcontinent. The El Reno Group microfloras have been extensively described by Wilson (1962), Morgan (1967), and Clapham (1970), and a comprehensive synthesis of North America Permian palynology has recently been published by Barker, Wood, and Eames (1991).
2. Kazanian of the cis-Uralian eastern European platform,

to which Sivertseva (cited in Andreeva et al., 1966) refers it on the basis of the presence of *Ullmannia* spores.

3. Kazanian of the Salt Range, Wargal Formation. In Pakistan, parts of the Ufimian and Kazanian, corresponding to the stratigraphic gap between the Amb and Wargal formations, are absent, and there *L. virkkiae* is present starting in the Upper Kazanian (Balme, 1970).
4. "Saxonian" of the "Facies de Leouvé" (Dôme de Barrot, southern France), in which Visscher, Huddleston Slater-Offerhaus, and Wong (1974) refer to the presence of *L. virkkiae.* It is the oldest known record of that species in the French Permian, but because of the absence of a continuous fossiliferous sequence in that area, this correlation is not so strong as a direct equation.

A correlation between Tregiovo and Esterel (Le Muy) (Visscher, 1968) seems not to be suitable; in fact, the Esterel microflora probably will be found to be younger than that of Tregiovo if the presence of *Perisaccus granulatus* Klaus and *Playfordiaspora hexareticulata* (Klaus) can be confirmed. The microflora reported by Schaarschmidt (1980) from Wetterau (Germany) seems to be of the same age as that of Tregiovo.

Regarding comparisons with Lodèvian microfloras, the microfloras reported by Doubinger (1974) and Doubinger, Odin, and Conrad (1987) have a strong "Autunian" or Lower Permian character, particularly the associations LO1 and LO2.

The LO2 association fits with the Wolfcampian North American midcontinent associations, where *Costapollenites ellipticus* and pollens of the *Nuskoisporites* complex are present in the Admire Group, which corresponds to the basal Wolfcampian (Tschudy and Kosanke, 1966; Barker et al., 1991).

The LO3 association, which we have compared with the Zechstein Kupferschiefer, does not fit with the biostratigraphy of the ichnofauna. Moreover, it does not find a place in a sporomorph-distribution framework for western, central, and southern Europe; neither can it be compared with the North American midcontinent Guadalupian microfloras (Kazanian and Midian, according to the Archbold and Dickins stratigraphy) (Wood, Foster, and Barker, 1991), because the *Lueckisporites virkkiae* palynofacies is widely represented there. It perhaps could fall in an interval extending from the Artinskian to the Kungurian, in being similar to that of the Collio Formation. Similarity indices between those two stratigraphic levels include *microgranulatus* Klaus and *L. globosus* Klaus; the remaining LO3 association has a strong older character than the Collio microflora.

The importance of Tregiovo derives from the presence there of numerous levels with tetrapods and rich microfloras. At our present state of knowledge, it allows correlations with the Kazanian of eastern Tethys and very likely with the eastern European platform. According to Cassinis and Doubinger (1991a,b), Tregiovo could be of an older age, and it has recently been referred to the Kungurian; given its current sporomorph

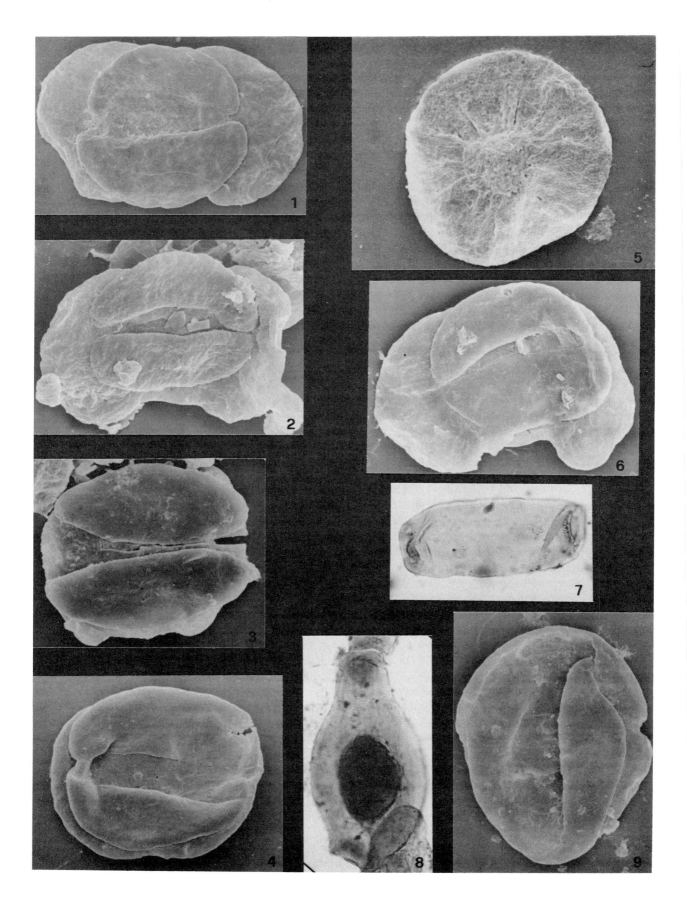

list, and pending a revision of Russian stratotypes, the more reliable age assignment would seem to be Kazanian.

Dolomitic Basin

Throughout the area going from Bolzano on the west to Cadore on the east and from Trento in the south to Val Pusteria in the north, the numerous microfloras observed in the Val Gardena Sandstone and the Bellerophon Formation show same characters and analogies in composition, both qualitatively and quantitatively, as well as occurrences of certain taxa that have proved to be useful for intrabasinal and interregional correlations. The events recognized are the following:

1. the first appearance of *Playfordiaspora hexareticulata* (Klaus 1963) Pittau n. comb.
2. the first appearance of *Lueckisporites parvus* Klaus 1963 (*partim*) (= *Lueckisporites virkkiae,* norm Bc, of Visscher, 1971 = *Lueckisporites virkkiae,* variant B, of Clarke, 1965)
3. the beginning of fungi blooming

All three are always recognizable in the II and III sedimentary cycles (Figure 6.4) inclusive of the Val Gardena Sandstone and the Bellerophon Formation (Massari et al., 1988).

1. The first bioevent is, in our opinion, the most striking one from which to establish correlations within the European Upper Permian and, through indirect correlations, with the marine stages of the standard scale. The taxonomic position of this species has been a matter of disagreement among several palynologists; a correct species assignment will come after revision and comparative studies of the sexine ultrastructure of the three species *Guthoerlisporites cancellosus* Playford and Dettmann 1965, *Endosporites hexareticulatus* Klaus 1963, and *Endosporites velatus* Leschik 1956. We believe that *Nuskoisporites crenulatus* Wilson 1963 is a species distinct from both *G. cancellosus* and *E. hexareticulatus,* because *N. crenulatus* possesses a thick, dense protosaccus, whereas the latter two have a thin pseudosaccus. A similar opinion has recently been expressed by Wood et al. (1991). Thus, there is no basis for continued use of Foster's synonymy. According to Farabee, Taylor, and Taylor (1989), the Triassic species *Playfordiaspora cancellosa* (Playford and Dettmann) Maheshwari and Banerji (1975) must be regarded as a species distinct from the Permian *E. velatus* and *E. hexareticulatus,* unless comparative studies demonstrate otherwise. It is our opinion that the Australian, South Alpine, and Zechstein specimens are quite comparable, but pending a revision of Leschik's material, we prefer to keep *E. hexareticulatus* and *E. velatus* formally distinct, even if the Alpine species is common in the German Zechstein. We formally define the synonymy, because at the present time the proper

name for Klaus's species is *Playfordiaspora hexareticulata* (Klaus) Pittau n. comb. [Basionym: *Endosporites hexareticulatus* Klaus 1963 (plate 4, fig. 9). Synonyms: *Playfordiaspora crenulata* (Wilson) Foster (plate 18, figs. 6–8).] *Playfordiaspora hexareticulata,* because of its wide geographic distribution, together with its limited stratigraphic range, seems to be more than merely favorable for use in correlation. It is already known in many areas: in the Zechstein Basin in Germany, cycles I–IV (Leschik, 1956; Schaarschmidt, 1963; Grebe and Schweitzer, 1964; Ecke, 1984; Rebelle and Doubinger, 1988); in England in the Magnesian Limestones and the Hilton Plant Bed (Clarke, 1965); in sedimentary cycles II and III of the Val Gardena Sandstone and the Bellerophon Formation of the Southern Alps (Pittau, cited in Massari et al., 1988; Klaus, 1963); in the Upper Permian of Spain (Cassinis, Toutin-Mourin, and Virgili, 1993); in the Dzhulfian of Australia (*Playfordiaspora crenulata* Zone of Foster, 1979); in Israel in the Arqov Formation of the Upper Permian (Eshet, 1990); at the top of the Permian in Kenya in sediments equivalent to the *Protohaploxypinus microcorpus* and *Lunatisporites pellucidus* palynozones of Australia (Hankel, 1992); in India and Madagascar, where it is known as *Guthoerlisporites cancellosus* (Playford and Dettman 1965) and extends up to the Triassic (reworking?); near the Permian–Triassic boundary in Tanzania (Wopfner and Kaaya, 1991).

2. First appearance of *Lueckisporites parvus* Klaus 1963 (*partim*) (plate 12, fig. 59, and plate 13, fig. 61). We take this taxon in a restricted sense; that is, we are considering only specimens with oval-to-well-rounded bodies, sacci reduced to almost the diameter of the body, and thick proximal sexine, with the outer surface almost levigate. It corresponds also to *Lueckisporites,* norm Bc, of Visscher (1971), to *Lueckisporites virkkiae,* variant B, of Clarke (1965), and to the specimens included in the open name "*Lueckisporites* n. sp." by Pittau (cited in Massari et al., 1988). This taxon is common to both the Zechstein and the Dolomitic Basin (Visscher, 1971; Massari et al., 1988). It is also present in the Upper Permian part of the Buntsandstein in the Iberian Range in Spain (M. A. Conti, unpublished data) and in the Upper Zechstein in Poland (Dibovà Jachowicz, 1974). In the Bletterbach section it appears at the same level as the local disappearance of *Playfordiaspora hexareticulata,* but in the Dolomitic Basin they are found together, because *P. hexareticulata* reaches the upper part of the Bellerophon Formation. It has never been found in the lower part of the Val Gardena Sandstone. Its occurrences are useful for intrabasinal correlations, and for the time being its wide distribution in central and western Europe makes it one of the useful elements for interregional correlations; there is no evidence of its presence in eastern Tethys or in Gondwana, but it is never abundant, except in very peculiar environments, such as in Ireland at the top of the Permian sedimentation. Visscher

---

Plate 6.1 (opposite). figs. 1, 2, 6. *Lueckisporites virkkiae* Potonié and Klaus, 1954. Cortiana section, Dolomitic Alps. Magnifications × 1100. figs. 3, 4, 9. *Lueckisporites parvus* Klaus, 1963. Bletterbach section, Dolomitic Alps. Magnification of fig. 3 and 4: × 1400; of fig. 9: × 1300.

fig. 5. *Playfordiaspora hexareticulata* (Klaus) n. comb. Seceda Mountain, Dolomitic Alps. Magnification × 650. fig. 7, 8. Fungi: *Tympanicysta* cells. Bletterbach section, Dolomitic Alps. Magnification × 1000.

(1971) linked this morphon to the evolutionary lineage of *Lueckisporites virkkiae,* establishing biostratigraphic steps. We can debate whether or not this taxon is connected with that hypothetical evolutionary line, but its biostratigraphic value as morphon or as species still remains. Our opinion is that the numerous morphologic varieties represent different species of *Lueckisporites.* The first occurrence of *Lueckisporites parvus* Klaus is intra-Dzhulfian in the Southern Alps, and it extends up to the end of the Bellerophon Formation, where it is concurrent with the fusulinid genera *Reichelina* and *Palaeofusulina.* Thus far it has not been reported in the Tesero horizon at the Permian–Triassic boundary. If the presence of these forams is indicative of the Changxingian Stage, as has been reported by Ouyang and Utting (1990) from the Changxingian stratotype at Meishan (southern China), *L. parvus* extends into that age.

3. The fungi blooming event must be regarded as one of the great events at the end of the Permian, because occurrences of fungal cells in marine sediments of late Permian and early Triassic age are worldwide. The importance of those cells has been highlighted by several authors (Balme, 1979; Foster, 1979; Visscher and Brugman, 1988; Massari et al., 1988), but as to their meaning, we do not know if they were of marine or terrestrial origin. A great increase in moisture and a hot climate caused the blooming of fungi, which likely affected plants or marine organisms, eventually causing them to die. As they developed, the fungi were not influenced by the type of vegetation they encountered, as they were spread throughout the Italian Alps, Israel (Eshet, 1990), Australia (Foster, 1979), and Greenland (Balme, 1979), all of which belonged to different palaeobioprovinces, although the fungi probably were controlled by climate. In the Southern Alps they appear in the sediments of the second major cycle, where the marine Bellerophon Formation is interfingered with the continental Val Gardena Sandstone. Their presence through the remaining sequence is episodic and reaches its maximum in the Tesero horizon, in layers containing the very impoverished marine fauna, the so-called mixed fauna (Broglio Loriga et al., 1988). The unfavorable lithologies of units between the Mazzin member and the Campil member of the Werfen Formation do not permit verification of the presence of that phenomenon through the Alpine Lower Triassic. In the Italian Southern Alps, that event spanned the Dzhulfian to the Permian–Triassic boundary. In the central and northern Zechstein Basin, the fungi blooming event has not been detected (even though marine sedimentation occurred in the six cycles), very likely because of the incomplete record of the Permian–Triassic sequences. In the French Permian the event is unknown because of the absence of such young sediments.

Interbasinal correlations

Comparing our microfloras and ichnofaunas with those of the Lodève Basin in southern France, it seems to us that even the youngest LO3 association, assigned by Rebelle and Doubinger (1988) to the Zechstein, is not younger than the Alpine Collio

Formation; it may even be a little older. Also, Rabejac's ichnofauna assemblage is comparable to the Pulpito assemblage of the Collio unit. Because of the close analogies in microflora compositions between the Lodève and St. Afrique basins (southern France), we can also extend the age of the former to the red sediments of St. Afrique, at least for the palynologically investigated part. Thus, the youngest microflora thus far known in the French Permian is that described by Visscher (1968) at Esterel (southern France).

Comparisons with German basins can be made using the Tregiovo and Wetterau microfloras, the latter described by Schaarschmidt (1980), in which the presence of *Luecksporites virkkiae* and the still abundant early Permian sporomorphs makes it older than the Zechstein microflora, but younger than the Lodèvian and Autunian. Those described in the German Rotliegendes of the Kuseler and Waderner groups are quite similar and can be compared to each other, both referable to the Lower Permian; thus the lithostratigraphic units (Unterrotliegendes and Oberrotliegendes) cannot be used in a chronostratigraphic sense.

Larger-scale correlations

The previously analysed situation concerning the stratigraphy of the Italian continental Permian and the possible correlations with other basins could be a key to sorting out the difficulties typically encountered with continental sediments in other European regions. From this viewpoint, and because of the nature of the selected taxa (both ichnofossils and microfloras), which have wide geographical distributions, we can try to fix a subdivision of the Southern Alpine Permian sediments into a few discrete intervals, thus to pass successively to wider correlations using the same intervals.

The Permian units that crop out in our sections can be divided into three intervals: an interval following the *D. didactylus* FAD, an interval characterized by a relict fauna, and, after the *L. virkkiae* FO, a third interval, after the *Rhynchosauroides* FAD and the *Playfordiaspora hexareticulata* FAD, with a diversified fauna and the fungi blooming event.

Considering the low levels of evolutionary change between the first two examined assemblages, a short time interval might be assumed for the association preceding the *D. didactylus* FAD that could encompass most of the faunas found within the Rotliegendes (both German and eastern European) and within most of the French basins; thus, for all of these we can assume the same very short time interval.

In this framework, parts of the "Assoziation" of Kozur (1980) and the "Tetrapodenfahrtenzone" of Holub and Kozur (1981) seem to be largely speculative, being based only on inconsistent systematics, as are at least two of the zones used by Boy and Fichter (1988a).

The first short interval could be set below the Collio FsU, which on the basis of the data of Doubinger (cited in Cassinis and Neri, 1990) is Artinskian to Ufimian p.p. in age. The Collio FsU, in turn, precedes the impoverished Tregiovo FsU, which on the basis of its sporomorphs (Doubinger, cited in Cassinis and Neri, 1990) could be Kungurian–Ufimian in age.

Because of the brevity of the first interval, the lifetime of the Bletterbach FU was too short to show any evolutionary change. For that faunal unit, a time interval could be assigned as the interval succeeding, or just starting with, the Illawarra reversal event (Italian IGCP-203 group, 1988), successive to the Kubergandinian fusulinids very recently found reworked in the breccias at the base of the Val Gardena Sandstone, preceding and interfingering with the marine levels ascribed to the Dorashamian. On the basis of faunal diversity, it was ascribed to Bakker's III Dynasty, Dzhulfian in age. The same age was assumed on the basis of the strong correspondence to Pitrat's (1973) Permian diversity acme (Ceoloni et al., 1988a). Pitrat's work (1973) was recently revised by King (1991), without significant change in the timing of the faunal-diversity acme.

Other data more or less directly connected with the standard scale are available. They can be used for larger-scale correlations, together with tetrapods and sporomorphs, and are as follows:

1. The lower portion of the Collio FU, the Pulpito FsU, on the basis of its faunal composition, can be ascribed by comparison to the reptile II Dynasty (Bakker, 1977).

2. The presence of the Kubergandinian fusulinid *Misellina,* found in the subsurface of the Adriatic Sea, in the uppermost portion of the Goggau Limestone (Cassinis and Doubinger, 1991a). The Goggau Limestone was later reworked into the Tarvisio Breccia before the beginning of deposition of the Val Gardena Sandstone.

3. The appearance of *Ullmannia* spores in the Akt'yubinsk area of the Kazanian stratotype (Andreeva et al., 1966), which includes spores of *Lueckisporites virkkiae* and *Jugasporites.*

4. The Illawarra reversal event, which according to various authors, as reviewed by Cassinis and Doubinger (1991a), is registered in the basal part of the Val Gardena Sandstone 10 m from the base, below both the *Rhynchosauroides* FAD and the *Playfordiaspora hexareticulata* FAD.

5. The correspondence between Bakker's III Dynasty and the Bletterbach FU is assumed on the basis of the degree of faunal diversity.

6. The presence of the fusulinids *Reichelina* and (?)*Palaeofusulina,* which are indicative of the Changxingian at Meishan in southern China (Ouyang and Utting, 1990), in the upper Bellerophon Formation (Conti et al., 1986).

7. The presence of an Upper Dorashamian brachiopod fauna within the Bellerophon Formation (Broglio Loriga et al., 1988) and the presence of a mixed Permian–Triassic fauna in its uppermost part.

The main points summarized here are shown schematically in Figure 6.5. After an integrated analysis of the foregoing data, we suggest that the hiatus between the Permian lower cycle and upper cycle in the Southern Alps lasted from the Ufimian to the Midian, or perhaps only from the Kazanian p.p. to the Midian (if the appearance of *L. virkkiae* should be confirmed in the Kazanian).

Suitability of previously used terms

Before discussing our results regarding larger-scale correlations, we shall briefly consider the meanings of some French and German terms currently used in describing the stratigraphical subdivisions of continental sediments of Permian age.

The French trinity is still frequently used with chronostratigraphic and geochronologic meanings, but the "Autunian" in the type area consists only of sediments lying disconformably on the substratum; moreover, the upper boundary is either unclear or unknown (Chateauneuf and Pacaud, 1991). The "Saxonian" and "Thuringian" are also quite mysterious entities. All three terms seem difficult to apply (perhaps because of the impossibility of defining stratotypes), and their calibration appears at present an uncertain venture.

The "Rotliegendes" and "Zechstein" of the German tradition are simply lithostratigraphic terms; they could be considered as supergroups, including a lot of formations. Moreover, there are frequent gaps within them; the whole succession, having been discontinuously deposited in separate small basins, is difficult to calibrate to the main marine events or to describe by physical stratigraphy. The use of such terms is suitable only in the context of their original meanings and in their type areas.

Moreover, both the German terms and the French terms have been used extensively by tradition, which has meant a century of inappropriate uses (e.g., names in quotation marks, used as if they were ages), as well as a wide range of different attributed meanings. As witness to that, we frequently see that authors have published papers with multiple scales, using the same names, with the boundaries shifting, some by millions of years. Consequently, we suspect that even if they were to be revised, such terms would continue to be cryptic. In conclusion, such terms ("Rotliegendes" and "Zechstein"; "Autunian," "Saxonian," and "Thuringian"), when used with chronostratigraphic or geochronologic meanings, seem not only to be invalid according to the ISG rules but also to be detrimental, as they can lead to erroneous conclusions. On the whole, it also seems that their future suitability should be questioned, either for chronostratigraphic or for geochronological purposes. On the other hand, definitions of new stages or ages within continental sediments are to be discouraged.

Faunal-age concept versus age/stage

After our literature analyses and our personal experiences, we find that we have completely negative feelings toward the previously used chronostratigraphic, biostratigraphic, and geochronological terms, and we suggest application of the faunal-

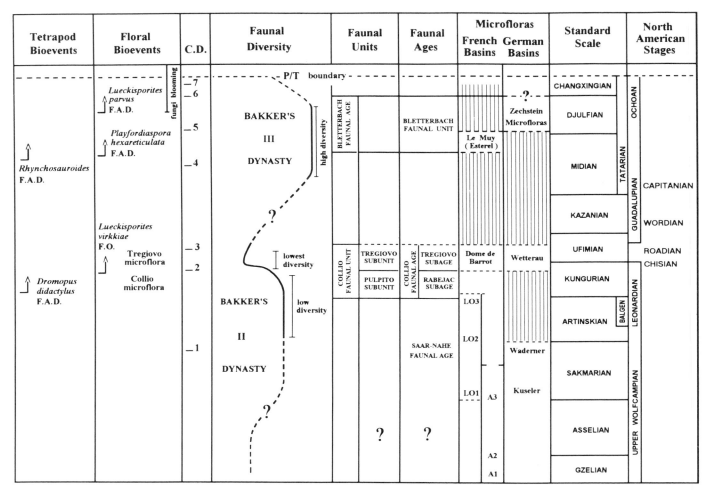

Figure 6.5. Correlation scheme. In the C.D. column (correlation data) are represented some main points: 1, Bakker's II Dynasty; 2, *Misellina*; 3, *Ullmannia*; 4, Illawarra reversal event; 5, Bakker's III Dynasty; 6, *Reichelina*; 7, late Permian brachiopods. These correlation data are discussed in the text.

age concept for continental sequences, instead of the stage/age concept, at least as a step in the right direction while waiting for a better-defined framework.

A first conclusion of our analysis is that we should shorten our estimate of the time interval during which the typical sediments of the Rotliegendes, the Saar-Nahe, and portions of the French basins were deposited. Thus, all the assemblages observed at those outcrops should be time-compressed and might represent a single faunal complex, to which we ascribe a position lower than the Rabejac fauna. Such a complex, typical of the Thuringian Forest, is here named the Saar-Nahe FA, to avoid confusion with the abused term "Thuringian." (Probably, in successive refinings, such a faunal age might be further subdivided; it might start with the appearance of *Ichniotherium cottae,* which is between the Manebach and Goldlauter faunas.) It is followed by the faunas present in the Rabejac levels, here named the Rabejac FsA. They are usually followed by a very long gap. It seems that only in the Central Alps was there a standstill of the relict

Tregiovo FsU, in its turn taking the name Tregiovo FsA. This peculiarity was previously recognized by Remy and Remy (1978) and Cassinis and Neri (1990). The Rabejac FsA and the Tregiovo FsA are here joined into a single unit named the Collio FA, on the basis of the same evolutionary level shown by tetrapods and sporomorphs; they could be subdivided only on the basis of a disappearance event.

A second point is the singularity of the faunal complex ascribed to the Bletterbach FU among the continental deposits of Europe; at the moment, only the trackway found in the Zechstein near Gera reaches a comparable evolutionary level. Such a fauna, which unfortunately cannot be traced later than the Dzhulfian because of changes in facies, is indicative of a very long gap in the record, because no evolutionary form transitional from more ancient tetrapods is known. This complex is named the Bletterbach FA.

Thus the suggested provisional succession of European Permian faunal ages is, at this time, limited to only four subdivisions, which will take their names from the most characteristic faunas:

(C)   Bletterbach FA
(B)   Collio FA
    (B2)   Tregiovo FsA
    (B1)   Rabejac FsA
(A)   Saar-Nahe FA

This faunal-age succession will allow correlations, in a relatively less precise, but easygoing and creditable way, of sedimentary units and faunas. According to the stated principle that the only chronostratigraphic units to be used are those based on marine successions, faunal ages can be compared chronostratigraphically only with standard-scale terms. Correlations are shown in Figure 6.5.

The resulting framework, which provides two very short sedimentation intervals, separated by a long gap, seems well fitted to the schemes of Gebhardt, Schneider, and Hoffmann (1991), Cassinis and Doubinger (1991b), and, in part, Kozur (1988). Thus, the framework suggested by this study seems to have overcome a first limitation.

This study is a first attempt to solve certain problems of subdivisions and correlations of Permian continental deposits, and further studies will be necessary. For the construction of a new and better system for subdivisions and nomenclature, we chose to take as our main basis only the peri-Mediterranean successions, as they are characterized by very well exposed sections, by relatively frequent footprint- and sporomorph-bearing levels, and by frequent volcanic deposits (which might be of help in absolute dating), and also they are free of the burden of traditional names. Moreover, studies of stratigraphic sequences are being carried out, and within the Upper Permian marginal deposits there are scattered marine faunas that can facilitate calibrations with the standard scale.

References

Andreeva, E. M. M., Kruchinina, N. V., Lyuber, A. A., Oshurkova, M. V., Panova, T. A., Pokrovskaja, I. M., Romanovskaja, G. M., Sivertseva, I. A., and Stel'mak, N. K. 1966. Assemblages of spores, pollen and other plant microfossils characteristic of various stratigraphical subdivisions from the Upper Precambrian to the Oligocene in the URSS (in Russian). In *Paleopalynology,* vol. 2, ed. I. M. Prokovskaja, pp. 296–497. Leningrad: Trudy V.S.E.G.E.I. (English translation, 1970, National Lending Library for Science and Technology.)

Bakker, R. T. 1977. Tetrapod mass extinctions – a model of the regulation and speciation rates and immigration by cycles of topographic diversity. In *Patterns of Evolution as Illustrated by the Fossil Record,* ed. A. Hallam, pp. 439–68.

Balme, B. E. 1970. Palynology of Permian and Triassic strata in the Salt Range and Surghar Range, West Pakistan. In *Stratigraphic Boundary Problems: Permian and Triassic of West Pakistan,* ed. B. Kummel and C. Teichert, pp. 305–453. Special publication **4**. University Press of Kansas.

Balme, B. E. 1979. Palynology of Permian–Triassic boundary beds at Kap Stosch, East Greenland. *Meddelelser om Gronland* **202**:1–37.

Barker, G. W., Wood, G. D., and Eames, L. E. 1991. Palynomorphs from the Lower Permian midcontinent and southwest United States and comparison to assemblages from Western Europe and the U.S.S.R. (Donets River basin). Unpublished report.

Boy, J. A., and Fichter, J. 1988a. Zur Stratigraphie des hoheren Rotliegend im Saar-Nahe-Becken (Unter-Perm; SW-Deutschland) und seiner Korrelation mit anderen Gebieten. *Neues Jahrbuch für Geologie und Paläontologie, Abhandlungen* **176**:331–94.

Boy, J. A., and Fichter, J. 1988b. Ist die stratigraphische Verbreitung der Tetrapodenfährten im Rotliegend ökologisch beeinflusst? *Zeitschrift für geologische Wissenschaften* **16**:877–83.

Broglio Loriga, C., Neri, C., Pasini, M., and Posenato, R. 1988. The upper Bellerophon Fm. and the P/T boundary, northern slope of the Sass de Putia Mt. In *Field Conference on the Permian and the Permian–Triassic Boundary in the South-Alpine Segment of the Western Tethys,* ed. Italian IGCP-203 group, pp. 82–8. Excursion guidebook, Società Geologica Italiana.

Cassinis, G. 1966. La Formazione di Collio nell'area tipo dell'Alta Val Trompia (Permiano inferiore bresciano). *Rivista Italiana di Paleontologia e Stratigrafia* **72**:507–88.

Cassinis, G., and Doubinger, J. 1991a. On the geological time of the typical Collio and Tregiovo continental beds in the South Alpine Permian (Italy), and some additional observations. *Atti Ticinensi di Scienze della Terra* **34**:1–20.

Cassinis, G., and Doubinger, J. 1991b. Artinskian to Ufimian palynomorph assemblages from the central southern Alps, Italy, and their regional stratigraphic implications. In *Contributions to Eurasian Geology: Papers Presented at the International Congress on the Permian System of the World, Perm, Russia, 1991,* part 1, pp. 9–18. Occasional publications, new series, 8B. ESRI.

Cassinis, G., Massari, F., Neri, C., and Venturini, C. 1988. The continental Permian in the Southern Alps (Italy). A review. *Zeitschrift für geologische Wissenschaften* **16**:1117–26.

Cassinis, G., and Neri, C. 1990. Collio and Tregiovo Permian continental basins (Southern Alps, Italy): a general comparison. *Atti Ticinensi di Scienze della Terra* **33**:11–15.

Cassinis, G., Toutin-Mourin, N., and Virgili, C. 1993. Permian and Triassic events in the continental domains of Mediterranean Europe. In *Permo-Triassic Events in the Eastern Tethys, Stratigraphy, Classification, and Relations with the Western Tethys,* ed. W. C. Sweet et al., pp. 60–77. Cambridge University Press.

Ceoloni, P., Conti, M. A., Mariotti, N., Mietto, P., and Nicosia, U. 1987. Tetrapod footprints from Collio Fm. (Lombardy, Northern Italy). *Memorie di Scienze Geologiche* **39**:213–33.

Ceoloni, P., Conti, M. A., Mariotti, N., Mietto, P., and Nicosia, U. 1988a. Tetrapod footprint faunas from Southern and Central Europe. *Zeitschrift für geologische Wissenschaften* **16**:895–906.

Ceoloni, P., Conti, M. A., Mariotti, N., and Nicosia, U. 1988b. New late Permian tetrapod footprints from the Southern Alps. *Memorie della Società Geologica Italiana* **34**:45–65.

Chateauneuf, J. J., and Pacaud, G. 1991. *Le Bassin Permien d'Autun Stratotype de l'Autunien.* Livret-guide de l'excursion organisée pour l'Association des Geologues du Permien (6 Juillet, 1991), pp. 1–12.

Clapham, M. B., Jr. 1970. Permian miospore from the Flowerpot

Formation of western Oklahoma. *Micropaleontology* **16**:15–36.

Clarke, R. F. A. 1965. British Permian saccate and monosulcate miospores. *Palaeontology* **8**:322–54.

Clement Westerhof, J. A. 1974. In situ pollen from gymnospermous cones from the Upper Permian of the Italian Alps. A preliminary account. *Review of Paleobotany and Palynology* **17**:63–74.

Conti, M. A., Fontana, D., Mariotti, N., Massari, F., Neri, C., Nicosia, U., Pasini, M., and Pittau, D. 1986. The Bletterbach-Butterloch section (Val Gardena Sandstone and Bellerophon Formation). In *Field Conference on the Permian and the Permian–Triassic Boundary in the South-Alpine Segment of the Western Tethys*, ed. Italian IGCP-203 group, pp. 99–109. Excursion guidebook, Società Geologica Italiana.

Conti, M. A., Leonardi, G., Mariotti, N., and Nicosia, U. 1977. Tetrapod footprints of the "Val Gardena Sandstone" (North Italy). Their paleontological, stratigraphic and paleoenvironmental meaning. *Paleontographia Italica, nuova serie* **40**:1–91.

Conti, M. A., Mariotti, N., Mietto, P., and Nicosia, U. 1989. Correlation elements in the Northern Italy Permian sediments. In *Résumés II Reunion S.F.G.P., 22 June 1988*. Paris: S.F.G.P.

Conti, M. A., Mariotti, N., Mietto, P., and Nicosia, U. 1991. Nuove ricerche sugli icnofossili della Formazione di Collio in Val Trompia (Brescia). *Natura Bresciana. Annali Museo Civico di Storia Naturale, Brescia* **26**:109–19.

De Boer, J. 1963. The geology of the Vicentinian Alps (NE Italy). *Geologica Ultraiectina* **11**:1–178.

Dibovà-Jachowicz, S. 1974. Analise palynologique des sédiments rouges saliferès du Zechstein supérieur ("zouber rouge") a Klodawa, Pologne. *Review of Paleobotany and Palynology* **17**:57–61.

Doubinger, J. 1974. Etudes palynologiques dans l'Autunien. *Review of Paleobotany and Palynology* **17**:21–38.

Doubinger, J., Odin, B., and Conrad, G. 1987. Les associations polliniques du Permien continental du bassin de Lodève (Herault, France): characterization de l'Autunien supérieur, du "Saxonien" et du Thuringien. *Annales de la Société Géologique du Nord* **106**:103–9.

Ecke, H. 1984. Palynostratigraphic data from an evaporite solution breccia at the eastern margin of the Leinetal Graben near Northeim (Southern Lower Saxony, F.R.). *Neues Jahrbuch für Geologie und Paläontologie, Abhandlungen* **170**:167–82.

Eshet, Y. 1990. The palynostratigraphy of the Permian–Triassic boundary in Israel: two approaches to biostratigraphy. *Israel Journal of Earth Sciences* **39**:1–15.

Farabee, M. J., Taylor, T. N., and Taylor, E. L. 1989. Pollen and spore assemblage from the Falla Formation (Upper Triassic), Central Transantarctic Mountains, Antarctica. *Review of Paleobotany and Palynology* **61**:101–38.

Fichter, J. 1976. Tetrapodenfährten aus dem Unterrotliegenden (Autun, Unterperm) von Odernheim/Glan. *Mainzer geowissenschaftliche Mitteilungen* **5**:87–109.

Fichter, J. 1982. Aktuopaläontologische Untersuchungen an den Fährten einheimischer Urodelen und Lacertilier. Teil 1: Die Morphologie der Fährten in Abhängigkeit von der Sedimentbeschaffenheit. *Mainzer naturwissenschaftlicher Archiv* **20**:91–129.

Fichter, J. 1983a. Tetrapodenfährten aus dem saarpfälzischen Rotliegenden (?Ober-Karbon–Unter-Perm; Südwest-Deutschland). Teil 1: Fährten der Gattungen *Saurichnites,*

Limnopus, Amphisauroides, Protritonichnites, Gilmoreichnus, Hyloidichnus und *Jacobiichnus. Mainzer geowissenschaftliche Mitteilungen* **12**:9–121.

Fichter, J. 1983b. Tetrapodenfährten aus dem saarpfälzischen Rotliegenden. (?Ober-Karbon–Unter-Perm; Südwest-Deutschland). Teil 2: Die fährten der Gattungen *Foliipes, Varanopus, Ichniotherium, Dimetropus, Palmichnus, Phalangichnus,* cf. *Chelichnus,* cf. *Laoporus* und *Anhomoiichnium. Mainzer naturwissenschaftlicher Archiv* **21**: 125–86.

Fichter, J. 1984. Neue Tetrapodenfährten aus den saarpfälzischen Standenbühl-Schichten (Unter-Perm; SW-Deutschland). *Mainzer geowissenschaftliche Mitteilungen* **12**:123–58.

Fichter, J., and Kowalczyk, G. 1983. Tetrapodenfährten aus dem Rotliegenden der Wetterau und ihre stratigraphische Auswertung. *Mainzer naturwissenschaftlicher Archiv* **22**: 211–29.

Foster, C. B. 1979. *Permian Plant Microfossils of the Blair Athol Measures, Baralaba Coal Measures and Basal Rewan Formation of Queensland*. Paleontological paper **45**. Publications of the Geological Survey of Queensland (Australia).

Gand, G. 1988. Les traces de vertebres Tetrapodes du Permien français. Paleontologie, stratigraphie, paleoenvironments. Thèse, Université de Bourgogne.

Gand, G., and Haubold, H. 1988. Permische Tetrapodenfährte in Mitteleuropa, stratigraphische und paläontologische Aspekte. *Zeitschrift für geologische Wissenschaften* **16**: 885–94.

Gebhardt, U., Schneider, J., and Hoffmann, N. 1991. Modelle zur Stratigraphie und Beckenentwicklung im Rotliegenden der Norddeutschen Senke. *Geologisches Jahrbuch* **A127**: 405–27.

Grebe, H., and Schweitzer, H. J. 1964. Die Sporae dispersae des niederrheinischen Zechsteins. *Fortschritte in der Geologie von Rheinland und Westfalen* **12**:201–24.

Hankel, O. 1992. Late Permian to Early Triassic microfloral assemblages from the Maji ya Chumvi Formation, Kenya. *Review of Paleobotany and Palynology* **72**:129–48.

Haubold, H. 1971. Ichnia Amphibiorum et Reptiliorum fossilium. In *Handbuch der Palaeoherpetologie,* vol. **18**, ed. O. Kuhn.

Haubold, H. 1984. *Saurierfährten. Die Neue Brehm-Bucherei,* Wittenberg Lutherstadt: A. Ziemsen Verlag.

Haubold, H., and Katzung, G. 1975. Die Position der Autun/Saxon-Grenze in Europa und Nordamerika. *Schriftenreihe für geologische Wissenschaften* **3**:87–138.

Haubold, H., and Sarjeant, W. A. S. 1974. Tetrapodenfährte aus den Keele und Enville Groups (Permokarbon: Stefan und Autun) von Shropshire and South Staffordshire, Grossbritannien). *Zeitschrift für geologische Wissenschaften* **1**:895–933.

Hedberg, H. D. (ed.) 1976. *International Stratigraphic Guide: A Guide to Stratigraphic Classification, Terminology, and Procedure*. International Subcommission on Stratigraphic Classification, IUGS Commission on Stratigraphy.

Hedlund, R. W. 1965. Palynological assemblage from the Permian Wellington Formation, Noble County, Oklahoma. *Oklahoma Geological Survey, Oklahoma Geological Notes* **25**:236–41.

Holub, V., and Kozur, H. 1981. Revision einiger Tetrapodenfährten des Rotliegenden und biostratigraphische Auswertung der Tetrapodenfährten des obersten Karbon und Perm. *Geologisch-paläontologische Mitteilungen* **11**: 149–93.

Italian IGCP-203 group. 1988. *Field Conference on the Permian and the Permian–Triassic Boundary in the South-Alpine Segment of the Western Tethys.* Excursion guidebook, Società Geologica Italiana.

King, G. M. 1991. Terrestrial tetrapods and the end Permian event: a comparison of analyses. *Historical Biology* 5:239–55.

Klaus, W. 1963. Sporen aus dem Südalpinen Perm. *Jahrbuch der geologischen Bundesanstalt* 106:229–361.

Kozur, H. 1980. Beiträge zur Stratigraphie des Perm. Teil III(2): Zur Korrelation der überwiegend kontinentalen Ablagerungen des obersten Karbons und Perms von Mittel- und Westeuropa. *Freiberger Forschungshefte, Paleontologie* C348:69–172.

Kozur, H. 1988. The Age of the Central European Rotliegendes. *Zeitschrift für geologische Wissenschaften* 16:907–15.

Leonardi, P., Conti, M. A., Leonardi, G., Mariotti, N., and Nicosia, U. 1975. *Pachypes dolomiticus* n. gen. n. sp.; pareiasaur footprint from the "Val Gardena Sandstone" (Middle Permian) in the Western Dolomites (N. Italy). *Accademia Nazionale dei Lincei, Rendiconti Classe Scienze matematiche, fisiche e naturali, serie 8* 57:221–32.

Leonardi, G., and Nicosia, U. 1973. Stegocephaloid footprint in the Middle Permian sandstone (Groedener Sandsteine) of the Western Dolomites. *Annali Università di Ferrara, nuova serie* 9:1116–249.

Leschik, G. 1956. Sporen aus dem Salzton des Zechstein von Neuhof (bei Fulda). *Palaeontographica* B100:122–42.

Lützner, H. 1987. *Sedimentary and Volcanic Rotliegendes of the Saale Depression. Symposium on Rotliegendes in Central Europe (Erfurt).* Excursion guidebook.

Maheshwari, H. K., and Banerji, J. 1975. Lower Triassic palynomorphs from the Maitur Formation, West Bengal, India. *Palaeontographica* B152:149–90.

Massari, F., Conti, M. A., Fontana, D., Helmold, K., Mariotti, N., Neri, C., Nicosia, U., Ori, G. G., Pasini, M., and Pittau, D. 1988. The Val Gardena Sandstone and Bellerophon Formation in the Bletterbach Gorge (Alto Adige, Italy): biostratigraphy and sedimentology. *Memorie di Scienze Geologiche* 40:229–73.

Mietto, P. 1981. Una grande impronta di pareiasauro nel Permiano di Recoaro (Vicenza). *Rendiconti della Società Geologica Italiana* 4:363–4.

Morgan, B. E. 1967. Palynology of a portion of the El Reno Group (Permian), southwest Oklahoma. Ph.D. dissertation, University of Oklahoma.

Müller, A. H. 1959. Die erste Wilbertierfährte (*Paradoxichnium problematicum* n. g., n. sp.) aus dem terrestrischen Zechstein von Thüringen. *Monatsberichte der Deutschen Akademie der Wissenschaften* 1:613–23.

NASC. 1983. North America Stratigraphic Code. *Bulletin of the American Association of Petroleum Geologists* 67:841–75.

Ori, G. G., Dalla, S., and Cassinis, G. 1988. Depositional history of the Permian continental sequence in the Val Trompia–Passo Croce Domini area (Brescian Alps, Italy). *Memorie della Società Geologica Italiana* 34:141–54.

Ouyang, S., and Utting, J. 1990. Palynology of Upper Permian and Lower Triassic rocks, Meishan, Changxing county, Zhejiang province, China. *Review of Paleobotany and Palynology* 66:65–103.

Pitrat, C. W. 1973. Vertebrates and the Permo-Triassic extinction. *Palaeogeography, Palaeoclimatology, Palaeoecology* 14:249–64.

Playford, G., and Dettman, M. E. 1965. Rhaeto-Liassic plant microfossils from the Leigh Creek Coal Measures, South Australia. *Senckenbergiana Lethaea* 46:127–81.

Rebelle, M., and Doubinger, J. 1988. Etudes palynologique dans le bassin évaporitique du Zechstein (Permien supérieur): aspects stratigraphiques, paléoécologiques et paléoclimatologiques. *Cahiers de Micropaléontologie* 3:5–17.

Remy, W., and Remy, R. 1978. Die Flora des Perms im Trompia-Tal und die Grenze Saxon/Thuring in der Alpen. *Argumenta Palaeobotanica* 5:57–90.

Schaarschmidt, H. 1963. Sporen und Hystrichosphaerideen aus dem Zechstein von Budingen in der Wetterau. *Palaeontographica* B113:38–91.

Schaarschmidt, H. 1980. Pollen flora and vegetation at the end of the Lower Permian. In *Proceeding of the IV International Palynology Conference, Lucknow (1976–77),* vol. 2, pp. 750–2.

Schmidt, H. 1959. Die Cornberger Fährten im Rahmen der Vierfüssler-Entwicklung. *Abhandlungen des Hessischen Landesamtes für Bondenforschung* 28:1–137.

Schneider, J. 1989. Basic problems of biogeography and biostratigraphy of the Upper Carboniferous and Rotliegendes. *Acta Museii Reginaehradecensis, Series A: Scientiae naturales* 22:31–44.

Tschudy, R. H., and Kosanke, R. M. 1966. Early Permian vesiculate pollen from Texas, USA. *Paleobotanist* 15:59–71.

Visscher, H. 1968. On the Thuringian age of the Upper Paleozoic sedimentary and volcanic deposits of the Esterel (Southern France). *Review of Paleobotany and Palynology* 6:71–83.

Visscher, H. 1971. *The Permian and Triassic of the Kingscourt Outlier, Ireland – A Palynological Investigation Related to Regional Stratigraphical Problems in the Permian and Triassic of Western Europe.* Geological Survey of Ireland, special paper 1.

Visscher, H., and Brugman, W. A. 1988. The Permian–Triassic boundary in the Southern Alps: a palynological approach. *Memorie della Società Geologica Italiana* 34:121–8.

Visscher, H., Huddleston Slater-Offerhaus, M. G., and Wong, T. F. 1974. Palynological assemblages from "Saxonian" deposits of the Saar-Nahe Basin (Germany) and the Dome de Barrot (France) – an approach to chronostratigraphy. *Review of Paleobotany and Palynology* 17:39–56.

Wilson, L. R. 1962. Permian plant microfossils from the Flowerpot Formation, Greer County, Oklahoma. *Oklahoma Geological Survey, Circular* 49:1–50.

Wood, G. D., Foster, C. B., and Barker, G. W. 1991. Permian palynological assemblages from Gondwana: intra- and inter-regional correlations and geological consequences. Unpublished report.

Wopfner, H., and Kaaya, C. Z. 1991. Stratigraphy and morphotectonics of Karoo deposits of the northern Selous Basin, Tanzania. *Geological Magazine* 128:319–34.

7 Permian chronostratigraphic subdivisions and events in China

YANG ZUNYI, ZHAN LIPEI, YANG JIDUAN, and WU SHUNBAO

The Permian system is well developed in various facies over wide areas in China, and it has been intensively studied by Chinese geologists since the 1930s. In his study of the Permian of South China, involving marine as well as paralic facies, Huang Jiqing (T. K. Huang, 1932) adopted a tripartite subdivision, namely, the Lower Permian (Chuanshan), the Middle Permian (Yanghsin), and the Upper Permian (Loping). In the same year Li Siguang (J. S. Lee) and Zhu Sen (1932) were dividing the Permian in the Longtan district, Nanjing, into the Lower Permian Chuanshan Limestone, the Middle Permian Chihsia beds and Kufeng beds, and the Upper Permian Longtan coal series. The Lower Permian was then regarded as correlative to the Lower Permian of international usage. Sun Yunzhu (Y. C. Sun, 1939) subsequently suggested a bipartite subdivision for the Permian of South China: the Lower Permian Yanghsin Series and the Upper Permian Loping Series. He considered the Chuanshan Series to be Upper Carboniferous, based in part on the regional disconformity recognized between the Chuanshan and Yanghsin series, marking what is called the Qiang-Gui or Yunnan movement. For a long time, such a twofold division of the Permian system was followed by all workers in China, a usage endorsed by two national stratigraphic congresses in China, held in 1959 and 1979. Since the later part of the 1970s, richly fossiliferous and continuous Carboniferous–Permian sequences have been discovered in Yunnan, Guangxi, and Guizhou provinces, and detailed biostratigraphic studies have been undertaken, involving fusulinids, nonfusulinid foraminifers, rugose corals, brachiopods, conodonts, and sponges. That work has enabled a number of scholars (Zhang Zhuqi, 1985; Huang Jiqing and Chen Bingwei, 1987; Zhou Tieming, Sheng Jinzhang, and Wang Wujing, 1987; Zhang Zhenghua, Wang Zhihao, and Li Changjin, 1988; Fan Jiasong et al., 1990) to reconsider the biostratigraphic subdivision of the Permian in China.

Classification scheme for the marine and paralic Permian in South China

This classification is according to Huang Jiqing and Chen Bingwei (1987):

Lower Triassic	*Otoceras* bed
Upper Permian (Lopingian)	
Changxingian	*Palaeofusulina* Zone
Wujiapingian	*Codonofusiella* Zone
Middle Permian (Yangxinian)	
Maokouan	*Yabeina* Zone
	Neoschwagerina Zone
Qixiaian	*Cancellina* Zone
	Misellina Zone
Lower Permian (Qiannanian*)	
Longyin Formation	
Upper	*Chalaroschwagerina-Pamirina* Zone
	Robustoschwagerina-Sphaeroschwagerina Zone
Lower	*Pseudoschwagerina* Acme Zone
Upper Carboniferous (Mapingian, *s.s.*)	*Montiparus-Triticites* Acme Zone

In light of their detailed study of the Carboniferous–Permian boundary strata and fusulinid zonation in Xiaodushan, Guangnan, southeast Yunnan, Zhou Tieming et al. (1987) adopted the following tripartite subdivision of the Permian (conformably overlying the Upper Carboniferous Xiaodushanian Stage): the Lower Permian Mapingian and Changmeian stages, the Middle Permian Qixiaian and Maokouan stages, and the Upper Permian Wujiapingian and Changxingian stages. They also suggested that the first appearance of pseudoschwagerinids be taken as marking the beginning of the Permian. Zhang Zhenghua et al. (1988) also adopted a tripartite scheme: the Lower Permian (Qiannan Series), involving the Zisongian and Yangchangian stages; the Middle Permian (Yangxinian Series), embracing the Qixiaian and Maokouan stages; and the Upper Permian (Lopingian Series), with the Wujiapingian and Changxingian stages. In the

*Newly coined, meaning "South Guizhou," where the Lower Permian is well exposed.

66

latter scheme, the Zisongian and Yangchangian are newly named units, and the base of the *Pseudoschwagerina* Range Zone is considered the lower limit of the Permian. The tripartite division advocated here differs slightly from that of Huang Jiqing and Chen Bingwei (1987) in the naming of the Lower Permian series and the erection of zones, namely, the Lower Permian Chuanshanian Series, with the Mapingian and Longyinian stages, the Middle Permian Yangxinian Series, with the Chihsia (Qixiaian) and Maokouan stages, and the Upper Permian Lopingian Series, with the Wujiapingian and Changxingian stages (Table 7.1).

Chronostratigraphic units of the marine and paralic Permian in South China

The Permian is exceptionally well developed in the upper Yangtze River area, covering the adjoining districts of Yunnan, Guizhou, and Guangxi, and the lower Yangtze area (central Hunan and central Jiangxi), where various faunas such as fusulinids, corals, brachiopods, ammonoids, and conodonts are represented. Successions of these taxa serve as appropriate criteria for establishing zonations and a relative chronology (Figure 7.1, Tables 7.1–7.3).

Mapingian Stage

The Mapingian Stage is based on the Maping Limestone succession, which is well developed around Liuzhou (the ancient city of Maping), Guangxi province. According to Sheng Jinzhang (1962), the Maping Limestone or Formation (originally thought to be late Carboniferous) has two fusulinid zones, a *Triticites* Zone in the lower part and a *Pseudoschwagerina* Zone in the upper. Those zones have been widely accepted. Yang Jingzhi and Wu Wangshi nominated the limestone as the type sequence for the Mapingian Stage, which embraces three fusulinid zones, in ascending order: *Triticites* Zone, *Sphaeroschwagerina* Zone, and *Robustoschwagerina schellwieni* Zone. The last two fall within the broader *Pseudoschwagerina* Zone of Dunbar and Skinner. Hou Hongfei et al. (1982) also subdivided the Mapingian into three zones, but in a different way:

1. *Montiparus* Zone
2. *Triticites* Zone
3. *Pseudoschwagerina-Zellia* Zone

In that scheme, the *Pseudoschwagerina* Zone refers to an acme zone or assemblage zone, without specifying its range or the locations of its base and top. Hence the stage was not accurately defined, causing confusion in our understanding of the *Pseudoschwagerina* Zone, particularly its upper limit. So there arose two usages for the name of this zone, a broad sense and a strict sense. For example, the Upper Mapingian *Sphaeroschwagerina moelleri* Zone (equivalent to *Pseudoschwagerina* Zone) in the Ningzheng area, Jiangsu province (strict sense), is not equivalent to the Upper Mapingian *Pseudoschwagerina* Acme Zone and the *Schwagerina* Range Zone (both equal to the *Pseudoschwagerina*

Range Zone) (broad sense) in the Qinglong area, Guizhou province. In our opinion, the range of the *Pseudoschwagerina* Zone (*s.s.*) is correlative to that of the Russian (Urals) Asselian Stage, whereas the range of the *Pseudoschwagerina* Zone (*s.l.*) covers the Russian (Urals) Asselian and Sakmarian stages. It was therefore necessary to define the Mapingian more precisely. We concur with Zhou Tieming et al. (1987), who redefined the Mapingian as a stage to include only the original Upper Mapingian *Pseudoschwagerina* Zone (or *Sphaeroschwagerina* Zone). The original Lower Mapingian *Triticites* Zone was given a new status as the Xiaodushanian Stage of late Carboniferous age (Zhou Tieming et al., 1987). The late Carboniferous–early Permian zonation can be summarized as follows:

Early Permian	
(Chuanshan Series)	
Longyinian	*Pamirina chinlingensis* Zone
(Changmeian)	*Chalaroschwagerina tumentis* Zone
	Pseudofusulina vulgaris – *Laxifusulina iniqua* Zone
Mapingian	*Pseudoschwagerina robusta–* *Zellia chengkungensis* Zone
	Pseudoschwagerina morsei– *Robustoschwagerina* *xiaodushanica* Zone
Late Carboniferous	
Xiaodushanian	*Triticites*

However, the Mapingian, as redefined by Zhou Tieming et al. (1987), lacks clear lower and upper limits, and its distribution as well as lateral continuity remain to be ascertained. According to Xiao Weiming et al. (1986), the Mapingian in southern Guizhou contains distinct and persistent biozones that may yet serve to mark the units of the stage. Here it is proposed to use the first appearance of *Pseudoschwagerina uddeni* to mark the beginning of the Mapingian, and the disappearance of *Sphaeroschwagerina moelleri* to mark the end. Hence the Mapingian in the carbonate facies of the Yunnan-Guizhou-Guangxi area yields the following succession of fusulinid zones:

1. *Pseudoschwagerina uddeni* Range Zone
2. *Robustoschwagerina kahleri* Range Zone
3. *Sphaeroschwagerina moelleri* Range Zone

In addition to fusulinids, the Mapingian is rich in rugose corals (the *Streptophyllidium-Diversiphyllum* assemblage), brachiopods (*Choristites jigulensis–Protadanthus elegans* assemblage or *Anidanthus enagardi–Buxtonia–Linoproductus* assemblage, consisting mainly of *Eomarginifera pusilla*, *Kutorginella*, *Eliva mignon*, *Enteletes lamarcki*, *Buxtonia magna*, and *Linoproductus sinensis*), and conodonts (in ascending order, *Streptognathodus elongatus–S. simplex* Zone, *S. wabaunsensis–S. fuchengensis* Zone, and *Neogondolella bisselli* Zone). These taxa readily establish correlation of the redefined Mapingian Stage with the Asselian of the Russian Urals.

Table 7.1. Correlation of marine Permian faunal sequence in China

Period and series	Stage	Fusulinids	Corals	Brachiopods	Ammonoids	Conodonts
Upper Permian (Lopingian Series)	Changxingian Stage (Changhsingian)	Palaeofusulina sinensis–Reichelina changhsingensis Z	Waagenophyllum–Huayunophyllum AZ	Spinomarginifera chengyaoyenensis–Pugnax pseudouah ASZ Peltichia zigzag–Spinomarginifera kueichouensis ASZ	Rotodiscoceras–Pleuronodoceras SZ Tapanshanites–Pseudo-stephanites SZ Shevyrevites–Paratirolites SZ	Neogondolella subcarinata subcarinata–N. subcarinata changxingensis Z
	Wujiapingian Stage (Wuchiapingian)	Codonofusiella kwangsiana Z	Liangshanophyllum-Lophophyllidium AZ	Orthotetina ruber–Permophricodothyris grandis ASZ Edriosteges poyangensis–Tyloplecta yangtzeensis ASZ	Pseudotirolites Z Sanyangites Z Araxoceras–Konglingites Z Anderssonoceras–Prototoceras Z	Neogondolella liangshanensis–N. bitteri AZ
Middle Permian (Yangxinian Series)	Maokouan Stage	Neomisellina-Codonofusiella AZ Neomisellina multivoluta Z Afghanella schenki–Yabeina gubleri Z Neoschwagerina simplex Z Cancellina Z	Huangophyllum Z Ipciphyllum–Iranophyllum AZ Polythecalis yangtzeenis AZ	Urushtenoidea–Neoplicatifera huangi ASZ Cryptospirifer striata ASZ Chaoiella reticulata–Tyloplecta grandicosta ASZ	Shouchangoceras Z Waagenoceras Z Kufengoceras Z	Sweetognathodus hanzhongensis–Neogondolella asserata AZ Neogondolella idahoensis–N. serrata AZ
	Qixianian Stage (Chihsian)	Shengella-Maklaya Z Misellina claudiae Z Brevaxina dyhrenfurthi–Misellina termieri Z	Hayasakaia elegantula AZ Wentzellophyllum volzi AZ	Orthotichia chekiangensis–Liraplecta richthofeni ASZ	Popanoceras–Neocrinites–Metaperrinites AZ	Neogondolella aserrata AZ Neogondolella idahoensis–N. gujioensis AZ Neostreptognathodus pequopensis AZ
Lower Permian (Chuanshanian Series)	Longyinian Stage	Pamirina darvasica Z Robustoschwagerina ziyunensis Z	Wentzelastraea Z Kepingophyllum irregulare AZ	Orthotichia magnifica–Choristites pavlovi ASZ Choristites-Leptodus ASZ	Propinacoceras Z Propopanoceras Z	Sweetognathus whitei Z
	Mapingian Stage	Sphaeroschwagerina moelleri Z Robustoschwagerina kahleri Z Pseudoschwagerina uddeni Z	Streptophyllidium-Diversiphyllum AZ Nephelophyllum simplex Z	Choristites jigulensis–Protanadanthus elegans ASZ	Properrinites–Eoasianites AZ	Neogondolella bisselli Z Streptognathodus fuchengensis Z S. wabaunsensis Z S. elongatus Z

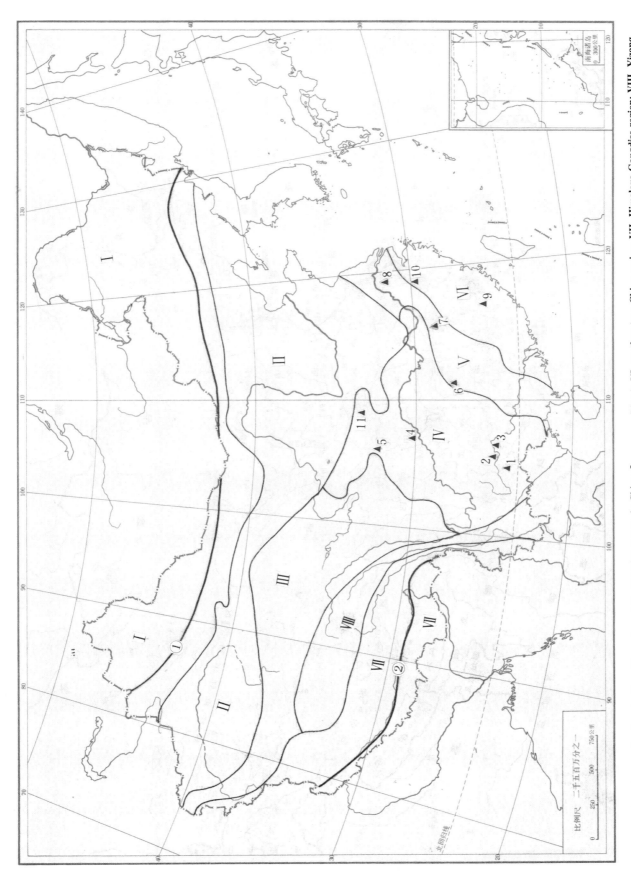

Figure 7.1. Sketch map showing stratigraphic regions of the Permian system in China: I, Junggar-Hinggan region; II, Tarim–northern China region; III, Kunlun–Bayan Har–Jinshajiang region; IV, Qinling-Yangtze region; V, Jiangnan region (south of the Yangtze River); VI, southeastern China region; VII, Himalaya-Gangdise region; VIII, Yizang (Tibet)–western Yunnan region; ①, Ebinur Lake–Juyan Lake–Solon Yar–Moron River fault zone; ②, Yarlung-Zangbo fault zone.

Longyinian Stage

The Longyin Formation is proposed here as the stratotype section for the Longyinian Stage and is situated in Longyin, Pu'an, Guizhou province. According to Wu Wangshi and Zhang Linxin (1979), the succession starts in the late Carboniferous Shazitang Formation (limestone and argillaceous limestone, 350 m), represented by the *Triticites* and *Sphaeroschwagerina constans* zones, and is overlain by the Longyin Formation (sandy shale, siltstone intercalated with marlstone lenses, 350 m), represented by the *Sphaeroschwagerina glomerosa* Zone (the same bed in Huagong, Qinglong, also yielding *Sphaeroschwagerina moelleri* and the ammonoid *Propopanoceras* sp.). The Baomeshan Formation (shale and limestone) overlies the Longyin Formation and is represented by the *Robustoschwagerina schellwieni* Zone, with associated corals (*Wentzelastraea* Zone, plus *Iranophyllum* and *Szechuanophyllum*), and in Huagong, Qinglong, the same bed contains the fusulinid *Pseudofusulina kraffti*. The Longyinian Stage is proposed as follows, using the Longyin Formation and the Baomeshan Formation as stratotypes:

Lower Permian (Chuanshan
   Series)
  Longyinian
    Upper (Baomeshan   *Robustoschwagerina*
      Formation)   *schellwieni* Zone
    Lower (Longyin   *Sphaeroschwagerina*
      Formation)   *glomerosa* Subzone
  Mapingian
    Shazitang Formation   *Sphaeroschwagerina*
    (middle and upper   *constans* Subzone
    parts)   *Pseudoschwagerina*
                   Acme Zone

Upper Carboniferous
  Xiaodushanian
    Shazitang Formation   *Triticites* Acme Zone
    (lower part)   *Montiparus* Acme
                   Zone

In their detailed study of the Longlinian Stage, a new Lower Permian unit between the *Pseudoschwagerina*-bearing Upper Carboniferous Mapingian Stage and the "Lower Permian" Chihsian Stage in the type locality (Longlin village, Longlin, Guangxi), Huang Zhixun, Shi Yan, and Wei Muchao (1982) indicate that the new stage is characterized by a peculiar fusulinid faunule composed of *Pamirina* (*Nanpanella*)–*P.* (*Pamirina*)–*Nagatoella* (*Darvasites*) that spread widely over western, central, and eastern Tethys.

According to Xiao Weiming et al. (1986), the Longyinian fusulinids in platform-marginal carbonate facies can be divided into (1) the *Robustoschwagerina ziyunensis* Range Zone and (2) the *Pamirina darvasica* Range Zone. The former is characterized by its abundance of highly evolved *Robustoschwagerina*, associated with *Sphaeroschwagerina*, *Zellia*, *Paraschwagerina*, *Eoparafusulina*, and corals (*Streptophyllum*, *Diversiphyllum*, *Antheria*,

and *Kepingophyllum simplex*), whereas the latter is typified by *P. darvasica* plus newly appearing elements such as *Chalaroschwagerina*, *Laxifusulina*, *Toriyamaia*, *Nagatoella*, *Schwagerina cushmani*, *Pseudofusulina kraffti*, and corals (*Wentzelastraea*, *Prowentzelellites*, *Iranophyllum*, and *Kepingophyllum*). In certain areas (Pu'an, for instance), the lower part of the Longyinian contains ammonoids (*Propopanoceras*, *Propinacoceras*, and *Popanoceras*), whereas in other localities (Yangchang, Ziyun) the upper part of the Longyinian yields *Parapronorites* and *Popanoceras*. In Nasui, Luoding, the Longyinian is rich in conodonts of the *Sweetognathus whitei* Zone (Xiong Jianfei and Zhai Zhiqian, 1985). Judging from the fusulinids and the stratigraphic sequence, the Lower Longyinian *Robustoschwagerina ziyunensis* Zone correlates with the topmost Ziyunian *Eoparafusulina contracta–Zellia heritschi* Subzone (Zhang Zhenghua et al., 1988), whereas the Upper Longyinian (Baomeshan Formation *Pamirina darvasites* Zone) correlates with the Yangchangian *Chalaroschwagerina-Pamirina-Staffella* Zone (Zhang Zhenghua et al., 1988).

The lower and upper limits of the Longyinian (Table 7.2) are marked respectively by the beginning of the *Sphaeroschwagerina glomerosa* Zone or *Robustoschwagerina ziyunensis* Zone and the termination of the *Pamirina darvasica* Zone or *Robustoschwagerina schellwieni* Zone. This stage is widespread over parts of Yunnan, Guizhou, and Guangxi, as well as in southern Qinling, but it probably is lacking over most parts of South China.

Because *Sphaeroschwagerina glomerosa* appears in the Sakmarian Stage of the Darvas area, and *Pseudofusulina moelleri* is a zonal fusulinid present in the Sakmarian Tastub Substage in the Russian southern Urals (where *Propopanoceras* is also an important index), the Lower Longyinian is correlated with the foregoing two horizons. *Pamirina darvasica*, *Chalaroschwagerina*, and *Pseudofusulina kraffti* appear in the Yakhtashian of the Darvas area, permitting correlation of the Upper Longyinian with the Sterlitamak Stage. In short, the Longyinian is correlative with the Sakmarian of the southern Urals.

Qixiaian Stage

The Qixia Limestone is the stratotype for the Qixiaian Stage in the Qixia Mountains, east of Nanjing. Lithologically it consists of limestone, siliceous rock, and limestone with chert nodules. Faunally it yields fusulinids, corals, brachiopods, and some bryozoans and ostracods. The Qixia Limestone, named by F. von Richthofen in 1882, was studied in the early 1930s by Li Siguang and others. Summarizing the early fusulinid biostratigraphic research, Sheng Jinzhang (1962) presented the following subdivision:

Qixiaian Formation
  Qixia Limestone   *Parafusulina multiseptata* Zone
  member   *Nankinella inflata* Zone
  Swine Limestone   *Misellina claudiae* Zone
  member   *Schwagerina tschernyschewi*
                  Zone

Table 7.2. Correlation of Lower–Middle Permian Subdivisions and Fossil Zones in China and Two Other Regions

Stratigraphy	Pu'an, Guizhou Wu and Zhang (1979)	Ziyun-Ceheng, Guizhou Xiao et al. (1986)	Yangchang, Ziyun Guizhou Zhang et al. (1988)	Guangnan, Yunnan Zhou et al. (1987)	Ningzheng, Jiangsu Zhang (1983)	SE Hunan Zhou (1982)	Darvasi Leven (1928–1980)	S Urals Rauser (1965)
Middle Permian (Yangxin Ser) Qixian / Oixia Fm	Oixia Fm (Qixian)	*Misellina* Range-zone: *Maklaya elliptica* Subz; *Shengella simplex* Subz; *Misellina claudiae* Subz; *M. termieri* Subz; *Brevaxina dyhrenfurthi* Subz	Oixialan: *Armenina* Az; *Misellina claudiae* Subz / *Misellina* Acme z; *M. termieri* – *Brevaxina* Subz; *Brevaxina dyhrenfurthi* Subz	Oixialan: *Misselina ovalis* Z; *Brevaxina otakiensis* Z	Oixia Fm: *Misellina claudiae* Z	Oixia Fm: *Misellina claudiae* Z; *Staffella vulgaris* Z; *Schwagerina cushmani* Z	Bolorian: *Misellina ovalis* Z; *M. parvicostata* Z; *Brevaxina dyhrenfurthi* Z	Artinskian: *Pseudofusulina concessa* Z; *P. schellwieni* Z
Lower Permian (Chuanshan Series) Longyinian (Baomeshan Fm., Longyin Fm)	Baomeshan Fm. / Longyin Fm (Longyinian): *Robustoschwagerina schellwieni* Z	*P. darvasica* Chronozone: *Pamirina darvasica* Z	Yangchangian: *Chalaroschwagerina* – *Staffella* Az	Changmeian: *Pseudofusulina vulgaris* – *Laxifusulina iniqua* Z; *Chalaroschwagerina tumentis* Z; *Pamirina chinlingensis* Z	Swine Ls Fm: *Darvasites ordinatus* Z		Yakhtashian: *Chalaroschwagerina vulgaris* – *Pamirina* Z; *Chalaroschwagerina solida* Z	Sterlitamak (Sakmarian): *Pseudofusulina callos* Z
Mapingian (Shazitang Fm)	Shazitang Fm (Mapingian): *Sphaeroschwagerina* zone — *S. glomerosa* Subz; *S. constans* Subz	*Sphaeroschwagerina* Range-zone: *Robustoschwagerina ziyunensis* Subz; *Sphaeroschwagerina moelleri* Subz; *Robustoschwagerina kahleri* Subz; *Pseudoschwagerina uddeni* Z	Zisongian: *Sphaeroschwagerina* Rz	Mapingian: *Sphaeroschwagerina* – *Pseudoschwagerina* Z	Chuanshan Fm: *Sphaeroschwagerina moelleri* Z	Chuanshan Fm: *Pseudoschwagerina* Z	Sakmarian: *Robustoschwagerina schellwieni* – *Paraschwagerina mira* Z (*Sphaero. glomerosa*); *Sphaeroschwagerina sphaerica* – *Pseudofusulina firma* Z. Asselian: *Sphaero. moelleri* – *Pseudofusulina fecunda* Z; *Pseudoschwagerina vulgaris* – *P. fusiformis* Z	Tastub (Sakmarian): *Pseudofusulina verneuili* Z; *P. moelleri* Z
Carboniferous Upper	*Triticites* Z	*Triticites* Z	Guoyanian: *Pseudoschwagerina uddeni* – *Pstexana* Z; *Triticites paramonti* – *Montiparus weiningica* Z	Xiaodushanian: *Triticites* – *Protriticites* Z	*Triticites* Z	*Triticites* Z	Gzhelian: *Daixina sokensis* Z; *Jigulites jigulensis* Z	

That subdivision was revised by Wang Jianhua (1978), who proposed the following:

Qixia Formation   *Parafusulina multiseptata* Zone
                    Nankinella orbicularia Zone
                    Misellina claudiae Zone

Wang pointed out that because *Schwagerina tschernyschewi* is associated with *Sphaeroschwagerina* and *Pseudoschwagerina* (both belonging to the Mapingian), it should no longer be treated as one of the Qixiaian units. As regards the *Parafusulina multiseptata* Zone, it is associated with *Chusenella*, *Codonofusiella*, and brachiopods (*Urushtenoidea, Dictyoclostoidea*), all of which display the Maokouan affinity; hence it correlates with the *Cancellina* Zone (Zhan's idea) (Zhan Lipei et al., 1982), leaving only two zones (*Misellina claudiae* and *Nankinella orbicularia*) for the Qixiaian (Table 7.3). Zhang Linxin (1983) redefined the Qixia Swine Limestone to include *Darvasites ordinatus*, *Schwagerina tschernyschewi*, and *S. cushmani*, which form the *Darvasites ordinatus* fauna or zone named *Schwagerina cushmani* (Zhou Zuren, 1982) in southeastern Hunan and the *Schwagerina cushmani–Darvasites* fauna in central Jiangxi. Consequently, the Qixiaian in the Nanjing-Zhenjiang area contains three fusulinid zones, in ascending order: the *Darvasites ordinatus* Zone, the *Misellina claudiae* Zone, and the *Nankinella orbicularia* Zone.

According to Xiao Weiming et al. (1986) (Table 7.2), the *Darvasites ordinatus* Zone also contains *Schwagerina cushmani* and *Pseudoschwagerina huananensis,* which appear in southern Guizhou in the *Misellina termieri* Subzone that underlies the *Misellina claudiae* Subzone. Hence, the *D. ordinatus* or *Schwagerina cushmani* Zone corresponds to the Qixiaian basal *Misellina termieri* Subzone in southern Guizhou, but not with the *Pamirina darvasica* or *Robustoschwagerina schellwieni* Zone of the Baomeshan Formation (Table 7.2). It should be noted here that all over the Nanjing-Zhenjiang area, central Hunan, and central Jiangxi, the Longyinian lacks the *Sphaeroschwagerina glomerosa* or *Robustoschwagerina ziyunensis* Zone. The parachomata-bearing *Misellina* Range Zone (Tables 7.1 and 7.3) is here regarded as the range of the Qixiaian Stage, which contains the following subzones:

1. *Misellina termieri–Brevaxina dyhrenfurthi* Range Subzone
2. *Misellina claudiae* Acme Subzone
3. *Shengella-Maklaya* Acme Subzone

Lower and upper limits of the Qixiaian Stage

The first appearance of Misellininae (*Misellina, Brevaxina*) is now generally taken to mark the beginning of the Qixiaian (i.e., the lower limit of the stage can be drawn at the base of the *Darvasites ordinatus* Zone or *Schwagerina cushmani* Zone). The disappearance of *Misellina* and the first occurrence of *Cancellina* mark the upper limit of the Qixiaian Stage.

Maokouan Stage

The name Maokou Limestone was given by Yue Shengxun (1929) to the rocks exposed by the Maokou River, Langdai, Guizhou, but the stage stratotype section is located in Bali, Langdai, Qinglong county, Guizhou, where the rocks consist chiefly of grey to greyish white limestone, moderately well bedded to massive, and bioclastic limestone, 67 m thick, richly fossiliferous, especially with fusulinids, with less abundant corals and brachiopods. The three fusulinid zones were recognized by Zhan Lipei et al. (1982) as (1) the *Cancellina* Zone, (2) the *Neoschwagerina margaritae* Zone, and (3) the *Yabeina-Neomisellina* Zone. Those zones are closely correlated with the zones established by Sheng Jinzhang (1962): *Cancellina, Neoschwagerina, Yabeina.* Yang Zhengdong (1985) carried out a phylogenetic study of the Maokouan fusulinids from the Datieguan section, Langdai. Studies of the Maokouan faunas in the neighbouring areas by Xiao Weiming et al. (1986) and Zhang Zhenghua et al. (1988) established the biostratigraphic successions of fusulinids, corals, brachiopods, and conodonts, and zonations were erected (in ascending order): (1) *Cancellina liuzhiensis* Range Zone, (2) *Neoschwagerina multivoluta* Zone, (3) *Afghanella schenki–Yabeina gubleri* Range Zone, and (4) *Neomisellina multivoluta* Zone. These roughly reflect the four evolutionary stages of the Maokouan fusulinids. Characteristic Maokouan corals are represented by *Ipciphyllum, Iranophyllum,* and *Allotropiophyllum,* forming the *Ipciphyllum-Iranophyllum* Assemblage Zone, which also contains the genera *Wentzelellites, Wentzelloides, Parawentzelella, Parawentzellophyllum,* and *Thomasiphyllum.*

In the upper Yangtze area there is a well-known disconformity between rocks of the Maokouan Stage and the Wujiaping or Loping (or Longtan) Formation, reflecting the Dongwu movement (epeirogeny). So the Maokouan fusulinid zones mentioned earlier do not provide a complete record of the Maokouan fusulinids. At present, the most nearly complete record can be found in eastern China (South Jiangsu; Tonglu, Zhejiang; and Longyan, Fujian), where the Middle and Upper Permian form a continuous sequence. The Yanqiao Formation, for instance, in the upper part of the Middle Permian in southern Jiangsu has rich fusulinid faunas that include chiefly *Neomisellina compacta, N. lepida, N. multivoluta,* and *Codonofusiella wusiana.* In Tonglu, Zhejiang, the upper part of the Middle Permian Lengwu Formation yields such fusulinids as *Polydiexodina, Condonofusiella schubertelloides, Neomisellina multivoluta,* and *N. douvillei.* Those faunas are referred to as the *Neomisellina-Codonofusiella* Assemblage Zone, which is regarded as the uppermost fusulinid zone of the Maokouan (Table 7.1). It is not represented in the upper Yangtze area.

In facies favorable for development of corals, two coral zones can be recognized in the Maokouan: the lower *Hayasakaia-Polythecalis* Acme Zone and the upper *Ipciphyllum-Allotropiophyllum* Assemblage Zone. Associated brachiopods also permit determination of the biostratigraphic succession: (1) *Chaoiella reticulata–Tyloplecta grandicosta* Acme Zone (oldest),

Table 7.3. Correlation of Middle-Upper Permian Subdivisions and Fossil Zones in China and four other countries

Series / Stage	Ziyun-Ceheng, Guizhou	Huayingshan, Sichuan	S. Jiangsu	Longyan-Yong'an, Fujian	Trans caucasus	Central Iran	SW Japan	N. America
U. Pm (Lepingian Ser.) — Changxingian	*Palaeofusulina sinensis-* / *Reichelina changhsing-ensis* AsZ (Changxing Fm)	*Palaeofusulina sinensis* Z / *Huayinophyllum irregulare* Z / *Pelichia zigzag* Z (Changxing Fm)	*Rotodiscoceras-Pleuronodoceras* AsZ / *Palaeofusulina sinensis* Z / *Pseudotirolites* Z / *Tapanshanites-Pseudostephanites* AsZ / *Palaeofusulina minima* Z / *Sanyangites* Z (Dalong Fm, Changxing Fm)	*Pseudotirolites asiatica* Z / *Pleuronodoceras* Z / *Palaeofusulina fusiformis* Z / *Paryphella sinuata* Z (Dalong Fm)	Dorashamian	Dorasham-ian	*Palaeofusulina sinensis* Z	Ochoan
Wujiapingian	*Codonofusiella kwangsiana* Z / *C. kweichowensis* Z / *C. lui* Z (Wujiaping Fm)	*Codonofusiella liui* Z / *Gallowayinella meitienensis* Z / *Edriosteges poyangensis* Z (Longtan Fm) / Emeishan basalt	*Araxoceras-Konglingites* AsZ / *Anderssonoceras-Prototoceras* AsZ / *Edriosteges poyangensis* Z / *Gigantopteris* A / *Codonofusiella-Neomisellina* AsZ (Longtan Fm, Yangqiao Fm)	*Gigantopteris nicotianae-folia* AZ / *Compsopteris* AZ / *Neuropteridium* AZ / *Shouchangoceras* Z (Cuipingshan Fm, Tongziyan Fm)	Dzhulfian	Dzhulfian	*Nanlingella simplex* Z / *Lepidolina kumaensis* Z / *Lepidolina multiseptata* Z (Maizuru Group)	
M. Pm (Yangxinian Ser.) — Maokouan	*Neomisellina multivoluta* Z / *Afghanella schenki-* / *Yabeina gubleri* AsZ / *Neoschwagerina simplex* Z / *Cancellina* Z (Maokou Fm)	*Chusenella conicocylindrica* Z / *Yabeina gubleri* Z / *Ipciphyllum ipci* Z (Maokou Fm)	*Altudoceras* Z / *Paragastrioceras* Z / *Paracelites* Z / *Parafuslina multiseptata* Z (Gufeng Fm)	*Uncisteges crenulata* Z / *Altudoceras* Z / *Waagenoceras* Z / *Paragastrioceras* Z / *Cancellina* Z (Wenbishan Fm)	Khachik / Gnishik	Abadehian / Guadalu-pian	*Colania douvillei-Neo. haydeni* AsZ / *Neoschwagerina craticulifera* Z / *Parafusulina kaenmizensis-Maklaya* sp. AsZ (Akiyoshi Limestone Group)	Capitan-ian / Wordian / Roadian (Guadalupian)
Qixian	*Misellina* Rz (Qixia Fm)	*Cryptospirifer striata* Az / *Nankinella orbicularia-Pisolina excessa* AsZ (Qixia Fm, Liangshan Fm)	*Misellina claudiae* Z / *Schwagerina tschernyschewi* Z or *Darvasites ordinatus* Z (Qixia Fm)	*Misellina claudiae* Z / *Staffella moellerana* Z / *Schwagerina cushmani* Z (Qixia Fm)	Artiskian		*Misellina claudiae* Z / *M. dyhrenfurthi* Z / *Pseudofusulina kraffti* Z	Leonardian

(2) *Cryptospirifer striata* Acme Zone (also *Tyloplecta nankingensis, T. grandicosta*), (3) *Urushtenoidea–Neoplicatifera huangi* Acme Zone (*Monticulifera, Tyloplecta nankingensis, Urushtenoidea,* and *Permundaria*). In platform basinal facies (e.g., Ziyun and Luoding, in Guizhou), where siliceous limestone, chert, and limestone are developed, conodonts and fusulinids are abundant, with some corals. The Maokouan conodonts recognized are *Mesogondolella serrata, Sweetognathus hangzhongensis, Mesogondolella postserrata,* and *Clarkina "bitteri."*

Lower and upper limits of the Maokouan Stage

It is now generally accepted that the base of the *Cancellina* Range Zone is the lower limit of the Maokouan Stage, and the termination of the verbeekinids and neoschwagerinids (i.e., the top of the *Neomisellina-Codonofusiella* Acme Zone) is the upper limit of the Maokouan Stage (Sheng Jinzhang, 1963; Yang Zhengdong, 1985).

According to Jin Yugan, Mei Shilong, and Zhu Zili (1993) and Mei Shilong, Jin Yugan, and Wardlaw (1993), a new stage (not yet named) marked by three conodont zones (*Mesogondolella altudaensis, M. prexuanhanensis, M. xuanhanensis*) can be erected for the upper "Kuhfeng" beds lying between the Maokou Formation and the Wujiaping Formation. So far as is known, this is the most nearly complete conodont succession known from the upper part of the Middle Permian in China. Whether or not the same will be found elsewhere remains to be determined (Table 7.1).

Wujiapingian Stage

The Wujiapingian Stage is based on the Wujiaping Limestone succession, named by Lu Yanhao (1956), with its stratotype section located in Liangshan, Hanzhong, southern Shaanxi. The Wujiaping Limestone is about 252 m thick, rich in *Codonofusiella, Reichelina,* and *Liangshanophyllum,* and it overlies the Wangbo Shale, both of which, according to Sheng Jinzhang (1962), form the Wujiaping Formation (*s.l.*), representing the lower part of the Upper Permian in South China. Sheng also erected the *Codonofusiella* Zone (*C. kwangsiana* Zone, according to Zhan Lipei et al., 1982; Zhan Lipei and Li Li, 1984) for the Wujiapingian. Restudy of the Liangshan section in southern Shaanxi enabled Rui Lin et al. (1984) to propose the following succession for the Wujiaping (Tables 7.1 and 7.3); *Codonofusiella* Zone, *Liangshanophyllum lui–L. wuae* Assemblage Zone (coral), and *Mesogondolella liangshanensis–M. bitteri* Zone (conodont).

In paralic facies (e.g., Huayingshan, Sichuan) the Longtan (Loping) Formation is characterized by abundant brachiopods (*Edriosteges poyangensis, Tyloplecta yangtzeensis*), with associated fusulinids (*Codonofusiella lui, C. schubertelloides*), bivalves, and plant fossils (*Gigantopteris nicotianaefolia, Lobatannularia*). In the lower Yangtze region (Jiangsu, Nanjing, central Jiangxi) the Loping (Longtan) Formation is divisible into three members: (1) a lower coal-bearing member comprising argillite and sandstone, with coal seams that contain *Gigantopteris, Lobatannularia,* and *Pecopteris;* (2) a middle marine member marked by argillite, sandstone, and siliceous rock, whose lower part yields

brachiopods (*Edriosteges poyangensis, Tyloplecta yangtzeensis*) and whose upper part contains ammonoids (*Anderssonoceras, Prototoceras, Konglingites,* and *Sanyangites*); (3) an upper clastic member comprising neritic argillite, sandstone, siliceous rock or sandstone, argillite, and carbonaceous shale intercalated with coal seams.

Lower and upper limits and events of the Wujiapingian Stage

It is widely accepted that the base of the *Codonofusiella kwangsiana* Zone is the lower limit of the stage, and the termination of this zonal species marks the upper limit of the Wujiapingian Stage.

Changxingian Stage

The Changxingian Stage is based on the Changxing Limestone succession erected by A. W. Grabau for the rocks exposed in Changxing county, Zhejiang province. That limestone was regarded by Huang Jiqing (1932) as representative of the upper part of the Upper Permian in South China, and during the first national stratigraphic congress of China in 1959 it was formally named the Changxing Formation by Sheng Jinzhang (1962), who recognized its characterization with the *Palaeofusulina* Zone. Zhao Jinke, Liang, and Zheng (1978) made a detailed study of the Changxing type section, dividing it into a lower part, the Baoqing member, and an upper part, the Meishan member. Zhao Jinke et al. (1981) designated it the stratotype section of the Changxingian, with well-defined lower and upper limits. The older Baoqing member contains fusulinids (*Gallowayina meitienensis* Zone) and conodonts ("*Neogondolella subcarinata–N. subcarinata elongata*" Zone). The younger Meishan member yields fusulinids (*Palaeofusulina sinensis* Zone), ammonoids (*Pseudotirolites-Pleuronodoceras* Zone, *Rotodiscoceras* Zone), and conodonts ("*Neogondolella subcarinata changxingensis*" Zone). The disappearance of *Codonofusiella* and the abundant occurrences of *Palaeofusulina* (*P. minima*) and *Gallowayinella* define the base of the Changxingian Stage. Further studies by Sheng Jinzhang et al. (1984) subdivided the *Palaeofusulina* Zone into a lower *P. minima–Gallowayinella* Subzone and an upper *P. sinensis* Subzone. Zhan Lipei et al. (1982) took the first appearance of the highly evolved *P. sinensis* as marking the beginning of the Changxingian. This differentiates the faunal succession from the older primitive palaeofusulinids and *Gallowayinella*-bearing rocks of the Upper Wujiapingian. The genus *Gallowayinella* first appears within the Middle or Lower Wujiapingian (e.g., Huayingshan, Sichuan, and Lianxian, Guandong) (Table 7.3).

Classification and correlation of the nonmarine Permian in China

Nonmarine Permian beds are also widespread in China. Both terrestrial and paralic facies predominate north of a line joining the Kunlun, Qinling, and Dabie mountains, whereas south of that line, in parts of South China, one finds chiefly paralic or terrestrial deposits of middle and late Permian age only; early Permian nonmarine beds are rarely present. On the basis of their lithologic and floral characteristics, the Permian nonmarine rocks can be divided into four regions: the Northern China Border Region, the North China Region, the South China Region, and the Himalayan Region (Figure 7.2).

Stratigraphic sequence (Table 7.4)

Northern China Border Region

This region, north of the Tianshan-Beishan and south of the Hinggan Mountains, includes the Junggar Basin, the western part of the Turfan Basin, and the Songhuajiang Plain, where the Lower Permian is characterized by terrestrial facies with marine intercalations, and in some areas is made up of volcanic rocks. The Middle and Upper Permian are terrestrial in nature, containing the Angara flora. The Permian sequence, which is best developed in the Junggar Basin, has been described (Yang Jiduan et al., 1986) in detail as follows:

Lower Permian Series

1. Aortu Formation: greyish black sandstone, siltstone intercalated with limestone lenses; contains invertebrate fossils, such as brachiopods (*Marginifera pusilla*), corals (*Cyathocarinia pararegulata*), and ammonoids (*Somoholites glomerosus, Neopronorites carboniferus, Prouddenites primus*), plants (*Calamites* sp.), and palynomorphs; 228 m thick.

Middle Permian Series

2. Shirenzigou Formation: grey and greyish black sandstone, argillite, limestone, siliceous rocks, and basal conglomerate; contains fossil plants (*Walchia* sp.), brachiopods (*Neospirifer* sp., *Echinoconchus* sp.), and palynomorphs; 250 m thick.

3. Tashikula Formation: greyish black fine-grained sandstone, argillite intercalated with siliceous rock and limestone; contains plant fossils (*Walchia, Cordaianthus curvatus, C. volkmanii, Noeggerathiopsis* sp.), brachiopods (*Neospirifer* sp., *Chonetes* sp.), and palynomorphs; 718–3,000 m thick.

4. Wulabo Formation: an alternation of greyish black sandstone and argillite plus limestone; contains plant fossils (*Paracalamites* sp., *Walchia* sp.), palynomorphs, and bivalves (*Palaeoanodonta*); 1,400 m thick.

5. Jingjingzigou Formation: greyish green sandstone, argillite, and tuffite; contains fossil plants (*Cordaites* sp., *Calamites* sp.) and palynomorphs; 1,650 m thick.

6. Lucaogou Formation: black shale, oil shale, siltstone, and dolomitic limestone; contains bivalves (*Urmoia* sp., *Anthraconauta*), trilobites (*Pseudophillipsia*), fish (*Tienshaniscus longipterus*), and palynomorphs; 1,100 m thick.

7. Hongyanchi Formation: greyish green fine sandstone, argillite, and limestone; contains bivalves (*Palaeoanodonta pseudolongissima, Anthraconauta* spp.),

plant fossils (*Calamites, Prynadacopteris anthriscifolia, Noeggerathiopsis* sp.), and palynomorphs; 730 m thick.

Upper Permian Series

8. Quanzijie Formation: purplish red, greyish green conglomerate, sandstone, argillite; contains fossil plants (*Callipteris zeilleri, Comia partita, Iniopteris sibirica, Prynadacopteris anthriscifolia, Noeggerathiopsis angustifolia*), bivalves (*Palaeoanodonta* cf. *longissima, P. solonensis*), ostracods (*Darwinula* spp., *Vymella xinjiangensis, Panxiania ovata, Kunpania scopulosa*), and palynomorphs; 61–223 m thick.

9. Wutonggou Formation: greyish green sandstone, conglomerate, argillite intercalated with a thin coal seam and marl; contains plant fossils (*Callipteris zeilleri, Comiopteris* sp., *Prynadacopteris anthriscifolia, Noeggerathiopsis angustifolia*), bivalves (*Palaeoanodonta* cf. *longissima, Palaeomutella keyserlingi*), ostracods (*Darwinula* spp., *Panxiania ovata*), vertebrate fossils (Dicynodontia), and palynomorphs; 87–286 m thick.

10. Guodikeng Formation (Middle Permian–Lower Triassic) (P_2–T_1): dark red, purplish red, and greyish green (variegated) argillite and sandy argillite intercalated with greyish green thin–medium-bedded fine limestone and marlite lenses; contains plant fossils (*Lepidostrobophyllum* sp., *Paracalamites* sp., *Zamiopteris* cf. *glossopteroides, Walchia* sp., *Viatscheslavia* cf. *vorcuntensis*), bivalves (*Palaeoanodonta* spp.), ostracods (*Darwinula* spp., *Darwinuloides* spp., *Vymella subglobosa, Panxiania xinjiangensis*), vertebrate fossils (*Striodon magnus, Jimusaria sinkiangensis, Dicynodon tienshanensis, Lystrosaurus* sp.), and palynomorphs; 68–128 m thick.

Lower Triassic Series

11. Jiucaiyuan Formation: a lower member consisting of greyish green sandstone and argillite, and an upper member of brick-red-to-dark-red argillite intercalated with greyish green sandstone; contains plants (*Pecopteris* sp.), ostracods (*Darwinula rotundata, D. trassiana,* other *Darwinula* spp.), vertebrate fossils (*Lystrosaurus broomi, L. hedini, L. weidenreichi, L. shichanggouensis, L. robustus, L. youngi, Chasmatosaurus yuani*), and palynomorphs; 170–376 m thick.

North China Region

This region includes the southern part of northeastern China, North China, the Qaidam Basin, the northern Qilian Mountains, and the Tarim Basin, over which the Permian–Triassic succession consists chiefly of terrestrial beds, intercalated with littoral deposits only in the lower part; it carries Cathaysian flora. The typical Permian–Triassic sequence in Shaanxi province is as follows:

Lower Permian Series

1. Taiyuan Formation: black shale, sandy shale, greyish

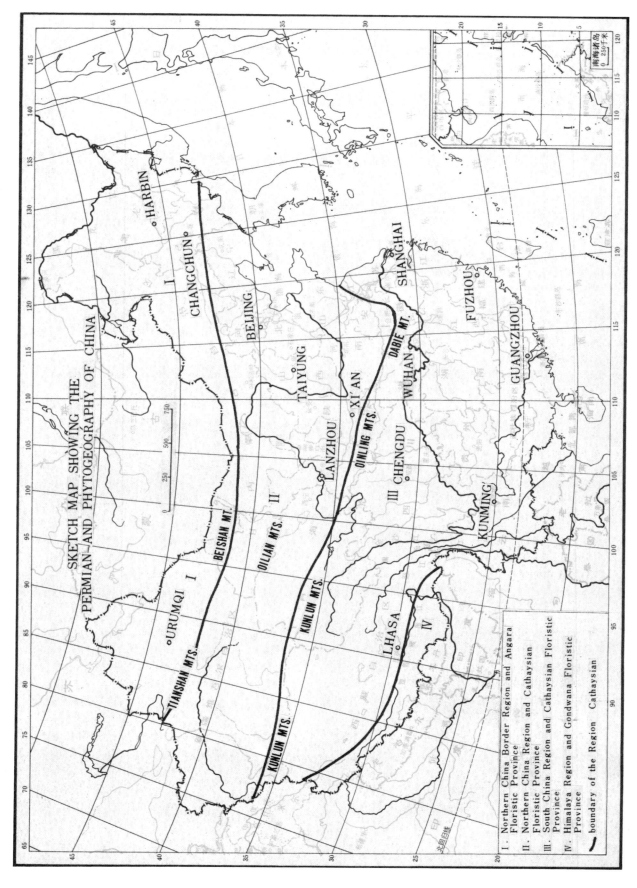

SKETCH MAP SHOWING THE
PERMIAN AND PHYTOGEOGRAPHY OF CHINA

I. Northern China Border Region and Angara
   Floristic Province
II. Northern China Region and Cathaysian
    Floristic Province
III. South China Region and Cathaysian Floristic
     Province
IV. Himalaya Region and Gondwana Floristic
    Province
━━ boundary of the Region Cathaysian

HARBIN
CHANGCHUN
BEIJING
TAIYÜNG
XI'AN
QINLING MTS.
LANZHOU
BEISHAN MT.
QILIAN MTS.
KUNLUN MTS.
TIANSHAN MTS.
URUMQI
KUNLUN MTS.
LHASA
CHENGDU
WUHAN
DABIE MT.
SHANGHAI
FUZHOU
GUANGZHOU
KUNMING

I
II
III
IV

Figure 7.2. Sketch map showing Permian regions and phytogeography.

Table 7.4. *Simplified Correlation Chart of Upper Carboniferous-Lower Triassic (Non-Marine and Partly Marine)*

Series		Northern China Border Region — Junggar Basin Xinjiang	North China Region — Shanxi-Hebei-Neimongol Province	South China Region — Fujian-Guangdong Province	South China Region — Yunnan-Guizhou-Sichuan Province	South China Region — Jiangsu-Zhejiang Province	Himalaya Region — Xizang	Russia
Lower Triassic	T₁	Jiucaiyuan Fm.	Liujiagou Fm.	Xikou Fm.	Kayitou Fm.	Yingken Fm.		Veteuger Fm.
Permian	P₃	Guodikeng Fm.	Sunjiagou Fm.	Talung Fm. (Marine)	Xuanwei Fm.	Changxing Fm. (Marine)		Tartarian
		Wutonggou Fm. Quanzijie Fm.	Upper Shihhotze Fm.	Cuipingshan Fm.	Omeishan Basalt Fm.	Longtan Fm.	Qubuerga Fm. (Marine)	
	P₂	Hongyanchi Fm. Lucaogou Fm.		Tongtzeyan Fm.	Maokou Fm. (Marine)	Yanchiao Fm.		Kazanian Ufimian
		Jingjingzigou Fm.	Lower Shihhotze Fm.	Wenbishan Fm.	Chihsia Fm. (Marine)			Kungurian
		Wulabo Fm. Tashikula Fm. Shirenzigou Fm.	Shanxi Fm.	Chihsia Fm. (Marine)	Liangshan Fm.	Chihsia Fm. (Marine) Liangshan Fm.	Qubu Fm.	Artinskian
	P₁	Aortu Fm.		Chuanshan Fm.	Baomeshan Fm. (Marine)	Chuanshan Fm. (Marine)	Jilong Fm.	Sakmarian Asselian
U. Carb.	C₂	Qijiagou Fm. (Marine)	Taiyuan Fm.	(Marine)	Longyin Fm. (Marine) Maping Fm. (Marine)			Gzhelian

white sandstone plus limestone and coal; contains plant fossils (*Lepidodendron posthumii, Neuropteris ovata, Sphenophyllum oblongifolium, Pecopteris feminae-formis, Emplectopteridium* sp., *Callipteridium* sp.), brachiopods (*Choristites* cf. *pavlovi*), fusulinids (*Rugosofusulina* sp., *Pseudoschwagerina texana, Triticites* spp.), and palynomorphs; 92–118 m thick.

Middle Permian Series

2. Shanxi Formation: dark grey, greyish white, and black argillite, sandy argillite, sandstone; contains plant fossils (*Emplectopteridium alatum, Callipteridium koraiense, Emplectopteris triangularis, Alethopteris ascendens, Lepidodendron posthumii, Odontopteris subcrenulata, Tingia partita, Taeniopteris* spp.) and palynomorphs; 40–170 m thick.

3. Lower Shihhotze Formation: yellowish green and greyish green medium and fine sandstone, argillite, silty argillite, and carbonaceous argillite; top part intercalated with purplish red or variegated argillite and one or two layers of aluminous argillite and iron concretions, locally containing pebble conglomerates and coal seams; contains plant fossils (e.g., *Emplectopteris triangularis, Cathaysiopteris whitei, Sphenophyllum neofimbriatum, Procycas densinervis, Pterophyllum daihoense, P. cutelliforme, Lepidodendron tachingshanensis, Taeniopteris* spp.) and palynomorphs; 130–260 m thick.

Upper Permian Series

4. Upper Shihhotze Formation: purplish red and yellowish green argillite and sandstone; contains plant fossils (*Gigantopteris dictyophylloides, Gigantonoclea hallei, Rhipidopsis parvi, Chiropteris reniformis, Pseudorhipidopsis brecicaulis, Psygmophyllum multipartitum, Annularia shirakii, Lobatannularia multifolia, Sphenophyllum sinocoreanum, S. kobense*) and palynomorphs; 40–700 m thick.

5. Sunjiagou Formation: a lower part consisting of yellowish green and purplish red pebbly sandstone, and an upper part that is an alternation of red sandstone and argillite; contains fossil plants (*Ullmannia bronnii, U. frumentaria, Yuania magnifolia, Pseudovoltzia leibeana, Callipteris martinsis, Sphenobaiera micronervis, Scytophyllum sunjiagouensis, Taeniopteris longifolia, Quadrocladas solmsii, Shihtienfenia permica*), vertebrate fossils (*Shansisaurus xuecunensis*), and palynomorphs; 100–200 m thick.

Lower Triassic Series

6. Liujiagou Formation: purplish red sandstone; contains plant fossils (*Pleuromeia jiaochengensis*) and palynomorphs; 360–600 m thick.

South China Region
This is an extensive region lying between the North China Region and the Himalayan Region (Figure 7.2). The Permian

succession is marine except in a small number of areas where coal-bearing deposits developed during the early Qixiaian Stage. Elsewhere, paralic facies prevailed during the Maokouan–Longtanian (Lopingian), and in some areas terrestrial or paralic facies dominated during the Changxingian and early Triassic. The Cathaysian flora is well known in this region.

Himalayan Region
Lying south of the Gangdise and Nienqin-Tanggula and north of the Chinese border, the terrestrial Permian successions are not well developed. Here, the Permian is marine. It is only in south Xizang that beds correlatable with the paralic Qixiaian and Maokouan (Qubu Formation) can be found. The Qubu Formation is a sequence of greenish yellow and greyish white fine sandstone, marl, and carbonaceous shale; it is 20 m thick and contains plant fossils (*Sphenophyllum speciosum, Raniganjia qubuensis, Glossopteris angustifolia, G. communis, G. indica*) that are characteristic of the Gondwana flora.

Permian boundaries and events in terrestrial–paralic facies

A tripartite division is proposed for the nonmarine Permian in China, based chiefly on plants and palynomorphs, as well as on faunistic elements associated with paralic facies, plus nonbiotic criteria such as distinctive lithology and palaeomagnetism.

Broad Permian correlations linking China with Japan, Iran, Russia, and North America are shown in Tables 7.2 and 7.3.

Lower Permian boundary and Carboniferous–Permian events

This boundary is difficult to describe, because there is no consensus on its placement. Here we consider the boundary to lie above the marine and paralic sequence in South China (i.e., the boundary between the *Pseudoschwagerina* and *Triticites* zones). In the North China Region, the base of the Permian is drawn between the Maoergou (Middle Taiyuan Formation), with *Pseudoschwagerina*, and the Jinci Formation, with *Triticites*. In the Northern China Border Region it is at least temporarily considered as passing between the paralic Aortu Formation and the marine Qijiagou Formation. There are different opinions about the age of the Jiujiakou Formation, as no fusulinids have been found – only ammonoids, such as *Neopronorites carboniferus, Somoholites glomerosus, Eoasianites* cf. *millsi, Prouddenites* cf. *primus, Glaphyites pararegulatus,* and *G. qijiagouensis,* which have variously been regarded as late Carboniferous (Sheng Huaibin, 1981), late Carboniferous (Asselian) to early Permian (Sakmarian to Artinskian), or middle–late Mapingian. Besides, the Aortu Formation yields palynomorphs, with predominant pteridophytic spores (mainly *Cyclogranisporites* spp., *Verrucosisporites* sp., and some *Endosporites* sp.), and a few pollen species (mainly *Cordaitina* sp., *Vittatina* sp., *Florinites* sp.), as well as more numerous striate disaccate pollen (chiefly *Protohaploxypinus* sp., *Striatoabietes*

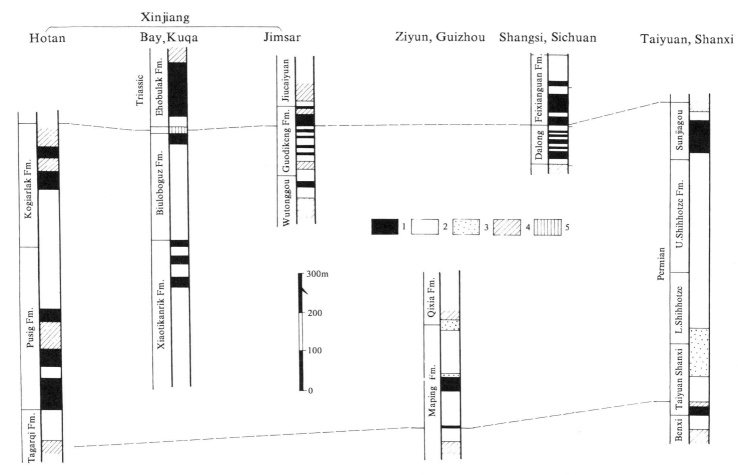

Figure 7.3. Correlation of Permian–Triassic columnar magnetic sections in China: 1, normal; 2, reversed; 3, mixed; 4, unknown; 5, lacuna.

sp., *Hamiapollenites* sp.) and much richer nonstriate disaccate pollen (mainly *Pityosporites*). Those palynomorphs had been considered to be of late Carboniferous age. It should be pointed out, however, that the foregoing assemblage is similar in character to the Asselian floras of the Russian Urals region (Faddeeva, 1990). Hence the Aortu Formation is here regarded as early Permian.

Marked changes took place in both biotic and nonbiotic factors around the Carboniferous–Permian boundary. First, faunas, including fusulinids, corals, brachiopods, and bivalves, increased their numbers of taxa in the earliest Permian (e.g., the first appearance of *Pseudoschwagerina*). Second, the terrestrial flora changed; *Linopteris,* which had flourished in the Carboniferous, declined sharply at the boundary and was replaced by lepidodendrons such as *L. posthumii, L. szelanum,* and *Cathaysiodendron incertum.* Among the pteridosperms, characteristic Permian taxa such as *Tingia, Emplectopteridium,* and *Callipteridium* suddenly appeared. Third, not only was there an increase in pteridophytic spores above the boundary, but also numerous gymnospermous and pteridospermous pollen forms

appeared (e.g., *Florinites, Cordaitina, Hamiapollenites, Striatoabietes,* and *Pityosporites*), reflecting a Permian aspect (Gao Lianda, 1985; Hou Jingpeng and Wang Zhi, 1990). Fourth, at the beginning of the Permian, the sedimentary facies and palaeoenvironment experienced great changes. Within the Northern China Border Region, in the Junggar Basin, for example, there was a distinct regression by the early Permian, so that marine conditions prevailed only along the south margin of the basin, where the paralic Aortu Formation overlies the Carboniferous Qijiagou Formation. In a number of localities within the Tarim Basin, marine facies persisted, as shown by the presence of the *Pseudoschwagerina* Zone, but below it the *Triticites* Zone is lacking, thus reflecting a gap in deposition. Fifth, palaeomagnetically, there was normal polarity at the boundary, but immediately above it the polarity was chiefly reversed (e.g., in the basal Permian) (Figure 7.3) (Lin Wanzhi, Shao Jian, and Zhao Zhangyuan, 1984; Li Yongan, Li Qiang, and Zhan Hui, 1991; Zhang Wenzhi, 1992). Consequently, certain major events must have occurred near the Carboniferous–Permian (C–P) boundary.

Lower–Middle Permian boundary and events

In the northern China Border Region, the Lower–Middle Permian boundary lies between the Aortu Formation and the Shirenzigou Formation; in the North China Region it is between the Taiyuan Formation and the Shaanxi Formation; in the South China Region it is between the Baomeshan Formation and the Liangshan Formation (Table 7.2). There are distinctly sharp changes in biotic and nonbiotic features. In the Northern China Border Region, Middle Permian floras are chiefly composed of *Cordaianthus, Cordaites, Walchia, Noeggerathiopsis,* and *Paracalamites.* Among palynomorphs, gymnospermous pollen predominates, at 50–95%, with major elements such as *Crustaesporites speciosus, Pseudocrustaesporites wulaboensis, Hamiapollenites viriosus, H. limbalis,* other *Hamiapollenites* spp., *Protohaploxypinus verus, P. perfectus, Protohaploxypinus,* other *Protohaploxypinus* spp., *Striatoabietes parviextensisaccus, Vittatina striata, V. vittifera, Cordaitina* spp., *Florinites* spp., *Pityosporites* spp., and *Cycadopites* sp. (Hou Jingpeng and Wang Zhi, 1990). Of these, *Hamiapollenites, Cordaitina,* and *Vittatina* increase in abundance, being distinctly more abundant than in the Lower Permian (Table 7.1). The palynomorph assemblage is remarkably similar to that of the Artinskian–Kazanian in Russia (Faddeeva, 1990). In the North China Region, there are significant floristic differences between the Middle and Lower Permian: The Cathaysian taxon *Lobatannularia sinensis* first appears in the Lower Shihhotze Formation, pteridopterids and pteridospermopsids dominate, *Taeniopteris* is well developed and widely distributed, and *Emplectopteridium* and *Emplectopteris* show signs of having begun to flourish. For the Middle Permian, the palynomorph assemblage is still dominated by pteridopterid spores, but the typical early Permian *Reinschospora* and *Murospora* have disappeared, and pteridospermous pollen has increased in content, reaching 30% (e.g., *Florinites* spp., *Gulisporites cochlearius, Anticapipollis tornatilis, Potonieisporites* spp., *Limitisporites* spp., *Pityosporites* spp., and *Alisporites* spp.).

Nonbiotic changes may be indicated by signs of crustal movement: uplift of land in many areas, and a clear regression seen in the Northern China Border Region. Near Urumchi, in the Junggar Basin, the Lower Permian paralic strata are unconformably overlain by the Middle Permian Shirenzigou Formation, of terrestrial origin, with marked and widespread unconformities within the basin. In the North China Region, regression was also marked in the early Permian, as the Taiyuan Formation is unconformably overlain by the terrestrial Shanxi Formation. In the northeastern part of the Tarim Basin, Lower Permian marine deposits are overlain by Middle Permian terrestrial strata. In South China the regression is distinctly seen, with the Lower Permian disconformably overlain by paralic Middle Permian coal-bearing deposits (the Liangshan Formation). These observations indicate the widespread influence of some event (crustal movement, sea-level change) between the early and middle Permian.

Middle–Upper Permian boundary and events

In the Northern China Border Region, the Middle–Upper Permian boundary is drawn between the Middle Permian Hongyanchi Formation and the Upper Permian Quanzijie Formation; in the North China Region it is between the Lower and Upper Shihhotze Formation; in South China it is between the Yanchiao Formation and the Longtan Formation or between the Maokou Formation and the Xuanwei Formation. This boundary is also marked by biotic and nonbiotic events. In the Northern China Border Region the Upper Permian is dominated by the late Angara flora (chiefly *Callipteris, Comia, Iniopteris, Zamiopteris, Noeggerathiopsis*), and the palynomorph assemblage is dominated by gymnospermous pollen. In the Middle Permian, *Cordaitina* and pleurosaccate pollen decrease appreciably, whereas smooth disaccate pollen and pteridophytic spores increase in number. Also present are *Apiculatisporis spinosus, A. xiaolongkouensis, Kraeuselisporites spinullosus, Tuberculatosporites homotubercularis, Limatulasporites fossulatus, Endosporites multirugulatus, Alisporites sublevis, Sulcatisporites potoniei, Vitreisporites pallidus, Pteruchipollenites reticorpus, Klausipollenites schaubergeri, Falcisporites zapfei, Protohaploxypinus samoilovichii, Lueckisporites virkkiae, Striatopodocarpites* sp., and *Taeniaesporites* sp. (*Limatulasporites-Alisporites-Lueckisporites* assemblage) (Hou Jingpeng and Wang Zhi, 1990) (Table 7.5). This assemblage is quite similar to that of the Tatarian in Russia.

In the North China Region the flora in the lower member of the Upper Permian (Upper Shihhotze) differs sharply from that of the Middle Permian Lower Shihhotze. *Gigantonoclea, Lobatannularia,* and *Sphenophyllum,* which were present in the middle Permian, flourished in the late Permian. There were also more numerous *Rhipidopsis, Pseudorhipidopsis, Psygmophyllum, Chiropteris,* and *Yuania* (*Gigantonoclea hallei–Lobatannularia heianensis* assemblage) (Table 7.5). Within the palynomorph assemblage, gymnospermous pollen increase in total content (40–50%), represented by *Potonieisporites neglactus, Nuskoisporites dulhuntyi, Vitreisporites pallidus, Anticapipollis gibbosa, A. elongata, A. tornatilis, Limitisporites, Sulcatisporites, Vecicaspora, Vetigisporites, Protohaploxypinus, Chordasporites,* and *Vittatina,* and there are important spores, such as *Punctatisporites palmipedites, Patellisporites meishanensis, Triquitrites tribullatus, T. paraporatus, Microtorispora gigantea, M. media, M. cathayensis, Tuberculatosporites medius, Gulisporites cochlearius, Torispora securis* (*Patellisporites-Macrotospora-Nuskoisporites* assemblage).

In the South China Region, the Upper Permian Longtan Formation and Xuanwei Formation are similar in character to the Upper Shihhotze. They are especially rich in *Gigantonoclea* and *Gigantopteris* and also yield numerous *Pecopteris, Otofolium,* and *Lobatannularia multifolia* (*Gigantopteris nicotianaefolia–Lobatannularia multifolia* assemblage). The palynomorph assemblage contains 90% spores, among which are *Gulisporites cochlearius, Waltzispora strictus, Nixispora sinica, Neoraistrickia spanie, Proterispora sparsus, Triquitrites sinensis, Tripartites cristatus,*

Table 7.5. Permian–Lower Triassic floras and palynoflora assemblages in China

Region / Fossil Series	Northern China Border Region		North China Region		South China Region		Himalaya Region
	Flora	Palynoflora	Flora	Palynoflora	Flora	Palynoflora	Flora
Triassic — T₁	*Pecopteris* sp.	*Limatulasporites-Lundbladispora-Taeniaesporites-Equisetosporites* Ass.	*Pleuromeia-Crematopteris-Neocalamites*	*Lundbladispora-Taeniaesporites-Cycadopites* Ass.	*Neuropteridium-Voltzia* Ass.	*Aratrisporites-Lundbladispora* Ass.	*Sphenophyllum speciosum* *Raniganjia qubuensis* *Dichotomopteris qubuensis* *Glossopteris angustifolia* *G. communis* *G. indica*
Permian — P₃	*Callipteris-Comia-Iniopteris* Ass.	*Limatulasporites-Alisporites-Lueckisporites* Ass.	*Ullmannia bronnii-Yuania magnifolia* Ass. *Gigantonoclea hallei-Lobatannularia heianensis* Ass.	*Lueckisporites-Jugasporites-Protohaploxypinus* Ass. *Patellisporites-Macrotospora-Nuskoisporites* Ass.	*Ullmannia* cf. *bronnii-Gigantonoclea guizhouensis* Ass. *Gigantopteris nicotianaefolia-Lobatannularia multifolia* Ass.	*Yunnanospora-Proterisispora-Gradenasporites* Ass. *Patellisporites-Macrotospora-Anticapipollis* Ass.	
Permian — P₂	*Paracalamites* *Noeggerathiopsis* *Walchia* *Cordiates* *Cordaianthus volkmanii* *Walchia bipinnata* *Neoggerathiopsis*	*Corduitina* *Hamiapollenites* *Vittatina* *Protohaploxypinus* *Striatoabietes* *Calamospora* *Florinites* *Hamiapollenites* *Protohaploxypinus*	*Cathaysiopteris whitei-Emplectopteris triangularis* Ass. *Emplectopteridium alatum-Taeniopteris multinervis-Emplectopteris triangularis* Ass.	*Patellisporites-Gulisporites-Disaceites* Ass. *Gulisporites* *Sinulatisporites* *Florinites* *Anticapipollis* *Pityosporites*	*Gigantonoclea fukienensis-Tingia carbonica* Ass. *Emplectopteris triangularis-Taeniopteris multinervis* Ass.	*Patellisporites* *Cossisporites* *Triquirites* *Neoraistrickia* *Knoxisporites* *Sinulatisporites* *Lycospora* *Laevigatosporites* *Torispora* *Florinites*	
Permian — P₁	*Calamites* sp.	*Cyclogranisporites* *Verrucosisporites* *Corduitina* *Florinites* *Protohaploxypinus* *Striatoabietes* *Hamiapollenites* *Pityosporites*	*Neuropteris ovata-Lepidopteris posthumi* Ass.	*Glisporites* *Kaipingispora* *Spinosporites* *Florinites* *Striatosporites* *Pityosporites*			

Crassispora kosankae, Patellisporites meishanensis, Microtorispora media, M. cf. *gigantea, Yunnanospora radiata,* and *Taeniaetosporites yunnanensis.* Gymnosperms are rare, represented only by *Vitreisporites pallidus, Anticapipollis tornatilis, Abeneaepollenites* sp., *Bacrosporites* sp., *Cordaitina* sp., and *Protopinus* sp. (*Patellisporites-Macrospora-Anticapipollis* assemblage) (Ouyang Shu, 1986). This assemblage closely resembles that in the Upper Shihhotze. Nonbiotic changes can also be observed in various regions, such as the Northern China Border Region and North China Region, where the Upper Permian is almost entirely composed of terrestrial purplish red clastics, locally intercalated with pyroclastics. In the South China Region, nonmarine deposition went on further as a result of marine regression. In Sichuan, Guizhou, and Yunnan, above the Middle Permian strata there is a series of variegated nonmarine coal-bearing deposits overlying basal basalt flows, with a maximum thickness of several hundred metres (Table 7.1). In the Himalayan Region, no Upper Permian is represented, testifying to crustal uplift between the middle and late Permian, and the climate then became much drier and hotter.

Permian–Triassic boundary and events

This boundary was settled in 1992 in the Northern China Border Region, where it is known to pass through the Middle and Upper Guodikeng Formation. In the North China Region it is drawn between the Sunjiagou Formation and the Liujiagou Formation, in the South China Region between the Xuanwei Formation and the Kaylitou Formation. This is the most distinct boundary marked by the biotic and nonbiotic events:

1. In regard to the extinction or impoverishment of flora, in the Northern China Border Region the late Permian Angara flora (*Callipteris-Iniopteris-Comia* flora) had laregly perished by the end of the Permian; so plant fossils have rarely been found in the earliest Triassic. In the North China Region the *Gigantonoclea hallei–Lobatannularia heianensis* assemblage (early part of the late Permian) and the *Ullmannia bronnii–Yuania magnifolia* assemblage (late part of the late Permian) also became extinct by the early Triassic, and it was not until the late part of the early Triassic that the simple *Pleuromeia-Voltzia* assemblage appeared. In the South China Region the *Gigantopteris nicotiannaefolia–Lobatannularia multifolia* assemblage (early part of the late Permian) and the *Ullmannia* cf. *bronnii–Gigantonoclea guizhouensis* assemblage (late part of the late Permian) were extinct by the end of the Permian, replaced by the *Neuropteridium-Voltzia* assemblage by the late part of the early Triassic (Table 7.5).

2. The bivalves experienced a sudden change near the Permian–Triassic boundary (e.g., the *Palaeoanodonta-Palaeomutella* assemblage was limited within the Permian).

3. There was a sharp decrease in ostracods. More varied ostracods flourished in the late Permian (chiefly the *Vymella-Panxiania-Darwinuloides-Darwinula-Sulchonella* assemblage), and the early Triassic is represented by only a single *Darwinula* assemblage.

4. Marked changes also took place in palaeovertebrates. Dicynodonts were present in the late Permian in the Northern China Border Region, and a mixed fauna of Pareisauridae and Dicynodontidae died out by the end of the Permian, whereas the *Lystrosaurus* fauna appeared again in the early Triassic in Xinjiang. But in the basal Lower Triassic strata there appear individual dicynodonts. Such a mixed record has also been observed in southern Africa.

5. The palynomorph assemblages are marked by both sharp differences and transitional characters (Table 7.5). In the North China Region, Upper Permian gymnospermous pollen predominates, whereas in the Lower Triassic, pteridophytid spores are dominant. Within the two assemblages just mentioned, there were also great changes in type and number (e.g., the Palaeozoic type, such as *Vittatina,* which occurred widely in the late Permian, died out in the early Triassic). Another type, *Cordaitina,* which flourished in variety and abundance in the late Permian, is represented by a single species (*Cordaitina uralensis*) in the Lower Triassic. Other forms, such as *Lueckisporites virkkiae, Protohaploxypinus,* and *Striatopodocarpites,* also decreased sharply. Conversely, *Taeniaesporites,* which appears in small numbers in the Upper Permian, rapidly increases in the Lower Triassic; for example, *Limatulasporites* increases from 13% (average) in the upper part of the Upper Permian to 30% (average) in the Lower Triassic; *Lundbladispora* increases not only in number but also in the variety of types (from one type to five types) (Yang Jiduan et al., 1984, 1986, 1992) (Table 7.5). That tendency toward palynomorph change is evident in South China as well as in other parts of the world.

6. There were clear changes in the colours of sediments: In the later part of the late Permian, the sediments were mostly coloured red, and less often yellowish green and grey, whereas in the early Triassic, redbeds developed as a universal phenomenon.

7. Environment and climate changes (due to stress) are easily noticed, for many late Permian floras and the increasing *Lundbladispora* within the palynomorph assemblage all point to a worsening of the ecological environment and the development of an arid climate, all of which were unfavourable to plant life.

8. Palaeomagnetically (Figure 7.3), an abnormal change can be seen at the Permian–Triassic boundary, where the Upper Permian strata are marked mainly by reversed polarity, with a rapid change into a mixed belt of normal and reversed polarities at the top of the Permian, and farther up, in the Lower Triassic, a mixed belt typified by normal polarity is registered. That palaeomagnetic signature is present not only in the terrestrial P_3–T_1 sequence in Jimsar, Xinjiang (Northern China Border Region), but also in the marine P_3–T_1 sequence over South China (Shangsi, Guangyuan, and Sichuan provinces) (Figure 7.3) (Li Huamei et al., 1989; Li Yongan et al., 1991). This is further evidence of an abnormal event at the Permo–Triassic junction.

Summary of Permian events

The Permian sequences in China are punctuated by a number of geologic and biotic events: sea-level changes related to transgressions and regressions on various scales, global or regional, and from varied causes; geochemical anomalies; volcanism and biotic mass extinctions and/or biotic regeneration (or renewal) at geologic boundaries. All the Permian events discussed earlier have been considered in relation to four boundaries (C–P, P_1–P_2, P_2–P_3, P_3–T_1).

The Permian events in South China include three main mass extinctions and regenerations: one between the early and late Longyinian (equivalent to the early and late Sakmarian of the Urals in Russia), another between the middle Permian Maokouan and the late Permian Wujiapingian, and a third between the Permian and the Triassic, which is well known.

Early–late Longyinian biotic mass extinction and regeneration

In a number of areas there is a disconformity between the Lower Permian Mapingian and the Middle Permian Qixiaian, resulting from the Qiang-Gui or Yunnan uplift, as evidenced by deposition of the basal Qixiaian coal-bearing clastics on top of the Mapingian limestones. Even in southern Guizhou and northern Guangxi, where Lower and Middle Permian sequences form an apparently continuous carbonate succession, the basal Qixiaian differs markedly from the underlying Mapingian; in other words, there is a distinct lithologic boundary between the two. Similarly, there is a sharp biotic change separating those two units. The fusulinid subfamily Pseudoschwagerininae disappears abruptly at the top of the Mapingian and is replaced by the parachomata-bearing Misellininae in the Qixiaian.

Sudden changes also took place in the corals, for new genera such as *Wentzellophyllum*, *Polythecalis*, and *Hayasakaia* flourished in the Qixiaian. Among the brachiopods, the *Choristites* of the Mapingian were replaced by the *Tyloplecta nankingensis* group in the Qixiaian. Consequently, it appears that there is a line of biotic demarcation between the Mapingian and the Qixiaian. However, in the Qiang-Gui area, where the Carboniferous and Permian are continuous, such a line should lie within the Longyinian itself, between the Lower and Upper Longyinian (i.e., equivalent to the Sakmarian Tastub and Sterlitamak substages of the southern Urals in Russia). According to Xiao Weiming et al. (1986), in southern Giuzhou, pseudoschwagerinids flourished below that boundary (*Pseudoschwagerina*, *Sphaeroschwagerina*, *Zellia*, *Occidentoschwagerina*), along with schwagerinids (*Triticites* and *Quasifusulina*), whereas above that boundary the pseudoschwagerinids entirely died out, and the latter almost so. They were replaced by abundant new elements, such as *Pamirina*, *Nagatoella*, *Chalaroschwagerina*, *Laxifusulina*, *Parafusulinella*, *Toriyamaia*, *Minojapanella*, *Nankinella*, and *Sphaerolina*. Both *Schwagerina* and *Pseudofusulina* had advanced to a new stage of development, with significant changes in shape and septal structures. The much simpler Lower Longyinian *Streptophyllidium-Diversiphyllum* assemblage was replaced by the more complicated Upper Longyinian *Szechuanophyllum*, *Iranophyllum*, *Wentzelastraea*, *Parawentzelellites*, *Prowentzelellites*, *Anfractophyllum*, *Lons-*

daleiastraea, and *Laophyllum*, all having septa of three or more grades. Above that boundary, brachiopods also increased, with some new elements, such as *Orthotichia magnifica*. Similar changes occurred in other fossil groups, such as ammonoids, nonfusulinid foraminifera, and ostracods. In short, there was a distinct biotic mass extinction within the Longyinian.

In other parts of the Tethys, such a sharp boundary of biotic extinction is also seen, as, for example, in Darvas and the Pamirs in central Asia, where the Sakmarian *Pseudoschwagerina*, *Sphaeroschwagerina*, and *Zellia* are replaced by the Yakhtashian *Pamirina darvasica*, *Chalaroschwagerina*, and *Pseudofusulina kraffti*.

Maokouan–Wujiapingian biotic mass extinction and renewal

It is generally believed that the Dongwu movement was the cause of the disconformity and stratigraphic gap between the Maokouan and the Wujiapingian, with attendant changes in biotic makeup. Massive extinctions had occurred in fusulinids, corals, brachiopods, and ammonoids by the terminal Maokouan, and a number of new groups appeared in the Wujiapingian. That provided the basis for the distinct biotic demarcation between the Yangxinian and the Lopingian.

The rich Maokouan fusulinids (Verbeekinidae and Neoschwagerinidae) were almost totally exterminated, along with a number of Schubertellidae and Staffellidae, and were succeeded by the Wujiapingian *Codonofusiella*, *Reichelina*, *Nankinella*, *Sphaerolina*, and *Staffella*, in addition to the newly appearing *Chenia*, *Haoella*, and *Leella*. The replacement of corals is reflected in the new taxonomic makeup; generally, there were sharp decreases in the numbers of families and genera, and apparent increases in diversity and abundance among such compound corals as *Liangshanophyllum* and *Waagenophyllum*. The brachiopod succession shows that the once-flourishing Maokouan *Vediproductus*, *Permundaria*, *Monticulifera*, *Neoplicatifera*, *Urushtenoidea*, and *Uncisteges* had disappeared by the end of the Maokouan, with 30% of all genera becoming extinct. They were replaced by new forms in the Wujiapingian, including *Perigeyerella*, *Transannatia*, *Alatoproductus*, *Edriosteges*, *Oldhamina*, *Gubleria*, *Tschernyschevia*, and *Semibrachythyris* (with new genera amounting to 30% of the total genera). At the same time, there appeared characteristic species such as *Tyloplecta yangtzeensis* and *Permophricodothyris grandis*. Ammonoids that dominated the middle of the late Maokouan were *Altudoceras*, *Paragastrioceras*, *Paracelitifera*, *Strigogoniatites*, *Kufengoceras*, and *Shouchangoceras*, belonging to the Shouchangoceratidae, Paragastrioceratidae, and Cyclolobidae. Those had all disappeared by the end of the Maokouan and were replaced in the Wujiapingian by the Anderssonoceratidae and Araxoceratidae, represented by *Anderssonoceras*, *Prototoceras*, *Araxoceras*, *Konglingites*, and *Sanyangites*. In conclusion, there is considerable faunal evidence of a major extinction event at the end of the Maokouan, with significant new taxa appearing in the Wujiapingian.

Mass extinction at the Permian–Triassic boundary

This is one of the best-known mass extinctions. It occurred not only in China, but throughout the Tethys. It is now generally clear that important Palaeozoic invertebrate groups, such as fusulinids, tetracorals, tabulates, productids, spiriferids, trilobites, and treptostromatous bryozoans, died out altogether, either at or below the Permian–Triassic boundary, as described by Yang Zunyi et al. (1987, 1991, 1993) and Sweet et al. (1992).

Causes of mass extinctions

Quoting from Yang Zunyi et al. (1993), "the world-wide regression including the Late Changxingian regression and the terminal Changxingian transgression is considered the main cause of a group of catastrophes (catastrophe group), which was aggravated by frequent volcanism; from both of them there were triggered tributary (associated) events such as ecotope changes, oxygen deficiency, salinity oscillation, temperature rise and fall, marine acidity and pollution by poisonous substances. The superposition or combination in time and space of these tributary events constituted an unusually gigantic 'catastrophe group', which led finally to a great event, mass-extinction"

Catastrophe group in the Permian–Triassic transitional period in South China

Leading events
    Transgressions and regressions
    Volcanic eruptions
Tributary events
    Ecotope changes
    Oxygen deficiency
    Salinity oscillations
    Temperature fluctuations
    Marine acidity
    Pollution by poisonous substances
Ultimate event
    Mass extinction

Volcanic event

Extensive volcanism broke out during the early part of the Wujiapingian, thus bringing about the Emeishan Basalt eruption, which spread over the western part of the Yangtze Platform, covering the vast Sichuan-Guizhou-Yunnan area. That basalt is up to 600 m thick (40 m at Huayingshan, near Chongqing, Sichuan; 247 m in Yongshan, northern Yunnan; 527 m in Shuicheng, western Guizhou). That gigantic extrusion was caused by a strong extensional rupture within the sea trough lying west of the Yangtze Platform, and it reflected the Dongwu movement, regarded generally as an episode of the Hercynian (*s.l.*) crustal movement.

Conclusion

A threefold subdivision of the Permian in China is advocated here, in place of the twofold subdivision generally adopted by Chinese geologists for many years. It consists of the Lower Permian (Chuanshanian), with the Longyinian and Mapingian stages, the Middle Permian (Yangxinian), with the Qixiaian and Maokouan stages and an unnamed stage, and the Upper Permian (Lopingian), with the Wujiapingian and Changxingian stages. The three series are separated by biotic and nonbiotic events, with the latter largely having greater influence than the former. A yet-unnamed stage in the upper part of the Middle Permian, lying betweeen the Maokou Formation and the Wujiaping Formation, is inserted here in recognition of the contributions by Jin Yugan et al. (1993) and Mei Shilong et al. (1993).

This first attempt at a threefold subdivision for the terrestrial and paralic Permian in China is based on analyses of various biotic and geologic events discussed herein.

Acknowledgment

The work of this project (IGCP-272) was supported financially by the State Natural Science Foundation of China.

References

Ding Wenjie, Xia G. Y., Xu S. Y., Zhao S. Y., Li L., and Zhang Y. X. 1991. *The Carboniferous–Permian Boundary in China.* Beijing: Geological Publishing House.

Eshet, Y. 1992. The palynofloral succession and palynological events in the Permo–Triassic boundary interval in Israel. In *Permo–Triassic events in the Eastern Tethys*, pp. 134–45. Cambridge University Press.

Faddeeva, I. Z. 1990. Palaeopalynology of Permian. In *Practical Palynostratigraphy* (in Russian), ed. L. A. Panova, M. V. Oshurkova, and G. M. Romanovskaya, pp. 59–80. Moscow: G.M.-NGEI Ministry of Geology, USSR.

Fan Jiasong, Qi Jingwen, Zhou Tieming, Zhang Wei, and Zhong Xiaolin. 1990. *Permian Reefs in Longlin, Guangxi* (in Chinese, with English summary). Beijing: Geological Publishing House.

Gao Lianda. 1985. Carboniferous and early Permian spore assemblages of north China region and the boundary of the Carboniferous and Permian. In *Proceedings of the 10th International Congress on the Stratigraphy and Geology of the Carboniferous,* vol. 2, pp. 409–24.

Gao Lianda, Shen Zhida, and Qin Dianxi. 1989. Discovery of early Permian palynomorphic assemblages from Kaili county, Guizhou, and their stratigraphic significance (in Chinese, with English summary). *Geology of Guizhou* 6:1–13.

Geoscience Department, Academia Sinica, and Xinjiang Petroleum Administration. 1989. *The Evolution of the Junggar Basin and Petroleum* (in Chinese). Beijing: Science Press.

Hou Hongfei, Wang Zengji, Wu Xianghe, and Yang Shipu. 1982. The Carboniferous system of China. In *Stratigraphy of China,* vol. 1 (in Chinese, with English summary). Beijing: Geological Publishing House.

Hou Jingpeng and Wang Zhi. 1990. *Permian Palynomorph Assemblages of Northern Xinjiang* (in Chinese, with English summary). Beijing: China Environmental Science Press.

Huang Jiqing (T. K. Huang). 1932. *The Permian Formations of South China* (in Chinese). Memoirs, Geologica Sinica, A(10).

Huang Jiqing and Chen Bingwei. 1987. *The Evolution of the Tethys in China and Adjacent Regions* (in Chinese and English). Beijing: Geological Publishing House.

Huang Zhixun, Shi Yan, and Wei Muchao. 1982. A new Permian stratigraphic unit – the Longlinian stage (in Chinese, with English abstract). *Quarterly, Chengdu College of Geology* 4:63–73.

Iranian-Japanese Research Group. 1981. The Permian and Lower Triassic system in Abadeh region, central Iran. *Memoirs Faculty of Sciences, Tokyo University, Ser. Geol. Min.* 47:61–133.

Jin Yugan, Mei Shilong, and Zhu Zili. 1993. *The Potential Stratigraphic Levels for Guadalupian/Lopingian Boundary.* Permophiles, no. 23, IUGS.

Kozur, H. 1978. Beitrage zur Stratigraphie der Perms. Teil II: Die Conodontenchronologie des Perms. *Freiburg. Forschunger. H.c.* 334:85–161.

Lapkin, I. T., Blom, G. I., Grigoryev, N. V., Entsova, F. I., Zamerenov, A. K., Kalantar, I. Z., Kisnerius, Y. L., Kuchtinov, D. A., Lutkevich, E. M., Movshovich, E. V., Sokolova, E. I., Sterlin, B. P., Suveizdis, P. I., and Tverdochlebov, V. P. 1973. The Permian–Triassic boundary on the Russian platform, *Memoirs, Canadian Society of Petroleum Geologists* 2:150–7.

Leven, E. Y. 1979. The Permian Bolorskyi formation: strata, characteristics and correlation (in Russian). *Izv. Akad. Nauk CCCP* 1:53–65.

Leven, E. Y. 1980. The Permian Yachtashskyi formation: strata, characteristics and correlation (in Russian). *Tom.* 8:50–609.

Li Huamei and Wang Junda. 1989. *Palaeomagnetism of Permian–Triassic Boundary Section at Shangsi (Guangyuan County, Sichuan Province).* Geological Memoirs, 2(9). Beijing: Geological Publishing House.

Li Siguang (J. S. Lee) and Zhu Sen. 1932. *A Geological Guide to the Lungtan District, Nanking* (in English, with Chinese summary). Beijing: Nat. Res. Inst., Geol. Acad. Sinica.

Li Yongan, Li Qiang, Liu Yuliang, Zhang Zhengkun, and Jing Xing 1989a. Study on palaeomagnetism of Tarim craton since late Palaeozoic. *Xinjiang Geology* 7:2–78.

Li Yongan, Li Qiang, and Zhan Hui. 1991. Palaeomagnetism and several geologic tectonics of Xinjiang (in Chinese, with English summary). *Xinjiang Geology* 9:23–9.

Li Yongan, Li Yanpin, Gao Zhenjia, Li Qiang, Zhai Yongjian, Zhang Zhengkun, Sharps, R., McWilliams, M., and Cox, A. 1989b. Sinian palaeomagnetic research of Aksu-Keping region, Xinjiang, China (in Chinese, with English summary). *Xinjiang Geology* 7:79–88.

Li Zishun and Zhan Lipei. 1989. Permian biostratigraphy and event stratigraphy in N. Sichuan and S. Shaanxi provinces (in Chinese, with English summary). Geological Memoirs, 2(9). Beijing: Geological Publishing House.

Lin Wanzhi, Shao Jian, and Zhao Zhangyuan. 1984. Palaeomagnetic features of Sino-Korea plate in late Palaeozoic era (in Chinese, with English summary). *Geophysical and Geochemical Exploration* 8:5.

Lu Yanhao. 1956. The Permian in Liangshan, Hanzhong, with a discussion on the subdivision and correlation of the Permian in S. China (in Chinese, with English abstract). *Acta Geological Sinica* 36:2.

Mei Shilong, Jin Yugan, and Wardlaw, B. R. 1993. Succession of conodont zones from the Permian "Kuhfeng" formation, Xuanhan, Sichuan and its implication in global correlation. *Acta Palaeontologia Sinica* 42:1–21.

Ouyang Shu. 1986. Palynology of Upper Permian and Lower Triassic strata of Fuyuan district, E. Yunnan (in Chinese, with English summary). *Acta Palaeontologia Sinica* 169: 1–122.

Rauser-Chernousova, D. M., et al. 1979. The Carboniferous–Permian boundary in the U.S.S.R. In *The Carboniferous of the U.S.S.R.*

Richthofen, F. von. 1882. *Das Chinas,* vol. 2. Berlin.

Rui Lin. 1981. The Permian system of Yangzi stratigraphic province (in Chinese, with English summary). *Journal of Stratigraphy* 5:263–75.

Rui Lin. 1983. On the *Lepidolina kumaensis* fusulinacean fauna (in Chinese, with English summary). *Bulletin Nanjing Institute of Geology and Palaeontology, Acad. Sinica* 6:249–70.

Rui Lin. 1986. On the Carboniferous–Permian boundary (in Chinese, with English summary). *Journal of Stratigraphy* 10:249–61.

Rui Lin, Zhao Jiaming, Mu Xinan, Wang Keliang, and Wang Zhihao. 1984. Restudy of the Wujiaping limestone in Liangshan, Hanzhong, Shaanxi province. *Journal of Stratigraphy* 8:179–93.

Sheng Huaibin. 1981. Late Carboniferous ammonites from N. Tianshan, Xinjiang (in Chinese, with English summary). *Bulletin, Institute of Geology, CAGS* 3:83–95.

Sheng Jinzhang. 1962. *The Permian System of China* (in Chinese, with English summary). Beijing: Science Press.

Sheng Jinzhang. 1963. Permian fusulinids in Guangxi, Guizhou, and Sichuan. *Palaeontologia Sinica, n.s.* 10.

Sheng Jinzhang, Chen Chuzhen, Wang Yigang, Rui Lin, Liao Zhuoting, Yuji Bando, Kenichi Ishii, Keiji Nakazawa, and Koji Nakamura. 1984. Permian–Triassic boundary in middle and eastern Tethys. *Journal of the Faculty of Science, Hokkaido University, ser. IV* 21:133–81.

Sun Yunzhu (Y. C. Sun). 1939. The uppermost Permian ammonoid fauna from Kuangxi and its stratigraphic significance (in Chinese). Paper presented at 40th anniversary, Peking University.

Sweet, W. C., Yang Zunyi, Dickins, J. M., and Yin Hongfu. 1992. Permo-Triassic events in the Eastern Tethys: stratigraphy, classification, and relations with Western Tethys. In *World and Regional Geology,* vol. 2, ed. W. C. Sweet, Yang Zunyi, J. M. Dickins, and Yin Hongfu. Cambridge University Press.

Wang Jianhua. 1978. Boundaries and fossil zones of Qixia formation in the Nanjing area (in Chinese). *Journal of Stratigraphy* 2:67–73.

Wang Zengji, Hou Hongfei, and Yang Shifu. 1990. *The Carboniferous System of China* (in Chinese). Stratigraphy of China, vol. 8. Beijing: Geological Publishing House.

Wu Wangshi and Zhang Linxin. 1979. The upper Carboniferous and its upper limit of Pu'an, Qinglong, Guizhou. In *Carbonate Strata of the Southwest Region* (in Chinese). Beijing: Science Press.

Xiao Weiming, Wang Hongdi, Zhang Linxin, and Dong Wenlan. 1986. *Early Permian Stratigraphy and Its Biotas* (in Chinese, with English summary). Guizhou: People's Publishing House.

Xiong Jianfei and Zhai Zhiqian. 1985. Biostratigraphy of

Carboniferous conodonts and fusulinids in the Black area (Ruya–Wanmo–Namu–Luoding), Guizhou (in Chinese). *Geology of Guizhou* 3:269–87.

Yang Jiduan, Qu Lifan, Zhou Huiqin, Cheng Zhengwu, Zhou Tongshun, Hou Jingpeng, Li Peixian, and Sun Suyin. 1984. Late Permian and early Triassic continental strata and fossil assemblages in northern China (in Chinese, with English summary). *Scientific Papers for Geological International Exchange* 1:87–9.

Yang Jiduan, Qu Lifan, Zhou Huiqin, Cheng Zhengwu, Zhou Tongshun, Hou Jingpeng, Li Peixian, and Sun Suyin. 1986. *Permian and Triassic Strata and Fossil Assemblages in the Dalongkou Area of Jimsar, Xinjiang* (in Chinese, with English summary). Geological Memoirs, 3. Beijing: Geological Publishing House.

Yang Jiduan, Qu Lifan, Zhou Huiqin, Cheng Zhengwu, Zhou Tongshun, Hou Jingpeng, Li Peixian, and Sun Suyin. 1988. Continental Permian–Triassic boundary and events (in Chinese, with English summary). *Geological Science* 2:366–74.

Yang Jiduan, Qu Lifan, Zhou Huiqin, Cheng Zhengwu, Zhou Tongshun, Hou Jingpeng, Li Peixian, and Sun Suyin. 1989. Nonmarine Triassic biostratigraphy and boundaries in China. In *Progress in geosciences of China (1985–1988): Papers of the 28th IGC,* vol. 3, pp. 21–4.

Yang Jiduan, Qu Lifan, Zhou Huiqin, Cheng Zhengwu, Zhou Tongshun, Hou Jingpeng, Li Peixian, and Sun Suyin. 1992. Classification and correlation of nonmarine Permo–Triassic boundary in China. In *Permo–Triassic Events in the Eastern Tethys,* pp. 56–9. Cambridge University Press.

Yang Zhengdong. 1985. Restudy on fusulinds in the Maokou limestone at Datieguan, Langdai, Guizhou (in Chinese, with English summary). *Acta Micropalaeontologia Sinica* 2:307–88.

Yang Zunyi, Yin Hongfu, Yang Fenqqing, Ding Meihua, and Xu Guirong. 1987. *Permian–Triassic Boundary Stratigraphy and Faunas of South China* (in Chinese, with English summary). Beijing: Geological Publishing House.

Yang Zunyi, Wu Shunbao, Yin Hongfu, and Zhang Kexin. 1991. *Geological Events of the Permo–Triassic Transitional Period in South China* (in Chinese, with English summary). Beijing: Geological Publishing House.

Yang Zunyi, Wu Shunbao, Yin Hongfu, Xi Guirong, and Bi Xianmei. 1993. *Permo–Triassic Events of South China.* Beijing: Geological Publishing House.

Yue Shengxun. 1929. *Geological Reconnaissance of West Kweichow.* Geological Survey of China, bulletin 12.

Zhan Lipei, Chen Yuling, Li Li, Chen Bingwei, et al. 1982. *The Permian System of China. 1: An Outline of the Stratigraphy of China* (in Chinese, with English summary). Beijing: Geological Publishing House.

Zhan Lipei and Li Li. 1984. On some problems of the Permian in China (in Chinese, with English summary). *Bulletin, Institute of Geology, CAGS* 9:169–81.

Zhang Linxin, 1983. On the Swine limestone (in Chinese, with English summary). *Journal of Stratigraphy* 7:84–90.

Zhang Shanzhen, Yao Zhaoqi, Mo Zhuanguan, and Li Xingwue. 1982. Subdivision and correlation of the Permian continental strata in China. In *Stratigraphical Correlation Chart of China with Explanatory Text,* pp. 171–93 (in Chinese). Beijing: Science Press.

Zhang Wenzhi, 1992. Magnetostratigraphy (in Chinese). In *The Carboniferous–Permian Boundary in China,* pp. 93–8. Beijing: Geological Publishing House.

Zhang Zhenghua, Wang Zhihao, and Li Changjin. 1988. *Permian Strata of South Guizhou* (in Chinese, with English summary). Guizhou: People's Publishing House.

Zhang Zhiming and Wu Shaozu. 1991. The Permian of Xinjing (in Chinese). In *The Palaeozoic of Xinjiang. 2: Stratigraphic Summary of Xinjiang,* pp. 329–482. Urumqi: Xinjiang People's Publishing House.

Zhang Zhuqi. 1985. *The Permian System of South China.* China Mining Institute, bulletin 1.

Zhao Jinke, Liang X. L., and Zheng Z. G. 1978. Late Permian cephalopods of South China. *Palaeontologica Sinica, n.s.* B12:1–194.

Zhao Jinke, Sheng J. C., Yao Z. Q., Liang X. L., Chen C. C., Rui L., and Liao Z. T. 1981. The Changsingian and Permian–Triassic boundary of South China (in Chinese, with English abstract). *Bulletin of the Nanjing Institute of Geology and Palaeontology, Academia Sinica* 1:1–85.

Zhao Xijin. 1980. Mesozoic vertebrates and strata in northern Xinjiang (in Chinese). *Bulletin of the Institute of Palaeontology and Palaeoanthropology, Academia Sinica* A15:1–119.

Zhou Tieming, Sheng Jinzhang, and Wang Wujing. 1987. Carboniferous–Permian boundary strata and fusulinid zonation of Xiaodushan, Guan'an, Yunnan (in Chinese, with English summary). *Acta Micropalaeontologica Sinica* 4:123–45.

Zhou Zuren. 1982. Early Qixiaian fusulinid *Schwagerina cushmani* fauna of early Permian in SE Yunnan (in Chinese, with English summary). *Acta Palaeontologica Sinica* 21:225–48.

8 Indosinian Tectogeny in the geological correlation of Vietnam and adjacent regions

J. M. DICKINS and PHAN CU TIEN

The concept of "Indosinian" was first introduced by Fromaget (1931, 1934) for folding in Indochina that took place during the Anthracolithic (a term, no longer commonly in use, for the Carboniferous and Permian together) and the Triassic. It is now just over 50 years since that concept was elaborated in an account of the work of the Service Géologique de l'Indochine (Fromaget, 1941).

The term "Indosinian" has been widely used, with many different meanings, for folding events in southern, central, southeastern, and eastern Asia that took place during the Permian and Triassic, or during various parts of that time interval. More recently it has been coupled (Dickins, 1985, 1988a,b) with the term "Hunter-Bowen" from eastern Australia to describe a period of worldwide folding and compression, beginning at the mid-Permian, involving a twofold division of the Permian in the Russian type sequence, and continuing to the end of the Triassic. There have been many recent studies of the geology of Indochina, with abundant new data. Within the framework of IGCP Project 272, and with related information from neighbouring countries, we here return to the issue of the Indosinian Tectogeny. This chapter describes some newly collected data related to the issue and presents a preliminary geological correlation for that noteworthy period on a fairly large scale in Southeast Asia. The authors express their sincere thanks to Professor K. Mori, Professor Vu Khuc, Dr. I. Metcalfe, Professor K. Nakazawa, Professor Yin Hongfu, Dr. Y. D. Zakharov, and others for their valuable recommendations and for providing information related to the topic of study.

Indosinian Tectogeny and some related studies

The first descriptions of the geological structure and lithosphere of Indochina were provided by the work of E. Suess, E. Blondel, and many other French geologists in the early years of the twentieth century. They advanced the concepts of the "Annamia" and the South China stabilized platform, the Cambodia Geoblock, and so forth. The Indosinian Tectogeny and "Indosinias" (the "tectono-stratigraphic units" of the Indosinian structural element) were considered by Fromaget as reflecting the tectonic characteristics of Indochina during the "Neotriassic." The structural plan of the Neotriassic is conspicuously expressed in the basic structural framework of the region.

According to Fromaget, the Indosinian Tectogeny, which formed the Indosinides, began with pulsating movements in the middle Anthracolithic, between the early and late Moskovian, postdating the Hercynian, to which he ascribed the Annamides. It consisted of a phase of undulating folding in the Permian, a second phase between the middle Triassic and Carnian, and a folding phase that began at the end of the Carnian, considered the first maximal orogeny. The fourth phase, together with the transgression in the Norian, is also called the "charriage" (thrust) phase, corresponding with the second maximal orogeny that completed the formation of the Indosinian folds before the beginning of the Jurassic (Fromaget, 1931). Recent mapping work (Phan Cu Tien, personal communication) has identified the main part of Fromaget's Upper Moskovian limestone in the northwestern part of Vietnam as Middle Triassic, so that the pulsating movements of Fromaget apparently refer to Permian and Triassic movements. The Indosinian Tectogeny was described by Fromaget (1941, figs. 16–18) as the result of compression between Gondwanaland* and Eurasia. That process led to compression of the internal elements of the structural frame, creating folds in geosyncline zones and causing the development of charriages in some locations, especially the Songda structural element.

The concepts of Fromaget were developed further in subsequent works. Nevertheless, many of his studies were overlooked by Saurin and Fontaine, who wrote during a period of rapid development of geosynclinal and mobilist theories.

Saurin accepted nearly all the conclusions of Fromaget, but he described the geostructural frame of Indochina in a more simple way. According to Saurin, the extensive complex folding of the

*Fromaget used "Continent de Gondwana." The term "Gondwana" was first used for the sequence in peninsular India extending apparently from low in the Permian to the mid-Cretaceous. The term "Gondwana-land" ("Gondwanaland" in this chapter) was used later by Suess to describe a continental area that he regarded as extending along the southern margin of a Mesozoic southern Asian sea that he named Tethys. The land to the north he called Angara-land. There seems considerable merit in retaining that usage to avoid the confusion of two distinctly different meanings for the term "Gondwana" (Dickins, 1994).

basement played the main role, but it was complicated by thrusts as a result of lateral movements.

There have been some differences of opinion regarding the role of the geosynclinal folding complexes, but all later studies of the geology of Indochina and its regions have remarked on the specific character of the early Mesozoic (Indosinian). Those studies include the works by Dovjikov (1965), Ngo Thuong San and Rezanov (1965), Tran Duc Luong (1975), and others, who referred to a Mesozoic (Indosinian) Orogeny and Mesozoic formations – formations of revived platforms or reworked platforms or formations belonging to particular structures of eastern Asia – and to rifts of different origins. In some works, the Indosinian geosynclinal regime was denied or was limited to the north-central, western, and northwestern parts of Vietnam or to the Songda Depression (Nguyen Xuan Tung, 1982; Gatinsky et al., 1984; Le Duy Bach, 1985).

In the period 1965–1975, geologic studies were conducted in the northern part of the territory, where the undifferentiated Carboniferous–Permian formations consist of limestones of considerable thicknesses.

With such characteristics, the Carboniferous–Permian strata were considered to reflect a stable period in the geological development of Vietnam. The timing of the origin of the Triassic depressions was determined to have been at the beginning of the Triassic or at the end of the early Triassic, and the Triassic marine sedimentary accumulation was considered to have been completed by the time of the pre-Norian folding, on the basis of the fossils collected (Dovjikov, 1965). Since 1975, with the extension of geological investigations to South Vietnam, and because of the effect of a new concept – global tectonics – various works on geology have taken a different direction.

There has been a tendency to consider the Indosinian Tectogeny as corresponding with the rifting process of the continental crust in the early Devonian or in the late Permian (Van Duc Chuong, 1983; Le Duy Bach, 1985). There have been analyses of the Indosinian Tectogeny and geological formations according to the concepts of global tectonics (Murphy, 1975; Le Thac Xinh, 1981) and palinspastic reconstruction of the Indochina continent (Nguyen Xuan Tung, 1982). The nature of the Indosinian Tectogeny and the geological correlations between the Indosinian Tectogeny and orogenic folding in the adjacent territories have been discussed by Dickins (1985), Metcalfe (1986), and Workman (1977).

In Vietnam, correlated studies of the geodynamic regime, palaeomagnetism, and palinspastic reconstructions of the late Palaeozoic–early Mesozoic structural elements are being promoted in order to clarify, step by step, the outstanding problems.

Geological events during the late Palaeozoic–early Mesozoic in Vietnam

The geological events during the late Palaeozoic–early Mesozoic in Vietnam are described for each terrane. The terranes are understood as accumulation basins or parts of accumulation basins formed in the period studied. The differentiation is based on the lateral distribution of geological formations of the same age, their accumulation conditions, and environment, taking into account the characters of geological development in earlier and later periods. The following terranes, formed during the late Palaeozoic–early Mesozoic, are delineated: the Northeast Vietnam Segment, a component part of the South China Terrane; the Indosinian Geoblock, comprising foldbelts and volcanic belts as well; the Muong Te Zone, the eastern extension of the Yunnan-Malaysia Terrane; and the Songda structural element, whose boundaries and formations will be partially discussed later (Figure 8.1).

Northeast Vietnam Segment

In the Northeast Vietnam Segment, the Carboniferous–Permian formations are characterized by homogeneous, thickly bedded limestones rich in faunas; in some places, in the lower parts, there are intercalated thin beds of siliceous rocks (Figures 8.2 and 8.3). The sedimentary interruption between the Carboniferous–Permian limestones and the older formations was determined mainly on the basis of the fossils collected. In Ha Lang Cat Ba, late Devonian sediments consisting of limestones and manganese-bearing siliceous rocks unconformably underlie the Carboniferous limestones. In the underlying formation, the fossils reported include *Quasiendothyra communis*, *Q. kobeitusana*, *Palmatolepis gracilis*, and *Spathognathodus inortatus*, of the late Fammenian. In the overlying formations, the main fossils are foraminifera pertaining to the *Dainella* or *Chernyshinella-Tournayella* and *Endothyranopsis* zones of the Lower Carboniferous, the *Profusulinella* and *Fusulinella* zones of the Middle Carboniferous, the *Obsoletes*, *Triticites*, and *Schwagerina* of the Upper Carboniferous, and the *Robustoschwagerina*, *Misellina*, *Cancellina*, and *Neoschwagerina* zones of the Permian. Those faunas are widespread, although to different degrees, in the various outcrop areas of the Carboniferous–Permian in the Northeast Vietnam Segment.

A major geological event has been clearly identified at the beginning of the late Permian (Dzhulfian stage), when a new cycle of volcano-sedimentary accumulation began. That filled up the depressions of Songhien and Anchau and covered the older, folded basement strata in the peripheries of those depressions. An Upper Permian bauxite bed, a few metres to 10 m thick, composed mainly of gypsite and diaspore, rests unconformably on the weathered and eroded surface of the Upper Palaeozoic limestone. The bauxite bed is overlain by siliceous terrigenous rock, with limestone and siliceous limestone on top, containing *Leptodus*, *Palaeofusulina*, and *Reichelina*, which characterize the upper part of the Permian in many areas of Southeast Asia.

The Triassic sediments overlying the Upper Permian limestone consist of siltstone, schist, and calcareous schist, containing *Claraia*, *Eumorphotis*, and *Glytophyceras*. Passing upward, the sediments have flyschoid characteristics, with intercalated siltstone, sandstone, and shale, and marl and limestone in some places. Volcanic rocks are noted in many of the outcrop areas of the Lower and Middle Triassic, consisting mainly of rhyolite,

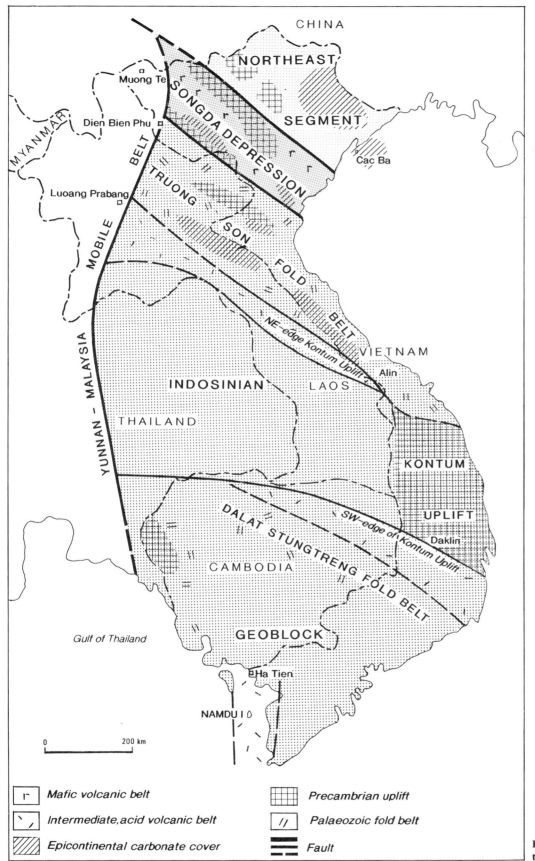

Figure 8.1. Late Palaeozoic structures and palaeogeography.

Mafic volcanic belt

Intermediate, acid volcanic belt

Epicontinental carbonate cover

Precambrian uplift

Palaeozoic fold belt

Fault

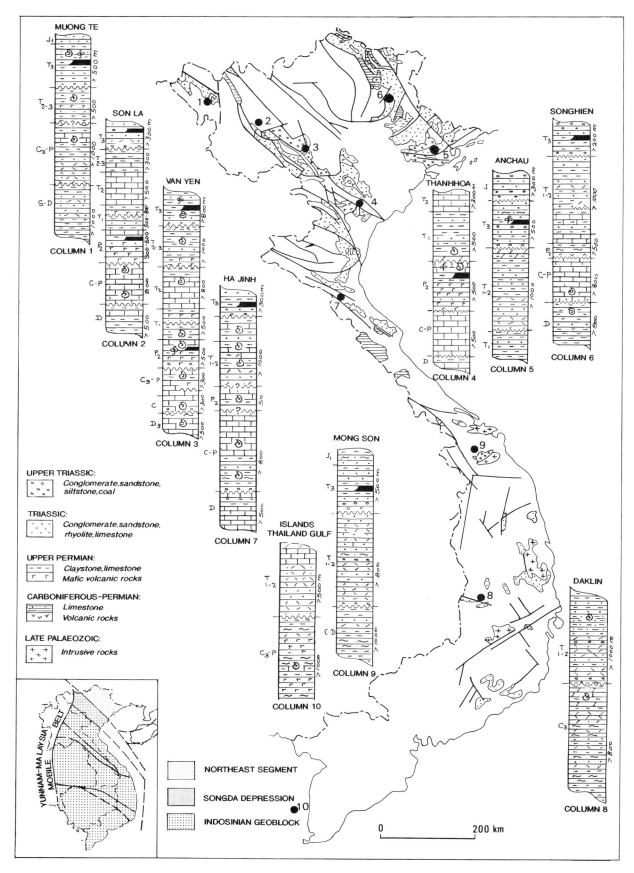

Figure 8.2. Upper Palaeozoic–Lower Mesozoic formations in Vietnam.

dacite, rhyolite-porphyry, and tuff. Along the Thatkhe-Langson deep-seated faults, some basalt is encountered. The Lower Triassic is differentiated on the basis of fossil complexes of "*Otoceras,*" *Flemingites-Owenites,* and *Tirolites.* In the Middle Triassic, fossils corresponding to the *Kellnerites samneuensis* complex of the upper part of the Anisian and the *Trigonodus-Costatoria* complex of the Ladinian are noted in both the Songhien and Anchau depressions. In those depressions, shale and marl containing fresh-water fauna and fish overlie the Ladinian sediments and are presumed to be Carnian in age. The regular distributions of volcanics in different outcrops, as well as the changes in their volume in cross sections, serve as a basis for delineating geologic events during the Olenekian of the Lower Triassic and the Anisian and Ladinian of the Middle Triassic in the northeast (Tran Van Tri, 1979). On the basis of newly reported data, the second prominent geological event during the Permian–Triassic period in the Northeast Vietnam Segment occurred at the end of the Carnian, with the appearance of gabbro and high-alumina granite associations and coal-bearing continental formations filling premontane depressions and intracontinental grabens.

The intrusions of gabbro complexes are in the form of small massifs distributed along faults. Beside olivine gabbro, which is the main component of the complexes, there are also gabbro pegmatites, anorthosites, and pyroxenites. In their composition (SiO_2 49.1–49.3%, total alkali 2–3%), sodium (Na) dominates over potassium (K). The high-alumina granite association is variably developed and also is distributed along faults. The main phase that has been distinguished consists of granite-aplite that contains garnet and pegmatite that contains tourmaline. The complex is characterized by a moderate aluminium (Al) content and a total alkali content of 6.9%, with K_2O clearly predominating over Na_2O.

The Norian–Rhaetian coal-bearing formations have thicknesses varying from a few hundred metres in the intracontinental grabens to 4,000–5,000 m in the premontane depressions. Within the latter there are seams of coarse clastic sediments and coal seams that alternate with each other in alluvial and proluvial rhythmic groups. The number of coal seams sometimes reaches 30, and their thicknesses vary from a few tens of centimetres to some tens of metres. Besides *Gervillia praecursor* and *Bakewellia,* the sediments are rich in flora, with the well-known Hongai assemblage, which marks the continental environment of the Northeast Vietnam Segment in the Mesozoic. The coal-bearing sediments are typified by a new structural character, unconformably transgressing over the older formations, including the granite association described earlier.

Indosinian Geoblock

In the Indosinian Geoblock, the volcano-sedimentary accumulation during the late Palaeozoic–early Mesozoic was differentiated laterally into the Truongson and Dalat foldbelts on two sides and the Kontum Uplift in the centre. In the Truongson Foldbelt, the Upper Palaeozoic terrigenous-carbonate and carbonate sediments have increasing contents of limestone from the edge of the Kontum Uplift to the north. In the Lakhe area, the Lower Carboniferous sediments are composed of basal conglomerate layers that also overlie the Upper Devonian manganese-bearing sediments and older sediments similar to those of the Northeast Vietnam Segment. In some places, the terrigenous sediments are up to 700–800 m thick, consisting of shale, chert, and sandstone. Along the strike, in the territory of Laos in the corresponding stratigraphical position, there are some coal seams. In the Lower Carboniferous there is also limestone, with thicknesses varying according to the variations of the terrigenous rocks. In the Lakhe and Quydat areas, the Lower Carboniferous fossils include *Posidonia (Posidonia) protobecheri, Archegonus (Phillibole) priscus, Megachonetes* cf. *zimmermari,* and *Rugosochonetes hardrensis* in the terrigenous sediments of the lower part of the section, and *Dainella chomatica, Endothyranopsis crassasphaerica, Plectogyra prisca, Polygnathus bischoffi,* and *Neopriniotus tulensis* in the limestones of the upper part. In the coal-bearing sequences in Laos, the fossils reported are *Stigmaria ficoides, Sigillaria brardi, Calamites goepperti,* and *Pecopteris dentata* (in Salavan), and *Asterophyllites longifolius* and *Sigillaria brardi* (in Namlik).

The Carboniferous–Permian limestones rest discordantly on the foregoing sequences and contain foraminifera belonging to the *Profusulinella, Fusulinella, Triticites, Schwagerina,* and *Neoschwagerina* zones – middle Carboniferous to middle Permian in age.

In the Upper Permian, terrigenous rocks with brachiopods and bivalves predominate. In the small areas of shale outcrops there are *Leptodus* (Tran Thi Chi Thuan and Fontaine, 1968), and in the limestones, *Nankinella* and *Colaniella.*

The recently obtained data do not confirm the role of the Lower Indosinian delineated by Fromaget within the Kontum Uplift. The Upper Carboniferous–Permian volcano-terrigenous formations are manifest only in the Alin Band at the northern edge of this uplift, and the Daklin Band at the southern edge. In the same strike direction, those volcano-terrigenous formations have considerable thicknesses and extensive exposures in Khanngkhay and Xiengkhouang, in Laos, as well as Strungtreng and Tbenmeanchay, in Cambodia. The characteristic rock association consists of polymictic sandstone containing pebbles of andesite. The effusive rocks described have thicknesses varying from 300–400 m to more than 1,000 m. The faunas in the interbeds include *Schwagerina, Pseudofusulina,* and *Verbeekina* in the limestone, together with brachiopods and bryozoans in the schist, pertaining to the late Carboniferous–Permian. The effusive rocks usually are characterized by high contents of iron (Fe) (especially FeO, at 5.5%) and Na_2O, and low contents of SiO_2 and Al_2O_3, belonging to the calc-alkaline group of andesite-basalts. As distinguished from the Northeast Vietnam Segment, in the Indosinian Geoblock the geological events of the late Carboniferous–Permian are rather clearly noted, according to the appearance of volcanics associated with gabbro-diorite/diorite and granodiorite-granite, developed both on the foldbelts and on the Kontum Uplift. The intrusions have moderate-to-low acidities; total alkali reaches 4.5–5.5% in

		Northeast Vietnam	Northwest Vietnam (Songda Depression)	Northern Central Vietnam		Southern Central & South Vietnam
UPPER TRIASSIC		HONGAI and VANLANG SUITES: *Coal-bearing sediments containing Floras and some Bivalves* / *Basal conglomerates*	SUOIBANG SUITE: *Coal-bearing sediment: containing Juravites, Halobia, Burmesia, Parathisbites, Floras* / *Basal conglomerates*	NONGSON, DONGDO SUITES: *Coal-bearing red coloured formation containing Bivalves, Floras* / *Basal conglomerates*	?	
MIDDLE TRIASSIC		MAUSON SUITES: *Sandstones, shales, marls containing Freshwater Bivalves* — NAKHUAT SUITE: *Volcanics, tuffs, siltstones, shales, marls containing Kellnerites, Costatoria, Trigonodus* ?	NAMMU, SONGBOI FORMATIONS: *Siltstones, limestones, volcanics containing Halobia, Daonella, Discotropites* — NAMTHAM SUITE: *Shale, marl, limestones containing Daonella, Posidonia* — DONGGIAO SUITE: *Limestones containing Cuccoceras, Paraceratites*	? QUYLANG SUITE: *Siltstones, limestones containing Costatoria, Trigonodus* — DONGTRAU SUITE: *Sandstones, shales, volcanics, limestones, Achrochordiceras, Costatoria*		CHAUTHOI and HONNGHE FORMATION: *Limestone, shales, volcanics containing Balatonites, Daonella* ? — MANGGIANG FORMATION: *Volcanics, tuffs, sandstones, siltstones containing Entolium*
LOWER TRIASSIC		LANGSON and SONGHIEN SUITES: *Siltstones, sandstones, marls volcanics, tuffs, shales, containing Claraia, Glyptophiceras, Flemingites, Kashmirites*	CONOI SUITE: *Sandstones, siltstones, limestones, volcanics, tuffs containing Claraia, Lytophiceras, Eumorphotis, Tirolites, etc.*	*Basal conglomerates* ?		SONGSAIGON SUITE: *Sandstones, shales, marls containing "Otoceras," "Ophiceras," Claraia, Eumorphotis, Gyronrites*
UPPER PERMIAN		DONGDANG SUITE: *Limestones, shales containing Palaeofusulina, Leptodus* / *Basal beds of bauxite, breccias,* / *Eroded surface*	YENDUYET SUITE: *Cherts shales containing Leptodus, Gigantopteris limestones containing Palaeofusulina, Reichelina* — CAMTHUY FORMATION: *Basalts, porphyrites*	*Limestones containing Codonofusiella, Pachyphloia, shales, Leptodus, Meekella*		TATHIET FORMATION: *Limestones containing Palaeofusulina, Neoendthyra, Reichelina*
LOWER–MIDDLE PERMIAN		Upper part of BACSON SERIES: *Limestones containing Neoschwagerina, Verbeekina*	Upper part of BACSON SERIES and BANDIET FORMATION: *Limestones containing Triticites, Verbeekina basalts, porphyrites, tuff in Songda Depression*	*Limestones containing Robustoschwagerina, Verbeekina, Neoschwagerina*		HATIEN FORMATION, DAKLIN FORMATION: *Limestones containing Neoschwagerina, Parafusulina, and andesites, rhyolites, tuffs*
UPPER CARBONIFER.	?	Lower and Middle part of BACSON SERIES: *Limestones, containing Dainella, Endothyranopsis, Millerella, Profusulinela Fusulinella, Obsoletes, Triticites, Schwagerina.*	Lower and Middle part of BACSON SERIES: *Limestones containing Endothyranopsis, Millerella, Profusulinella, Fusulina, Obsoletes, Triticites, Schwagerina*	? MUONGLONG FORMATION: *Limestones containing Endothyranops, Profusulinella, Fusulinella, Obsoletes* ?		?
MIDDLE CARBONIFER.						
LOWER CARBONIFER.		*Basal conglomerates*	*Basal conglomerates*	LAKHE SUITE: *Limestones, cherts, coal shales containing Dainella etc.* / *Basal conglomerates*		*Cherts, sandstones, shales*
UPPER DEVONIAN		TOCTAT SUITE: *Manganese bearing limestones, cherts containing Quasiendothyra, Palmatolepis*	BANCAI SUITE: *Manganese limestones, cherts containing Quasiendothyra, Palmatolepis*	*Limestones containing Palmatolepis, Polygnatus, Cyrtospirifer*		

the gabbro-diorite, 5.5–7.3% in the diorite-granodiorite, and 0.5–8.6% in the granitoid, with Na_2O sharply predominating over K_2O. U-Pb and K-Ar radiochronological ages from nearly 30 samples vary from 250 to 360 Ma.

The Upper Permian formations have been described only from small exposures of limestone rich in *Palaeofusulina, Reichelina,* and *Neoendothyra* along the upper course of the Saigon River, where they show conformable relationships, the Triassic shale containing fossils of the *"Otoceras (Metotoceras)"* assemblage (Tran Duc Luong and Nguyen Xuan Bao, 1980). In the Indosinian Geoblock, the stratigraphical relationships of the Upper Permian limestone with the Carboniferous–Permian formations, as well as the Triassic volcanics, are not as clear as in the Northeast Vietnam Segment, despite the undoubted similarities in sedimentation and in some fossils in the two regions. The Triassic formations, perhaps mainly the Lower and Middle Triassic, consist of basal conglomerate, sandstone, siltstone, rhyolite, dacite, and tuff. The limestone lenses have variable thicknesses, from some tens of metres in the south to some hundreds of metres in the north-central area. The volcanic rocks usually are in the form of interbeds some 10 m thick and sometimes can be seen in the whole cross section, as in the Kontum Uplift. In the terrigenous and carbonate sediments, the fossils belong to the assemblages of *Balatonites balatonicus, Paraceratites trinodosus,* and *Costatoria-Trigonodus* or *Daonella.* There are also Middle Triassic foraminifera, with *Glomospirella* and *Ammodiscus* in limestone lenses. These seem to be the last formations of the Triassic that contain marine fauna recognized in the Indosinian Geoblock (Tran Duc Luong and Nguyen Xuan Bao, 1980). A geological event that occurred during the early–middle Triassic was the appearance of volcanics, belonging to the dacite-rhyolite association, with small amounts of calc-alkaline andesite and latite. They are characterized by normal Fe and K contents, similar to the subalkaline, high-alumina rhyolite-trachyte. The characteristic extrusions described are widespread in the Kontum Uplift and are scattered in the foldbelts on both sides. The intrusive association consists of biotite-granite, granite-granophyre, granosyenite, and granodiorite-porphyry, as well as dykes and veins of granite, aplite, pegmatite, and lamprophyre. Isotope dating indicates an interval of 177–228 m.y. The most characteristic intrusive activity in the Indosinian Geoblock took place in the Carnian, bringing associations of pyroxenite-norite and high-alumina granitoids. The first association, of gabbro, gabbro-diorite, diorite, and pyroxenite, usually is in the form of small massifs distributed along the faults. In the second association, the intrusions are multiform. The principal phase includes adamellite and melanocratic biotite-granite, and the vein phase includes granite-aplite that contains garnet and pegmatite that contains tourmaline. The intrusive rocks are characterized by moderate alumina content (13%) and total alkali content (6.9%), with K_2O clearly dominating over Na_2O.

On the basis of field observations, these intrusions seem to be late Triassic in age; however, isotope dating indicates ages concentrated in the range of 232–242 Ma, thus conflicting with some of the data mentioned earlier.

The Norian–Rhaetian coal-bearing formations lie unconformably on the formations described earlier, including the intrusions. That sequence also contains the floras of the Hongai assemblage and is characterized by red-coloured coarse clastic formations intercalated with coal seams and having thicknesses up to 1,000 m.

Songda Depression

In the Songda Depression, the geological events that took place were not completely similar to those in the two terranes described earlier, although the Carboniferous limestone is also underlain by late Devonian manganese-bearing formations in the two margins of the depression. The first event recognized in the depression involved the appearance of a mafic volcanic formation in a narrow band at the centre of the depression (Fromaget, 1927; Phan Cu Tien, 1977). The formation is composed of alkaline olivine basalt, albitized basalt-porphyrite, and tuff intercalated with sandstone in the lower part, then passing into limestone. The clastic rocks contain *Propinacoceras actubense* and *Dytomopyge,* together with bivalves, whereas the fauna of the limestone includes *Triticites schwageriniformis, Misellina,* and *Neoschwagerina.* The thickness of the formation reaches 500–800 m. That Permian geological event is recognized on the basis of the intrusive association of serpentinite, dunite, diorite-granodiorite, and aplite. Belonging to the first association, the dunites, usually serpentinized, occur in the form of veins and lenses that range from 1–2 m to some tens of metres thick, closely related to the mafic volcanics described earlier. They also penetrate conformably with the structural trends in the exposed Palaeozoic formations. Belonging to the second association, the intrusions are also in the form of small massifs and intrude the Palaeozoic formations in the southeastern part of the depression and along its northern edge, where it is bounded by the Northeast Vietnam Segment. The intrusions consist of diorite, granodiorite, and granite-aplite characterized by siliceous saturation and a low alkali content in the granodiorite, but a high content in the granite-aplite. Isotope ages are in the range of 280–320 Ma for the second association.

Corresponding to the geological events in the Dzhulfian, in the Songda Depression notable amounts of mafic volcanics appeared. They covered the Carboniferous–Permian formations in the centre of the depression, and even the old formations on both sides of the depression.

The Upper Permian consists of two parts. The lower part contains tuffaceous conglomerate, olivine basalt, tholeiite, and tuff, passing upward into chert and shale, with thin beds of coal,

Figure 8.3 (opposite). Geologic events in the late Palaeozoic–early Mesozoic in Vietnam.

whereas the upper part consists mainly of limestone. Fossils reported from the limestone include *Palaeofusulina, Reichelina,* and *Codonofusiella,* similar to those in the sediments of the same age in the Northeast Vietnam Segment. In the shale, besides the plants *Gigantopteris nicotianaefolia, Pecopteris* sp., and *Lobatannularia multifolia,* there are *Leptodus, Oldhamina,* and *Strophomenida.* The thicknesses of the Upper Permian sediments vary from 300–500 m to more than 1,000 m.

The Upper Permian volcanics in the Songda Depression are characterized by low contents of SiO_2 and Al_2O_3 and high contents of TiO_2 (2.9–3.2%), Fe, MgO, and alkali. Mafic volcanics are intercalated with the terrigenous-carbonate sediments of the early, middle and late (Carnian) Triassic. With thicknesses of 3,000–4,000 m, the pre-Norian formations of the Songda Depression usually are schistose and folded, and thrusting can be seen, accompanied by faults along the fold axis in the centre of the depression.

The fossils found in the lowest part of the Triassic are *Claraia, Eumorphotis,* and *Lyptophiceras.* In the upper part there are abundant *Tirolites seminudas, Paraceratites binodosus,* and *Protrachyceras costulatum.* The abundance of *Halobia* together with *Torquistites* in the coral-reef limestone perhaps indicates that it represents the last part of the pre-Norian marine deposition in the Songda Depression. In the Triassic, besides the intrusive association of peridotite-gabbro-diabase, diabase, and granophyre that is related to the volcanic rocks, there is an association of melanogranite and two mica-granites of late Triassic age cropping out at the southeast of the depression. The first association is similar in character to the related volcanics, and the second is characterized by oversaturation of SiO_2 and a medium content of alkali. The isotope age for the fine-grained melanogranite is 228 Ma.

The Norian–Rhaetian coal-bearing formations in the Songda Depression transgressively overlie the Palaeozoic and older Triassic formations. In the lower part, the formations consist of layers of marl and shell limestone intercalated with shale and coal seams and contain *Gervillia praecursor, Halobia distincta,* and *Paratibetites tornquisti.* The youngest marine Triassic formation in Vietnam is found in the Songda Depression. In the middle part of that formation, the sediments consist of coarse rocks; coal seams are intercalated with the sediments in the upper part. The thickness ranges from 300–400 m to nearly 1,000 m.

Muong Te Zone

In the eastern extension of the Yunnan-Malaysia Terrane – in the Muong Te area – the Upper Palaeozoic formations consist only of terrigenous sediments and intermediate and acidic volcanics similar to those in the south of the Kontum Uplift, with thicknesses of 1,500–2,000 m; in the upper part, limestone lenses contain *Misellina ovalis* and *Pseudofusulina complicata* (Dovjikov, 1965). An intrusive association of diorite and granodiorite-granite is distributed in various exposures. The intrusive rocks intersecting the Carboniferous–Permian formations are covered by Upper Triassic–Jurassic coal-bearing red-coloured sediments.

Those sediments have a monoclinal occurrence, spread in mountain reliefs and extending toward Laos.

Indosinian Tectogeny in the recent geological correlation in Vietnam and adjacent regions

In the Upper Palaeozoic–Lower Mesozoic sequences in Vietnam, different terranes are distinguishable on the basis of petrographic structural analysis and the regularity of their lateral distributions. The earliest geological event is recognizable throughout the whole region. Although the new data have not revealed any of the Moskovian intrusions described by Fromaget, the Lower Carboniferous terrigenous formations and subsequently the Upper Carboniferous–Permian volcanics and Permian intrusive associations have been confirmed. In the Northeast Vietnam Segment, the Carboniferous–Permian accumulations characterize the continental-shelf formation. Impressions of a late Carboniferous–Permian event are recognized only in some depressions related to deep-seated faults. In the Songda Depression, that event was marked by strong basaltic activity.

The effects of the Indosinian Tectogeny are apparent in a major event associated with the development of the Dzhulfian Stage (Upper Permian of Vietnam). That event is widely manifested in almost the whole territory of Vietnam and adjacent countries, although to different degrees. It is marked by widespread unconformity and volcanic activity. In the Northeast Vietnam Segment it began with the formation of intracontinental depressions (e.g., Songhien, Anchau, Samneua), accompanied by compression in the Dzhulfian. Terrigenous and volcano-terrigenous formations covered the whole of the areas of Proterozoic and Palaeozoic basement. Formation of the new depressions began prior to the biological changes associated with the Permian–Triassic boundary in Vietnam and South China. According to the fossils, large-scale volcanism seems to have occurred in the early Triassic, or even at the beginning of the middle Triassic. Marine transgression perhaps reached its maximum in the middle Triassic, coinciding with the widespread development of the *Paraceratites* and *Balatonites* assemblages in the thick-bedded limestone and limestone intercalated with black shale. By the end of the Carnian, an important orogenic phase of the Indosinian Tectogeny accompanied the wide intrusive activities and the deposition of coal-bearing molasse, mainly of continental origin, filling the late Permian–Triassic depressions in Vietnam, the southeastern part of Tethys. A continental regime had formed by the end of the Triassic in most of Vietnam and adjacent regions, eliminating the previous westerly marine connection with Tethys. The early Jurassic marine environment in Ban Don, Huunien, a narrow inlet from the southern ocean, was the only exception in Indochina.

The Songda Depression had special features during the late Palaeozoic–early Mesozoic that distinguished it from the terranes on either side. Especially important was the development of intense mafic (and ultramafic) intrusive and extrusive activities in the Permian and Triassic. The boundary of the depression in the late Carboniferous–Permian can be delineated on the basis of the different distribution of mafic volcanics along the Songma and Songhong faults. The faunas provide some indication of distinctive tectonic activities at about the Carboniferous–Permian boundary and in the mid-Permian (of the twofold subdivision of the Russian type sequence). As in other parts of Vietnam, the Dzhulfian event is well reflected and is distinguished by mafic igneous activitiy, as well as more acidic activity, as registered in other parts. Because of its distinctive characteristics, the depression has been regarded as a suture, but in both the Permian and the Triassic, as described herein, all the terranes of Vietnam not only had floral and faunal features in common but also had many common features of geological development. The Songda structure can be correlated with some structures in South China that show parallel biological developments.

Considering its entire development (from the Cambrian apatite-bearing formations at the northern border with China, through the medial thermodynamic metamorphic type of the Devonian sediments in the Takhoa Anticline, to the formations described in this chapter), the Songda structure seems to have distinctive characteristics that could represent allochthonous features, but that remain to be satisfactorily explained.

In comparing the geological developments in adjacent regions, we see that the clearly apparent, first distinctive event of the late Palaeozoic–Triassic, the Dzhulfian event described earlier, which involved all of Vietnam, coincided with the boundary of the Maokou Formation and the Emeishan Basalt Formation in China (Jin Yugan, 1993; Yang Zunyi, Liang Diny, and Nie Zetong, 1993), as well as the boundary between the Chuping and Kodiang Limestones in the western basin of Malaysia, or the boundary between the Mersing Formation and the Mursan and Ledili Volcanics in the eastern depression of that country (Tan and Khoo, 1978). It corresponds to a major event in Japan that marked the boundary between the Middle and Upper Permian, as those terms are traditionally used in Japan. In the Okiyoshi area of southwestern Honshu, the carbonate platform that had existed through the late Carboniferous and early and middle Permian in Japan was tectonically broken up and in the late Permian was deluged with a thick blanket of volcano-clastic and volcanic material (Dickins, 1991; Sano and Kanmara, 1991). In the Kitakami Mountains of northeastern Honshu, at the beginning of the Dzhulfian, the middle Permian carbonate platform was destroyed by strong tectonic activity at the boundary of the Kanokura Series and the Toyoma Series, and that was followed by a clastic sequence (probably with a strong volcanic component) in which lenticular limestone was developed. Much debris, including large blocks from the Kanokura, was incorporated in the overlying clastic sequence.

In the Kitakami Mountains, two earlier distinctive events in the Permian are apparent. An unconformity is found between the Lower Permian and the underlying sequence, in this case the Lower Carboniferous, and there is an unconformity between the Kanokura and the underlying Sakamotozawa Series (Lower and Middle Permian, as traditionally used in Japan). That corresponds closely to the mid-Permian of the twofold Russian subdivision, and in the Kitakami Mountains the influx of

volcanic material interrupted the carbonate development. Some information, especially from the Songda Depression, suggests that those two developments may be reflected in Vietnam. The mid-Permian event in eastern Australia has been described by the name Hunter-Bowen (Indosinian) Orogenic Phase and is recognized also in New Zealand, southern Asia, Japan, southern Europe, and many other parts of the world (Dickins 1985, 1987, 1988a,b, 1992). The beginning of that phase was associated with a major worldwide transgression. The Midian–Dzhulfian boundary is only now being recognized as associated with other important worldwide tectonic, magmatic, and regressive-transgressive events, and the Vietnam sequences offer important and distinctive information. The Triassic seems to have had more orderly transgressive and orogenic activities.

References

Blondel, F. 1929. Les mouvements tectoniques de l'Indochine français. In *Comptes Rendus 4e Congress Science Pacifique,* vol. 2B, pp. 613–20.

Dickins, J. M. 1985. Late Palaeozoic and Mesozoic "Orogeny" in eastern Australia. In *Advances in the Study of the Sydney Basin. Proceedings of the 19th Symposium,* pp. 8–9. University of Newcastle, Department of Geology.

Dickins, J. M. 1987. Tethys – a geosyncline formed on continental crust? In *Shallow Tethys,* vol. 2, ed. K. G. Mckenzie, pp. 149–56. Rotterdam: Balkema.

Dickins, J. M. 1988a. The world significance of the Hunter-Bowen (Indosinian) orogenic phase. In *Advances in the Study of the Sydney Basin. Proceedings of the 22nd Symposium,* pp. 69–74. University of Newcastle, Department of Geology.

Dickins, J. M. 1988b. The world significance of the Hunter/Bowen (Indosinian) mid-Permian to Triassic folding phase. *Memorie della Società Geologica Italiana* 34:345–52.

Dickins, J. M. 1991. Permian of Japan and its significance for world understanding. In *Shallow Tethys 3,* ed. T. Kotaka, J. M. Dickins, K. G. Mckenzie, K. Mori, K. Ogasawara, and G. D. Stanley, pp. 343–51. Sendai: Saito Ho-on Kai. Special publication 3.

Dickins, J. M. 1992. Permo–Triassic orogenic, paleoclimate, and eustatic events and their implications for biotic alteration. In *Permo–Triassic Events in Eastern Tethys – Stratigraphy, Classification and Relations with the Western Tethys,* ed. W. C. Sweet, Yang Zunyi, J. M. Dickins, and Yin Hongfu, pp. 169–74. Cambridge University Press.

Dickins, J. M. 1994. What is Pangaea? In *Pangea: Global environments and resources,* ed. A. F. Embry, B. Beauchamp, and D. J. Glass, pp. 67–80. Canadian Society of Petroleum Geologists Memoir 17. Calgary.

Dovjikov, A. E. (ed.) 1965. *Geology of North Vietnam* (in Russian and Vietnamese). Hanoi: General Department of Geology.

Fromaget, J. 1927. *Etudes géologiques dans le Nord de l'Indochine centrale.* Bulletin 16. Hanoi: Service Géologique de l'Indochine.

Fromaget, J. 1931. *L'Anthracolique en Indochine après la régression moscovienne ses transgression et sa stratigraphie.* Hanoi: Service Géologique de l'Indochine. Bulletin 19.

Fromaget, J. 1934. Observations et reflexions sur la géologie stratigraphique et structurale de l'Indochine. *Bulletin de la Societé Géologique de France* 5:101–64.

Fromaget, J. 1941. *L'Indochine français, sa structure géologique ses roches ses mines et leurs rélations possibles avec tectonique*. Bulletin 26. Hanoi: Service Géologique de l'Indochine.

Gatinsky, Y., Hutchinson, C. S., Minh, N. M., and Tri, T. V. 1984. Tectonic evolution of Southeast Asia. In *Tectonic of Asia. Proceedings of the 27th International Geological Congress*, vol. 5, pp. 225–41.

Jin Yugan. 1993. Pre-Lopingian benthos crisis. In *Douzième Congres International de la Stratigraphie et Géologie du Carbonifère et Permien, Comptes Rendus*, vol. 2, ed. S. Archangelsky, pp. 239–78.

Le Duy Bach. 1985. Tectonic evolution of the earth's crust in Indochina (in Vietnamese). *Tap chi Dia chat* 176–177:25–39.

Le Thac Xinh. 1981. Mineralization of the continental rifting epoch, North Vietnam. In *Proceedings of the 4th Conference on the Geology of Southeast Asia (GEOSEA). Manila*, pp. 633–9.

Le Thac Xinh and Ta Hoang Tinh. 1975. Attempt at analysis of structure and metallogenic epochs of Indochina by the plate tectonic concept (in Vietnamese). *Dia chat* 118:1–20.

Metcalfe, I. 1986. Late Paleozoic paleogeography of Southeast Asia: some stratigraphical palaeontological and palaeomagnetic constraints. *Geological Society of Malaysia Bulletin* 19:153–64.

Murphy, R. W. 1975. Tertiary basins of Southeast Asia. *Southeast Asia Petroleum Exploration Society Proceedings* 2:1–36.

Ngo Thuong San and Rezanov, I. A. 1965. General features of the history of vertical movements in North Vietnam (in Vietnamese). *Sinh vat – Dia hoc* 4:1–10, 72–83, 163–80.

Nguyen Xuan Tung. 1982. Geodynamic evolution of Vietnam and neighbouring regions (in Vietnamese). *Dia chat va Khoang san* 1:179–217.

Phan Cu Tien. 1977. On the subdivision of Upper Permian and Lower Triassic in Northwest Bacbo (in Vietnamese). *Dia chat* 131:1–7.

Phan Cu Tien (ed.). 1989. *Geology of Kampuchea, Laos and Vietnam*. Hanoi.

Sano, H., and Kanmara, K. 1991. Collapse of organic reef complex. *Journal of the Geology Society of Japan* 97:113–33.

Tan, B. K., and Khoo, T. T. 1978. Review of the development of the geology and mineral resources of Malaysia and Singapore. In *Proceedings of the 3rd Conference on the Geology of Southeast Asia (GEOSEA), Bangkok*, pp. 655–73.

Tran Duc Luong. 1975. On the geotectonics of Northern Vietnam (in Vietnamese). *Ban do dia chat (Hanoi)* 19:41–62.

Tran Duc Luong and Nguyen Xuan Bao. 1980. Introduction of the geological map of Vietnam at the scale of 1:500,000 (in Vietnamese). *Ban do dia chat (Hanoi)* 46:4–11.

Tran Duc Luong, Nguyen Xuan Bao, Dao Dinh Thuc, Tran Tat Thang, and Nguyen Van Quy. 1985. On the newly compiled tectonic map of Vietnam at 1:1,500,000 scale (in Vietnamese). *Ban do dia chat* 63:5–11.

Tran Thi Chi Thuan, and Fontaine, H. 1968. Note sur la province de Quang Tri et description de quelques brachiopods. *Archieve géologique Vietnam (Saigon)* 12:55–105.

Tran Van Tri. 1979. *Geology of Vietnam (North Part)*. Hanoi: Institute of Geology and Mineral Resources.

Tran Van Tri, Nguyen Dinh Uy, and Dam Ngoc. 1986. The main tectonic features of Vietnam. In *Proceedings of the 1st Congress of the Geology of Indochina (CIG), Hanoi*, vol. 1, pp. 361–6.

Van Duc Chuong. 1983. Processes in the formation of the continental crust in Vietnam and adjacent regions (in Vietnamese). *Dia chat* 161:1–8.

Workman, D. R. 1977. *Geology of Laos, Cambodia, South Vietnam and the Eastern part of Thailand*. London: Institute of Geological Sciences. Overseas Geology and Mineral Resources.

Yang Zunyi, Liang Diny, and Nie Zetong. 1993. On two Permian submarine extension sedimentation events along the north margin of the Gondwana and the west margin of the Yangtze Massif. In *Douzième Congres International de la Stratigraphie et Géologie du Carbonifère et Permien, Comptes Rendus*, vol. 2, ed. S. Archangelsky, pp. 467–74. Buenos Aires.

9   Sitsa flora from the Permian of South Primorye

VERA G. ZIMINA

The Sitsa flora was first described by M. D. Zalessky in 1929. He investigated the collection sampled by M. A. Pavlov, I. A. Klock, and A. V. Martynov from the Malaya Sitsa River basin in the Partizansk region. Of the forms he described, only *Pecopteris anthriscifolia* Goeppert had been known earlier. The other species (*Pecopteris maritima* Zalessky, *Callipteris sahnii* Zalessky, *C. orientalis* Zalessky, *C. congermana* Zalessky, *Thinnfeldia pavlovi* Zalessky, *Odontopteris ussuriensis* Zalessky, *Scapanophyllum sitzense* Zalessky, *Ctenis renaulti* Zalessky) were described as being new species. There were also some species (*Annularia schurowskii* Schmalhausen, *Phyllotheca* cf. *deliquescens* Eichwald, and *Noeggerathiopsis aegualis* Goeppert) that were reported only in text. Those data suggested a late Permian age for the Sitsa flora. At the same time, Zalessky found some forms in his collection "that foretold the beginning of Mesozoic time" (Zalessky, 1929a, p. 124).

Later, Zalessky (1929b, 1930) described four other species (*Synopteris demetriana* Zalessky, *Sitzia kloki* Zalessky, *Sitzopteris superba* Zalessky, and *Rhipidopsis elegans* Zalessky). He concluded that the presence of *Pecopteris anthriscifolia* Goeppert in the Sitsa flora made it comparable with the late Permian floras of Pechora (Pechora Formation) and Kuznetsk (Kolchuginskian Formation).

When late Permian plant fossils were found in other localities of South Primorye, the question arose as to the position of the Sitsa Suite in the Upper Permian section of the region (the type of the suite is located in the type region of the Sitsa flora). In 1939, Kryshtofovich described flora with *Gigantopteris nicotianaefolia* Schenk that was found by V. Z. Skorokhod near the village of Poltavka (Fadeevka) in West Primorye. It was recognized in sequences occurring under limestone with Kazanian fossils. Kryshtofovich noted that the flora from Poltavka correlated with the flora of the Malaya Sitsa River basin, though it contained lesser amounts of Cathaysian elements.

Conclusions about the age of the Sitsa flora in South Primorye have changed repeatedly. In 1948, Maslennikov placed the sand-schist deposits of the Sitsa Suite in the uppermost part of the Permian. In the stratigraphic schemes by Belyaevsky and Prynada (1949), Eliseeva and Sosnina (1956), Organova et al. (1961), and Eliseeva and Radchenko (1964), deposits character-ized by the Sitsa flora lie above the Chandalaz, Barabash, and Kaluzin marine strata (limestones and some other rocks).

When new localities with late Permian floras were found, it was shown that nonmarine deposits with Sitsa-type fossils often alternated with marine deposits (Vasilyev and Likht, 1961). Moreover, one of the Sitsa flora elements (*Callipteris sahnii* Zalessky) was found in the lowermost part of the Upper Permian (Upper Pospelov Suite) (Zimina, 1969). Researchers began to consider the beds with the Sitsa flora more ancient – coeval with the late Midian deposits of the Chandalaz and Barabash suites (Vasilyev and Nevolin, 1969; Burago, 1986). At the same time, Radchenko (1956) contended that the Sitsa Suite correlated with the Erunakovskian Suite, which crowns the section of the Permian in the Kuznetsk Basin.

The lower part of the "Barabash" Suite on the Muravyev-Amursky Peninsula has been distinguished by N. G. Melnikov as a new Vladivostok Suite. That was supported by V. I. Burago, and Radchenko's study (1956) showed that the plant assemblage from those deposits was "intermediate between the Pospelov (Early Permian) one and Sitsa one (which is typical of the second half of Late Permian)" (Vasilyev and Nevolin, 1969, p. 152). Later it was assumed that the Vladivostok assemblage corresponded to the *Monodiexodina sutschanica* Zone of the Chandalaz Suite and to the Sitsa assemblage of the *Metadoliolina lepida* Zone of the same suite (Nikitina, Kiseleva, and Burago, 1970; Burago, 1973; Burago et al., 1974). That point of view regarding the sequence of Permian plant assemblages in South Primorye is reflected in the unified stratigraphic scheme for Russia, though the question of correlation of the two assemblages mentioned earlier is open to discussion. So it was necessary to study in detail the plant remains from both the type locality of the Sitsa flora and neighbouring localities in the Partizansk River basin.

Flora of the Malaya Sitsa (Tigrovaya) River basin

As noted earlier, the main flora described by Zalessky was found in the Tigrovaya River (former Malaya Sitsa River) basin, but his sampling locations are not known with certainty. I sampled the plant remains from the right bank of the Tigrovaya River in the

Partizansk municipal region. They occur in a sandy-shaley sequence, with thick conglomerates at the base. The sequence is characterized by alternating dark grey and greyish brown middle-to-fine-grained sandstones, dark grey siltstones, and greenish grey tuff-siltstones, with thin interbeds of coal-bearing siltstones and coals. In the upper part of the section, porphyrites are sometimes found (but they may be sills). The total thickness of the sequence is about 550 m.

Because there is no continuous section of the Permian along the river, the relationship between the sandy-shaley sequence (containing the fossil flora) and the marine sediments remains unclarified in this region. To improve our estimates of the age of those deposits, in-depth study of the plant assemblages was required. It turned out that it was necessary to revise the taxa described by Zalessky. The problem was that in his publications, Zalessky provided drawings of only the new species and genera. Most of the drawings had been made from specimens in M. A. Pavlov's collection, which has been lost. However, some drawings had been made from samples in A. V. Martynov's collection, which is now kept in the Geological Institute of the Russian Academy of Sciences (Moscow), catalogue number 2124. I have looked through that collection with S. V. Meyen and have taken photographs of the most important fossil imprints. In addition to Martynov's collection, I have used material that I have collected from the Tigrovaya River and some other locations with Sitsa-type floras in the Primorye region in an effort to determine its systematic composition.

Table 9.1 shows that some taxa (*Callipteris sahnii* Zalessky, *C. orientalis* Zalessky, *Odontopteris ussuriensis* Zalessky, *Sitzopteris superba* Zalessky, *Rhipidopsis elegans* Zalessky) can be inserted into the common list of Sitsa flora without any changes. I have also left unchanged the species *Pecopteris maritima* Zalessky, in spite of the fact that Zalessky (1934) later included that species in the synonymy of *Pecopteris synica* Zalessky. I do not share that opinion, because the diagnostic signs that were used as the basis for *P. synica* are not known, because of the absence of photos and holotype definition. Zalessky showed the imprint of a pinna with elongate pinnules, mucronate cone, and limbate rachis (Zalessky, 1934, fig. 13) and attributed it to *P. synica* Zalessky. On the fossil imprint from the Pechora River basin (Zalessky, 1934, fig. 10), those signs are not observed, just as they are not observed on the imprint in Plate 9.1(1), which was originally described by Zalessky (1929a) as *P. maritima* Zalessky. It seems to me more likely that Zalessky was correct when he considered the latter species to be similar to *P. anthriscifolia* Goeppert, but it is not appropriate to make any change in nomenclature without additional material on *P. maritima*.

Table 9.1. *Revision of the Permian plants from the Malaya Sitsa (Tigrovaya) River described by Zalessky*

Original description	Revised name
Pecopteris maritima Zalessky (Zalessky, 1929a, fig. 1)	*Pecopteris maritima* Zalessky
P. anthriscifolia Goeppert (Zalessky, 1929a, figs. 3 and 6)	*P. anthriscifolia* Goeppert
P. anthriscifolia Goeppert (Zalessky, 1929a, fig. 5)	*Pecopteris?* sp.
P. anthriscifolia Goeppert (Zalessky, 1929a, fig. 7)	*Prynadaeopteris* cf. *alifera* Fefilova
P. synica Zalessky (Zalessky, 1934)	*Pecopteris synica* Zalessky
Callipteris sahnii Zalessky (Zalessky, 1929a)	*Callipteris sahnii* Zalessky
C. orientalis Zalessky (Zalessky, 1929a, fig. 10)	*C. orientalis* Zalessky
C. congermana Zalessky (Zalessky, 1929a, fig. 11)	*Comia congermana* (Zalessky) Zimina (Zimina, 1977) [Figure 9.2(2) herein])
Odontopteris ussuriensis Zalessky (Zalessky, 1929a)	*Odontopteris?* ussuriensis Zalessky
Scapanophyllum sitzense Zalessky (Zalessky, 1929a)	?
Sitzia kloki Zalessky (Zalessky, 1930)	Original material was not investigated
Sitzopteris superba Zalessky (Zalessky, 1930)	*Sitzopteris superba* Zalessky
Siniopteris demetriana Zalessky (Zalessky, 1929b)	*Psygmophyllum demetrianum* (Zalessky) Burago (Burago, 1982)
Thinnfeldia pavlovi Zalessky *Thinnfeldia* sp. (Zalessky, 1929a, fig. 15)	? *Brongniartites* sp.
Ctenis renaulti Zalessky (Zalessky, 1929a)	Apparently absent from the Permian
Rhipidopsis elegans Zalessky (Zalessky, 1930)	*Rhipidopsis elegans* Zalessky
Ullmannia longifolia Zalessky (Zalessky, 1929a)	Apparently absent from the Permian

Even a cursory examination of the drawings that accompanied the description of *P. anthriscifolia* Goeppert will show that Zalessky's (1929a) approach to that species was too broad. Martynov's collection contains the originals on which Zalessky based his figures 3 and 5–7. I believe that only the imprints

---

Plate 9.1 (opposite). (1) *Pecopteris maritima* Zalessky, GIN-2124/17 (Martynov's collection), a specimen illustrated by Zalessky (1929a, fig. 1); 1a, full size; 1b (a part of the imprint), ×2; Tigrovaya (Malaya Sitsa) River, Sitsa Suite. (2–4) *Pecopteris anthriscifolia* Goeppert: 2, GIN-2124/25, a specimen illustrated by Zalessky (1929a, fig. 3); 2a, full size; 2b, ×2; 3, GIN-2124/3, a specimen illustrated by Zalessky (1929a, fig. 6); ×2; 4, GIN-2124/26, a specimen to which Zalessky referred (1929a, p. 128); full size; Tigrovaya River, Sitsa Suite. (5) *Pecopteris?* sp., GIN-2124/23, a specimen illustrated by Zalessky (1929a, fig. 5); full size (originally it was attributed to *P. anthriscifolia* Goeppert); Tigrovaya River, Sitsa Suite. (6) *Callipteris sahnii* Zalessky, GIN-2124/18; full size; Tigrovaya River, Sitsa Suite. (7) *Zamiopteris* cf. *dubia* Zimina (in association with fragment of *Cordaites?* sp.), GIN-2124/1; full size; Tigrovaya River, Sitsa Suite.

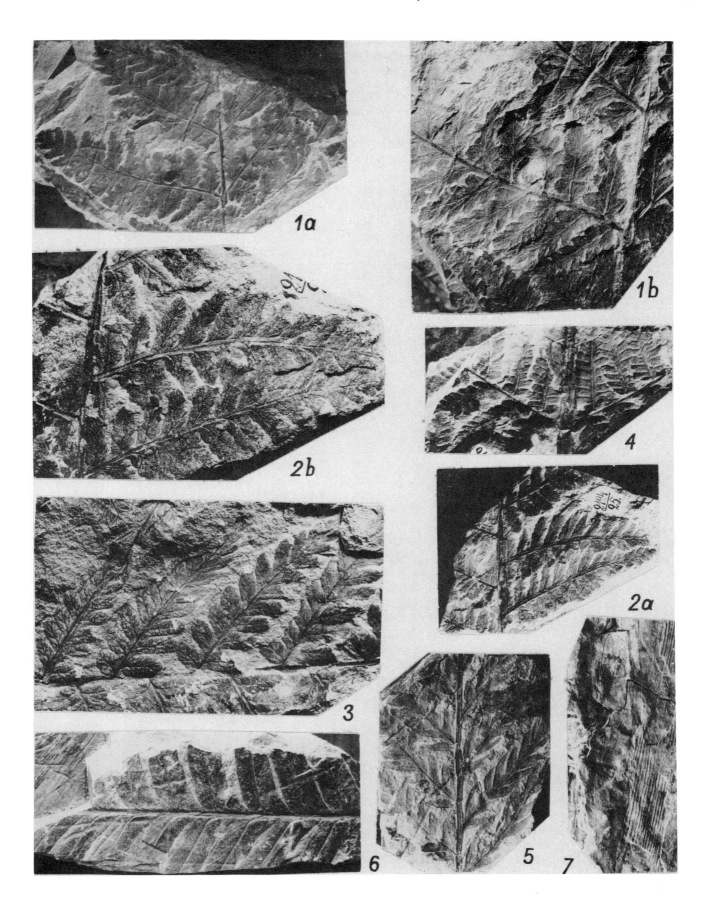

shown in Plate 9.1(2 and 3) (Zalessky's figs. 3 and 6) can be attributed to *P. anthriscifolia* Goeppert. The remaining specimens must be excluded from *P. anthriscifolia*. This is true for the imprint shown in Plate 9.1(5) (Zalessky, 1929a, fig. 5), which differs from *P. anthriscifolia* in strongly mucronate pinnules, with subsilate edges, and for the imprint of the spore-bearing pinna shown in Plate 9.2(1) (Zalessky's fig. 7), which differs from *P. anthriscifolia* in its widely limbate rachis.

Various researchers differ on the problem of the generic positioning of fern-like leaves, described by Zalessky as *P. anthriscifolia* Goeppert. Radchenko (1956) included them in *Prynadaeopteris*. Fefilova (1973) inserted those imprints into the synonymy of *Pecopteris anthriscifolia* (Goeppert) f. *anthriscifolia* Zalessky. I attributed similar imprints from South Primorye to *Prynadaeopteris*. But the spore-bearing pinna attributed earlier to *P. anthriscifolia* Goeppert (Zalessky, 1929a, fig. 7), as noted earlier, differs from the sterile pinnae of this species in its widely limbate rachis, and so it is more closely related to *Prynadaeopteris alifera* Fefilova. The fact that the leaves of the *P. anthriscifolia* Goeppert type show the presence of fructification is supported by material from the Primorye and Pechora (Fefilova, 1973) regions. Thus the assignment of this species to *Prynadaeopteris* appears to be premature.

Earlier I had noted (Zimina, 1977) that Zalessky (1929a) had erred when illustrating and defining plant remains as *Callipteris congermana* Zalessky. One can see that by looking at the picture of the original, Plate 9.2(2). The pattern of pinna venation shown by Zalessky (1929a, fig. 11) indicates that the plant belongs to *Comia*. Judging from the finding of two other species of *Comia* in the Partizansk area, examples of this species are common enough in the Sitsa flora.

I question the presence of representatives of *Thinnfeldia, Scapanophyllum,* and *Ctenis* in the Sitsa flora of the type locality. According to Zalessky, *Thinnfeldia* in the Tigrovaya River basin is represented by two species: *T. pavlovi* Zalessky and *Thinnfeldia* sp. Martynov's collection contains the specimen that was used for description of the species defined as *Thinnfeldia* sp., though that imprint is fragmental and does not give an indication of the pinnule's nature [Plate 9.2(6)]. Unlike Zalessky, I define it as *Brongniartites* cf. *salicifolius* (Fischer) Zalessky. My assignment is supported by the finding of the remains of an identical plant fragment, but with well-preserved pinnules, in the neighbouring region of the Muravyev-Amursky Peninsula (Lazurnaya River). These imprints are indistinguishable from those in the Urals area (Melchak River), which were classified by Zalessky (1927) as *B. salicifolius* (Fischer) Zalessky. It is unlikely

that the imprint illustrated by Zalessky (1929a, fig. 12) and defined as *Thinnfeldia pavlovi* Zalessky belongs to *Brongniartites*, because of the pinnule shape. But that it is not *Thinnfeldia* is supported by the finding of the plant in Upper Permian deposits, near the village of Shevelevka, similar in pinnule shape to the species described by Zalessky from the material sampled in the Tigrovaya River basin. At the same time, this plant belongs rather to a new genus, as the unusual arrangement of pinnules (in spiral order) is observed in the lower part of the plant.

Zalessky, who described a single representative of *Scapanophyllum* (*S. sitzense* Zalessky), allowed for contradiction (Zalessky, 1929a, fig. 14). Judging from the note to the original description of the type species, it was initially described as *Ganganopteris*. However, Zalessky then verified that the structures he considered anastomoses were "the imprints of diagonal joints in narrowed mesophile of the leaf" (Zalessky, 1929a, p. 133), and he decided to use the material available to distinguish a new genus, *Scapanophyllum*. At the same time, when describing that genus, he used the term "anastomosis" (sometimes without quotation marks), probably assuming that sign as diagnostic, thus contradicting his own earlier note. In Martynov's collection there is the imprint no. 2124/19, which is marked *Scapanophyllum sitzense* Zalessky, but the author of that species, in his original paper (Zalessky, 1929a), did not refer to Martynov's specimen. I have not found any anastomoses on Martynov's specimen. It is conceivable that it belongs to *Zamiopteris*. Detailed study of another specimen from Martynov's collection [Plates 9.1(7) and 9.3(2)] has shown that representatives of this genus are present in Sitsa. I have found imprints with the same structure in the Partizansk municipal area. In spite of the fragmentary nature of the material, we can conclude that both the specimen from Martynov's collection and the fossil imprints I have found belong to the same species – *Zamiopteris* cf. *dubia* Zimina.

Determining the age of the imprint of *Ctenis renaulti* Zalessky (Zalessky, 1929a, fig. 16) is a complicated problem. That the drawing was made from a *Ctenis* imprint is beyond question; however, the presence of representatives of this genus in the Permian strata in Primorye has not been demonstrated until now. There is some alarm that the specimen with the *Ctenis* imprint may derive from the Lower Cretaceous, which is widespread in that region, as it was with specimens bearing the imprints of *Ullmannia longifolia* Zalessky and *U. frumentaria* (Schlotheim) Goeppert (Zalessky, 1929a).

It should be noted that the imprint described by Zalessky

Plate 9.2 (opposite). (1) *Prynadaeopteris* cf. *alifera* Fefilova, GIN-2124/24, spore-bearing pinna (the second pinnule from the bottom was illustrated by Zalessky (1929a, fig. 7); ×2; Tigrovaya River, Sitsa Suite. (2, 3) *Comia congermana* (Zalessky) Zimina: 2, GIN-2124/13; 2a, pinna of next to last (second) order; full size; 2b, second pinna from bottom of last order, probably illustrated by Zalessky (1929a, fig. 11); ×2; 3, GIN-2124/6, the upper part of the pinna; ×2; Tigrovaya River, Sitsa Suite. (4) *Odontopteris? ussuriensis* Zalessky, GIN-2124/1, a specimen to which Zalessky referred (1929a, p. 133); full size; Tigrovaya River, Sitsa Suite. (5) *Callipteris orientalis* Zalessky, GIN-2124/8, a specimen to which Zalessky referred (1929a, p. 130); ×2; Tigrovaya River, Sitsa Suite. (6) *Brongniartites* cf. *salicifolius* (Fischer) Zalessky, GIN-2124/11, a specimen illustrated by Zalessky (1929a, fig. 15); ×2; Tigrovaya River, Sitsa Suite. (7) *Brongniartites* sp., GIN-2124/20, a specimen illustrated by Zalessky (1929a, fig. 9) (originally it was attributed to *Callipteris sahnii* Zalessky); full size; Tigrovaya River, Sitsa Suite.

(1929a, fig. 9) as the pinna top part of *Callipteris sahnii* Zalessky is closer to that of *Brongniartites*. In Plate 9.2(7) it is well seen that the pinnules of the illustrated plant are not arranged in pairs and are overgrown in the cone only at the base. Typical representatives of *C. sahnii* Zalessky have pinnules approaching each other or nearly opposed [Plate 9.1(5)], and on the pinna cone they almost completely intergrow or are half intergrown. The range of pinna variability in *C. sahnii* Zalessky is well seen from the imprint of an excellently preserved frond of this species that I found on the Muravyev-Amursky Peninsula (Shevelevka River).

The morphology of the *Sitzopteris superba* Zalessky frond from the Tigrovaya River had, until recently, been shown in only a single drawing (Zalessky, 1930, fig. 10). The finding of two fragmentary imprints of this species on the Muravyev-Amursky Peninsula [Plate 9.3(4)] provided additional information on nervation and variability in the frond's upper part.

In addition to the plants earlier described from Martynov's collection, I have distinguished in it *Pterophyllum* aff. *slobodskiense* Schvedov [Plate 9.3(1)]. The remains of that plant were found in the Shaiga River basin.

In the section along the right bank of the Tigrovaya River, in the Partizansk municipal area, I have distinguished two main floristic assemblages. The assemblage from the lower part of the section includes *Paracalamites* sp., *Pecopteris* (*Asterotheca*) *orientalis* (Schenk) Potonie, *Callipteris orientalis* Zalessky, *C. adzvensis* Zalessky, *Comia enisejevensis* Schwedov [Plate 9.3(3, 5, 8)], and undeterminable remains of cordaites.

In the upper part of the section there are *Paracalamites* sp., *Pecopteris* (*Asterotheca*) *orientalis* (Schenk) Potonie, *P. andersonii* Halle, *P. anthriscifolia* Goeppert, *Callipteris sahnii* Zalessky, *C. orientalis* Zalessky, *C. adzvensis* Zalessky, Callipteris n. sp., *Comia* aff. *laceratifolia* (Halle) Zimina, *Comia* sp., *Zamiopteris* cf. *dubia* Zimina, *Cordaites?* cf. *buragoi* Zimina, *Rufloria derzavinii* (Neuburg) S. Meyen, *R.* cf. *ulannurica* Durante, *Crassinervia?* cf. *neuburgiana* Zimina, *Crassinervia* sp. 1, *Lepeophyllum* sp., *Samaropsis piramidalis* Neuburg, *S. khalfinii* Suchov, and *S. irregularis* Neuburg [Plate 9.3(6, 7, 9–12)].

In spite of the presence of the forms found in common with floras from the Tigrovaya River basin, described by Zalessky, the plant remains from the sequences of the Partizansk municipal area significantly supplement the total list of Sitsa flora. Common species are *Pecopteris anthriscifolia* Goeppert, *Callipteris sahnii* Zalessky, and *C. orientalis* Zalessky, as well as *Zamiopteris* cf. *dubia* Zimina, which I determined from Martynov's collection.

Conclusion

As mentioned earlier, the Sitsa flora (Sitsa assemblage) was long thought to include abundant Mesozoic elements, and so the deposits (Sitsa Suite) hosting it were placed in the uppermost part of the Permian or in the Midian, just above the location of the Vladivostok assemblage.

I had earlier reported that in the Sitsa flora of the Partizanskaya River basin, together with late Permian species, there are some early Permian elements (Zimina, 1977). That conclusion was confirmed by the presence of the following species: *Paracalamites frigidus* Neuburg (Krasnoarmeysky Station), *P. similis* Zalessky and *Rufloria derzavinii* (Neuburg) Meyen (Senkina Shapka Mountain), and *Crassinervia?* cf. *neuburgiana* Zimina (Tigrovaya River). The fact that early Permian elements are also common in the Vladivostok Suite suggests an age the same as or close to that of the Sitsa and Vladivostok suites, especially as the species in common were then also found among late Permian forms. Additional information on the Sitsa flora from the Tigrovaya River and recent data (Burago, 1990) on flora from the Vladivostok Suite in southwestern Primorye confirm my earlier supposition. In the composition of flora from the stratotype section of the Vladivostok Suite, along the Pervaya Rechka River (that part of the section was earlier attributed to the lower part and the very beginning of the upper part of the Barabash Suite), there are the following species in common with Sitsa flora: *Callipteris adzuensis* Zalessky, *C. sahnii* Zalessky, *Comia congermana* Zalessky, *Rufloria derzavinii* (Neuburg) S. Meyen, and *Psygmophyllum demetrianum* (Zalessky) Burago. If we take into account the data on flora from deposits recently attributed to the Vladivostok horizon (Lower and Middle Barabash Subsuites) in West Primorye, the list can be supplemented by the following species: *Pecopteris maritima* Zalessky, *P. synica* Zalessky, *P. anthriscifolia* Goeppert, *Zamiopteris dubia* Zimina, and *Rhipidopsis elegans* Zalessky.

It is interesting that *Zamiopteris dubia* Zimina, a typical species of Sitsa flora from the Partizanskaya River basin, is known in West Primorye only from sequences in the watershed of the Barkhatnaya and Bogataya rivers, sequences that are placed in the lower part of the Middle Barabash Subsuite (upper part of the Vladivostok horizon, Upper Murgabian). The Middle Barabash Subsuite is overlain there by limestone, with early Midian *Monodiexodina* (Burago, 1990). It is possible that the sandy-schist sequences hosting the Sitsa flora in the type region should be correlated precisely with that part of the Permian in West Primorye, where we have the most representative section

---

Plate 9.3 (opposite). (1) **Pterophyllum** aff. **slobodskiense** Schvedov, GIN-2142/2; ×2; Tigrovaya River, Sitsa Suite. (2) **Zamiopteris** cf. **dubia** Zimina, GIN-2124/1; ×3; Tigrovaya River, Sitsa Suite. (3) **Pecopteris** (**Asterotheca**) **orientalis** (Schenk) Potonie, DVGI-407/113; ×2; Tigrovaya River, Sitsa Suite. (4) **Sitzopteris superba** Zalessky, DVGI-408/285; ×1.5; Shevelevka village, Vladivostok Suite. (5) **Callipteris adzvensis** Zalessky, DVGI-408/292; full size; Tigrovaya River, Sitsa Suite. (6) **Callipteris** n. sp., DVGI-408/293; full size; Tigrovaya River, Sitsa Suite.

(7) **Samaropsis khalfinii** Suchov, DVGI-408/289; ×5; Tigrovaya River, Sitsa Suite. (8) **Comia enisejevensis** Schwedov, DVGI-407/148; ×1.5; Tigrovaya River, Sitsa Suite. (9, 10) **Rufloria derzavinii** (Neuburg) S. Meyen: 9, DVGI-408/291; full size; 10, DVGI-408/290 microstructure; ×20; Tigrovaya River, Sitsa Suite. (11) **Samaropsis piramidalis** Neuburg, DVGI-408/287; ×5; Tigrovaya River, Sitsa Suite. (12) **Samaropsis irregularis** Neuburg, DVGI-408/286; Tigrovaya River, Sitsa Suite.

of the Murgabian–Midian boundary beds in the Far East. It should be noted that the previously mentioned beds in the village of Fadeevka that yield *Gigantopteris* are also placed at the level of the lower part of the Middle Barabash Subsuite in West Primorye.

New data on the Sitsa flora from the type region have allowed more accurate correlations with the remote Upper Permian sections of Angara (Kuzbass, Urals area, and Mongolia). Thus, the prevalence of *Callipteris* and the presence of representatives of *Rufloria* in the Sitsa assemblage from the type locality show that it is most closely comparable to the floras of the Uskatskian and Lenin (lower part) horizons of the Kuznetsk Basin, correlated with the Kazanian and the lowermost part of the Tatarian (Betekhtina et al., 1988). The presence of the species similar to *Brongniartites salicifolius* Zalessky allows correlation of the Sitsa flora in its type locality with the flora of the Belebeev Suite in the Melchak River (Urals region), dated now as Upper Kazanian (Vladimirovich, 1986). At approximately the same level is the coal-bearing and tuffogenic sedimentary Tabun-Tologoya member of southern Mongolia, where some forms similar to *Callipteris sahnii* Zalessky and *Zamiopteris dubia* Zimina (i.e., species typical of the Sitsa flora of South Primorye) have been found.

References

Belyaevsky, N. A., and Prynada, V. D. 1949. New data on stratigraphy of the south Sikhote-Alin (in Russian). *Trudy Vsesoyuznogo Nauchno-Issledovatel'skogo Geologicheskogo Instituta (VSEGEI), Obschaya Seriya* **1949**(9):131–8.

Betekhtina, O. A., Gorelova, S. G., Batyaeva, S. K., Dryagina, L. L., and Tokareva, P. A. 1988. Palaeontological characteristic of the regional horizons of the Kuznetsk Basin (in Russian). *Trudy Instituta Geologii i Geophiziki* **707**:9–12.

Burago, V. I. 1973. Cathaysian elements in Permian flora of South Primorye (in Russian). *Geologiya i Geophizika* **1973**(11):54–61.

Burago, V. I. 1982. The morphology of the leaf in *Psygmophyllum* (in Russian). *Paleontologicheskii Zhurnal* **1982**(2):128–36.

Burago, V. I. 1986. On the problem of the boundary between Angaran and Cathaysian vegetable kingdoms (in Russian). In *Permo-Triasovye Sobytiya v Razvitii Organicheskogo Mira Severo-Vostochnoi Azii* [*Permian-Triassic Events in the Development of the Organic World of North-East Asia*], ed. Y. D. Zakharov and Y. I. Onoprienko, pp. 6–23. Vladivostok: Dal'nevostochnoye Otdeleniye Akademii Nauk SSSR.

Burago, V. I. 1990. Vladivostok horizon of the Upper Permian in South Primorye (in Russian). In *Novye Dannye po Biostratigrafii Paleozoya i Mezozoya Yuga Dal'nego Vostoka* [*New Data on Palaeozoic and Mesozoic Biostratigraphy of South Far East*], ed. Y. D. Zakharov and Y. I. Onoprienko, pp. 81–103. Vladivostok: Dal'nevostochnoye Otdeleniye Akademii Nauk SSSR.

Burago, V. I., Kiseleva, A. V., Kotlyar, G. V., Nikitina, A. P., Sosnina, M. I., and Tashchi, S. M. 1974. Palaeontological characteristics of Permian deposits of South Primorye (in Russian). In *Paleozoi Dal'nego Vostoka* [*The Paleozoic of*

the Far East], ed. L. I. Krasnyi and L. I. Popeko, pp. 214–35. Khabarovsk: Dal'nevostochnyi Nauchnyi Centr.

Eliseeva, V. K., and Radchenko, G. P. 1964. Stratigraphy of Permian continental and volcanogene formations of South Primorye (in Russian). *Trudy Vsesouznogo Nauchno-Issledovatel'skogo Geologicheskogo Instituta (VSEGEI)* **107**:31–52.

Eliseeva, V. K., and Sosnina, M. I. 1956. Stratigraphy of Upper Palaeozoic sediments of Primorye (abstract) (in Russian). *Soveschaniye po Unifitsirovannym Stratigraphicheskim Skhemam Dal'nego Vostoka. Tezisy* [*Meeting on Unified Stratigraphic Schemes of the Far East, Abstracts*], pp. 29–32. Khabarovsk.

Fefilova, L. A. 1973. *Paporotnikovidnye Permi Severa Pre-dural'skogo Progiba* [*Permian Ferns of the North Urals Trough*]. Leningrad: Nauka.

Kryshtofovich, A. N. 1939. New findings of the fossil flora in Far East, as landmarks of stratigraphy (in Russian). In *Akademiku V. A. Obruchevu k 50-letiyu Nauchnoi i Pedagogicheskoi Deyatel'nosti* [*To the 50-Years Scientific and Pedagogical Activity of Acad. V. A. Obruchev*], pp. 277–313. Moskva-Leningrad: Izdatel'stvo Akademii Nauk SSSR.

Maslennikov, D. F. 1948. New data on stratigraphy of Upper Palaeozoic deposits of the south Far East (in Russian). *Materialy Vsesoyuznogo Nauchno-Issledovatel'skogo Geologicheskogo Instituta (VSEGEI), Obschaya Serya* **1948**(8):79–84.

Nikitina, A. P., Kiseleva, A. V., and Burago, V. I. 1970. A scheme of biostratigraphic subdivision of the Barabash Suite of the Upper Permian in South-West Primorye (in Russian). *Doklady Akademii Nauk SSSR* **191**:187–9.

Organova, N. M., Kirvolutsky, V. N., Petrachenko, E. D., and Zimina, V. G. 1961. Stratigraphy of Palaeozoic sediments of South-West Primorye (abstract) (in Russian). In *Tezisy Dokladov na Sessii Soveta Dal'nevostochnogo Filiala Sibirskogo Otdeleniya Akademii Nauk SSSR po Itogam Nauchnykh Issledovanii 1960 Goda* [*Session of the Far East Division, Siberian Branch, USSR Academy of Sciences on Scientific Investigations in 1960, Abstracts*], pp. 27–9. Vladivostok.

Radchenko, G. P. 1956. Leading forms of fossil plants of coal-bearing deposits of the Kuznetsk Basin (in Russian). In *Atlas Rukovodyaschikh Form Iskopaemoi Fauny i Flory Permskikh Otlozhenii Kuznetskogo Basseina* [*Atlas of Leading Forms of Fossil Fauna and Flora from Permian Deposits of the Kuznetsk Basin*], ed. V. I. Yavorsky, pp. 110–206. Moskva: Gosgeoltekhizdat.

Vasilyev, B. I., and Likht, F. R. 1961. New data on the stratigraphy of the Upper Permian of South Primorye (in Russian). *Soobshcheniya Dal'nevostochnogo Filiala Sibirskogo Otdeleniya Akademii Nauk SSSR* **1961**(14):27–30.

Vasilyev, B. I., and Nevolin, L. A. 1969. Permian system (in Russian). In *Geologiya SSSR. Primorskii Krai* [*Geology of the USSR. Primorye Region*], vol. **32**, ed. I. I. Bersenyev, pp. 120–85. Moskva: Nedra.

Vladimirovich, V. P. 1986. High plants (in Russian). *Trudy Vsesoyuznogo Nauchno-Issledovatel'skogo Geologo-Razvedochnogo Instituta (VNIGRI), n. ser.* **331**:32–8.

Zalessky, M. D. 1927. Permian flora of Ural region of Angara (in Russian). *Trudy Geologicheskogo Komiteta* 176:1–52.

Zalessky, M. D. 1929a. Permian plants from the Malaya Sitsa River in the Sutchan region (in Russian). *Izvestiya Akademii Nauk SSSR* **1929**(2):123–38.

Zalessky, M. D. 1929b. Sur le *Syniopteris nestriana* n. g. et sp.

noveaux vegetaux permiens. *Izvestiya Akademii Nauk SSSR* **1929**(8):729–36.

Zalessky, M. D. 1930. Distribution of fossil flora relative to Gondwana one in North Eurasia (in Russian). *Izvestiya Akademii Nauk SSSR* **1930**(9):913–30.

Zalessky, M. D. 1934. Observations sur les vegetaux permiens du Bassin de la Petchora. I. *Izvestiya Akademii Nauk SSSR* **1934**(2–3):24–9.

Zimina, V. G. 1969. On the age of the Pospelov Suite and the time of appearance of Gondwana elements in the Permian flora of South Primorye (in Russian). *Doklady Akademii Nauk SSSR* **189**(5):1073–6.

Zimina, V. G. 1977. *Flora Rannei i Nachala Pozdnei Permi Yuzhnogo Primor'ya* [*Flora of the Early and the Beginning of the Late Permian of South Primorye*]. Moskva: Nauka.

10 Late Permian bimodal volcanism in South Primorye

ALEXANDER A. VRZHOSEK

The effects of the late Permian volcanism in Primorye are evident primarily along the southern margin of the Khanka Massif, a crystalline massif representing the extreme eastern block of the Sino-Korean Platform. Volcanism developed there under a variety of geodynamic conditions. In the western part of South Primorye, Upper Permian volcanics and associated terrigenous and carbonaceous rocks settled in the sea basin, which was formed through the destruction of the ancient sialic basement by the Hercynian orogenic movement. When that orogenesis was completed, that depression turned into a foldbelt, which is mapped as the West Primorye (or Khasan-Grodekovo) structural-facies zone (Nazarenko and Bazhanov, 1987).

In the eastern part of South Primorye, Upper Permian volcanic rocks formed under the platform conditions directly on the area of the Khanka Massif in separate volcano-tectonic depressions. Their occurrence immediately on the ancient Precambrian rocks shows that they and accompanying terrigenous deposits belong to the complex of the platform mantle.

This chapter describes the structure and composition of the Upper Permian volcanogenic complex of South Primorye.

Position and structure of the volcanogenic complex

The Upper Permian volcanogenic formation is best studied in the West Primorye Fold Zone. It is a marginal eastern member of the meridional Laoelin-Grodekovo Depression, which wedges from the south to the Sino-Korean Platform, covering the adjacent territories in Korea and China (Smirnov, 1963). Deep faults outline the depression and separate it from ancient western and eastern crystalline massifs. In the south, the depression is open toward the Japanese Sea, which suggests that it is an aulacogen that resulted from rift genesis.

The depression is 7.5–8 km deep (to the crystalline basement) and is composed of Permian deposits; however, its lithologic and facies compositions are diverse. At the base of the section there are Lower Permian volcanic-terrigenous strata of the Dunai and Pospelov suites. The Dunai Suite (Sakmarian and Jaktashian stages) is made up of coarsely fragmented pyroclastic sediments, with rare porphyritic flows. The overlying Pospelov Suite (Bolorian and Kubergandinian stages) is composed predomi-

nantly of alternating sandstones and siltstones. However, those deposits do not crop out directly in the area of the West Primorye Zone, but have been found only along the eastern margin of the depression at its junction with the Khanka Massif, in the Muravyev-Amursky Peninsula and along the coast and subjacent islands of Ussuri Bay.

Within the depression, erosional shear removed the younger Permian deposits that had accreted the stratigraphical section upward without hiatuses and unconformities. Those deposits are most common in the southern part of the fold zone, where they can be divided into two series: volcanogenic sequences and black shale. The black-shale deposits are near the central part of the depression, and the volcanogenic sequences extend along its slopes, tracing deep faults and forming a system of conjugate-facies subzones.

The most comprehensive sections of the Upper Permian sequences from the volcanogenic subzone crop out within the Barabash Anticlinorium (Narva-Amba Interfluve), where they are known as the Barabash Suite (Murgabian and Midian stages). Its thickness is 3,300–4,700 m. The suite is separated into two subsuites representing a typical bimodal series: effusives of basic composition (spilites, basaltic porphyrites, and tuffs) predominate in its lower part, and a volcanogenic member of acidic composition (tuffs of quartz-keratophyres, with rare horizons of lavas) occupies the upper part. Both parts contain interbeds of sandstones and siltstones, the amounts of which increase gradually in the upper part of the section in each subsuite, which indicates slow waning of each cycle of volcanism and the predominance of terrigenous sedimentation during the final stages of volcanic activity.

During the sedimentation process, carbonate accumulation took place, with the formation of shelf reefs on the submerged continental margins along the coast of a rifting depression. The close association of effusives (including lavas) with biogenic limestones containing marine faunas, the discovery of benthic organisms in baked tuffs, the unusual patterns of occurrence of some flows (pillow-lava presence), and other features indicate that the volcanogenic series was formed under marine conditions. However, the effusives appear to have erupted at shallow depths, and in some places lavas and accompanying agglomerates of acidic composition erupted in a subaerial environment. The formation conditions, the composition, and the character of the secondary alterations that resulted in greenstone transformation of volcanic rocks show the Barabash Suite to be an analogue of spilite-keratophyre formation, where basic and acidic effusives are connected paragenetically.

The ratios of those different rocks change along the fold zone from south to north (i.e., in the direction of aulacogen wedging). Basaltic porphyrites disappear from the volcanogenic sequences because of the significantly increasing thicknesses (from 200 to 3,800 m) of the formations of quartz-keratophyre composition, which in their turn are replaced by black-shale deposits developed in the area of the adjacent terrigenous subzone. The changes in composition and volume of the riftogenic depression probably were related to the basement, primarily to the increased thickness of the earth's crustal granitic layer to the north (Argentov, Ospanov, and Popov, 1970).

Comparisons of the sections across the depression's strike to the east show sharp decreases in the thicknesses of Upper Permian deposits and rapid changes from deep-sea conditions of sedimentation to subplatform deposition and then platform deposition in the Khanka Massif margin. Volcanogenic formations are related to the aulacogen's restriction of the flange fault zone and form a structural suture on the boundary with the massif. Along that deep fault and the adjacent dislocations caused by breaks were formed the magmatic centres that supplied magma to the volcanic apparatuses.

To the east of that fault, volcanic facies change: There are increases in pyroclastic and sedimentary rocks that contain fossilized plant remains. Agglomerate tuffs in the basement sections of the volcanogenic series in the Muravyev-Amursky Peninsula and the Russky and Popov islands contain fragments of early Palaeozoic granites, which indicates proximity to a crystalline basin. Facies changes in the junction zone with the Khanka Massif prompted the distinction of new suites, with the lower part of the section attributed to the Vladivostok Suite (Murgabian and Lower Induan stages), and the upper part to the Chandalaz Suite (Midian Stage), though the two suites, on the basis of numerous and diverse organic remains, compose a full stratigraphic analogue of the Barabash Suite. To the east of the junction zone, the volcanogenic series often is almost completely replaced by tuffaceous-terrigenous deposits that are slightly dislocated and thin and are represented predominantly by psephitic facies, which allows us to consider them a complex of the platform mantle.

It should be noted that basic volcanism was replaced by acid volcanism ubiquitously within a single age interval. In the West Primorye Zone, that change took place in the early Midian Stage. In the eastern regions (Partizanskaya River basin), that boundary is traced in the Chandalaz Suite at approximately the same stratigraphic level. In the junction zone with the Khanka Massif, in the Muravyev-Amursky Peninsula, it is confined to the upper part of the Vladivostok Suite, which according to palaeomagnetic data comprises the Lower Midian deposits (Zakharov and Sokarev, 1991).

Petrographic and petrochemical features of the volcanogenic formation

The Upper Permian volcanogenic formation, like most bimodal series, is composed of two rock groups: (1) basalts and andesite-basalts and (2) dacites and liparites, with rocks of intermediate compositions being minor components.

The first group is represented by aphyric or fine-porphyric diabasic and basaltic porphyrites that form numerous thin flows, with horizons of tuff and lava clasts. Pyroclastic formations account for 15–20% of the total rock within the fold zone, and 40–50% in the junction zone with the Khanka Massif and over the massif area. The lavas have anamygdaloidal texture; sometimes pillow cleavage is observed.

The lavas are composed of fine impregnations and laths of plagioclase and grains of clinopyroxene that are submerged in volcanic glass during magmatic eruptions. In rare cases, fine porphyritic segregations of olivine can be found that were completely replaced by serpochlorite. All the rocks have undergone the processes of albitization and greenstone alteration up to the formation of typical spilites.

The second rock group is represented by effusives of acidic composition. In contrast to the basalts, the lavas are rare and form isolated thin flows. Most of the rock (not less than 75–80%) is composed of pyroclastic formations, represented by both psephytic and unusual fine-grained ash tuffs alternating with streaky cherty tuffites.

The lavas have a porphyritic structure and in the glassy microfelsitic groundmass contain fine phenocrysts of plagioclase, potassium-feldspar, and quartz. Dark-coloured minerals are completely absent or have been completely replaced. The plagioclase, like that in spilites, is most often represented by albite. In their composition, the lavas correspond to albitized dacites and liparites and form a rock series transitional to typical quartz-keratophyres.

Among the volcanogenic formations of the Khasan-Grodekovo Fold Zone, rocks of intermediate composition are not found. Andesite-basalts and rare andesites appear only in the junction zone with the Khanka Massif (Muravyev-Amursky Peninsula), where they are not intermediate members, but form an independent group instead of basalts. Those rocks are most common in the area of the Khanka Massif within an Upper Permian volcano-tectonic depression, where they displace basalts and are represented by typical porphyry-like andesite-basalts and andesites with high contents of potassium. However, on the whole, the contrast in formation compositions is preserved, because the chemical compositions of lavas and pyroclastic rocks change (silica content increases) in parallel in basic and acidic rocks. In the junction area, dacites are altered to liparite-dacites, and on the Khanka Massif the liparites represented by quartz-keratophyres lie above andesite-basalts and andesites.

The evolution of the basaltic volcanism was completed with the formation of intrusive varieties of granophyric diabases. That fact and the chemical compositions of basalts and the rare-

Table 10.1. *Average chemical compositions of rocks in the Upper Permian volcanogenic formation*

Oxide	1[a]	2	3	4	5	6	7	8
SiO_2	49.38	50.51	49.98	50.14	56.96	65.53	70.25	74.40
TiO_2	1.20	1.77	2.15	1.71	1.15	0.68	0.56	0.29
Al_2O_3	16.40	16.94	15.24	16.48	16.11	15.38	13.79	12.54
Fe_2O_3	3.52	2.58	3.09	2.90	3.45	1.98	1.65	1.37
FeO	9.21	7.76	8.71	8.29	6.16	2.78	2.54	1.81
MnO	0.19	0.22	0.18	0.20	0.19	0.17	0.11	0.10
MgO	6.24	5.34	4.83	5.46	2.23	1.12	0.59	0.38
CaO	9.14	5.99	7.44	7.01	4.33	2.54	1.13	0.79
Na_2O	3.46	4.14	3.13	3.79	4.52	4.96	4.94	4.27
K_2O	0.35	0.60	0.90	0.60	1.51	2.22	2.91	2.80
P_2O_5	0.14	0.38	0.23	0.30	0.34	0.07	0.10	0.04
LOI	0.71	3.75	4.20	3.11	3.25	2.35	1.49	0.81
Total	99.94	99.98	100.08	99.99	100.20	99.78	100.06	100.00

[a]Columns 1–5, effusives of basic composition: 1, albitized basalts from the southern part of the Khasan-Grodekovo Zone; 2, spilites from the central part of the same zone; 3, albitized basalts from the northern part of the same zone; 4, spilites from the Khasan-Grodekovo Zone (average); 5, albitized andesite-basalts from the Muravyev-Amursky Peninsula. Columns 6–8, effusives of acidic composition: 6, albitophyres of the Khasan-Grodekovo Zone; 7, keratophyres of the same zone; 8, quartz keratophyres of the Muravyev-Amursky Peninsula.

element contents (Table 10.1) indicate that they belong to the tholeiitic series of rocks (Vrzhosek, 1980).

The principal petrographic and petrochemical features of the spilite-diabasic and quartz-keratophyric associations of late Permian volcanic rocks suggest genetic independence of those two rock types. As volcanism developed and basaltic centres differentiated, magma-generation zones were removed to higher horizons in the earth's crust, which resulted in melting of the sialic shell and widespread development of acid volcanism. Thus, the manifestations of late Permian bimodal volcanism and subsequent intrusions of large granitoid massifs, on which the Lower Triassic eroded deposits lie (Induan Stage), support the idea that at the end of the late Permian epoch, South Primorye was an active continental margin of the Khanka Massif.

References

Argentov, V. V., Ospanov, A. B., and Popov, A. A. 1970. Earth crust structure of South-West Primorye (in Russian). *Trudy Sakhalinskogo Kompleksnogo Nauchno-Issledovatel'skogo Instituta (SakhKNII)* **1987**(3):77–82.

Nazarenko, L. F., and Bazhanov, V. A. 1987. *Geologiya Primorskogo Kraya. III. Osnovnye Cherty Tektoniki i Istoriya Razvitiya* [*Geology of Primorye Region. III. The Main Features of Tectonic and Geological Evolution*] (preprint). Vladivostok: Dal'nevostochnyi Nauchnyi Centr Akademii Nauk SSSR.

Smirnov, A. M. 1963. *Sochleneniye Kitaiskoi Platformy s Tikhookeanskim Skladchatym Poyasom* [*The Transition Zone from the China Platform to the Pacific Mobile Belt*]. Moskva-Leningrad: Izdatel'stvo Akademii Nauk SSSR.

Vrzhosek, A. A. 1980. Bimodal volcanogenic complex of the Khasan-Grodekovo Fold Zone (in Russian). In *Petrokhimiya Magmaticheskikh Formatsii Vulkanitcheskikh Zon Dal'nego Vostoka* [*Petrochemistry of Magmatic Formations in Far Eastern Volcanic Belts*], ed. V. G. Sakhno and N. S. Nikolsky, pp. 3–16. Vladivostok: Dal'nevostochnyi Nauchnyi Centr Akademii Nauk SSSR.

Zakharov, Y. D., and Sokarev, A. N. 1991. *Biostratigraphiya i Paleomagnetizm Permi i Triasa Evrazii* [*Permian–Triassic Biostratigraphy and Palaeomagnetism of Eurasia*]. Moskva: Nauka.

11 Syngenetic and epigenetic mineral deposits in Permian and Triassic sequences of the Primorye region

LEV N. KHETCHIKOV, IVAN V. BURIJ, VITALIY G. GVOSDEV, VADIM G. KHOMICH, VLADIMIR V. IVANOV, VERA A. PAKHOMOVA, and VLADIMIR V. RATKIN

In the Primorye region, various syngenetic and epigenetic minerals are known from Permian and Triassic deposits. The main co-sedimentary mineral deposits are bituminous coals, limestones, glass sandstones, and rocks used for building stones. There has long been interest in the possibility of recovering syngenetic gold from the basal Triassic beds. Permian and Triassic deposits enclosing epigenetic endogenous minerals are known from South Primorye (tungsten, gold, and other polymetallic minerals) and Sikhote-Alin (tungsten, boron-silicate, and other polymetallic mineralizations).

Triassic coal accumulation in South Primorye

In Primorye, alternating Triassic marine and terrestrial fresh-water deposits are widely distributed. Coal deposits have been traced for great distances (more than 500 km), as from the coast of Peter the Great Bay in the south to the Bikin River basin in the north. Triassic coal deposits have been found in drill cores from depths of 1,510–2,100 m near the village of Borisovka (to the northwest of the town of Ussuriisk), indicating the possibility of significant deposits at great depths. Throughout the area there are numerous coal dumps alongside old mining sites, the documentation of which has not been preserved.

Coal deposits have been better studied in the Vladivostok coal region, where coals of Triassic, Cretaceous, and Tertiary age have long been known. Triassic coals attracted the attention of many miners at the end of the nineteenth century, when Vladivostok was built as a naval base. Spontaneously, without proper geological study, mining franchises were let, and small mines and adits were worked for coal. Because of the compli-cated tectonics of the region, and the variations in the quality of coals at different levels, most coal miners failed and gave up their mine workings, creating a bad impression of Triassic coal as a low-grade solid fuel. Also, the individual owners could not compete with the state mines in Sutchan (Partizansk), which mined Lower Cretaceous coals.

The firm of Briner & Co. was most successful in mining Triassic coals on the west coast of the Amur Gulf, in the Mongugai (Barabashevka) River basin, where at the end of the nineteenth century the engineer D. L. Ivanov's expedition discovered good coals. The firm mined semi-anthracite coal from two beds – "Inzhenernyi" and "Staryi." That coal had 30.8% ash and 8.22% volatiles. Coal was brought to the west coast of the Amur Gulf by 15 km of narrow-gauge railway, and then, by scows, it was brought to Vladivostok. Between 1908 and 1921, 56 tons of semi-anthracite were mined (Anhert, 1928). Eliashevich (1922), who worked at the Mongugai deposit in 1919–21, found coking coals and natural coke in the neighbouring Yulyevsky deposit. The mine closed in 1922 when the nationalization of mines took place in the Soviet Union.

Another important source is the Nadezhdinskoye deposit located along the upper reaches of the Knevichanka (Batalyanza) River, 16 km from the Nadezhdinskaya railway station. The coals that were found there when the Ussuri railway was built attracted the attention of a retired junior captain, who obtained the franchise for mining coal there. With the help of foreign financial capital and the participation of French geologists, general information on the deposit was gathered, and coal beds were sampled in small mining workings. Coal was brought to the Nadezhdinskaya station by winter means (sledge) and then by railway to Vladivostok. Tests for coal quality were carried out in boiler-houses and on board the ships of the navy, and the findings were encouraging. According to Anhert (1928), 7,687 tons of semi-anthracite were mined at the deposits from 1912 to 1923. Analysis of coal samples showed 5.12% volatiles and 12.85% ash. Later, a bed of coking coal was found. In succeeding years, exploration continued, in search of quality, low-ash coals. There were no positive findings, and the coals of that deposit were abandoned.

Interesting information, unfortunately never published, is available concerning the mine "Leonov" at the Lyanchikha deposit in the Vladivostok region (Sad-Gorod). I interviewed miners who had worked at old coal mines and found that in 1910 the firm of Lindholm & Co. had mined Cretaceous coals from the Podgorodnenka mines. They sank a vertical shaft to a depth of 120 m, with a ventilation gradient, and crosscuts reached six working beds from the Triassic. According to the data of Anhert (1928), that coal contained 15.36% ash and 25.09% volatiles. Having found coal of good quality only 30 km from Vladivostok, the firm prepared for mining operations, but in 1913 there was a fire at the mine, and it was closed down. Along the strike to the south, coal dumps are known near the former mine "Nikolaenka."

In 1945–48 the Daluglerazvedka trust explored those same coals. Twenty beds and interbeds of coal were found, but only 11 beds had economic thicknesses. Analysis of the coals showed an average ash content of 29.84%, volatiles 11.37–18.00%, and caloric capacity 7,839–8,591 kcal per ton. The deposit was not exploited.

The Surazhevka deposit of Triassic coals is found in the vicinity of the villages of Surazhevka and Radchikha, Shkotovo region. Coals found there at the end of the nineteenth century seemed to be interesting to many specialists. From 1930 to 1942, exploration of both Cretaceous and Triassic coals was carried out. About nine beds of finely fractured and crumpled Triassic coal (Burij, 1948) were revealed; five beds had economic thicknesses (0.8–2.2 m).

One of the reasons that Triassic coals came to be ignored was inadequate investigation of the stratigraphy of Triassic non-marine coal-bearing deposits. At present, extensive information is available that has allowed us to work out the stratigraphic details of the Triassic terrestrial and marine deposits of South Primorye and to show their interrelationships (Burij, Buryi, and Zharnikova, 1993).

In South Primorye, we distinguish the Ladinian, Carnian, early Norian, and Norian–Rhaetian stages of coal formation. The Ladinian coal-bearing deposits, represented by the Kiparisovka Suite (many geologists consider it to be early Carnian in age), 600 m thick, contain one to three coal beds, traced in the Lyanchikha, Nadezhdinskaya, and Mongugai deposits and in Borisovka hole no. 1 (Burij, 1959; Burij and Zharnikova, 1980). The Carnian coal deposits, about 600–720 m thick, are represented by the Sad-Gorod Suite (Lyanchikha, Nadezhdinskaya, Surazhevka, and Rakovka deposits) and Mongugai Suite (Mongugai deposit and Borisovka hole no. 1) (Burij, 1959). The Lower Norian coal sequences are represented by the Amba Suite (400 m thick), which we have recognized and studied on the Amba (Mongugai) River left bank, on the west coast of the Amur Gulf, though the coals still are poorly known. Analogous to the Amba Suite are the beds of the upper horizon of the Mongugai Suite in the Suifun Depression.

We believe that the strata of the Shkotovo deposit, conditionally correlated with the Shitukhe Suite (about 250 m thick) of debatable age (Krassilov and Shorokhova, 1973; Okuneva and Zheleznov, 1977), located on the Artemovka River left bank 9 km from the village of Shkotovo, belong to the Norian–Rhaetian coal sediments. The deposit was discovered in 1958 during a geologic survey. Detailed exploration by mine workings and drillholes has shown 10 coal beds and interbeds 0.2–0.4 m thick. The beds have complicated structures. They have been extensively dislocated by tectonic processes. Analysis of samples from two beds showed that they contain 30–60% ash and 20–30% volatiles, with caloric capacity of 7,836–8,689 kcal per ton.

Tin-tungsten, copper, gold-silver, and polymetallic mineralizations in southwestern Primorye

In southwestern Primorye, in the area known as the West Primorye structural-formation zone (SFZ), late Permian subaerial volcanic and sedimentary carbonate-terrigenous deposits are common, as described in detail by Vrzhosek in Chapter 10 in this volume.

The complex of granophyric granites is the youngest of the Permian intrusives. It is restricted mainly to the fields of volcanic accumulations. Petrologists consider it to be the intrusive co-magmatic to Upper Permian effusive-extrusive formations. The shape of the massif is irregular, stock-like, and discordant relative to the folded constructions. They crosscut the intrusives of early gabbroids and contact the granitoids of the tonalite-granite complex. Lower Triassic conglomerates (Induan Stage) lie transgressively on the rocks of this complex where it is contiguous to the West Primorye SFZ. The complex includes the rocks of two intrusive phases: (1) hornblende-biotite and biotite-granodiorites and (2) leucocratic granophyric granites. Dikes crosscutting the rocks of the complex are represented by aplite-like granites and rare spessartites. The rocks are characterized by leucocratic compositions, heterogeneous porphyry-like structures, and widely developed granophyric microstructures (Vrzhosek and Sakhno, 1990). Granophyric granites are depleted in calcium and potassium and enriched in sodium because of late magmatic (or autometasomatic) albitization of rocks.

In the areas of the Komissarovka, Pogranichnaya, and Tumannaya rivers various endogenetic mineralizations are known. In the areas of carbonate-rock occurrences, granitoids are associated with skarns, greisen-quartz, vein-streaked tin (cassiterite), molybdenum (molybdenite, chalcopyrite), tungsten (wolframite, scheelite), and gold (gold, bismuthine) mineralizations occurring in the zones of intrusive exocontact with terrigenous and volcanic rocks. Spatial and temporal correlations indicate that the endogenetic mineralization manifested in the West Primorye SFZ reflects a Hercynian metallogenic epoch. Its manifestations have not been adequately studied as late Palaeozoic ore-forming events are poorly known. In the West Primorye SFZ, heavy concentrations of aureoles and small placers of sheelite, cassiterite, wolframite, and gold (sometimes with platinum) are common. Within those aureoles are mineralization of native ores: cassiterite (quartz formations), molybdenum and tungsten, lead and zinc (skarn, quartz formations), molybdenum-copper-porphyry, and gold with rare metals (gold-bismuthine-scheelite).

Bodies with streaky veins of cassiterite, wolframite, and quartz, containing small amounts of muscovite, are rare finds in the West Primorye SFZ. Scheelite-quartz mineralization is concentrated in small vein bodies, in veinlet zones, and in clay shales and sandstones. In addition, scheelite, pyrite, chalcopyrite, arsenopyrite, bismuthine, and gold are present. Molybdenite-quartz mineralization is represented by small quartz and quartz-feldspar veins and veinlets. Sometimes molybdenite is associated with more abundant chalcopyrite. In a molybdenum-copper manifestation of the porphyry type in the central part of the West Primorye

SFZ, dispersed pyrite, chalcopyrite, bornite, and, more rarely, molybdenite have been recognized in gabbroids that have undergone biotitization, seritization, and silicification. In the contact part of the gabbroid intrusion, gold-bearing zones of hydrothermally altered rocks have been revealed. Sulphide (with arsenopyrite) mineralization, associated with granite-porphyry intrusions, is known there also.

In the zones of skarnization, mineralization often correlates with skarn composition. In garnet skarns, magnetite ores are common; pyroxene-garnet skarns contain scheelite mineralization associated with fine-scale molybdenite and minor chalcopyrite, galena, and sphalerite; pyroxene skarns show galena-sphalerite (sometimes with scheelite and molybdenite) mineralization. In pyroxene-garnet skarns, scheelite is associated with plagioclase, epidote, quartz, and pyrite. Usually those are seen as finely dispersed impregnations. Most likely they were deposited at the end of skarnization and at the beginning of the quartz-sulphide stages of mineralization of the superposed skarn.

Zones streaked with veins of quartz and rare metals (scheelite-bismuthine) and sulphidized terrigenous clay and black-shale sequences are considered the sources of native gold, which until recently was known only from placers and heavy concentrations of aureoles. Specialized explorations in the West Primorye SFZ in the Komissarovka area have revealed manifestations of gold (Au) and silver (Ag) mineralizations of epithermal type new for the subregion. Sulphide-poor Au and Ag mineralizations are localized in effusive-pyroclastic accumulations. In the volcanogenic and volcanogenic-sedimentary deposits of the latter, mica-andalusite-quartz and mica-quartz epigenetic changes are recognized. The ore bodies of the deposit are described only on the basis of sampling results. They are mineralized deposits of metasomatic rocks with rather irregular (bonanza) Au and Ag distributions. We believe that such ores are of the porphyry type, with finely impregnated and microveinlet-impregnated ore mineralizations. The large, Au-bearing veinlets (0.5–1.5 cm), which are rare in metasomatic rocks, are composed mainly of columnar-grained semitransparent quartz. The mineralization has Ag specialization, and in the ore the Au/Ag ratio varies from 0.03 to 0.3. Au can be associated with Ag, copper (Cu), lead (Pb), and tin (Sn), and occasionally with antimony (Sb) and arsenic (As). The cross sections of Au aureoles can reach 500 m; in their contours, the linear-anomaly width can be 5–80 m. The latter account for 10–20% of an area of high Au concentration. The main mineral concentrations of Au and Ag are due to Au-Ag solid solutions, native Ag, and Ag chalcogenides, visible only under a microscope. Similar fine, rare impregnations of other ore minerals are found in the deposit (sulphides, sulphoarsenides, arsenides, etc.).

The late Palaeozoic (Hercynian) age of the Au-Ag mineralization found in the West Primorye SFZ is demonstrated, on the one hand, by the spatial and structural restriction of some metasomatic mineral veins to the dacite and rhyolite-dacite extrusives belonging to the Upper Permian effusive-pyroclastic deposits and, on the other hand, by the location of epigenetically altered terrigenous and volcanic rocks in the exocontact zones of the late Permian granitoid intrusions.

Gold mineralization in South Primorye

We shall next discuss the few facts known about the higher Au content in Lower Triassic (Induan and Olenekian) deposits in South Primorye, which is distinguished by tectonists as an individual megablock. The first publication on the Au contents of Lower Triassic conglomerates in northeastern Askol'd Island was by Gudkov (1921). Half a century later, A. I. Burago, B. V. Tsoi, and others found a zonal geochemical aureole of Au on the island. Aureoles of that type have been mapped in some areas of Triassic deposits distributed along the northern coast of Peter the Great Bay. Within the latter, fluvial and littoral placers of Au and deposits of native Au-quartz mineralization are found. We shall consider in detail the geological position of Triassic deposits with significant Au contents.

On Askol'd Island, Lower Triassic deposits (Induan Stage) lie on the eroded surfaces of Palaeozoic gabbroids and granitoids and lie, with acute angular unconformity, on the intensively dynamothermally metamorphosed sedimentary-volcanogenic formations that are variously attributed to the Putyatin Suite (Silurian–Devonian?), the Dunai Suite (Lower Permian, P_1), or the Pospelov Suite (Upper Permian P_2). The total thicknesses of the Lower Triassic basal beds are 70–65 m, decreasing from east to west to 40 m. They feature alternations of conglomerates, gravelstones, and sandstones containing isolated lenses and interbeds of siltstones and sandy siltstones with bivalve and ammonoid remains. Psephitic material of conglomerates and gravelstone consists of rhyolite- and felsite-porphyries, hornfels, quartzites, fine-grained quartz, and, more rarely, porphyrites. Such rock associations in situ characterize the areas of Hercynian Au and Au-Ag mineralization that have recently been found in southwestern South Primorye. Sometimes the Triassic deposits and the overlying Jurassic sandy siltstones are intensively pyritized and silicified. Nonoxidized, thin, fine-impregnated pyrite accounts for 3–5% of the rock mass. The Au content in pyritized and silicified rocks is 0.2 g/t, and sometimes 1–1.3 g/t.

The Lower Triassic deposits of the South Primorye Continental Terrane, 24–25 km to the north-northeast of Askol'd Island, are represented by a thicker (150–170-m) series of sandstones and conglomerates overlying, with angular unconformity, all of the more ancient (P_1, P_2, and other) formations. Sandstones prevail over conglomerates and siltstones, with a 7:2:1 ratio. Basal deposits 10–30 m thick are composed of gravelstones and conglomerates, which are replaced upward by a consistent horizon of fine-to-medium-grained calcareous sandstones with coquina lenses.

The psephitic material of the basal deposits contains 70–80% siltstones and 20–30% sandstones and tuff-sandstones, and sometimes quartzite-like rocks and isolated magmatite pebbles (gabbro or granite). The areas of sulphidization, significant in size, are found in the basal deposits. There is evidence of sulphidization of epigenetic origin (e.g., the presence of sulphide microveinlets, rock lightening, signs of Au in matrix and pebble material). The sandstone member of the section is divided into three parts of comparable volumes. The thickness of the lower portion, of calcareous grey sandstones, varies from 30 to 40 m, and the upper portion (dirty-yellow calcareous sandstones with conglomerate interbeds 0.5 m thick) is 60 m thick. They are separated by a packet of obliquely bedded green-grey mica (in some areas poorly sorted), quartz, and quartz-feldspar sandstones containing siltstone interbeds (up to 10 m thick). In the upper portion of the calcareous sandstones, N. G. Suturin has collected the ammonoid *Bajarunia* (it belongs to the *Tirolites-Amphistephanites* Zone, according to Y. D. Zakharov). That has demonstrated the early Triassic (middle Olenekian) age of the layers based on the fossils, but it cannot be excluded that the basal deposits may be of Induan age, by analogy with the Askol'd and Russky islands. Epigenetic impregnation of sulphides and microveinlets of quartz occur between the deposits of the upper and lower portions of calcareous sandstones. They gravitate to the aureoles of silicification and argillization of rocks in the exocontact parts of magmatic bodies.

Sulphidization is seen locally in all irregularly altered lithotypes of Triassic sequences (conglomerates, coquina, sandstones, siltstones). In conglomerates, it is represented by pyrite and, more rarely, pyrrhotite, marcasite, arsenopyrite, and chalcopyrite.

Variations in the nature of impregnation are observed (0.001–3 mm), from poorly dispersed to veinlet-like and spot-clotted (concentrated). These are mainly xenomorphic and hemidiomorphic impregnations. Pyrite crystals, predominantly of cubic habit, are rare. Marcasite is characterized by case-shaped micrograins. Both the psephitic conglomerate material and the matrix have undergone sulphidization. In the areas of local sulphide manifestations, pyrites make up 3–5% of the rock volume.

The sulphides form a variety of aggregates. There is pyrite with microinclusions of pyrrhotite, marcasite, and chalcopyrite. In grain-marcasite masses, "porphyry-like" impregnations of pyrite metacrysts are found. Independent aggregations of pyrrhotite with chalcopyrite also occur. Fine inclusions of quartz, zircon, rutile, and other minerals are sometimes present in sulphide metagrains. In some cases, abundant magnetite is associated with sulphides. In coquina, pyrrhotite and marcasite predominate, and together with them and with pyrite and chalcopyrite, rare sphalerite, galena, and possibly troilite can occur. Pyrrhotite and marcasite form various isometric and vein-like aggregates. Most often, sulphides gravitate to siltstone hydroboudins. Sulphide grains are often corroded. They are closely associated with biotite and other minerals, indicating the supply and regrouping of the surrounding matter due to hornfels formation and hydrothermal metasomatosis of sedimentary series in the magmatite intrusions. In the areas where epigenetic alterations (including sulphidization) of terrigenous-sedimentary series with coquina interbeds were severe, Au mineralization is associated with mica-quartz metasomatites and quartz microveinlets.

Quantitative spectral analysis has shown a wide spectrum of admixed elements in iron sulphides from different horizons of Triassic sequences: Co, Ni, As, Sb, Cu, Pb, Zn, Sn, Au, and Ag. Their distribution variograms are complicated. The pyrite has higher Co and Ni concentrations than does pyrite from many other Au-bearing areas of the Russian Far East. Often, Co and Ni contents are more than 100 g/t. The contents of other admixed elements are comparable to those of pyrite from other Far East sites that had different geneses. Most samples of local pyrite and marcasite contain Au and Ag at less than 3–5 g/t. In individual samples the content of those metals can reach 73 and 86 g/t, respectively. On a distribution diagram, within the 10–50-g/t interval, Au has a poorly pronounced second modal class of concentration. Higher Au contents are seen in sulphide samples taken from areas where late gold mineralization is found in terrigenous-sedimentary rocks.

Sulphur-isotope compositions show the pyrites from conglomerates and sulphidized coquina to be analogous; in the former, $34_s \neq +7.1‰$ (one determination), and in the latter, $34_s \neq +5.3‰$ and $+7.0‰$. The sulphurs in pyrites from hydrothermal sulphide-quartz veins and epigenetically altered magmatic rocks have similar isotopic characteristics, though they are comparatively depleted of it in 34_s. In hydrothermal pyrite from veins and metasomatites, δS values vary from $+1.1‰$ to $+4.4‰$ (three determinations). The available data indicate different sulphur sources and physicochemical conditions for pyrite emplacement in some rocks, but they indicate the same source of sulphur but different physicochemical conditions for pyrite localization in some deposits of different facies, which is supported by the specific features of the sulphide mineral associations.

Studies of ore mineralizations in Triassic deposits on the flanks of Au-bearing areas show that the sulphidization zones in sedimentary series of different ages that are found beyond the ore-bearing areas require more detailed investigations. This is very important for assessing the nature of irregular, anomalous Au deposits in Permian–Triassic terrigenous sequences.

Skarn-scheelite mineralization in central Sikhote-Alin

Permian rocks have long been known in central Sikhote-Alin (Skorokhod, 1941). They have been well described on the basis of their fossils and often have been used to construct schemes to describe the distribution of tectonic events. They attracted even greater attention when the Vostok-2 and Lermontovka skarn-scheelite deposits were discovered. Those deposits are localized in Permian rocks that compose small allochthonous blocks in younger Mesozoic sequences. Studies of the deposits have shown that Permian rocks, limestones especially, greatly affected the formation of ore-containing magmatic systems. The Vostok-2 and Lermontovka deposits occur in different metallogenic zones (Ivanov, 1974). However, they are similar in terms of the nature of the host rocks, mineral composition, and history and time of ore formation. So the Vostok-2 deposit provides a good example of the role of Permian rocks in the formation of igneous rocks and skarn-scheelite ores.

According to recent ideas (Khanchuk, Kemkin, and Panchenko, 1989a), the Vostok-2 deposit is localized within the Samarka accretionary prism of Sikhote-Alin, consisting of turbidite and olistostrome series and exotic allochthonous blocks of different ages (middle Palaeozoic–Jurassic) that are fragments of oceanic and continental-margin deposits (Kemkin and Khanchuk, 1992). Ore bodies are restricted to the olistolith of Upper Permian rocks composed of sandstones, siltstones, and cherts, with limestone interbeds yielding brachiopods. Replacements of those rocks produced the main volume of ore bodies. The sandstones are polymictic and fine-grained and are composed mainly of plagioclase (30–50%), quartz (25–40%), potassium-feldspar (10–30%), and small fragments of siltstones and cherts. The cement is feldspar or quartz-feldspar-mica. The siltstones are made up of fragments of quartz and potassium-feldspar cemented with mica-clay material.

The limestones are light grey to black, with massive or banded textures (different colour bands), and are composed predominantly of calcite, with minor clay material. The total amount of silica, iron oxides, manganese, magnesium, phosphorus, potassium, and sodium is 2–3%. In the central part of the deposit, the limestones are marbled, the sandstones and siltstones have been altered to biotite-hornfels, and the cherts have been altered to quartzites (Stepanov, 1977). Those changes derived from the small ore-bearing central stock of granitoids (outcrop area 0.25 km^2; K-Ar age 112–114 Ma) that is composed of porphyry-like granodiorites, passing gradually into plagiogranites in some areas. In the central stock there is a pipe-like explosive breccia of granodiorite-porphyry, the fragments of which show granodiorites, biotite-granites, biotite-hornfels, and skarns.

Granodiorites are oversaturated with alumina and have high contents of calcium oxide, potassium, magnesium, chrome, vanadium, nickel, and cobalt. Chemical analyses of granodiorites have shown them to be I-type rocks. The ratio of the initial composition of strontium isotopes (^{86}Sr/^{87}Sr) in the granodiorites (Gladkov et al., 1984) is 0.7066, which excludes an upper-crust sialic source for the original melt, but that value is beyond the field of mantle values.

We believe that the rocks characterized by the chemical compositions and strontium-isotope ratio cited resulted from the melting of thick, extensive allochthonous blocks of oceanic crust, that are common in turbidite sequences and are composed of Palaeozoic ophiolites, Triassic basalts, and Jurassic picrites and basalts (Kemkin and Khanchuk, 1992). Those rocks have low ^{86}Sr/^{87}Sr ratios. The initial strontium-isotope ratios in granodiorites indicate that the melt was contaminated with a small amount of sialic material at the expense of Permian rocks, the xenoliths of which (biotite-hornfels) are often found in the granodiorites. Assimilation of Permian sandstones and siltstones, with abundant potassium-feldspar and fragments, by magmatic melts can result in potassium enrichment.

Studies of mineral-forming medium-size inclusions of granodiorites and plagiogranites in quartz show the possible influences of Permian rocks, especially limestones, on the compositions of granitoid melts. Melted inclusions, with accompanying fluid-crystal, gas–liquid, and solid phases, have been found in the granodiorites. The presence of melted inclusions with homogenization temperatures of 860–910°C indicates the melted nature of the granodiorites. We have found no melted inclusions, but only fluid inclusions, in plagiogranites, which suggests a metasomatic mode for their formation in peripheral parts of the magmatic chamber. The high-temperature fluids derived the necessary calcium from Permian limestones. The fluids trapped in inclusions and studied by cryometry and gas chromatography have been found to contain calcium chloride, sodium bicarbonate, and liquid carbon oxide, in addition to magnesium and sodium chlorides. In the gaseous phase, carbon dioxide predominates, and there are minor components of nitrogen, methane, and carbon oxide [the $(CO + CH_4)/CO_2$ ratio is 0.4]. The high contents of calcium, carbon dioxide, and bicarbonate ions in the fluids reflect the assimilation of Upper Permian limestones by a granitoid melt.

The ore bodies in the deposit are plate-like or vein-like beds, which were localized in Upper Permian limestones and biotite-hornfels on their contact with the granodiorite stock, controlled by the fault zone of northeastern strike. The ore bodies have complex inner structures and variable morphologies. All those features are defined by both the behaviour of the ore-controlling fault and the changes in rock composition, especially the regularities of limestone occurrences. Ore bodies are localized immediately on the southwestern extension of the limestone bed, 40 m thick. In the northeastern fault zone, which controlled the mineralization, the Upper Permian rocks were broken in a period prior to ore formation, and tectonic breccias of biotite-hornfels were formed, containing limestones as blocks, lenses, and fragments of different sizes, rather than as beds. The replacements of the brecciated rocks, having variable compositions, sizes, and shapes of fragments, were responsible for the variability in the morphology, composition, thickness, and inner structure of the skarn-ore deposit.

The process of deposit formation included several stages of mineralization (Stepanov, 1977): skarn, greisen, quartz-scheelite, sulphides, and carbonates. The skarn deposits are confined largely to the contacts between limestones and biotite-hornfels and to fractured zones among the biotite-hornfels, and, rarely, they occur among limestones blocks. The skarns are composed mainly of hedenbergite, plagioclase, and amphibole (cummingtonite) and contain minor amounts of grossularite, calcite, vesuvianite, and wollastonite. The skarn lodes have a zonal structure, and the zonality character is defined by the composition of the hosting Permian rocks. For example, skarns localized at the contacts between limestones and biotite-hornfels are characterized by the following metasomatic sequence: limestone-wollastonite skarn, pyroxene skarn, garnet-pyroxene skarn, pyroxene-plagioclase skarn, amphibole-plagioclase metasomatite, and biotite-hornfels. The total thickness of the skarn zone is 5–6 m. In the vein bodies in biotite-hornfels, the zonal composition is different. From the centre of a vein to its periphery, the following zones can be distinguished: pyroxene skarn, pyroxene-plagioclase (sometimes with garnet) skarn, amphibole-plagioclase metasomatite, and biotite-hornfels. The thicknesses of vein-like bodies there do not exceed 2–3 m, but when contiguous fractures are present, they can increase to 25–30 m. Skarn zones in limestones usually are vein-shaped and are composed of pyroxene, with rare vesuvianite and sometimes with wollastonite margins along the vein periphery. The skarn thickness there is less than 0.7 m.

After skarnization, the Permian rocks, together with skarns and granitoids, underwent greisenization. The composition of greisenized rocks and greisens is controlled by the nature of their source rocks. For example, after greisenization, the Permian rocks preserved many of their original unaltered minerals, such as quartz, potassium-feldspar, and plagioclase. Only the large plagioclase grains were sometimes completely replaced by newly formed albite, and the rock cement was sericitized. Biotitic hornfels was greisenized more intensively, and that process resulted in the formation of abundant fine-grained quartz-sericitic altered rocks composed of quartz (45%), muscovite-sericite (25%), chlorite, albite (18%), and calcite (4%), with rare arsenopyrite and pyrrhotite (7%) impregnations. Chemical analysis of the rocks has shown that during the transformation of biotitic hornfels to quartz-sericite metasomatic rocks, only silicon was added, whereas the aluminium, iron, magnesium, calcium, and potassium experienced intensive removal processes. So it appears not to have been accidental that in the solutions trapped in the quartz inclusions in quartz-muscovite metasomatic rocks, highly cationic contents of calcium (7.13 mol/kg H_2O), potassium (0.41 mol/kg H_2O), and magnesium (1.64 mol/kg H_2O) resulted, as determined by water-extract analysis, and of the anions, bicarbonate ions prevailed over chlorine ions (4.67 and 0.80 mol/kg H_2O, respectively).

The influences of the Permian rocks on the formation of quartz-scheelite and sulphide ores have not yet been determined with reasonable accuracy. Those rocks are localized in Permian rocks, not in the original ores, but in their skarnized or greisenized varieties. Indirect evidence suggests that such influences decreased over time, and the distribution was controlled by the tectonic activity in the ore field, aside from physicochemical parameters. The indirect evidence concerns the regular changes in the solution compositions trapped in the quartz at different stages of the process of ore formation. Analyses of water extracts from quartz (Stepanov, 1977) have shown that the calcium and sodium bicarbonate and sulphate types of solutions, with lower potassium, magnesium, and chlorine contents, were kept relatively constant during all stages of ore formation. However, when early quartz-scheelite mineralization passed through the quartz-scheelite-arsenopyrite stage to quartz-scheelite-chalcopyrite-pyrrhotite ore, in the solutions there were decreases in the contents of potassium (from 0.35 to 0.13 mol/kg H_2O), magnesium (from 0.53 to 0.06 mol/kg H_2O), calcium (from 3.44 to 1.7 mol/kg H_2O), and bicarbonate ion (from 2.80 to 2.38 mol/kg H_2O), and there were increases in sulphur (from 1.23 to 2.33 mol/kg H_2O) and chlorine (from 0.47 to 0.91 mol/kg H_2O). Those concentrations in the solutions were lowered at the expense of the components by which the solutions could be enriched through reactions with the Permian rocks. The roles of endogenous sulphur and chlorine increased gradually, and when

ore components accumulated in fluids and the thermodymamic parameters changed – in particular, when the temperature dropped from 400°C (quartz-scheelite ores) to 200°C and below (quartz-sphalerite ore) – that resulted in the formation of ore deposits of complex compositions.

Triassic limestone and boron-, lead-, and zinc-bearing skarns of the Dalnegorsk ore district (Sikhote-Alin accretionary fold system)

The Dalnegorsk ore district lies in the Sikhote-Alin area, 20 km from the coast of the Japan Sea. Lead (Pb) and zinc (Zn) deposits have been mined there for about 100 years. Four active mines, the Nikolaevsky, Partizansky, Verkhnij, and Sadovoe deposits, yield one-third of the Pb and Zn ores produced in Russia. In the middle part of the district is the Dalnegorsk borosilicate deposit, the largest in Eurasia, which totally covers Russia's needs for boron (B). All of those deposits, hosted in Triassic limestone, are of the post-accretionary skarn type.

Limestone in early Cretaceous olistostrome

According to a recent interpretation (Golozubov et al., 1992), the Mesozoic rocks of the Dalnegorsk ore district form the Taukha Terrane, an early Cretaceous accretionary wedge. It consists of packets of turbidite and olistostrome, with blocks of Triassic and Jurassic sandstone, chert, and Jurassic limestone. The olistostrome matrix consists of Berriasian to Valanginian siltstones and sandstones of shelf origin. Blocks of Triassic limestone reach several kilometres in length, with thicknesses of 100–150 m. These limestone bodies generally are accompanied by a train of sedimentary breccias, with small limestone fragments. Vesicular basalts up to 50 m thick are common at the bases of these bodies. Basaltic injections in limestone bodies are observed locally and often are used as evidence of a later origin for the basalts than for the limestone sills. However, cool stratigraphic contacts between the basalts and the overlying limestone are also observed. Limestone fills the vesicles in the underlying basalt. In the Sovetsky deposit, the limestone overlying the basalts carries abundant small basaltic fragments.

The limestone is very pure; clastic material is completely absent. Its content of carbonic calcium, by weight, is about 99%. Dolomitic limestone occurs randomly (2–4% MgO). The limestone is mostly organic and chemogenic. There are some bioherms, usually formed by calcareous algae, corals, bryozoans, and bivalves (see Chapter 20, this volume). There is also oolitic limestone, with abundant foraminiferal fossils clearly distinguishable under a microscope. Judging from the composition of oxygen and carbon isotopes, the limestones were formed in marine basins with normal salinity (Malakhov, Ignatyev, and Nosenko, 1987).

All of these compositional features and the geologic setting of the limestone units are in agreement with the suggestion by Khanchuk et al. (1989b) that at an early stage the limestones crowned the tops of guyots. During the early Cretaceous accretion, they were torn off their volcanic basement, together with some underlying basalts in the zone of subduction of the oceanic plate over the continental plate.

Limestone in the post-accretionary volcanic sequence

A volcanic sequence, overlying the folded olistostrome complex of the Dalnegorsk district, was formed at 80–60 Ma. Its lower part (80–70 Ma) consists predominantly of ignimbrite and tuff of felsic composition (Primorsky Formation), and its upper part (70–60 Ma) of lava and tuffs of rhyodacite and andesite (Dalnegorsk Formation). In the lower part of the section, rather small (20–100 m) limestone, chert, and sandstone blocks are locally present among the tuffs of the Primorsky Formation. The facies features of the host volcanic rocks exclude the possibility of their origin as explosive breccias. These allochthonous units are confined to the boundaries of volcanic depressions. Their occurrences among volcanic rocks resulted from the collapse of steep walls of ring synvolcanic faults at an early stage in the formation of volcano-tectonic depressions (i.e., early Cretaceous destruction of olistostrome created the limestone blocks).

Geologic setting of skarns

All the skarns in the Dalnegorsk district are of the infiltration type, having resulted from interactions between hydrothermal solutions and limestone, producing Pb-Zn- and B-bearing skarns. The distribution of Pb-Zn skarns clearly was controlled by the contacts between limestones and alumosilicate rocks. Only those limestone blocks that are close to or at the contacts with the late Cretaceous volcanic sequence overlying the olistostrome are mineralized. The largest skarn ore-bearing bodies of the Nikolaevsky, Verkhnij, and Partizansky deposits occur exactly at the contacts with felsic volcanic rocks. Allochthonous units of limestone in basal layers of the post-accretionary volcanic sequence also tend to turn into skarns. There, limestone blocks 30–50 m long often are completely replaced by skarns. At the places where the steeply dipping sheeted limestones in the olistostrome contact the siltstones and sandstones of an early Cretaceous matrix, skarns form relatively thin (0.5–3.0 m) bodies. At depths of 500–700 m from the modern surface, skarns often form a tabular unit at the contact surface; 100–200 m below there, they pinch out. Toward the surface, the unit splits into a series of finger-like small bodies. At depth, skarns consist of hedenbergite, ilvaite, and andradite. As to clastic rocks, skarns normally contain axinite. Closer to the surface, skarns consist of 95–100% hedenbergite. The MnO content in hedenbergite changes from 3.0%, by weight, at depth to 12.0% near the modern surface. Skarns consisting of hedenbergite change at the surface to vein-like quartz-calcite bodies.

The sulphides crystallized immediately after the skarn silicates. A skarn-conformable zoning is observed in the distribution of sulphides. Low-iron (~3.0% Fe by weight) sphalerite, with small mineral inclusions of galena, anomalously enriched in bismuth (Bi) (1–4%) and Ag (0.2–0.3%), is observed in the lower part, where sulphides are associated with ilvaite-garnet-hedenbergite-skarns. The Pb/Zn ratio in the ore is below 0.5. Above, galena-sphalerite ore is associated with manganhedenbergite (Pb/Zn ≠ 0.8). The Fe content in sphalerite can reach 7%, and galena contains Ag and Bi colors (25–30 g/t). In the uppermost part, where sulphides occur in a quartz-calcite aggregate, galena (up to 0.1%) rich in Ag and antimony (Sb) predominates (Pb/Zn > 1.0). The sphalerite has a maximum Fe content of 10.0–12.0%. Ag-Sb sulphosalts are always associated with sulphides.

The formation of skarn aggregates, slightly preceding sulphide deposition, resulted predominantly from direct limestone replacement at the contact with alumosilicate rocks. Again, pyroxene, garnet, and ilvaite crystallized at depth, filling small palaeohydrothernal cavities in the limestone and forming radial and banded spherical aggregates of skarn minerals up to 2.0 m across. Such rocks are typical of the Verkhnij and Pervyi Sovetsky deposits. We think their crystallization took place when true solutions turned into gels. B-bearing skarns, which occupy a giant zone (>20,000 m^2, and >1 km below the surface), were formed by intensive replacement of olistolith, a vertically dipping fragment of limestone bed. The composition of the skarns corresponds to that of the ilvaite- and radite-hedenbergite skarns of the root portions of polymetallic ore bodies. In addition to sulphide-poor mineralization (predominantly low-Fe sphalerite, with Bi-Ag-bearing galena), they carry abundant datolite and axinite. Abundant borosilicates in those skarns are caused by redeposition of earlier danburite ore. Several studies (Khetchikov, Gnidash, and Ratkin, 1990; Nosenko et al., 1990; Khetchikov et al., 1991) have shown that before the crystallization of the ilvaite- and radite-hedenbergite skarns, a section of hydrothermal rocks with a vertical span of more than 1 km was formed within the boundaries of the same hydrothermal zone. In the lower part of the section, in the range of depths of 500–1,500 m below the modern surface, intense interactions of hydrothermal solutions with limestones resulted in the formation of grossularite-wollastonite skarns. Above those skarns, from the surface to a depth of 500 m, long, pipe-like cavities up to 30 m across were formed by intensive circulations of solutions. A partial transition of true solutions into gels caused the growth of fine-banded mineral kidneys on the walls of those cavities, consisting of radial wollastonite (95–98%) and datolite, hedenbergite, and calcium-free pyroxene-ferrosilite (2–5%) (Ratkin et al., 1992). The growth of mineral kidneys was changed by extensive danburite crystallization, forming druses of large crystals 1–50 cm across. The danburite accumulations over aggregates of mineral wollastonite kidneys on the walls of cavities are 2–3 cm thick. Magma of an intermediate composition was injected into the cavities, filling the spaces that remained after danburite crystallization. That resulted in the formation of stock-like andesite-porphyry bodies, conformable to those cavities.

All those events preceded the origin of ilvaite-andradite-hedenbergite skarns. During their formation, andesite was turned into skarns, with replacement of plagioclase by orthoclase

and hedenbergite. The danburite was altered most intensively –
it was completely replaced by silicates of newly formed skarns,
datolite, quartz, and calcite. Only fine-zoned pseudomorphs
provide evidence of the former presence of danburite. Field
observations and boron-isotope compositions show that in the
redistribution during danburite replacement the boron concentrated at upper levels of the deposits as axinite and datolite
associated with ilvaite- and radite-hedenbergite skarns (Ratkin
and Watson, 1993).

Age of skarns

There has been no dating of the skarns in the Dalnegorsk
district. Age estimates are based on dikes and ore relationships.
Andesitic dikes intruded after danburite deposition in paleohydrothermal cavities; K/Ar dating indicates an age of about
70 Ma (Pustov, 1990). Field observations have shown that no
notable brecciation of danburite druses trapped in the andesite
framework preceded the intrusion of andesitic magma. This
could be interpreted as evidence for magma injection immediately after the termination of hydrothermal activity. It coincided
with the time of termination of catastrophic volcanic eruptions,
the products of which were ignimbrite fields of the Primorsky
Formation. This allows us to relate the origin of early skarn
boron-bearing (danburite) mineral complexes to the volcano-
plutonic origin of the Primorsky Formation.

Later skarns associated with Pb-Zn and regenerated boron ore
were formed after the intrusion of andesitic bodies (i.e., they are
younger than 70 Ma). Numerous dikes of basalt porphyry, with
K/Ar ages of 62–64 Ma, cut the ore-bearing skarns. The
distribution of later skarn accumulations was controlled by the
volcanic structures of the Dalnegorsk volcano-plutonic complex
(70–60 Ma). Ore-bearing zones occur in the peripheries of
volcano-tectonic depressions, following marginal ring faults,
which allows us to relate those skarns to the formation of the
Dalnegorsk volcano-plutonic complex, suggesting an age of
around 65 Ma.

In the eastern part of the district under discussion, in ore-
bearing volcanic vents, polymetallic ores were formed almost
simultaneously with the eruptive processes (Ratkin et al., 1990).
The ages of those ore-bearing vents are 63–66 Ma (K/Ar dating),
which is in complete agreement with our idea.

Concluding remarks

Units of Triassic limestone in the early Cretaceous olistostrome
and among the post-accretionary volcanic rocks are fragments of
guyots. They played a very important role in controlling the
distribution of post-accretionary boron and Pb-Zn skarn deposits. Areas where allochthonous bodies of Triassic limestone are
spatially combined with ore-bearing late Cretaceous–Palaeogene
volcano-plutonic complexes are the areas of highest ore concentrations. The presence of limestone as blocks, inserted into the
matrix of alumosilicate clastic or volcanic rocks, was a decisive
factor in the development of numerous skarn deposits, because

in that setting (as compared with layered clastic and carbonate
sections) the possibility of an interaction between limestone and
alumosilicate rocks to form skarns would have increased by a
factor of several tens.

The origin of long pillars (to depths of 1,500 m) of ore-bearing
hydrothermal rocks was due to the presence of long hydrothermal cavities in the limestone units. They caused intensive
crystallization of ore and skarn minerals at great depths, with
free removal of the CO_2 reaction product. When prospecting for
skarn mineralization, one should remember that allochthonous
limestone bodies in post-accretionary volcanic rocks provide
evidence for the presence of limestones in an underlying
olistostrome. They are also signs of synvolcanic faults, favorable
for emplacement of skarn deposits.

References

Anhert, E. E. 1928. *Bogatstvo Nedr Dal'nego Vostoka [Mineral
    Riches of the Russian Far East]*. Khabarovsk-Vladivostok:
    Knizhnoe Delo.
Burij, I. V. 1948. Geological report on Surazhevka coal deposit
    (in Russian). *Trudy Dalnevostochnoi Bazy Akademii Nauk
    SSSR, Geologiya* 1:43–44.
Burij, I. V. 1959. Stratigraphy of Triassic deposits of South
    Primorye (in Russian). *Trudy Dalnevostochnogo Po-
    litekhnicheskogo Instituta* 54:3–34.
Burij, I. V., Buryi, G. I., and Zharnikova, N. K. 1993.
    Stratigraphy and interrelations of marine and continental
    Triassic in South Primorye. In *Nonmarine Triassic*, ed.
    S. G. Lucas and M. Morales. *New Mexico Museum of
    Natural History and Science Bulletin* **1993**(3):47–9.
Burij, I. V., and Zharnikova, N. K. 1980. Flora-bearing layers of
    the Ladinian Stage (Middle Triassic) in South Primorye (in
    Russian). *Bulleten Moskovskogo. Obstchestva Ispytatelei
    Prirody, Otdel Geologicheskii* **55**:45–53.
Eliashevik, M. K. 1922. *Vozrast i Kachestvo Yuzhnoussuriiskikh
    Iskopaemykh Uglei [Age and Quality of the South-Ussuri
    Fossil Coals]*. Vladivostok: Izdanie Geologicheskogo
    Komiteta Dal'nego Vostoka.
Gladkov, N. G., Goltsman, Y. V., Borisova, E. D., Pavlov,
    V. A., and Rub, M. G. 1984. Strontium isotope composi-
    tion of some ore-bearing magmatic associations of
    Primorye as indicator of their genesis (in Russian).
    Doklady Akademii Nauk SSSR **275**:1164–9.
Golozubov, V. V., Khanchuk, A. I., Kemkin, I. V., Panchenko,
    I. V., and Simanenko, V. P. 1992. *Taukhinskii i Zhura-
    vlevskii Terreiny (Yuzhnyi Sikhote-Alin) [Taucha and
    Zhuravlevka Terranes (Southern Sikhote-Alin Area)]* (pre-
    print). Vladivostok: Dal'nevostochnoe Otdeleniye Ros-
    siiskoi Akademii Nauk.
Gudkov, P. P. 1921. Askol'd mine and other gold deposits on
    Ascol'd Island. In *Materialy po Geologii Poleznykh
    Iskopaemykh Dalnego Vostoka*, pp. 1–25. Vladivostok:
    Izdaniye Geologicheskogo Komiteta Dal'nego Vostoka.
Ivanov, Y. G. 1974. *Geokhimicheskie i Mineralogicheskie kriterii
    Poiskov Vol'framovogo Orudeneniya [Geochemical and
    Mineral Criteria of Searching for Tungsten Mineralization]*.
    Moskva: Nedra.
Kemkin, I. V., and Khanchuk, A. I. 1992. New data on the age
    of paraautochthone of the Samarka accretionary complex
    of South Sikhote-Alin (in Russian). *Doklady Akademii
    Nauk SSSR* **324**:847–51.

Khanchuk, A. I., Kemkin, I. V., and Panchenko, I. V. 1989a. Geodynamic evolution of Sikhote-Alin and Sakhalin in Paleozoic and Mesozoic time (in Russian). In *Geologiya Tikhookeanskoi Okrainy Azii* [*Pacific Margin of Asia Geology*], ed. A. D. Shcheglov, pp. 218–54. Moskva: Nauka.

Khanchuk, A. I., Nikitina, A. P., Panchenko, I. V., Buryi, G. I., and Kemkin, I. V. 1989b. Paleozoic and Mesozoic guyots of the Sikhote-Alin area and Sakhalin Island (in Russian). *Doklady Akademii Nauk SSSR* **307**:196–9.

Khetchikov, L. N., Gnidash, N. V., and Ratkin, V. V. 1990. Evolution of mineral forming environs on data of studding of pseudomorphs over danburite crystals in cavities of the Dalnegorsk borosilicate deposit (in Russian). *Doklady Akademii Nauk SSSR* **315**:1466–9.

Khetchikov, L. N., Ratkin, V. V., Gnidash, N. V., and Kiselev, V. I. 1991. *Fluidnyi Rezhim Formirovaniya Pozdnikh Produktivnykh Associacii Dal'negorskogo Borosilikatnogo Mestorozhdeniya* [*Fluid Regime of the Origin of Late Ore-Bearing Assemblages of the Dalnegorsk Borosilicate Deposits*] (preprint). Vladivostok: Dal'nevostochnoye Otdeleniye Akademii. Nauk SSSR.

Krassilov, V. A., and Shorokhova, S. A. 1973. Early Jurassic flora of the Petrovka River (Primorye) (in Russian). In *Iskopaemye Flory i Phitogeographiya Dal'nego Vostoka* [*Fossil Flora and Phytostratigraphy of Far East*], ed. V. A. Krassilov, pp. 13–27. Vladivostok: Dal'nevostochnoye Otdeleniye Akademii Nauk SSSR.

Malakhov, V. V., Ignatyev, A. V., and Nosenko, N. A. 1987. On the conditions of borosilicate mineralization of the Dalnegorsk deposit using data on chemical and isotopic composition of carbonates (in Russian). In *Novye Dannye po Mineralogii Dal'nego Vostoka Rossii* [*New Data on Mineralogy of the Far East Russia*], ed. V. K. Finashin, pp. 68–76. Vladivostok: Dal'nevostochnoye Otdeleniye Akademii Nauk SSSR.

Nosenko, N. A., Ratkin, V. V., Logvenchev, P. I., Polokhov, V. P., and Pustov, Y. K. 1990. The Dalnegorsk borosilicate deposit – a product of polychronous skarn replacement (in Russian). *Doklady Akademii Nauk SSSR* **310**:178–82.

Okuneva, T. M., and Zheleznov, A. A. 1977. On Upper Triassic age of the Shetukhe Suite (South Primorye) (in Russian). *Doklady Akademii Nauk SSSR* **232**:879–82.

Pustov, Y. K. 1990. *Skarnovo-Rudnye Mineral'nye Associacii Usloviya ikh Obrazovaniya i Osobennosti Raspredeleniya v Predelakh Partizanskoi Struktury (Dal'negorskii Rudnyi Raion)* [*Skarn-Ore Mineral Assemblages, Conditions of Their Growth, and Features of Their Distribution within the Partizansk Structure (Dalnegorsk Ore District)*]. *Dissertatsiya na Soiskaniye Uchenoi Stepeni Candidata Nauk.* Moskva: Institut Geologii, Mineralogii i Geokhimii (IGEM).

Ratkin, V. V., Khetchikov, L. N., Gnidash, N. V., and Dmitriev, V. E. 1992. On the role of colloids and paleohydrothermal cavities for the formation of rhythmically-banded ore of the Dalnegorsk borosilicate deposit (in Russian). *Doklady Rossiiskoi Akademii Nauk* **325**:1214–17.

Ratkin, V. V., Simanenko, L. F., Kuznetsov, D. N., and Korol, P. V. 1990. Tin-zinc mineralization of the East Sikhote-Alin Volcanic Belt (in Russian). *Geologiya Rudnykh Mestorozhdenii* **1990**(2):68–77.

Ratkin, V. V., and Watson, B. N. 1993. The Dalnegorsk skarn borosilicate deposit – geology and boron source based on isotopic data (southern Far East Russia) (in Russian). *Tikhookeanskaya Geologiya* **1993**(6):95–102.

Skorokhod, V. Z. 1941. *Osnovnye Cherty Geologicheskogo Stroeniya Yuzhnoi Chasti Sovetskogo Dal'nego Vostoka* [*Specific Features of Geological Structure of the South Far East*]. Vladivostok: Primorskii Filial Vsesoyuznogo Geographicheskogo Obstchestva.

Stepanov, G. N. 1977. *Mineralogiya, Petrographiya i Genezis Skarnovo-Sheelit-Sul'phidnykh Mestorozhdenii Dal'nego Vostoka* [*Mineralogy, Petrography, and Genesis of Skarn-Sulphide Deposits of the Far East*]. Moskva: Nauka.

Vrzhosek, A. A., and Sakhno, V. D. 1990. Late Permian granitoidal magmatism of the east activized margin of Sino-Korean Shield (in Russian). In *Problemy Magmatizma i Metamorphizma Vostochnoi Azii* [*Problems of Magmatism and Metamorphism of Eastern Asia*], ed. N. L. Dobretsov and B. A. Litvinovsky, pp. 126–35. Moskva: Nauka.

12 The mid-Permian: major changes in geology, environment, and faunas and some evolutionary implications

J. M. DICKINS

Much attention has recently been given to the major biological changes in the Phanerozoic. The emphasis has been on extinctions, whereas the nature of the recoveries by faunas and floras after the times of major extinctions has not been so fully examined. In this chapter, attention is directed to that latter aspect. Emphasis is also placed on relating the biological developments to the geological changes and examining how those relationships interacted with the environment.

The term "mass extinction" is avoided, because it does not accurately describe most, if not all, of the examples of important extinctions during the Phanerozoic. In a detailed context of time and stratigraphy, it is not easy to say that any change was "abrupt" – see, for example, even the Permian–Triassic extinction (Dickins, 1983; Sweet et al., 1992), often rated as the largest within the Phanerozoic (regarding the Precambrian–Cambrian change as outside the Phanerozoic).

Until recently, correlations involving the mid-Permian (in the twofold Permian subdivision based on the Russian type area) have proved difficult, especially because of the worldwide regression and the associated strong tectonic changes (Dickins et al., 1989). The difficulty in making correlations has obscured both the geological and biological changes. A more nearly satisfactory resolution has become possible by considering the geological and biological aspects together, and that has led to the synthesis that is the subject of this chapter.

Geologic changes

Regression

Earlier (Dickins et al., 1989), the mid-Permian regression was tabulated, *inter alia,* in the Russian Platform–Ural Mountain sequence (the type area of the Permian), in Spitzbergen, in the southern Tethyan region of western Europe, through the Middle East into the Himalayan region, in Japan, in the United States, in Arctic and western Canada, and in eastern and western Australia.

The regression and subsequent transgression have now been documented in many areas: Novaya Zemlya (Korago, Kovaleva, and Trufanov, 1989), Sicily (Catalano, Di Stefano, and Kozur, 1992), France and Germany (Gebhardt, Schneider, and Hoff-

mann, 1991), Greece (Grant et al., 1991), the vast area extending from Turkey through the Arabian Peninsula, Iraq, and Iran (Powers et al., 1966; Edgell, 1977; Kashfi, 1992), Vietnam (Phan, 1991), China (Yang Zunyi, Cheng Yugi, and Wang Hongzhen, 1986), the Russian Far East (Zakharov, Panchenko, and Khancuk, 1992), South America (Argentina, Bolivia, Perú, and Brazil) (Newell, Chronic, and Roberts, 1953; Rocha Campos and Rosler, 1978; Azcuy and Caminos, 1987; von Gosen, Buggisch, and Dimieri, 1990), and an extensive area in North America from Texas to the west-central and western United States and to the southern Canadian Rockies and Alaska (McGugan, 1984; Jenson, 1986; Wardlaw and Grant, 1990; Bamber et al., 1992). They have been documented in arctic Canada and Greenland and in South Africa (Beauchamp, Harrison, and Henderson, 1989a,b; Stemmerick and Hakansson, 1991; Dickins, 1992a). The culmination of the regression and the beginning of the subsequent transgression were associated with the Ufimian boundary, as reported by Dickins (1987a, 1988a, 1992a,b) and as shown in correlation charts (Dickins et al., 1989), corresponding to marked tectonic and magmatic-volcanic changes, as discussed later in this chapter. In some places, where the changes were marked by tectonic and magmatic-volcanic features, the marine sedimentation appears to have been virtually uninterrupted, but the mid-Permian gap can be quite variable, and in some places only the earliest and the latest of the Permian marine stages are represented. These features have caused considerable problems in making correlations, especially involving the middle part of the Permian, where marine deposition often was replaced by hiatus or nonmarine deposition. There are also problems caused by transgressive overlap and confusion in superpositional relationships. In the Tethyan region, however, the recent synthesis by Leven (1993) has extensively tabulated that regressive–transgressive event coinciding with the Bolorian–Kubergandinian boundary, corresponding to the mid-Permian of other regions.

Tectonic changes

In many parts of the world, the beginning of the Upper Permian was marked by the onset of strong folding, structural changes,

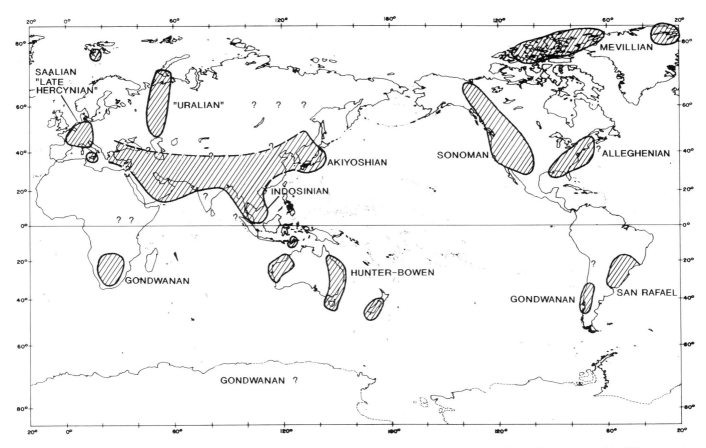

Figure 12.1. Areas with mid-Permian foldings and/or distinctive transgressive unconformities associated with the beginning of the Hunter-Bowen (Indosinian) Orogenic Phase. The terms used in different parts of the world for that tectonic phase are shown.

new basin formation, and substantial transgressive unconformities. Areas where the effects of those processes have been identified are shown in Figure 12.1, together with the names that have been applied to those orogenic events in different parts of the world. It is not possible to describe those features here in detail, but they have been discussed earlier (Dickins, 1987a, 1988a,b, 1989, 1991, 1992a,b, 1993a), along with the substantial supporting data. Such features have recently been described in Queensland by Korsch, Wake-Dyster, and Johnstone (1992). The term "Hunter-Bowen (Indosinian) Orogenic Phase" or folding phase is preferred to describe the worldwide phenomena. Nowhere are the features of that phase better illustrated than in eastern Australia. In the Bowen and Sydney basins and in the New England area, the Lower Permian strata are characterized by tensional faulting, and the mid-Permian strata show the effects of compressional folding, apparently accompanied in places by deep tensional crustal fracturing and transgressive overlap. An unconformity begins at about the Kungurian–Ufimian boundary (the beginning of the Hunter-Bowen). (Although historically the boundary between the Lower and Upper Permian in the Urals type area has been placed between the Kungarian and Ufimian stages, it seems likely that the important change took place during the Ufimian Stage, between the

Solikamsk and the Sheshminsk. That problem is discussed in Chapter 1 in this volume.) Parallel events occurred in southern and eastern Asia (Indosinian). In North America, the transgressive unconformity, accompanied by folding, extends from the western-central and western United States through western Canada as far north as Alaska (Wardlaw and Collinson, 1979, 1986; Bamber et al., 1992). The structural changes that in the United States led to the formation of the Phosphoria Basin are shown, for example, in the isopachs of Baars (1962). In some places, the folding has not yet been linked to the extensive transgressive unconformity, but the overall features leave little doubt that the tectonism was of a similar nature as that in, for example, arctic Canada, the southern part of western Europe, and the Middle East. In the Middle East, the Upper Permian strata in the Arabian Peninsula, Iraq, Turkey, and Iran transgress over pre-Permian strata for more than 1,000 km. In the southern Arabian Peninsula they rest on Lower Permian strata, probably with a hiatus, and in eastern Oman they rest on pre-Permian strata, with basalt at the base, perhaps indicating deep crustal fracturing, as folding appears to be suggested by the flysch-like turbidites and debris flows of the Upper Permian (Powers et al., 1966; Edgell, 1977; Blendinger, 1988; Kashfi, 1992).

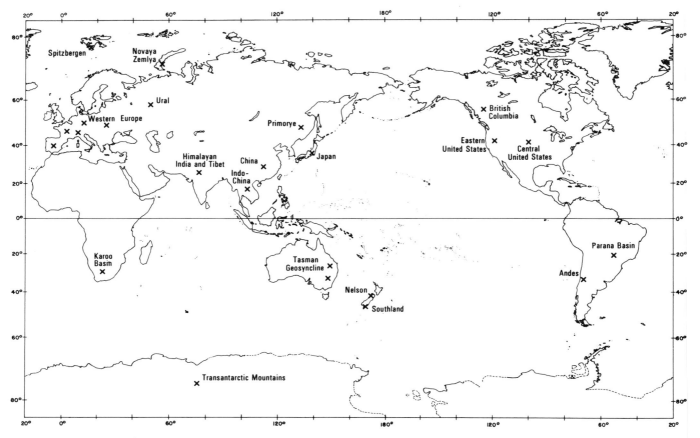

Figure 12.2. Areas of extensive ignimbritic and pyroclastic activities in the late Permian (marked by ×). (From Dickins, 1992a, with permission.)

Magmatic and volcanic activities

The Upper Permian was marked by the onset of a period of intensive, predominantly intermediate-to-acidic igneous activity. Large granitic batholiths and their volcanic equivalents are features of the Upper Permian (and Triassic) deposits on a worldwide scale. The extent of ignimbritic and pyroclastic activities is shown in Figure 12.2 (see also the references cited in the preceding sections of this chapter). The effects of that volcanism were more widespread than what we see at present. Its evidences are particularly strong in deposits of Ufimian age, as, for example, in Queensland and New South Wales, where acidic and intermediate volcanics are widespread and the appearance of such detritus is a feature in the sedimentary sequences.

Murchey and Jones (1992) described a mid-Permian chert event, reflected in siliceous rocks from 23 areas in the circum-Pacific and Mediteranean regions, coeval with the phosphate formation in the central and western United States and Canada. There can be little doubt that those rocks reflect the intensive volcanic and magmatic activities of the mid-Permian described in this chapter. They also lend additional credence to the special character of the "Boreal" fauna considered later in this chapter.

In eastern Australia, eastern Argentina, and Oman, a strong thermal event at the mid-Permian is indicated by the regional metamorphism of Lower Permian sediments. According to my observations, it probably is also reflected in Japan. It may be a widespread feature that in many places is obscured by overprinting of subsequent metamorphic events.

Climate changes

The main features of the climate are shown in Figure 12.3. During the Lower Permian the world climate became warmer, and after the mid-Permian a generally warm climate became universal, with, perhaps, at times, the exceptions of eastern Australia and northeastern Siberia, where warm temperate conditions may have prevailed (Dickins, 1985). A cold fluctuation, however, apparently was present right at the mid-Permian, and although there is no evidence of glaciation, it may have had consequences in relation to faunal changes and migrations.

Faunal changes

According to the information currently available, the faunal changes are seen most clearly in the pelecypods (bivalves) and

Lower Jurassic	*Warm, extensive humid zone* *Distinct arid zone*
Upper Triassic	*Hot distinct arid and humid zones*
Middle Triassic	*Hot distinct arid and humid zones*
Lower Triassic	*Hot and dry worldwide*
Upper Permian	*Warm and hot worldwide* *No glacials*
lower Upper Permian	*Cool fluctuation No moraine identified*
Remainder of Lower Permian	*No moraine identified* *"Dropstones" in Australia*
Sakmarian (Tastubian)	*Eustatic sea-level rise*
Asselian	*Main glaciation* *Widespread terrestrial glaciation*
Upper Carboniferous	*Glaciation limited* *Mainly montaine, some moraine*
Lower Carboniferous	*Warm and equable,* *no authenticated glaciation*
Upper Devonian	*Warm worldwide*

20–4/75

Figure 12.3. Summary of climatic conditions from the late Devonian to the early Jurassic. (From Dickins, 1993b, with permission.)

the tetrapods, although there were other important changes, especially in the fusulinids. The marine fauna of the Bowen Basin, Queensland, well illustrates the general features. Of the 83 species (mostly pelecypods, gastropods, and brachiopods) recorded from beds regarded as Ufimian and Kazanian by Dickins and Malone (1973), 38 species have been identified as also occurring below that level. In the underlying Upper Artinskian and Kungurian formations, 62 species are recorded, of which 25 do not range into the overlying formations. A feature of the Ufimian and Kazanian faunas, however, is that they show much more extensive worldwide relationships.

Pelecypods (bivalves)

Information on the pelecypods (and brachiopods) has been graphically tabulated by Astafieva and Yasamanov (1989). Marked changes occurred at the mid-Permian. The changes and proliferations at that time in Japan have been described by

Nakazawa and Newell (1968), and the worldwide changes have been discussed by Dickins et al. (1989) and Dickins (1993a). The new, diverse faunas proliferated in eastern Australia, New Zealand, the west-central United States, Spitzbergen, the northern Biarmian province of the Russian federation, including Novaya Zemlya, the Urals, Italy (Sosio), the Arabian Peninsula (Oman), Japan, and South America.

The new faunal elements are best known from the United States, where a large fauna of schizodids, with many new genera, has been described by Newell and Boyd (1975), and diverse faunas have been described by Chronic (1952) and Ciriacks (1963). A particular feature of that fauna was the appearance of a number of new genera of pectinoids that were very widespread. *Permophorus* first appeared, and other distinctive permophorids and the multivincular pterioid forms such as *Bakewellia* became well established. *Glytoleda,* previously known only from western and eastern Australia and New Zealand, appeared in Novaya Zemlya. A particularly rich development of that fauna can be found in the Sosio beds of Sicily.

Brachiopods

The distribution of the brachiopods is not as well established as that for the pelecypods, and the faunas from Oman (Lusaba Limestone) and Thailand (Rat Buri Limestone) remain to be assessed. Important groups that appeared included spiriferoids and productoids (Dickins et al., 1989). They were very widespread during the Ufimian and Kazanian in the Boreal, Austral, and Equatorial (Tethyan) regions.

Ammonoids

Fewer changes seem to have occurred in the ammonoids than in the pelecypods and the brachiopods, but the Ufimian faunas are yet to be adequately described. Hangovers from the Lower Permian, such as *Daubichites* and *Perrinites,* may also obscure the picture. Certainly by the Kazanian, there had been marked changes in the appearances of the *Waagenoceras* fauna – see the comment on the Oman fauna by Glenister and Furnish (1988, p. 62).

Fusulinids

Although some distinctive changes accompanied the rise of the *Cancellina* and *Armenina* faunas and the development of the *Misellina* fauna in the Ufimian, the major changes appear to have occurred in the Kazanian, with the development of the *Neoschwagerina* fauna. As in the case of the ammonoids, it is not clear whether that appearance results from ongoing difficulties with the Ufimian correlations or whether it reflects a recovery period after the rigorous conditions that existed in the late part of the early Permian, with the richest faunal evolution then accompanying the full development of the transgressive phase in the Kazanian (and Midian), after which the fauna was again affected by the regression and geological events of the Midian–Dzhulfian. Leven (1993) has tabulated the important changes in

	MAJOR REGRESSION–TRANSGRESSION	STRONG VOLCANISM	SIGNIFICANT CLIMATIC CHANGE	ORDER OF CHANGE IN BIOTA (1, 2 & 3)
CRETACEOUS–TERTIARY	X	X	X	1
TRIASSIC–JURASSIC	X	X	–	2
PERMIAN–TRIASSIC	X	X	X	1
MID–PERMIAN	X	X	–	3
PERMIAN–CARBONIFEROUS	X	–	?	3
DEVONIAN–CARBONIFEROUS	X	–	–	3
FRASNIAN–FAMENNIAN	X	?	–	2

20/0/22

Figure 12.4. Comparison of major faunal changes from the Frasnian–Fammennian to the Cretaceous–Tertiary.

the Tethyan in the Kubergandinian (= Roadian) and "Upper" Ufimian. (Although consideration of the Midian and Dzhulfian is outside the scope of this chapter, the Ufimian, Kazanian, and Midian deposits show, in terms of fauna and sedimentary character, relatively gradual changes, as compared with the more dramatic changes seen at the lower and upper boundaries of that group of stages, i.e., at the boundaries with the Kungurian below and the Dzhulfian above.)

Conodonts

Although in the Permian the conodont faunas were relatively few in numbers of genera and species, and the species tended to be rather persistent as compared with many other taxa, there was an important change in the Ufimian (Wardlaw and Grant, 1990) involving the replacement of *Neogondolella idahoensis* by *Neogondolella serrata* in the lower part of the Road Canyon Formation in Texas. The Road Canyon, according to the correlations accepted here, is regarded as Ufimian.

Tetrapods

A detailed examination of nonmarine tetrapods from the Permian and Lower Triassic has recently been published by Maxwell (1992). He concluded that a major period of extinction and diversification was associated with the Ufimian. That is clearly shown in his figures, and it parallels the available information based on marine invertebrates.

Synthesis and conclusions

There seems little doubt that the regression that culminated at the end of the Lower Permian had important deleterious effects on the faunas, marine and nonmarine, disrupting the existing environment, both in the sea and on land, and those effects were exacerbated by the tectonic changes that began at the mid-Permian. Together, those changes modified and/or destroyed and replaced the existing marine basins and the living areas on land.

It appears that large-scale volcanic activity that marked the beginning of the Upper Permian not only had deleterious effects on the faunas but also affected the ways in which the newly adapted faunas developed. No effect from the arrival of an extraterrestrial body is known.

Whether the cold fluctuations resulted from the volcanic activity or from another cause is not clear. No doubt it had effects on the faunas, and it may have facilitated the migration of more temperate elements across the equatorial regions, thus explaining, for example, the appearance of *Glytoleda* in Novaya Zemlya. After the cold fluctuation, apparently there was rapid warming, which allowed the newly developing faunas to spread rapidly over virtually the entire globe, diversifying and proliferating in the new, connected seas associated with the widespread marine transgression. A broad Tethys Sea ("Neotethys") was formed at that time, from southern Europe through the Middle East and across southern Asia to China, Japan, and Primorye, with links to Australia and New Zealand, and with a large seaway across the United States, from Texas through British Columbia, Alaska, arctic Canada, and Greenland to Spitzbergen, Novaya Zemlya, and the Urals. New faunas, both in the sea and on land, developed not in conditions of isolation but in conjunction with the further development of diverse faunas. This is shown, for example, in the diversification of the schizodids described by Newell and Boyd (1975) from the Phosphoria Formation of the United States.

Comparisons of the changes at the mid-Permian with other major faunal changes during the Phanerozoic are shown in Figure 12.4. The mid-Permian, for example, differed from the Permian–Triassic transition mainly in that it featured lesser numbers of extinctions. More taxa persisted longer, as reflected in the overlying beds, and the recovery period was not so traumatic. The tabulation suggests that the greater the number of factors that are involved in environmental changes, and the stronger those changes are, the greater will be the changes in the biota. Compared, for example with the Permian–Triassic transition, the pre-mid-Permian regression and tectonic activities appear to have been considerably less disruptive of the environ-

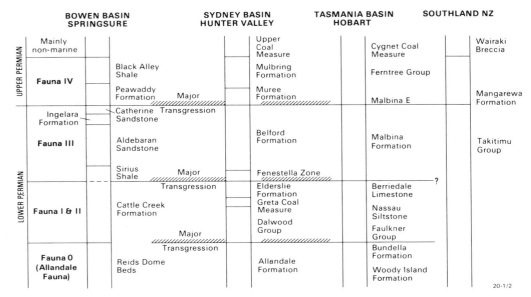

BOWEN BASIN SPRINGSURE			SYDNEY BASIN HUNTER VALLEY	TASMANIA BASIN HOBART	SOUTHLAND NZ

Figure 12.5. Marine faunal changes during the Permian in Australia and New Zealand. (From Dickins, 1984, with permission.)

ment and to have exerted less stress on the biotas, although, indeed, those were precursors leading up to the Permian–Triassic. At the Permian–Triassic boundary, the continental shelves had been largely or entirely obliterated, and the biotas were vulnerable to stress from other environmental factors operating at the time. Similarly with the recovery: After the mid-Permian, widespread favourable conditions were quickly established, leading to rapidly increasing diversity and proliferation, whereas conditions would remain relatively adverse throughout the Lower Triassic.

Some subtle features are, however, apparent. If the faunas from Queensland, New South Wales, and Tasmania are compared (Figure 12.5), differences that apparently reflect water temperatures can be detected (Dickins, 1978, 1984), and yet common morphological changes in pelecypods and brachiopods can allow reliable correlations. Presumably those morphological changes were sufficiently strong that they could override the effects of water-temperature differences, indicating better adaptation to the new conditions, perhaps reflecting changes in the chemical and sedimentary environments, such as the chemical effects of the particular volcanic activity and the effects of basin shape and atmospheric conditions, not at present obvious in the geological record. This is borne out by the development of the northern "Boreal" faunal province at the mid-Permian.

That province is characterized especially by a distinctive brachiopod fauna (Dickins et al., 1989) and sedimentologically by chert and evaporite and a restricted marine environment. Those conditions reflected the widespread type of basin development, apparently widespread acidic volcanic activity, and similar water temperatures and climatic conditions. The Boreal province occupied the northern part of the seaway described earlier, stretching from the west-central United States to Spitzbergen. Fusulinids were absent, and although that has been ascribed to cool water conditions, it more likely was due to the chemical and

physical conditions, which, along with the faunal relationships, suggest, to the contrary, warm water.

Summary

The faunal and geological changes associated with the mid-Permian have been described, and comparisons have been made with some other times of major changes in the Phanerozoic.

The late Lower Permian was marked by widespread regression, and the beginning of the Upper Permian (the twofold Permian subdivision) brought widespread tectonic movements and foldings associated with strong intrusive activity and explosive volcanic activity, which led to transgression and structural changes that resulted in extensive new basin development. Although extinction was not a particularly prominent feature, there was rapid spreading of the diversified faunas developed at the beginning of the Upper Permian, and distinctive assemblages occupied the new palaeogeographical provinces. A warm climate after the mid-Permian apparently encouraged migration.

Comparisons with the biological changes at the Carboniferous–Permian, Permian–Triassic, Triassic–Jurassic, and Cretaceous–Tertiary boundaries indicate that the greater the coincidence of major changes in the environment, the greater the magnitude of the biological changes.

The conclusion is that evolution is very sensitively and directly related to environmental change. Isolation is not necessary for specific and generic changes, which might indeed be encouraged by the course of diversification and proliferation.

Acknowledgments

I appreciate the help and collaboration of many individuals and organizations in many parts of the world. If there is any lasting merit in this study, it is largely because of the help I have received from friends and colleagues. Such collaboration has been a source of great pleasure and satisfaction. I accept sole responsibility for any mistakes or misconceptions that may remain in this report. I must especially thank Dr. N. W. Archbold, of Rusden Campus, University of Deakin, whose collaboration on Permian correlations has been invaluable, and Professor Yang Zunyi, of the China University of Geosciences, Beijing, who involved me in IGCP Project 203, dealing with the Permian–Triassic boundary sequences, which prompted me to look at other times of significant biologic and associated geologic changes in the Phanerozoic.

References

Astafieva, K. A., and Yasamanov, N. A. 1989. The evolution of late Paleozoic marine invertebrates in relation to changes in the environment (in Russian). *Transactions of the USSR Academy of Sciences* 309:231–3 (translation by Scripta Technica, Inc., 1991).

Azcuy, C. L., and Caminos, R. 1987. Diastrofisma. In *El Sistema Carbonifero en la Republica Argentina*, ed. S. Archangelsky, pp. 240–51. Cordoba, Argentina: Academia Nacional de Ciencias.

Baars, D. L. 1962. Permian system of the Colorado Plateau. *Bulletin of the American Association of Petroleum Geologists* 46:149–218.

Bamber, E. W., Henderson, C. M., Richards, B. C., and McGugan, A. 1992. Carboniferous and Permian stratigraphy of the Foreland Belt. In *Geology of Canada. Vol. 4: Geology of the Cordilleran Orogen in Canada*, ed. H. Gabrielse and C. J. Yorath, pp. 242–65. Geological Survey of Canada.

Beauchamp, B., Harrison, J. C., and Henderson, C. M. 1989a. *Upper Palaeozoic Stratigraphy and Basin Analysis of the Sverdrup Basin, Canadian Arctic Archipelago. Part 1: Time Frame and Tectonic Evolution.* Current research, part G, paper 89–1G, pp. 105–13. Geological Survey of Canada.

Beauchamp, B., Harrison, J. C., and Henderson, C. M. 1989b. *Upper Palaeozoic Stratigraphy and Basin Analysis of the Sverdrup Basin, Canadian Arctic Archipelago. Part 2: Transgressive–Regressive Sequences.* Current research, part G, paper 89–1G, pp. 115–24. Geological Survey of Canada.

Blendinger, W. 1988. Permian to Jurassic deep water sediments of the Eastern Oman Mountains: their significance for the evolution of the Arabian margin of the South Tethys. *Facies* 19:1–32.

Catalano, R., Di Stefano, P., and Kozur, H. 1992. New data on Permian and Triassic stratigraphy of western Sicily. *Neueus Jahrbuch für Geologie und Paläontologie, Abhandlungen* 184:25–61.

Chronic, H. 1952. Molluscan faunas from the Permian Kaibab Formation, Walnut Canyon, Arizona. *Geological Society of America Bulletin* 63:95–166.

Ciriacks, K. W. 1963. *Permian and Eotriassic bivalves of the Middle Rockies.* Bulletin 125. New York: American Museum of Natural History.

Dickins, J. M. 1978. Climate of the Permian in Australia: the invertebrate faunas. *Palaeogeography, palaeoclimatology, Palaeoecology* 23:33–46.

Dickins, J. M. 1983. Permian to Triassic changes in life. *Australasian Association of Palaeontologists Memoirs* 1:297–303.

Dickins, J. M. 1984. Evolution and climate in the Upper Palaeozoic. In *Fossils and Climate*, ed. P. Brenchley, pp. 317–27. New York: Wiley.

Dickins, J. M. 1985. Late Palaeozoic climate with special reference to invertebrate faunas. In *Compte Rendu Nuevieme Congres International de Stratigraphie et de Geologie du Carbonifere*, vol. 5, ed. J. T. Dutro, Jr., and H. W. Pfeffercorn, pp. 394–402. Carbondale: Southern Illinois University Press.

Dickins, J. M. 1987a. Tethys – a geosyncline formed on continental crust? In *Shallow Tethys*, vol. 2, ed. K. G. MacKenzie, pp. 149–58. Rotterdam: Balkema.

Dickins, J. M. 1987b. A history of research on Hunter Fault System or "Lineament." *Earth Sciences History* 6:205–13.

Dickins, J. M. 1988a. The world significance of the Hunter/Bowen (Indosinian) mid-Permian to Triassic Folding Phase. *Memorie della Società Geologica Italiana* 34:345–52.

Dickins, J. M. 1988b. The world significance of the Hunter-Bowen (Indosinian) Orogenic Phase. In *Advances in the Study of the Sydney Basin. Proceedings of the Twenty-Second Symposium*, pp. 69–74. Department of Geology, University of Newcastle.

Dickins, J. M. 1989. Major sea level changes, tectonism and extinctions. In *Compte Rendu Onzieme Congres International de Stratigraphie et de Geologie du Carbonifere, Beijing, China*, vol. 4, ed. Yin Yugan and Li Chien, pp. 135–44. Nanjing: Nanjing University Press.

Dickins, J. M. 1991. Permian of Japan and its significance for world understanding. In *Shallow Tethys 3*, ed. T. Kotaka et al., pp. 343–51. Special publication 3. Sendai: Saito Ho-on Kai.

Dickins, J. M. 1992a. Permian geology of Gondwana countries: an overview. *International Geology Review* 34:986–1000.

Dickins, J. M. 1992b. Permo–Triassic orogenic, paleoclimatic, and eustatic events and their implications for biotic alteration. In *Permo–Triassic Events in the Eastern Tethys*, ed. W. C. Sweet, Yang Zunyi, J. M. Dickins, and Yin Hongfu, pp. 169–74. Cambridge University Press.

Dickins, J. M. 1993a. Permian bivalve faunas – stratigraphical and geographical distribution. *Compte Rendu Douzieme Congres International de la Stratigraphie et Geologie du Carbonifere et Permien, Buenos Aires, Argentina*, vol. 2, pp. 523–36. Buenos Aires.

Dickins, J. M. 1993b. Climate of the Late Devonian to Triassic. *Palaeogeography, Palaeoclimatology, Palaeoecology* 100: 89–94.

Dickins, J. M., Archbold, N. W., Thomas, G. A., and Campbell H. J. 1989. Mid-Permian correlation. *Compte Rendu Onzieme Congress International de Stratigraphie et de Geologie du Carbonifere, Beijing, China*, vol. 2, ed. Yin Yugan and Li Chien, pp. 185–98. Nanjing: Nanjing University Press.

Dickins, J. M., and Malone, E. J. 1973. *Geology of the Bowen Basin, Queensland.* Bulletin 130. Canberra: Bureau of Mineral Resources.

Dickins, J. M., Shah, S. C., Archbold, N. W., Jin Yugan, Liang Dingyi, and Liu Benpei. 1993. Some climatic and tectonic implications of the Permian marine faunas of Peninsular India, Himalayas and Tibet. In *Gondwana Eight: Assembly, Evolution and Dispersal*, ed. R. H. Findlay, R. Unrug, M. R. Banks, and J. J. Veevers, pp. 333–42. Rotterdam: Balkema.

Edgell, H. S. 1977. The Permian System as an oil and gas reservoir in Iran, Iraq and Arabia. In *Second Iranian Geological Symposium,* pp. 161–87. Teheran.

Gebhardt, U., Schneider, J., and Hoffmann, N. 1991. Modelle zur stratigraphie und beckenentwicklung im Rotliegenden der Norddeutschen Senke. *Geologisches Jahrbuch, Reihe A* 127:405–27.

Glenister, B. F., and Furnish, W. M. 1988. Patterns in stratigraphic distribution of Popanocerataceae, Permian ammonoids. *Senckenbergiana Lethaea* 68:43–71.

Grant, R. E., Nestell, M. K., Baud, A., and Jenny, C. 1991. Permian stratigraphy of Hydra Island, Greece. *Palaios* 6:479–821.

Jenson, J. 1986. Stratigraphy and facies analysis of the Upper Kaibab and Lower Moenkopi Formations in southwest Washington County, Utah. *Brigham Young University Geology Studies* 33:21–42.

Kashfi, M. S., 1992. Geology of the Permian "super-giant" gas reservoirs in the Greater Persian Gulf Area. *Journal of Petroleum Geology* 15:465–80.

Korago, E. A., Kovaleva, G. N., and Trufanov, G. V. 1989. Formations, tectonics and history of geologic development of the Kimmerides of Novaya Zemlya. *Academy of Sciences of the USSR, Geotectonics* 23:497–514 (English translation by the American Geophysical Union and the Geological Society of America).

Korsch, R. J., Wake-Dyster, K. D., and Johnstone, D. W. 1992. Seismic imaging of Late Palaeozoic–Early Mesozoic extensional and contractional structures in the Bowen and Surat basins, eastern Australia. *Tectonophysics* 215:273–94.

Leven, E. Y. 1993. Main events in Permian history of Tethys and fusulinids. *Stratigraphy and Correlation* 1:51–65 (translation).

McGugan, A. 1984. Carboniferous and Permian Ishbel Group stratigraphy, North Saskatchewan Valley, Canadian Rocky Mountains, western Alberta. *Bulletin of Canadian Petroleum Geology* 32:372–81.

Maxwell, W. D. 1992. Permian and Early Triassic extinction of non-marine tetrapods. *Palaeontology* 35:571–83.

Murchey, B. L., and Jones, D. L. 1992. A mid-Permian chert event: widespread deposition of biogenic siliceous sediments in coastal, island arc and oceanic basins. *Palaeogeography, Palaeoclimatology, Palaeoecology* 96:161–74.

Nakazawa, K., and Newell, N. D. 1968. *Permian Bivalves of Japan.* Faculty of Science, Kyoto University Memoirs, Series of Geology and Mineralogy no. 25.

Newell, N. D., and Boyd, D. W. 1975. *Parallel Evolution in Early Trigoniacian Bivalves.* Bulletin 154. American Museum of Natural History.

Newell, N. D., Chronic, B. J., and Roberts, T. G. 1953. *Upper Paleozoic of Peru.* Memoir 58. Geological Society of America.

Phan, C. T. 1991. Stratigraphic correlation of Permian and Triassic in Vietnam. In *Shallow Tethys 3,* pp. 359–70. ed. T. Kotaka, J. M. Dickins, K. G. McKenzie, K. Mori, K. Ogasawara, and G. D. Stanley, Jr. Special Publication 3. Sendai: Saito Ho-on Kai.

Powers, R. W., Ramirez, L. F., Redmond, C. D., and Elberg, E. L., Jr., 1966. *Geology of the Arabian Peninsula.* Professional paper 560-D. United States Geological Survey.

Rocha Campos, A. C., and Rosler, O. 1978. Late Paleozoic faunal and floral successions in the Parana Basin, southeastern Brazil. *Boletim IG Instituto de Geosciencias Universidade de São Paulo* 9:1–16.

Stemmerick, L., and Hakansson, E. 1991. Carboniferous and Permian history of the Wandel Sea Basin, North Greenland. In *Sedimentary Basins of North Greenland,* pp. 141–51. ed. J. S. Pell and M. Sonderholm. Publication 160. Godthaab: Gronlands Geologiske Undersogelse.

Sweet, W. C., Yang Zunyi, Dickins, J. M., and Yin Hongfu (eds.) 1992. *Permo–Triassic Events in the Eastern Tethys,* Cambridge University Press.

von Gosen, W., Buggisch, W., and Dimieri, L. V. 1990. Structural and metaphoric evolution of the Sierras Australes (Buenos Aires Province/Argentina). *Geologische Rundschau* 79:797–821.

Wardlaw, B. R., and Collinson, J. W. 1979. Youngest Permian conodont faunas from the Great Basin and Rocky Mountain regions. In *Conodont Biostratigraphy of the Great Basin and Rocky Mountains,* ed. C. A. Sandberg and D. L. Clark, pp. 151–63. Studies no. 26. Provo, UT: Brigham Young University.

Wardlaw, B. R., and Collinson, J. W. 1986. Paleontology and deposition of the Phosphoria Formation. *Contributions to Geology, University of Wyoming* 24:107–42.

Wardlaw, B. R., and Grant, R. E. 1990. *Conodont Biostratigraphy of the Permian Road Canyon Formation, Glass Mountains, Texas.* Bulletin 1895A. U.S. Geological Survey.

Yang Zunyi, Cheng Yuqi, and Wang Hongzhen. 1986. *The Geology of China.* Oxford Monographs on Geology and Geophysics, no. 3.

Zakharov, Y. D., Panchenko, I. V., and Khancuk, I. (eds.) 1992. *A Field Guide to the Late Paleozoic and Early Mesozoic Circum-Pacific Bio- and Geological Events.* Publication 89. Far Eastern Geological Institute, Russian Academy of Sciences, Far Eastern Branch.

13 Variations in the disappearance patterns of rugosan corals in Tethys and their implications for environments at the end of the Permian

YOICHI EZAKI

Both before and after the biologic events near the Permian–Triassic boundary, large-scale extinctions and new appearances of organisms occurred at higher taxonomic levels. Each taxonomic unit has its own terminal phylogenetic history, showing a distinctive disappearance pattern in various ways. Permian corals have been reported from many regions of the world, but the Rugosa became extinct at the end of the Permian (Oliver, 1980; Ezaki, 1989), first disappearing in the Boreal province and finally in the Tethys province at the latest Changxingian. Although the local disappearance events of rugose corals obviously were heterochronous, all Rugosa had disappeared by the time of formation of the mixed-fauna beds described by Yin (1985). There were no Triassic representatives, which makes positive biostratigraphic determination of the Permian–Triassic boundary using Rugosa impossible. The occurrences of late Permian corals were restricted geographically, but specific groups survived in conditions that were suitable for them. Permian Rugosa provide important case examples of the patterns and processes of the end-Permian extinction and their relationships to environments.

Upper Permian stratigraphy and interregional correlations have been intensively studied in each Tethyan region. Stratigraphic schemes for the Upper Permian have been proposed, for example, by Kotlyar et al. (1989) and Nakazawa (1993), and the standard Permian stratigraphy is based on detailed biostratigraphic analyses, especially using fusulinids, conodonts, and ammonoids.

This chapter reviews the subject of extinction of Rugosa at the end of the Permian and summarizes the Permian faunas at several reference sections in terms of their succession from the middle Permian. The various faunal successions are compared thoroughly in order to extract a general pattern of rugosan disappearance and to understand the relationships to environments at the end of the Permian. I shall also discuss the importance of recognizing a "boundary difference" at the Permian–Triassic boundary, defined in various manners.

The specifics of local faunas are treated in the sources cited along with each explanation. A threefold Permian subdivision is adopted here, for convenience. The Lower–Middle Permian boundary is equivalent to the Kungurian–Ufimian boundary,

and the Upper Permian begins at the base of the *Codonofusiella-Reichelina* Zone (Midian–Dzhulfian boundary).

Extinction of Rugosa

The existence of Triassic Rugosa (Ruzhentsev and Sarycheva, 1965) and of direct rugosan descendants to the Scleractinia (Iljina, 1965, 1984) has been reported biostratigraphically and phylogenetically. Progress in biostratigraphy near the Permian–Triassic boundary has shown either that units considered lowermost Triassic should be placed in the uppermost Permian in the Transcaucasus (Teichert, Kummel, and Sweet, 1973) or that Permian corals were included in Triassic strata as reworked fossils in eastern Greenland (Teichert and Kummel, 1972). Rugose corals are no longer recognized in the mixed-fauna beds at the Permian–Triassic boundary (Yin, 1985), and the group had become extinct by the end of the Permian.

The possibility of direct rugosan origins for the Scleractinia has been argued many times. Iljina (1965, 1984), for example, proposed a theory of direct origin for the Scleractinia. Iljina described plerophyllid corals showing morphologic similarities to the Scleractinia, such as cyclic insertion of septa and radial arrangement of septa, from the Upper Permian of the Transcaucasus (Dzhulfa and Dorasham formations). Ezaki (1989) examined plerophyllids from the Upper Permian of Iran and confirmed the peculiar characters to which Iljina had attributed special phylogenetic importance. Ezaki showed that those characters were merely transitory, having resulted from spatial and structural restrictions in the inner corallite, and showed no phylogenetic relationship to the Scleractinia. Neither close intermediate forms between Rugosa and Scleractinia nor any rugosan groups ancestral to scleractinian corals have been found. The Rugosa became extinct without giving rise to any descendants.

Extinction patterns

Late Permian Rugosa are widely, though sporadically, distributed, from Slovenia and Hungary in the west, through Greece, Turkey, the Transcaucasus, Iran, Pakistan, South China, Japan, and South Primorye, to Omolon in the east. They were distributed notably in the areas of eastern Tethys, but even in western Tethys, such as presented-day Iran and the Transcaucasus, the Rugosa were abundant, though low in taxonomic diversity (e.g., Iljina, 1965; Flügel, 1971; Kropatcheva, 1989; Ezaki, 1991). Fedorowski (1989) summarized the overall Permian coral distributions and successions.

At present, it is difficult to discriminate late Permian faunal provinces, even within the Tethys area, owing to restrictions in occurrences and taxonomic diversity. However, we can describe the disappearance pattern for the Rugosa, based on regional late Permian faunas and their stratigraphic successions.

Iran (Abadeh and Julfa) and the Transcaucasus

The Permian has been thoroughly studied lithologically and biostratigraphically, using ammonoids, conodonts, fusulinids, and other groups [e.g., Ruzhentsev and Sarycheva, 1965; Stepanov, Golshani, and Stöcklin, 1969; Taraz, 1971, 1974; Teichert et al., 1973; Iranian-Japanese Research Group (IJRG), 1981]. Although the Upper Permian strata are paraconformably overlain by the Triassic shale beds in the Abadeh region (IJRG, 1981), the *Pleuronodoceras occidentale* Zone was found above the *Paratirolites* Zone in the Transcaucasus, validating the uppermost Permian as equivalent to the uppermost Changxingian (Zakharov, 1988). The $\delta^{13}C$ values are persistently high for lower Dorashamian strata in the Julfa region, but gradually decrease upward and drop remarkably in the Permian–Triassic boundary marls (Baud, Magaritz, and Holser, 1989).

Permian corals have been described by several workers (e.g., Iljina, 1965; Flügel, 1968, 1971; Kropatcheva, 1989; Ezaki, 1991). The middle Permian fauna included colonial waagenophyllids, such as *Ipciphyllum*, whereas the late Permian was dominated by highly endemic, solitary Rugosa (plerophyllids) (e.g., Iljina, 1965; Ezaki, 1991). Massive Wentzelellinae (having tertiary septa) and yatsengiids disappeared in the Midian, followed by Waagenophyllinae (having two orders of septa), and finally by non-dissepimented, solitary corals (*Pentaphyllum* and *Ufimia*) as survivors into the late Dzhulfian and the Dorashamian. Hence, the simpler corals disappeared higher in the stratigraphic record. The Permian rugose corals disappeared in succession, showing a distinctive disappearance pattern (Ezaki, 1991, 1993b), thus favouring the general trend indicated by Flügel (1970).

The benthic faunas, consisting of solitary rugose corals and brachiopods, flourished in the early Dzhulfian, but they were finally replaced by pelagic faunas, including cephalopods and conodonts (IJRG, 1981), in response to the prevalence of an open-marine, pelagic environment caused by the Dorashamian sea-level transgression (Altiner et al., 1980). The Dorashamian Rugosa in Iran are represented by paedomorphic forms (*Pentaphyllum breviseptum* and *P. minimum*), which were due to a peculiar habitat in latest Permian time (Ezaki, 1993a). The Rugosa were succeeded by pelagic organisms. The disappearance event was accompanied by major lithofacies and biofacies changes that reflected the deterioration of the marine environment, such as sea-level and salinity fluctuations on local and global scales (IJRG, 1981; Ezaki, 1993b).

Pakistan (Salt Range, including Surghar Range)

The Permian *Productus* Limestone (Zaluch Group) can be subdivided into three formations: the Amb, the Wargal, and the Chhidru formations, in ascending order (Teichert, 1966). The Chhidru Formation is paraconformably overlain by the lowermost part of the Mianwali Formation (Kathwai member), containing Permian survivors of brachiopods (Kummel and Teichert, 1970) or syn-depositional Permian brachiopods [Pakistani-Japanese Research Group (PJRG), 1985]. The Permian–Triassic boundary has been placed between the Chhidru and Mianwali formations or between the lower and middle units of the Kathwai member.

There is a gradual decline in $\delta^{13}C$ values from the upper part of the Wargal Formation to the Chhidru Formation, and a marked drop occurs in the uppermost Chhidru Formation (Baud et al., 1989). A conspicuous drop in boron content is also seen in the uppermost Chhidru Formation, due to fresh-water inflow into the basin, and that event was followed by a sea-level transgression to form the lower Kathwai member, which initiated another sedimentary cycle (PJRG, 1985).

Waagen and Wentzel (1886) described the Permian organisms, including corals, brachiopods, and bryozoans, from the Salt Range. Kato and Ezaki (1986) published a preliminary report on Permian coral biostratigraphy. The Permian corals are characterized by typical Tethyan elements and show a distinctive stratigraphic succession. No corals are found in the Amb Formation. Solitary and massive Wentzelellinae (e.g., *Iranophyllum*, *Wentzelella*, *Wentzelellites*) having tertiary and higher orders of septa disappeared in the lower part of the Wargal Formation. Fasciculate forms, such as *Waagenophyllum*, are abundant in the upper parts of the Wargal Formation (unit 4b at Chhidru, and the upper part of the Kalabagh member at Zaluch), which were deposited in a shallow-marine, reefal environment (PJRG, 1985). Those fasciculate corals survived to be deposited into the Chhidru Formation, 34 m below the top of the formation at Narmia Nala in the Surghar Range, where calcareous facies are more persistent than in the Salt Range. *Waagenophyllum* occurs in the dark-grey, bioclastic packstone in association with *Colaniella* and *Codonofusiella* (without *Palaeofusulina*), indicating a Dzhulfian age. No corals are found in the upper part of the Chhidru Formation, which is characterized by a high content of terrigenous sediments and by marked decreases in fusulinids, smaller foraminifera, and boron content, indicating considerable continental influence on the marine environment (PJRG, 1985).

Wentzelellinae disappeared first, followed by the platform-dwelling fasciculate forms (*Waagenophyllum*), not accompanied by the non-dissepimented, solitary corals. Corals disappeared earlier in the Salt Range than in the Surghar Range, reflecting a local variation in the marine environment.

The Zewan Formation in Kashmir shows similarities in faunal

composition and succession to the Wargal and Chhidru formations in the Salt Range, and therefore both regions have been assigned to peri-Gondwana (Nakazawa and Kapoor, 1977). A similar transgressive–regressive couplet is recorded near the Permian–Triassic boundary, but the Zewan Formation includes only the solitary coral *Euryphyllum cainodon,* and no fusulinids (Nakazawa et al., 1975b).

South China

Marine Permian deposits, showing a considerable variety of lithologies, are widely distributed on the Yangtze Block, which was situated in the equatorial region in the eastern part of Tethys (Scotese and McKerrow, 1990). The Permian–Triassic strata have been intensively studied lithologically, biostratigraphically, and geochemically, and the Permian–Triassic boundary is regarded as being continuous in places (e.g., Sheng et al., 1984; Yang et al., 1987, 1991; Li et al., 1989), although subaerial erosion also has been reported in the uppermost Permian (Reinhardt, 1988). Even in Upper Permian strata (Wujiapingian and Changxingian), bedded limestones are well developed and yield abundant fusulinids, smaller foraminifera, conodonts, and corals. The continuously developed Upper Permian carbonates were formed in a transgressive environment having some regressive phases of various magnitudes (Li et al., 1989, 1991). A gradual decline in $\delta^{13}C$ values has been reported near the Permian–Triassic boundary in the Shanxi section, Sichuan province, and the minimum values appear above that boundary (Yan et al., in Li et al., 1989).

The most reliable faunal data regarding the disappearance pattern for the Rugosa are available from the Permian of South China, because successive carbonate rocks containing abundant corals are better developed there (up to the uppermost Permian) than in any other region (Ezaki, 1994). Different lithologies include different types of corals, showing apparently varied patterns of disappearance from place to place. Coral biostratigraphic zones have been proposed for the Permian. In the Upper Permian, for example, Xu (1984) proposed, in Hunan and Hubei provinces, the *Plerophyllum guangxiense–P. xintanense* Assemblage Zone in the Wujiapingian and the *Waagenophyllum lui–Lophocarinophyllum lophophyllidum* Assemblage Zone in the Changxingian. Zhao (1984) proposed, at Xizang, Sichuan and Yunnan provinces, the *Liangshanophyllum wuae–L. sinense* Assemblage Zone below and the *Huayunophyllum longiseptatum–Waagenophyllum markamense* Assemblage Zone above.

Each Upper Permian fauna is characterized by different groups surviving from the Middle Permian, but they show similar patterns of disappearance. The faunas are generally represented by non-dissepimented, solitary corals, such as plerophyllids and lophophyllidiids (e.g., Zhao, 1976), in a basinal environment, but they are dominated by fasciculate waagenophyllids (*Waagenophyllum* and *Liangshanophyllum*) in places where carbonate rocks are developed up to the Upper Permian (e.g., Wu and Wang, 1974; Zhao, 1981). Even in such cases, the massive

Wentzelellinae (e.g., *Wentzelella* and *Polythecalis*) having tertiary septa and the yatsengiids disappeared in the Maokouan, leaving simply formed Waagenophyllinae having only two orders of septa.

Two apparent disappearance patterns can be recognized (Ezaki, 1994). One pattern is characterized by a general trend for earlier disappearances of more complex forms, including massive waagenophyllids, and the predominance of simple, solitary corals (e.g., plerophyllids and lophophyllidiids) as remnants. The other pattern is marked by the colonization of the platform-dwelling, fasciculate waagenophyllids until the latest Permian. The latter pattern is distinguished as a "truncation" of the general pattern, as a result of the persisting favourable conditions that permitted the colonization by the fasciculate waagenophyllids, followed by the suddenly adverse changes in environment at the end of the Permian. That deviation in disappearance pattern obviously was due to changes in environmental magnitudes and durations during the late Permian (Ezaki, 1994).

Calcareous algae and sponges flourished temporarily in the late Permian reefs (Flügel and Reinhardt, 1989), but most benthic organisms disappeared before the Permian–Triassic boundary, when the latest Changxingian transgression occurred (Li et al., 1991; Yang et al., 1991). Particular groups of brachiopods, ammonoids, and bivalves that could adapt to deeper-water facies survived across the boundary, forming mixed-fauna beds (Sheng et al., 1984; Yin, 1985).

Japan (South Kitakami Terrane)

The Japanese Permian corals have been studied since Yabe (1902) described *Lonsdaleia akasakensis* from the Permian of the Akasaka region in the Mino Terrane. No late Permian corals have been systematically described yet, in spite of the sporadic occurrences of corals. Minato (1975) and Kato (1990) summarized the Japanese Permian corals biostratigraphically and biogeographically.

The Permian strata are widely distributed in the South Kitakami Terrane, and they have been divided into three series: the Sakamotozawa Series, the Kanokura Series, and the Toyoma Series, corresponding to the Lower, Middle, and Upper Permian, respectively (Minato et al., 1979). The Lower and Middle Permian are well defined biostratigraphically on the basis of fusulinids (e.g., Choi, 1973).

The Toyoma Formation, in the Motoyoshi region, composed mainly of black shale, occasionally intercalated with thin sandy layers, includes coral fragments of simple, solitary form (precise taxonomy not yet determined). The underlying Iwaizaki Limestone, of Kanokuran age, is characterized by a carbonate buildup formed on a continental shelf, and corals of various growth forms occur, especially in its middle part, constructing a reef, in association with calcareous algae, bryozoans, and calcisponges (Machiyama, Kawamura, and Yoshida, 1992). Included are the waagenophyllid and yatsengiid corals such as *Waagenophyllum, Iranophyllum, Parawentzelella,* and *Yatsengia* (Morikawa et al.,

1958; Minato and Kato, 1965). The reef was drowned gradually, as indicated by the dominance of terrigenous mudstone and limestone slide blocks in the uppermost horizon of the Midian. Reef-dwelling, colonial corals disappeared first, together with fusulinids, leaving only non-dissepimented, solitary corals, such as *Lophophyllidium* and *Calophyllum,* in mudstone alternating with impure black limestone.

The fossiliferous Permian strata are also developed in the Kesennuma region, and some palaeontologic and biostratigraphic studies of that region have been published (Tazawa, 1973, 1976). The Sakamotozawa Series contains corals such as *Yatsengia, Waagenophyllum,* and *Protomichelinia* and subordinate numbers of non-dissepimented corals, and the overlying Kanokura Series includes *Parawentzelella, Lophophyllidium,* and *Sinopora* (Tazawa, 1976). The lower part of the Toyoma Series is characterized by black shale, whereas the upper part, yielding Permian foraminifera of Changxingian age (Tazawa, 1975; Ishii, Okimura, and Nakazawa, 1975), is marked by grey sandstone, shallow-marine limestone and conglomerate, and black shale. Very recently, a fragment of a simple, solitary coral (*Ufimia* sp.) was found in the upper part of the Toyoma Series (Y. Ezaki, unpublished data).

The Usuginu conglomerates (granitic-clast-dominated conglomerates) prevailed in the Kanokuran, initiating a new sedimentary regime (the granitic-clast provenance of Kawamura et al., 1990). The overlying Toyoma Formation is interpreted to have been deposited in a basinal condition sheltered from the open sea (Minato, 1944) or on a continental shelf or slope accompanied by upwelling (Kanisawa and Ehiro, 1986).

Waagenophyllid and yatsengiid corals flourished in the early Permian, but most disappeared in the middle Permian, obviously because of deterioration of habitat conditions, such as the prevalence of hinterlands and/or drowning of the reef. However, preferential conditions may have reappeared locally in the latest Permian, permitting habitation of only solitary corals. Finally, subaerial erosion and/or nondeposition (hiatus) occurred near the Permian–Triassic boundary.

Russia (South Primorye)

In South Primorye, the uppermost part of the Permian is represented by the Lyudyanza Formation. The formation has been subdivided biostratigraphically on the basis of ammonoids (Zakharov and Pavlov, 1986; Zakharov, 1992), and the *Cyclolobus kiselevae* beds have been placed in the uppermost Midian, although Kotlyar et al. (1989) regarded the immediately underlying *Xenodiscus subcarbonarius* beds as the base of the Dzhulfian. The *Liuchengoceras*-Pleuronodoceratidae beds are present in the upper Lyudyanza horizon (Zakharov, 1992), equivalent to the *Pleuronodoceras occidentale* Zone of Zakharov (1988) in the Transcaucasus.

The Senkina Shapka Mountain, along the Partizanskaya River, provides excellent Midian limestone containing numerous fusulinids of the Chandalaz horizon. The middle part of the bedded limestone (corresponding to the lower part of the

Metadoliolina lepida–Lepidolina kumaensis Zone) yields colonial waagenophyllids, including *Polythecalis chandalasensis* and *Lonsdaleiastraea* sp., and the *Szechuanophyllum kitakamiense–Wentzelloides (W.) ussuricus* Zone has been established (Kotlyar et al., 1989).

The Nakhodka Reef is a carbonate mound embedded in fine clastics in the upper Chandalaz horizon or in the lower Lyudyanza horizon of the Trudny Peninsula, and it contains abundant sphinctozoans, crinoids, and algae (Kotlyar et al., 1987, 1992). Non-dissepimented, solitary corals occur sporadically as subordinate bioclasts of the limestone. Kotlyar et al. (1989) regarded the *Pseudofavosites kotljarae–Calophyllum kabakovitchae* Zone, characterized by occurrences of *Calophyllum kabakovitchae, Pseudofavosites kotljarae, Paracaninia subtilis, P. ex. gr. simplex, P.* sp., and *Lophocarinophyllum* sp., as the topmost coral zone in the South Primorye region.

Colonial waagenophyllids flourished in the Midian, and they were succeeded by non-dissepimented, solitary corals in a carbonate mound. Such a faunal succession is typical and is commonly observed in the middle-to-late Permian corals in Tethyan regions.

Other areas

Other localities where late Permian corals occur are, for example, Zažar, Slovenia (Ramovš, 1958), Salamis, Greece (Nakazawa et al., 1975a), and Xizang, China (Wu, Liao, and Zhao, 1982). They are omitted from this discussion of the disappearance pattern of Rugosa because few biostratigraphically successive data are available.

Disappearance patterns and end-Permian environments

The late Permian Rugosa comprised different taxa and assemblages from one area to another, showing individual faunal successions. Their disappearance patterns, however, seem to have been similar to one another, and their extinction was global by the end of the Permian. What was common to all of those local and heterochronous disappearance events? The immediate causes of the disappearances may also have been variable, and plausible factors seem to have been present in each region and for each taxon.

The disappearance of the waagenophyllids was the most remarkable in the Tethyan faunal successions. Generally, the massive and solitary waagenophyllids disappeared before the middle–late Permian boundary. Most colonial, platform-dwelling waagenophyllids disappeared in the middle Permian because of the collapse of open-marine, shallow carbonate environments, as in Abadeh and Julfa, in the Salt Range, and in the South Kitakami Terrane. Afterward, only non-dissepimented, solitary corals could survive in the unfavourable habitats of those areas.

In the mid-Permian, the Hunter-Bowen Orogenic Phase began, accompanied by sea-level regression, vertical movements, intense acidic and intermediate volcanism, and a hot climate, and it prevailed toward the Permian–Triassic boundary

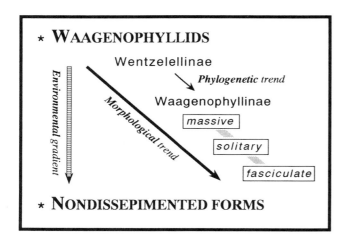

Figure 13.1. Successive disappearances of rugosan coral morphogroups, reflecting environmental gradients and a phylogenetic trend.

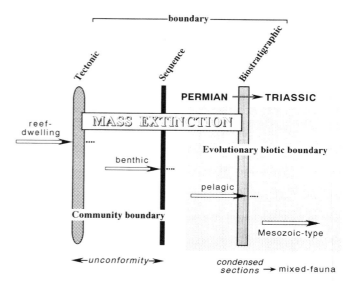

Figure 13.2. "Boundary differences" according to various criteria. Each boundary based on a different quality was accompanied by the disappearance of organisms with a specific mode of life, and boundaries can coincide. The erathemic boundary is ordinarily defined biostratigraphically.

(Dickins, 1989, 1992). Holser and Magaritz (1987) speculated that the geophysical regimes had already shifted in the Tatarian, before the Permian–Triassic boundary, and the extinction event may have been associated with rapid sea-level change and plateau volcanism.

The general disappearance trend was that the morphologically more complex corals disappeared earlier, and the last survivors were the non-dissepimented, solitary corals (Flügel, 1970). Fasciculate Waagenophyllinae could flourish on late Permian carbonate platforms in South China because of the continuation of exceptionally favourable habitats in a tropical climate for a long duration, but they disappeared at the end of the Permian. Such a disappearance pattern for Rugosa deviates from the general pattern as a "truncation" (Ezaki, 1994). The overall faunal succession obviously resulted from changes in the environment in the Permian. An apparent morphological trend from complex to simple depended on "environmental gradients" showing adverse directions (Figure 13.1).

The late Permian decline in the Wentzelellinae having tertiary septa should be explained also from a phylogenetic viewpoint. That is, the waagenophyllids, as a whole, had reached an acme in the middle Permian (Minato and Kato, 1965, 1970), and the late Permian corals were essentially characterized by survivors from the preceding ages. Wentzelellinae had already declined in the early part of the late Permian, and they were no longer diversified, despite the improvements in habitat conditions.

Upper Permian strata show marked effects from transgressive-regressive changes that produced a variety of lithofacies. A significant transgression occurred just prior to the Permian–Triassic boundary, after a regression that had formed a depositional boundary (Wignall and Hallam, 1992, 1993). Benthic faunas, consisting of rugose corals and fusulinids, may have been replaced by pelagic-type organisms such as ammonoids and conodonts in the latest Permian (e.g., Li et al., 1991; Yang et al., 1991). Particular taxonomic groups of conodonts, foraminifera, and brachiopods formed mixed-fauna beds at the Permian–

Triassic boundary (Sheng et al., 1984; Yin, 1985; Broglio Loriga et al., 1988). The Permian-type fossils are either *in situ* or derived, and the stratigraphic position of the fossil-bearing beds is either just below or just above the Permian–Triassic boundary.

A transgressive systems tract above a sequence boundary can form a condensed section represented by extinction events and resultant biostratigraphic stage boundaries and by concentrations of authigenic minerals and volcaniclastics (e.g., Donovan et al., 1988; Loutit et al., 1988). Most, if not all, Permian–Triassic boundary clays may be of volcanic origin [for reviews, see Yang et al. (1991) and Yin et al. (1992)], but the lithologic, geochemical, and biostratigraphic characteristics of the mixed-fauna beds are similar to those of the condensed sections.

The Permian–Triassic boundary is ordinarily defined biostratigraphically. The biostratigraphic boundary does not always coincide with the significant tectonic and depositional sequence boundaries, and it may be retarded (Figure 13.2). The widespread transgression near the Permian–Triassic boundary was a prelude to a new sedimentary regime that led to rather uniform lithofacies yielding *Otoceras* and *Claraia* from the Lower Triassic. *Otoceras* and *Claraia* themselves, however, were the Permian "relicts," from a phylogenetic viewpoint, and real Triassic representatives, such as the ammonoid meekoceratids and the conodont *Neospathodus*, originated later than the conventional Permian–Triassic boundary (Newell, 1988). The Permian–Triassic boundary therefore can be drawn differently (biostratigraphically, sedimentologically, tectonically, etc.), thus producing a "boundary difference" (Figure 13.2).

Some variation in the disappearance pattern for Rugosa was produced, in most cases by deviation from the general pattern, depending on the nature of the environmental conditions and the

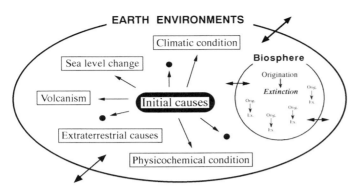

Figure 13.3. Schematic relationships among factors of various origins and extinction events of different magnitudes. Arrows indicate continuous interplay between individuals. Coeval biotic and abiotic factors are closely related to produce different or higher-order causes for each extinction event. Essential causes vary spatiotemporally and taxonomically.

sizes of their gradients, as recorded in each particular facies development. Of course, whether the trend shows merely environmental reflections or has phylogenetic connotations should be judged in terms of the main causes of the disappearance, through individual examinations of background palaeo-environmental and palaeoecological conditions.

Coeval factors interact to produce higher-order causes (Figure 13.3), and the multiple adverse causes of various origins selected corals repeatedly, ordinarily from complex forms to simple forms, according to their limits of tolerance. Although the late Permian corals were in places morphologically simple, in response to specific, unfavourable environments, they still remained genuine Rugosa. Not a single representative survived into the Triassic anywhere in the world. The Rugosa became extinct before the conventional Permian–Triassic boundary.

Acknowledgments

I would like to thank Dr. J. M. Dickins of the Australian Geological Survey Organisation for invaluable discussions. I am also grateful to Professor Emeritus K. Nakazawa of Kyoto University for generous guidance on the Permian–Triassic stratigraphy. My thanks also to Professor M. Kato of Hokkaido University and Dr. W. J. Sando of the U.S. Geological Survey for critical readings of the manuscript. This study was supported by the Scientific Research Fund of the Japanese Ministry of Education, Science and Culture (grants 04304009, 05740327, and 05304001).

References

Altiner, D., Baud, A., Guex, J., and Stampfli, G. 1980. La limite Permien–Trias dans quelques localités du Moyen-Orient: recherches stratigraphiques et micropaléontologiques. *Rivista Italiana di Paleontologia* 85:683–714.

Baud, A. Magaritz, M., and Holser, W. T. 1989. Permian–Triassic of the Tethys: carbon isotope studies. *Geologische Rundschau* 78:649–77.

Broglio Loriga, C., Neri, C., Pasini, M., and Posenato, R. 1988. Marine fossil assemblages from Upper Permian to lowermost Triassic in the Western Dolomites (Italy). *Memorie della Società Geologica Italiana* 34:5–44.

Choi, D. R. 1973. Permian fusulinids from the Setamai-Yahagi district, southern Kitakami Mountains, N.E. Japan. *Journal of the Faculty of Science, Hokkaido University, Series IV, Geology and Mineralogy* 16:1–132.

Dickins, J. M. 1989. Major sea level changes, tectonism and extinctions. In *Compte Rendu Onzieme Congres International de Stratigraphie et de Geologie du Carbonifere, Beijing, China,* vol. 4, pp. 135–44. Nanjing: Nanjing University Press.

Dickins, J. M. 1992. Permian–Triassic orogenic, paleoclimatic, and eustatic events and their implications for biotic alteration. In *Permo–Triassic Events in the Eastern Tethys,* ed. W. C. Sweet, Z. Y. Yang, J. M. Dickins, and H. F. Yin, pp. 169–74. Cambridge University Press.

Donovan, A. D., Baum, G. R., Blechschmidt, G. L., Loutit, T. S., Pflum, C. E., and Vail, P. R. 1988. Sequence stratigraphic setting of the Cretaceous–Tertiary boundary in central Alabama. In *Sea Level Changes: An Integrated Approach,* ed. C. K. Wilgus, B. S. Hastings, C. G. St. C. Kendall, H. W. Posamentier, C. A. Ross, and J. C. Van Wagoner, pp. 299–307. Special publication no. 42. Tulsa: Society of Economic Paleontologists and Mineralogists.

Ezaki, Y. 1989. Morphological and phylogenetic characteristics of Late Permian rugose corals in Iran. *Memoir of the Association of Australasian Palaeontologists* 8:275–81.

Ezaki, Y. 1991. Permian corals from Abadeh and Julfa, Iran, West Tethys. *Journal of the Faculty of Science, Hokkaido University, Series IV, Geology and Mineralogy* 23:53–146.

Ezaki, Y. 1993a. The last representatives of Rugosa in Abadeh and Julfa, Iran: survival and extinction. *Courier Forschungsinstitut Senckenberg* 164:75–80.

Ezaki, Y. 1993b. Sequential disappearance of Permian Rugosa in Iran and Transcaucasus, West Tethys. *Bulletin of the Geological Survey of Japan* 44:447–53.

Ezaki, Y. 1994. Pattern and paleoenvironmental implications of end-Permian extinction of Rugosa in South China. *Palaeogeography, Palaeoclimatology, Palaeoecology* 107:165–77.

Fedorowski, J. 1989. Extinction of Rugosa and Tabulata near the Permian/Triassic boundary. *Acta Palaeontologia Polonica* 34:47–70.

Flügel, E., and Reinhardt, J. W. 1989. Uppermost Permian reefs in Skyros (Greece) and Sichuan (China): implications for the Late Permian extinction event. *Palaios* 4:502–18.

Flügel, H. W. 1968. Korallen aus der oberen Nesen-Formation (Zhulfa-Stufe, Perm) des zentralen Elburz (Iran). *Neues Jahrbuch für Geologie und Paläontologie, Abhandlungen* 130:275–304.

Flügel, H. W. 1970. Die Entwicklung der rugosen Korallen im hohen Perm. *Verhandlungen der Geologischen Bundesanstalt* 1:146–61.

Flügel, H. W. 1971. Upper Permian corals from Julfa. *Geological Survey of Iran, Report* 19:109–39.

Holser, W. T., and Magaritz, M. 1987. Events near the Permian–Triassic boundary. *Modern Geology* 11:155–80.

Iljina, T. G. 1965. Late Permian and early Triassic tetracorals from the Transcaucasus area (in Russian). *Transactions of the Paleontological Institute, USSR Academy of Science* 107:1–104.

Iljina, T. G. 1984. Historical development of corals (in Russian). *Transactions of the Paleontological Institute, USSR Academy of Science* 198:1–184.

Iranian-Japanese Research Group. 1981. The Permian and the Lower Triassic systems in Abadeh region, central Iran. *Memoirs of the Faculty of Science, Kyoto University, Series of Geology and Mineralogy* 47:61–133.

Ishii, K., Okimura, Y., and Nakazawa, K. 1975. On the Genus *Colaniella* and its biostratigraphic significance. *Journal of Geosciences, Osaka City University* 19:107–38.

Kanisawa, S., and Ehiro, M. 1986. Occurrence and geochemical nature of phosphatic rocks and Mn-rich carbonate rocks in the Toyoman Series, Kitakami Mountains, northeastern Japan. *Journal of the Japanese Association of Mineralogists, Petrologists and Economic Geologists* 81:12–31.

Kato, M. 1990. Palaeozoic corals. In *Pre-Cretaceous Terranes of Japan*, ed. K. Ichikawa, S. Mizutani, I. Hara, S. Hada, and A. Yao, pp. 307–12. publication of IGCP-224. Osaka: IGCP.

Kato, M., and Ezaki, Y. 1986. Permian corals of Salt Range. *Proceedings of the Japan Academy* 62:231–4.

Kawamura, M., Kato, M., and Kitakami Paleozoic Research Group. 1990. Southern Kitakami Terrane. In *Pre-Cretaceous Terranes of Japan*, ed. K. Ichikawa, S. Mizutani, I. Hara, S. Hada, and A. Yao, pp. 249–66. Publication of IGCP-224. Osaka: IGCP.

Kotlyar, G. V., Beljaeva, G. V., Kiseleva, A. V., Kropatcheva, G. S., Kushnar, L. V., Nikitina, A. P., Pronina, G. P., and Chedija, I. O. 1992. Nakhodka reef. In *A Field Guide to the Late Paleozoic and Early Mesozoic Circum-Pacific Bio- and Geological Events,* ed. Y. D. Zakharov, I. V. Panchenko, and A. I. Khanchuk, pp. 59–62. Vladivostok: Russian Academy of Science, Far Eastern Branch, Far Eastern Geological Institute.

Kotlyar, G. V., Vuks, G. P., Kropatcheva, G. S., and Kushnar, L. V. 1987. Nakhodka reef-rock massif and the position of Ludjanza horizon within the stage scale of the Permian in the Tethys (in Russian). In *Problems of the Permian and Triassic Biostratigraphy of East USSR,* ed. Y. D. Zakharov and Y. I. Onoprienko, pp. 54–63. Vladivostok: Far Eastern Scientific Centre, USSR Academy of Science.

Kotlyar, G. V., Zakharov, Y. D., Kropatcheva, G. S., and Chedija, I. O. 1989. Upper Permian biostratigraphy (in Russian). In *Evolution of the Latest Permian Biota. Midian Regional Stage in the USSR,* ed. G. V. Kotlyar and Y. D. Zakharov, pp. 52–74. Leningrad: Leningrad Department of Publishing (Nauka).

Kropatcheva, G. S. 1989. Rugose corals (in Russian). In *Evolution of the Latest Permian Biota. Midian Regional Stage in the USSR,* ed. G. V. Kotlyar and Y. D. Zakharov, pp. 47–9, 107–17. Leningrad: Leningrad Department of Publishing (Nauka).

Kummel, B., and Teichert, C. 1970. Stratigraphy and paleontology of the Permian–Triassic boundary beds, Salt Range and Trans-Indus Range, West Pakistan. In *Stratigraphic Boundary Problems: Permian and Triassic of West Pakistan,* ed. B. Kummel and C. Teichert, pp. 1–110. Special publication 4, Department of Geology, University of Kansas. Lawrence: University Press of Kansas.

Li, Z. S., Zhan, L. P., Dai, J. Y., Jin, R. G., Zhu, X. F., Zhang, J. H., Huang, H. Q., Xu, D. Y., Yan, Z., and Li, H. M. 1989. *Study of the Permian–Triassic Biostratigraphy and Event Stratigraphy of Northern Sichuan and Southern Shaanxi* (in Chinese, with English summary). Beijing: Geological Publishing House.

Li, Z. S., Zhan, L. P., Yao, J. X., and Zhou, Y. Q. 1991. On the Permian–Triassic events in South China – probe into the end Permian abrupt extinction and its possible causes. In

Shallow Tethys 3, ed. T. Kotaka, J. M. Dickins, K. G. McKenzie, K. Mori, K. Ogasawara, and G. D. Stanley, Jr., pp. 371–85. Sendai: Saito Ho-on Kai.

Loutit, T. S., Hardenbol, J., Vail, P. R., and Baum, G. R. 1988. Condensed sections: the key to age determination and correlation of continental margin sequences. In *Sea Level Changes: An Integrated Approach,* ed. C. K. Wilgus, B. S. Hastings, C. G. St. C. Kendall, H. W. Posamentier, C. A. Ross, and J. C. Van Wagoner, pp. 183–213. Special publication no. 42. Tulsa: Society of Economic Paleontologists and Mineralogists.

Machiyama, H., Kawamura, T., and Yoshida, K. 1992. Two examples of Permian carbonate buildups in Japan. *Abstracts of the 29th International Geological Congress, Kyoto,* vol. 2, p. 332.

Minato, M. 1944. Stratigraphische Stellung der Usuginukonglomerate mit besonderer Brucksichtigung des Toyoma Meeres, eines Binnenmeeres der spateren permischen Zeit im Kitakami Gebirge, Japan. *Journal of the Geological Society of Japan* 51:166–87.

Minato, M. 1975. Japanese Palaeozoic corals. *Journal of the Geological Society of Japan* 81:103–26.

Minato, M., Hunashashi, M., Watanabe, J., and Kato, M. (eds.) 1979. *The Abean Orogeny.* Tokyo: Tokai University Press.

Minato, M., and Kato, M. 1965. Waagenophyllidae. *Journal of the Faculty of Science, Hokkaido University, Series IV, Geology and Mineralogy* 12:1–241.

Minato, M., and Kato, M. 1970. The distribution of Waagenophyllidae and Durhaminidae in the Upper Palaeozoic. *Japanese Journal of Geology and Geography* 41:1–14.

Morikawa, R., Sato, T., Shibazaki, T., Shinada, Y., Okubo, M., Nakazawa, K., Horiguchi, M., Murata, M., Kikuchi, Y., Taguchi, Y., and Takahashi, K. 1958. Stratigraphy and biostratigraphy of the "Iwaizaki Limestone" in the Southern Kitakami Mountainland (in Japanese, with English abstract). In *Jubilee Publication Dedicated to Professor H. Hujimoto, 60th Birthday,* pp. 81–90.

Nakazawa, K. 1993. Stratigraphy of the Permian–Triassic transition and the Paleozoic/Mesozoic boundary (in Japanese). *Bulletin of the Geological Survey of Japan* 44:425–45.

Nakazawa, K., Ishii, K., Kato, M., Okimura, Y., Nakamura, K., and Haralambous, D. 1975a. Upper Permian fossils from Island of Salamis, Greece. *Memoirs of the Faculty of Science, Kyoto University, Series of Geology and Mineralogy* 42:21–44.

Nakazawa, K., and Kapoor, H. M. 1977. Correlation of the marine Permian in the Tethys and Gondwana. In *Fourth International Gondwana Symposium, Calcutta, India,* pp. 1–18. Delhi: Hindustan Publishing Corp.

Nakazawa, K., Kapoor, H. M., Ishii, K., Bando, Y., Okimura, Y., and Tokuoka, T. 1975b. The Upper Permian and the Lower Triassic in Kashmir, India. *Memoirs of the Faculty of Science, Kyoto University, Series of Geology and Mineralogy* 42:1–106.

Newell, N. D. 1988. The Paleozoic/Mesozoic erathem boundary. *Memorie della Società Geologica Italiana* 34:303–11.

Oliver, W. A., Jr. 1980. The relationship of the scleractinian corals to the rugose corals. *Paleobiology* 6:146–60.

Pakistani-Japanese Research Group. 1985. Permian and Triassic Systems in the Salt Range and Surghar Range, Pakistan. In *The Tethys: Her Paleogeography and Paleobiogeography from Palaeozoic to Mesozoic,* ed. K. Nakazawa and J. M. Dickins, pp. 221–312. Tokyo: Tokai University Press.

Ramovš, A. 1958. Razvoj zgornjega perma v Loških in

Polhograjskih hribih. *Razprave Slovenska Akademija Znanosti in Umetnosti* 4:455–622.

Reinhardt, J. W. 1988. Uppermost Permian reefs and Permo–Triassic sedimentary facies from the southeastern margin of Sichuan basin, China. *Facies* 18:231–88.

Ruzhentsev, V. E., and Sarycheva, T. G. 1965. The development and change of marine organisms at the Paleozoic–Mesozoic boundary (in Russian). *Transactions of the Paleontological Institute, USSR Academy of Science* 108:1–431.

Scotese, C. R., and McKerrow, W. S. 1990. Revised world maps and introduction. In *Palaeozoic Palaeogeography and Biogeography,* ed. C. R. Scotese and W. S. McKerrow, pp. 1–12. The Geological Society memoir no. 12. London: The Geological Society Publishing House.

Sheng, J. Z., Chen, C. Z., Wang, Y. G., Rui, L., Liao, Z. T., Bando, Y., Ishii, K., Nakazawa, K., and Nakamura, K. 1984. Permian–Triassic boundary in Middle and Eastern Tethys. *Journal of the Faculty of Science, Hokkaido University, Series IV, Geology and Mineralogy* 21:133–81.

Stepanov, D. L., Golshani, F., and Stöcklin, J. 1969. *Upper Permian and Permian–Triassic boundary in North Iran.* Geological Survey of Iran, report no. 12.

Taraz, H. 1971. Uppermost Permian transition beds in Central Iran. *American Association of Petroleum Geologists Bulletin* 55:1280–94.

Taraz, H. 1974. *Geology of the Surmaq–Deh Bid Area, Abadeh Region, Central Iran.* Geological Survey of Iran, report no. 37.

Tazawa, J. 1973. Geology and the Kamiyasse area, Southern Kitakami Mountains (in Japanese, with English abstract). *Journal of the Geological Society of Japan* 79:677–86.

Tazawa, J. 1975. Uppermost Permian fossils from the Southern Kitakami Mountains, Northeast Japan. *Journal of the Geological Society of Japan* 81:629–40.

Tazawa, J. 1976. The Permian of Kesennuma, Kitakami Mountains: a preliminary report. *Earth Science* 30:175–85.

Teichert, C. 1966. Stratigraphic nomenclature and correlation of the Permian "Productus limestone," Salt Range, West Pakistan. *Geological Survey of Pakistan, Records* 15:1–19.

Teichert, C., and Kummel, B. 1972. Permian–Triassic boundary in the Kap Stosch area, East Greenland. *Bulletin of Canadian Petroleum Geology* 20:659–75.

Teichert, C., Kummel, B., and Sweet, W. C. 1973. Permian–Triassic strata, Kuh-e-Ali Bashi, northwestern Iran. *Bulletin of the Museum of Comparative Zoology* 145:359–472.

Waagen, W., and Wentzel, J. 1886. Salt Range fossils. I. Productus Limestone fossils. Coelenterata. *Palaeontologia Indica* 13:835–924.

Wignall, P. B., and Hallam, A. 1992. Anoxia as a cause of the Permian/Triassic mass extinction: facies evidence from northern Italy and the western United States. *Palaeogeography, Palaeoclimatology, Palaeoecology* 93:21–46.

Wignall, P. B., and Hallam, A. 1993. Griesbachian (earliest Triassic) palaeoenvironmental changes in the Salt Range, Pakistan and southeast China and their bearing on the Permo–Triassic mass extinction. *Palaeogeography, Palaeoclimatology, Palaeoecology* 102:215–37.

Wu, W. S., Liao, W. H., and Zhao, J. M. 1982. Palaeozoic rugose corals from Xizang (in Chinese, with English abstract). In *Palaeontology of Xizang,* ed. Nanjing Institute of Geology and Palaeontology, Academia Sinica, pp. 107–51. Beijing: Science Press.

Wu, W. S., and Wang, Z. 1974. Permian corals (in Chinese). In *Handbook of the Stratigraphy and Paleontology of Southwest China,* ed. Nanjing Institute of Geology and Palaeontology, Academia Sinica, pp. 296–9. Beijing: Science Press.

Xu, S. Y. 1984. The characters of the Permian coral faunas from Hunan and Hubei provinces (in Chinese, with English abstract). *Acta Palaeontologica Sinica* 23:605–16.

Yabe, H. 1902. Materials for a knowledge of the Anthracolithic fauna of Japan, 1. *Journal of the Geological Society, Tokyo* 9:1–5.

Yang, Z. Y., Wu, S. B., Yin, H. F., Xu, G. R., and Zheng, K. X. 1991. *Permo–Triassic Events of South China* (in Chinese, with English summary). Beijing: Geological Publishing House.

Yang, Z. Y., Yin, H. F., Wu, S. B., Yang, F. Q., Ding, M. H., and Xu, G. R. 1987. *Permian–Triassic Boundary Stratigraphy and Fauna of South China* (in Chinese). Geological memoirs, ser. 2, no. 6. Beijing: Geological Publishing House.

Yin, H. F. 1985. On the transitional bed and the Permian–Triassic boundary in South China. *Newsletters on Stratigraphy* 15:13–27.

Yin, H. F., Huang, S. J., Zhang, K. X., Hansen, H. J., Yang, F. Q., Ding, M. H., and Bie, X. M. 1992. The effects of volcanism on the Permo–Triassic mass extinction in South China. In *Permo–Triassic Events in the Eastern Tethys,* ed. W. C. Sweet, Z. Y. Yang, J. M. Dickins, and H. F. Yin, pp. 146–57. Cambridge University Press.

Zakharov, Y. D. 1988. Type and hypotype of the Permian–Triassic boundary. *Memorie della Società Geologica Italiana* 34:277–89.

Zakharov, Y. D. 1992. The Permo–Triassic boundary in the southern and eastern USSR and its international correlation. In *Permo–Triassic Events in the Eastern Tethys,* ed. W. C. Sweet, Z. Y. Yang, J. M. Dickins, and H. F. Yin, pp. 46–55. Cambridge University Press.

Zakharov, Y. D., and Pavlov, A. M. 1986. Permian cephalopods of Primorye region and the problem of Permian zonal stratification in Tethys (in Russian). In *Correlation of Permo–Triassic Sediments of East USSR,* ed. Y. D. Zakharov and Y. I. Onoprienko, pp. 5–32. Vladivostok: Far Eastern Scientific Centre, USSR Academy of Science.

Zhao, J. M. 1976. Late Permian rugose corals from Anshun, Luzhi and Qinglong, Guizhou Province (in Chinese, with English abstract). *Acta Palaeontologica Sinica* 15:213–22.

Zhao, J. M. 1981. Permian corals from Beichuan and Jiangyou of Sichuan and from Hanzhong of Shaanxi (in Chinese, with English abstract). *Memoirs of Nanjing Institute of Geology and Palaeontology* 15:233–81.

Zhao, J. M. 1984. Permian rugose corals from east Xizang, west Sichuan and north Yunnan (in Chinese, with English abstract). In *Stratigraphy and Palaeontology in W. Sichuan and E. Xizang, China. Part 4,* ed. Regional Geological Surveying Team of Geological Bureau of Sichuan and Nanjing Institute of Geology and Palaeontology, Academia Sinica, pp. 163–202. Chengdu: Sichuan Science and Technology Publishing House.

14 Fluctuations in pelagic environments near the Permian–Triassic boundary in the Mino-Tamba Terrane, southwest Japan

YOICHI EZAKI and KIYOKO KUWAHARA

Biostratigraphic and lithostratigraphic studies of strata near the Permian–Triassic (P–T) boundary, especially in carbonate and terrigenous facies, have been conducted for the Tethys, peri-Gondwana, and Boreal regions. Recently, environmental conditions near the P–T boundary, including those that caused the mass extinction, have received special attention through biostratigraphic, geochemical, and sedimentologic approaches (e.g., Sweet et al., 1992; Erwin, 1993).

Permian and Triassic marine successions are known to have been widely deposited, not only in near-shore situations but also in open-marine, pelagic environments (e.g., Ishiga and Imoto, 1980; Yao, Matsuda, and Isozaki, 1980). The P–T boundary is found in a sequence of pelagic sediments now incorporated into the Jurassic accretionary complexes in southwest Japan (e.g., Yamakita, 1987; Ishida, Yomashita, and Ishiga, 1992).

The bedded chert in the Mino-Tamba Terrane was deposited in a pelagic environment, with little continental influence, and the strata range from Upper Carboniferous to Middle Jurassic (Imoto, 1984a). Near the P–T boundary, the section is characterized by intergradings of bedded chert, siliceous claystone, black mudstone, and dolostone. Coarse-grained terrigenous components transported by currents are not included in those rocks. That pelagic sequence records secular and widespread oceanic changes in the Panthalassan Ocean near the P–T boundary.

In this chapter we attempt to clarify the Upper Permian–Lower Triassic successions in the Mino-Tamba Terrane. The palaeoceanographic implications of the pelagic sediments are considered in terms of P–T depositional and habitual fluctuations in the Panthalassan Ocean. The following sections have been selected for study: Sasayama (Ishida et al., 1992), Ubara (Kuwahara, Nakae, and Yao, 1991), Ryozen (Ishiga, Kito, and Imoto, 1982), Neo (Sano, 1988), Mt. Kinkazan (Sugiyama, 1992), Unuma (Yao et al., 1980), and Gujo-hachiman (Kuwahara et al., 1991), all sections in the Mino-Tamba Terrane (Figure 14.1). We adopt the Permian radiolarian zonation of Ishiga (1990) and the Lower Triassic zonation of Sugiyama (1992) for correlation.

Geologic setting

The Mino-Tamba Terrane is one of the Jurassic accretionary complexes in the inner zone of southwest Japan. Since the 1980s, radiolarian biostratigraphic analyses have established a well-documented oceanic-plate stratigraphy used to determine the genesis of the terrane and its later emplacement (Isozaki, Maruyama, and Furuoka, 1990; Matsuda and Isozaki, 1991). The terrane is characterized by Upper Carboniferous–Jurassic bedded cherts, Upper Carboniferous–Lower Permian greenstones, Permian shallow-water carbonates, P–T transition beds, and Jurassic clastics. The greenstones are mostly alkali basalts of seamount affinity; see Isozaki et al. (1990) for a review. Palaeomagnetic analysis of the Triassic bedded chert indicates that it was formed in an equatorial region (Shibuya and Sasajima, 1986). Terrigenous clastics of trench-fill sediments are mostly of Jurassic age. For a detailed geologic review, see Mizutani (1990).

The siliceous facies, composed mainly of bedded chert, include no coarse-grained terrigenous components, but they contain radiolarians, sponges, and conodonts, implying deposition in an open-marine, pelagic environment. Those pelagic sediments can grade laterally into shallow-water carbonates in seamount settings, showing a diverse algal and bryozoan biota that lived in a warm-water environment (Sano, 1988, 1989).

The Upper Permian–Lower Triassic successions are broken into several blocks by faults that locally show sheared boundary zones. Although it is difficult to conduct high-resolution stratigraphic analyses, we can reconstruct the original stratigraphy for each dislocated body on the basis of radiolarian and conodont biostratigraphy (Figure 14.2).

Lithology and fauna

The bedded chert shows a couplet structure consisting of vitric chert layers separated by clay partings (argillaceous layers). The nature of the bedded chert depended primarily on fluctuations in radiolarian productivity, and it was modified by various sedimentologic and diagenetic processes (Jenkyns and Winterer,

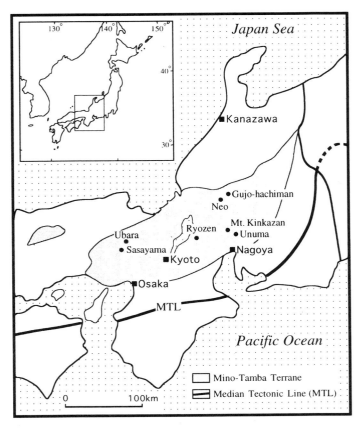

Figure 14.1. Index map showing the localities of our research in the Mino-Tamba Terrane, southwest Japan. The terrane distribution is adapted from the work of Ichikawa (1990).

1982). The Permian chert contains albaillellid, stauraxon, and spherical spumellarian radiolarians, whereas Triassic chert includes nassellarian and spumellarian radiolarians. Five units have been distinguished near the P–T boundary, based on the appearance of characteristic lithologies.

Folliculucullus-bearing bedded chert

The reddish, brownish-red, and chocolate-brown bedded cherts contain abundant radiolarians, ranging up to the *Follicucullus monacanthus* Assemblage Zone in Neo (Y. Ezaki, K. Kuwahara, unpublished data) or to the *Follicucullus scholasticus* Assemblage Zone in Sasayama (Ishida et al., 1992) (Figure 14.2). Those cherts are succeeded by grey bedded chert. Chert layers several centimetres in thickness are separated by thin clay interlayers. White chert several tens of centimetres in thickness is intercalated with the bedded chert in Sasayama (Ishida et al., 1992).

Neoalbaillella-bearing siliceous claystone beds

Direct contacts with the underlying bedded chert are not found. The claystone has a homogeneous smooth texture [Figure 14.1(3)] and consists mainly of microcrystalline quartz, clay minerals, opaque minerals dominated by pyrite, and radiolarian ghosts. It is variably altered to yellowish white in colour. The claystone, several centimetres in thickness, is interbedded with chert and/or finely laminated, black mudstone on a millimetre-to-centimetre scale, occasionally showing banded patterns [Plate 14.1(2)]. Microscopically, the claystone grades into radiolarian chert [Plate 14.1(3)] and fine-laminated mudstone with or without radiolarians. Cubic pyrite crystals are concentrated in

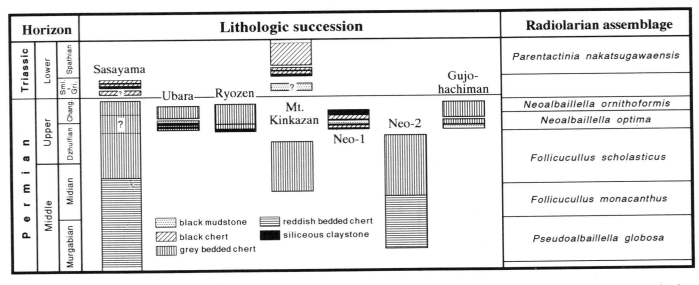

Figure 14.2. Lithologic successions near the P–T boundary in a pelagic environment in the Mino-Tamba Terrane. The lithologic and faunal data for Sasayama and Mt. Kinkazan are adapted from Ishida et al. (1992) and Sugiyama (1992), respectively. The radiolarian zonation is based mainly on the work of Ishiga (1990). Chang., Changxingian; Gri., Griesbachian; Smi., Smithian.

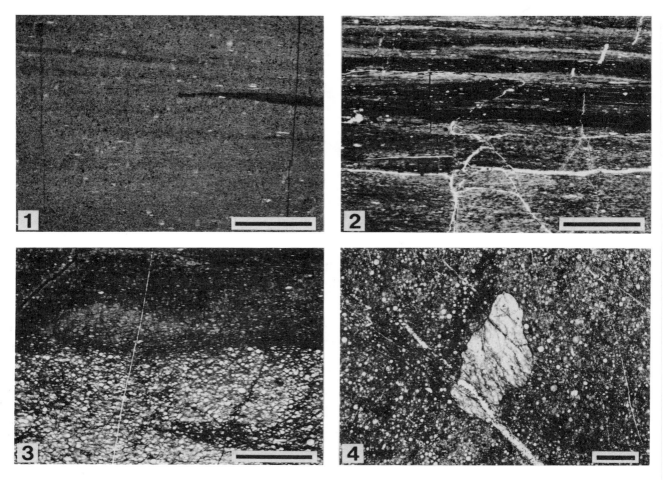

Plate 14.1. Siliceous claystone, black mudstone, and radiolarian chert near the P–T boundary (scale bar = 2 mm). (1) Siliceous claystone having black layers of pyrite concentration; *Neoalbaillella optima* Assemblage Zone in Ubara. (2) Siliceous claystone (lower part) and finely laminated beds of black mudstone and claystone (upper part);

N. optima Assemblage Zone in Neo. (3) Siliceous claystone and intercalating radiolarian chert in Neo. (4) Early Triassic muddy chert containing a chert grain of middle–late Permian age; *Parentactinia nakatsugawaensis* Assemblage in Mt. Kinkazan.

layers. Some radiolarians are flattened along the scaly foliation and dark-coloured seams. Conodonts (*Neogondolella* sp.) are found in the claystone and chert intercalations. Dolomite rhombs are scattered throughout in the grey claystone, but they are concentrated in a black, siliceous layer.

The claystone shows weak fissility and ductile deformation, and lenticular cherts are surrounded by crude clay partings.

Sheared surfaces are developed between different lithologies. The presence of lubricative rocks is one of the main reasons for the structural discontinuity at the P–T boundary (Nakae, 1993).

Radiolarians of the *Neoalbaillella optima* Assemblage Zone [Plate 14.2(9–12)] and sponge spicules occur in bedded chert and chert intercalations within grey siliceous claystone and black siliceous mudstone. The radiolarian chert alternates with clay-stone

Plate 14.2 (opposite). Radiolarian and conodont fossils from the Upper Permian in the Mino-Tamba Terrane (5, 9–12, and 16 from *Neoalbaillella optima* Assemblage Zone; 1–4, 6–8, and 13–15 from *N. ornithoformis* Assemblage Zone) (scale bar = 100 μm). (1) *Neoalbaillella optima* Ishiga, Kito, and Imoto; grey bedded chert in Gujo-hachiman. (2) *N. ornithoformis* Takemura and Nakaseko; grey bedded chert in Gujo-hachiman. (3) *Follicucullus falx* Caridroit and DeWever; grey bedded chert in Gujo-hachiman. (4) *F. scholasticus* Ormiston and Babcock; grey bedded chert in Gujo-hachiman. (5) *Albaillella excelsa* Ishiga, Kito, and Imoto; grey bedded chert in Gujo-hachiman. (6) *A. flexa* Kuwahara; grey bedded chert in Gujo-hachiman. (7) *A. levis* Ishiga, Kito, and

Imoto; grey bedded chert in Gujo-hachiman. (8) *A.* sp. aff. *A. levis* Ishiga, Kito, and Imoto; grey bedded chert in Sasayama. (9) *A. triangularis* Ishiga, Kito, and Imoto; alternation of grey siliceous claystone and chert in Ubara. (10) *A. triangularis* Ishiga, Kito, and Imoto; chert in black siliceous claystone in Gujo-hachiman. (11) *A. triangularis* Ishiga, Kito, and Imoto; alternation of chert and siliceous claystone in Neo. (12) *Entactinia* sp.; alternation of chert and siliceous claystone in Neo. (13) *Meshedea permica* Sashida and Tonishi; grey bedded chert in Gujo-hachiman. (14) *Ishigaum* sp.; grey bedded chert in Gujo-hachiman. (15) *Triplanospongos musashiensis* Sashida and Tonishi; grey bedded chert in Gujo-hachiman. (16) *Neogondolella* sp.; grey bedded chert in Ryozen.

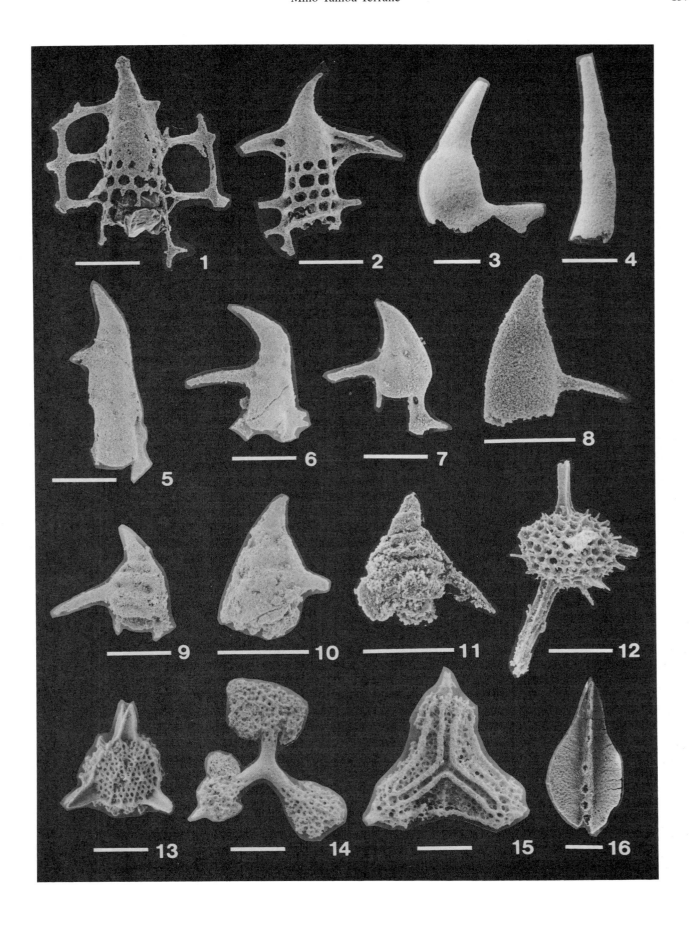

in the Ubara (Kuwahara et al., 1991), Neo (Sano, 1988), and Ryozen (Y. Ezaki, K. Kuwahara, unpublished data) sections, but it occurs as small lenses in black mudstone in Gujo-hachiman (Kuwahara et al., 1991). The *Neoalbaillella optima* Assemblage Zone is not detected in Sasayama (Ishida et al., 1992).

Grey bedded chert of the upper part of the Upper Permian

The foregoing claystone beds grade into bedded chert, with abundant radiolarians belonging to the fauna of the upper part of the *Neoalbaillella optima* Assemblage Zone and the *N. ornithoformis* Assemblage Zone in Ubara, Ryozen, and Gujo-hachiman [Figure 14.2(1–8, 13–15)]. The chert is dark grey, but occasionally it includes brownish-red layers in Gujo-hachiman and Ryozen, where the hematite content is very high. The radiolarian remains are filled with a mosaic, microcrystalline quartz and chalcedonic silica. The age of the *N. ornithoformis* Assemblage Zone is the late part of the late Permian, as indicated by fusulinids from contiguous limestones (Yamashita, Ishida, and Ishiga, 1992b). The bedded chert contains the conodont *Neogondolella changxingensis*, indicative of late Permian and other gondolellids [Plate 14.2(16)].

Permian–Triassic transition beds

The P–T transition beds in this study are marked by intergradings of chert, siliceous claystone, black mudstone, and dolostone. The transition beds span the latest Changxingian to the early Spathian, but a conformable P–T succession is not confirmed (Figure 14.2).

The average content of organic matter is 2–3% total organic carbon in the black mudstone in the Ashimi section (Yamashita et al., 1992a). The mudstone, alternating with chert and siliceous claystone, occasionally is greatly sheared.

Siliceous claystone beds grade upward into bedded chert (Imoto, 1984a). Rhythmic alternations (2–3 cm) of siliceous claystone and black chert are developed in Sasayama. Thin chert intercalations include flattened radiolarian ghosts. The chert is occasionally lenticular and nodular. In Unuma, intervening black mudstone and chert sometimes are amalgamated to show an apparently thick layer. Dolostone in black mudstone is also lenticular and nodular. Pyrite concentrations occur as layers in the claystone.

The conodonts *Neospathodus dieneri* and *Neospathodus waageni*, of Smithian age, occur in the claystone in Sasayama (Yamakita, 1993). No Griesbachian nor Dienerian fossils have been recovered (Ishiga, 1992).

Bedded chert of the upper part of the Lower Triassic

Black, muddy chert in Mt. Kinkazan contains radiolarians of the *Parentactinia nakatsugawaensis* Assemblage, of Spathian or older age, in which middle-to-late Permian radiolarians occur as derived fossils (Sugiyama, 1992) [Plate 14.1(4)]. The bedded chert is succeeded by widely distributed bedded chert of middle Triassic age.

Fluctuations in environment

Folliculullus-bearing bedded chert

The radiolarians were diversified and widespread. Few calcareous sediments were deposited where deposition was below the carbonate-compensation depth. The colour change of the bedded chert in the *Folliculullus* Zone from reddish to grey reflects major oceanic changes in degree of oxygenation (Noble and Renne, 1990). The reddish chert, as observed in the Sasayama and Neo sections, was deposited on an oxygenated bottom, whereas the overlying grey chert was formed on a stagnant, anoxic bottom (Ishiga, 1992).

Neoalbaillella-bearing siliceous claystone beds

The grey siliceous claystone with high clay content is related to deposition of more or less inorganic components, without showing high radiolarian productivity in the photic zone. The claystone is condensed, probably because of low sedimentation rates and the presence of hiatuses.

The siliceous claystone beds are not age-specific, but were formed repeatedly across the P–T boundary. Claystone deposition in the *Neoalbaillella optima* Assemblage Zone implies that Palaeozoic-type radiolarians were impoverished temporarily before the P–T extinction. Unfavourable habitat conditions appeared locally in the late Permian, and there were environmental perturbations, as shown by repeated, fine-scale intergradings of chert, claystone, and black mudstone.

Grey bedded chert of the upper part of the Upper Permian

Formation of Upper Permian claystone was succeeded by normal deposition of bedded chert in the upper part of the *Neoalbaillella optima* Assemblage Zone and/or the *N. ornithoformis* Assemblage Zone. Palaeozoic-type radiolarians recovered and flourished again up to the latest Permian, but almost all became extinct near the P–T boundary. The hematite layers in grey bedded chert suggest the occasional presence of an oxic environment.

Permian–Triassic transition beds

Widespread deposition of siliceous claystone resulted from a marked decline in radiolarians over a large area and for a long time. The deposition may have been related to very slow biogenic sedimentation in the early Triassic. However, the presence of chert intercalations indicates intermittent and episodic radiolarian blooming initiated by favourable environmental changes.

Carbonaceous mudstone shows a high organic-matter content,

because the matter was shielded from aerobic degradation on the sea bottom. Dolomite may have been formed *in situ* by inorganic precipitation. Some dolostones of lenticular and nodular form may have been reworked and transported as displaced carbonates from shallow water into a basin environment (Sano, 1988).

Bedded chert of the upper part of the Lower Triassic

Mesozoic-type radiolarians became diverse during the Spathian and later times when there were favourable conditions, and cyclic radiolarian blooming formed the bedded chert. Granules of Permian chert, reworked and transported by strong currents into the basins, are present in the Triassic chert.

Implications for environments in the Panthalassan Ocean

The Permian radiolarian chert was deposited in the western Panthalassan Ocean, whereas phosphorite and sponge-spicule-rich chert were formed along the western margin of Pangaea (Murchey and Jones, 1992), reflecting latitudinal, local variations in the oceanic environment. In the Panthalassan Ocean there may have been numerous seamounts and plateaus, judging from geochemical and mineralogic analyses of the greenstones (Isozaki et al., 1990) associated with Permian shallow-water carbonates. The volcanic masses form an irregular sea-floor topography and partition the basin, producing a marked variety of depositional settings in a pelagic environment (Imoto, 1983, 1984b). The shallow-water carbonates may represent slide blocks and breccias on the slopes and in the basin (Sano, 1988). Permian chert was also reworked and redeposited in the basin.

Permian chert with similar albaillellid radiolarian assemblages shows a wide latitudinal distribution (Murchey and Jones, 1992), indicating that the same faunal province had been established in the same climatic zone. Under the influence of westward-driven currents, strong upwelling of nutrient-rich waters may have occurred in the equatorial region (Parrish, 1982), providing conditions suitable for radiolarian blooming in the photic zone. The irregular sea-floor topography also provided sites favourable for radiolarian blooming (Imoto, 1984a,b), and at the same time it may have enhanced water stagnation.

Siliceous deposition was prevalent in the *Follicucullus scholasticus* Assemblage Zone, but it may have been succeeded by a widespread hiatus and disconformity in circum-Pacific and Mediterranean regions (Murchey and Jones, 1992). The colour change of chert from reddish to grey, reflecting the degree of oxygenation, is sharp and can be widely traced in the *Follicucullus monacanthus* Assemblage Zone and *F. scholasticus* Assemblage Zone, roughly Midian in age. The Hunter-Bowen Orogenic Phase became intensified after the Kungurian–Ufimian boundary, and it was accompanied by intensive acidic-to-intermediate volcanism. Large amounts of acidic-to-intermediate volcanic products are widely distributed in the Akiyoshi and Maizuru terranes in southwest Japan (e.g., Dickins, 1990; Kanmera, Sano, and Isozaki, 1990) and in South China in the form of volcaniclastic

rocks and P–T-boundary clays (Yang et al., 1987; Yin et al., 1992). Those volcanic events may have been closely related to widespread oceanic anoxia (Ishiga, 1992).

On the Yangtze Platform, fusulinids, calcareous algae, and sponges flourished up to the Changxingian reef (Flügel and Reinhardt, 1989), as did the radiolarian fauna in the Panthalassan Ocean. However, most benthic and pelagic organisms became impoverished because of deteriorating habitat conditions and became extinct near the P–T boundary (e.g., Yang et al., 1987; Ezaki, 1994, and Chapter 13, this volume), although little is known about any phylogenetic relation between, for example, the Palaeozoic-type and the Mesozoic-type radiolarians.

In near-shore environments, only specific organisms such as smaller foraminifers, conodonts, and brachiopods survived across the P–T boundary, forming mixed-fauna beds (Sheng et al., 1984). However, rather uniform lithofacies of muddy shales, containing species of the bivalve *Claraia*, were developed widely in the early Triassic. No lowermost Triassic cherts, except those of volcanic origin, accumulated in South China, where the latest Permian radiolarian chert was formed in some places. Microbial mats and algal stromatolites were formed at the time of organic impoverishment (Schubert and Bottjer, 1992).

The amount of CO_2 was increased by volcanism and oxidation of organic carbon on the widespread continental shelves during the late Permian regression, and it may have produced a warm climate and oceanic anoxia (Erwin, 1993). A series of events is reflected not only in the changes in chert colour but also in the deposition of black, carbonaceous mudstone alternating with black chert and claystone containing dolomite and pyrite crystals. The lithologic characteristics of the Upper Permian claystone beds and the P–T transition beds show marked decreases in the supply of biogenic silica, probably because of impoverishment of radiolarians and sponges, as well as an increase in inorganic materials produced by continental weathering and volcanism.

The Upper Permian claystone beds and the P–T transition beds, as a whole, were formed in reducing, ocean-bottom conditions, and they reflect devastating, unpredictable fluctuations in organic productivity, which may have been coupled with the end-Permian extinction events. A major change in the oceanic environment occurred near the P–T boundary on a global scale, producing peculiar lithologic successions.

Acknowledgments

We would like to express our sincere thanks to Dr. J. M. Dickins of the Australian Geological Survey Organisation, Professor A. Yao of Osaka City University, Dr. W. J. Sando of the U.S. Geological Survey, and Professor C. H. Stevens of San Jose State University for critical readings of the manuscript and invaluable discussions. We also thank Dr. H. Sano of Kyushu University and Mr. K. Kitao of Osaka City University for useful discussions and information. This study was supported by the Scientific Research Fund of the Japanese Ministry of Education, Science and Culture (grants 04304009, 05740327, and 05304001).

References

Dickins, J. M. 1990. Permian of Japan and its significance for world understanding. In *Shallow Tethys 3,* ed. T. Kotaka, J. M. Dickins, K. G. McKenzie, K. Mori, K. Ogasawara, and G. D. Stanley, Jr., pp. 343–51. Sendai: Saito Ho-on Kai.

Erwin, D. H. 1993. *The Great Paleozoic Crisis.* New York: Columbia University Press.

Ezaki, Y. 1994. Patterns and paleoenvironmental implications of end-Permian extinction of Rugosa in South China. *Palaeogeography, Palaeoclimatology, Palaeoecology* 107: 165–77.

Flügel, E., and Reinhardt, J. 1989. Uppermost Permian reefs in Skyros (Greece) and Sichuan (China): implication for the Late Permian extinction event. *Palaios* 4:502–18.

Ichikawa, K. 1990. Pre-Cretaceous Terranes of Japan. In *Pre-Cretaceous Terranes of Japan,* ed. K. Ichikawa, S. Mizutani, I. Hara, S. Hada, and A. Yao, pp. 1–12. Publication of IGCP-224. Osaka.

Imoto, N. 1983. Sedimentary structures of Permian–Triassic cherts in the Tamba district, southwest Japan. In *Siliceous Deposits in the Pacific Region,* ed. A. Iijima, J. R. Hein, and R. Siever, pp. 377–93. Amsterdam: Elsevier.

Imoto, N. 1984a. Late Paleozoic and Mesozoic cherts in the Tamba Belt, southwest Japan (part 1). *Bulletin of Kyoto University of Education, Series B* 65:15–40.

Imoto, N. 1984b. Late Paleozoic and Mesozoic cherts in the Tamba Belt, southwest Japan (part 2). *Bulletin of Kyoto University of Education, Series B* 65:41–71.

Ishida, K., Yamashita, M., and Ishiga, H. 1992. P/T boundary in pelagic sediments in the Tamba Belt, southwest Japan (in Japanese, with English abstract). *Geological Reports of Shimane University* 12:39–57.

Ishiga, H. 1990. Paleozoic radiolarians. In *Pre-Cretaceous Terranes of Japan,* ed. K. Ichikawa, S. Mizutani, I. Hara, S. Hada, and A. Yao, pp. 285–95. Publication of IGCP-224. Osaka.

Ishiga, H. 1992. Late Permian anoxic event and P/T boundary in pelagic sediments of southwest Japan. *Memoirs of the Faculty of Science, Shimane University* 26:117–29.

Ishiga, H., and Imoto, N. 1980. Some Permian radiolarians in the Tamba Belt, southwest Japan. *Earth Science (Chikyu Kagaku)* 34:332–45.

Ishiga, H., Kito, T., and Imoto, N. 1982. Late Permian radiolarian assemblages in the Tamba district and an adjacent area, southwest Japan. *Earth Science (Chikyu Kagaku)* 36:10–22.

Isozaki, Y., Maruyama, S., and Furuoka, F. 1990. Accreted oceanic materials in Japan. *Tectonophysics* 181:179–205.

Jenkyns, H. C., and Winterer, E. L. 1982. Palaeoceanography of Mesozoic ribbon radiolarites. *Earth and Planetary Science Letters* 60:351–75.

Kanmera, K., Sano, H., and Isozaki, Y. 1990. Akiyoshi Terrane. In *Pre-Cretaceous Terranes of Japan,* ed. K. Ichikawa, S. Mizutani, I. Hara, S. Hada, and A. Yao, pp. 49–62. Publication of IGCP-224. Osaka.

Kuwahara, K., Nakae, S., and Yao, A. 1991. Late Permian "Toishi-type" siliceous mudstone, Mino-Tamba Belt (in Japanese). *Journal of the Geological Society of Japan* 97:1005–8.

Matsuda, T., and Isozaki, Y. 1991. Well-documented travel history of Mesozoic pelagic chert in Japan: from remote ocean to subduction zone. *Tectonics* 10:475–99.

Mizutani, S. 1990. Mino Terrane. In *Pre-Cretaceous Terranes of Japan,* ed. K. Ichikawa, S. Mizutani, I. Hara, S. Hada, and A. Yao, pp. 121–35. Publication of IGCP-224. Osaka.

Murchey, B. L., and Jones D. L. 1992. A mid-Permian chert event: widespread deposition of biogenic siliceous sediments in coastal, island arc and oceanic basins. *Palaeogeography, Palaeoclimatology, Palaeoecology* 96:161–74.

Nakae, S. 1993. The Permo–Triassic boundary as a decollement zone within pelagic siliceous sediments with reference to Jurassic accretion of the Tamba Terrane, SW Japan (in Japanese, with English abstract). *Bulletin of the Geological Survey of Japan* 44:471–81.

Noble, P., and Renne, P. 1990. Paleoenvironmental and biostratigraphic significance of siliceous microfossils of the Permo–Triassic Redding section, eastern Klamath Mountains, California. *Marine Micropaleontology* 15:379–91.

Parrish, J. T. 1982. Upwelling and petroleum source beds, with reference to Paleozoic. *American Association of Petroleum Geologists Bulletin* 66:750–74.

Sano, H. 1988. Permian oceanic rocks of Mino terrane, central Japan. Part I. Chert facies. *Journal of the Geological Society of Japan* 94:697–709.

Sano, H. 1989. Permian oceanic rocks of Mino terrane, central Japan. Part IV. Supplements and concluding remarks. *Journal of the Geological Society of Japan* 95:595–602.

Schubert, J. K., and Bottjer, D. J. 1992. Early Triassic stromatolites as post-extinction disaster forms. *Geology* 20:883–6.

Sheng, J. Z., Chen, C. Z., Wang, Y. G., Rui, L., Liao, Z. T., Bando, Y., Ishii, K., Nakazawa, K., and Nakamura, K. 1984. Permian–Triassic boundary in Middle and Eastern Tethys. *Journal of the Faculty of Science, Hokkaido University, Series IV, Geology and Mineralogy* 21:133–81.

Shibuya, H., and Sasajima, S. 1986. Paleomagnetism of red cherts: a case study in the Inuyama area, central Japan. *Journal of Geophysical Research* 91:14105–16.

Sugiyama, K. 1992. Lower and Middle Triassic radiolarians from Mt. Kinkazan, Gifu prefecture, central Japan. *Transactions and Proceedings of the Palaeontological Society of Japan, New Series* 167:1180–223.

Sweet, W. C., Yang, Z. Y., Dickins, J. M., Yin, H. F. (eds.) 1992. *Permo–Triassic Events in the Eastern Tethys.* Cambridge University Press.

Yamakita, S. 1987. Stratigraphic relationship between Permian and Triassic strata of chert facies in the Chichibu Terrane in eastern Shikoku (in Japanese). *Journal of the Geological Society of Japan* 93:145–8.

Yamakita, S. 1993. Conodonts from P/T boundary sections of pelagic sediments in Japan (in Japanese). *Abstracts of the 100th Annual Meeting of the Geological Society of Japan,* pp. 64–5.

Yamashita, M., Ishida, K., and Ishiga, H. 1992a. Late Early to early Middle Triassic bedded cherts in the Tamba Belt and black organic mudstones in the P/T boundary, Southwest Japan (in Japanese, with English abstract). *Geological Reports of Shimane University* 11:87–96.

Yamashita, M., Ishida, K., and Ishiga, H. 1992b. *Palaeofusulina sinensis* age for Late Permian *Neoalbaillella ornithoformis* radiolarian zone, southwest Japan (in Japanese). *Journal of the Geological Society of Japan* 98:1145–8.

Yang, Z. Y., Yin, H. F., Wu, S. B., Yang, F. Q., Ding, M. H., and Xu, G. R. 1987. *Permian–Triassic Boundary Stratigraphy and Fauna of South China.* Geological memoirs, ser. 2, no. 6 (in Chinese, with English summary). Beijing: Geological Publishing House.

Yao, A., Matsuda, T., and Isozaki, Y. 1980. Triassic and Jurassic radiolarians from the Inuyama Area, central Japan. *Journal of Geosciences, Osaka City University* 23:135–54.

Yin, H. F., Huang, S. J., Zhang, K. X., Hansen, H. J., Yang, F. Q., Ding, M. H., and Bie, X. M. 1992. The effects of volcanism on the Permo–Triassic mass extinction in South China. In *Permo–Triassic Events in the Eastern Tethys,* ed. W. C. Sweet, Z. Y. Yang, J. M. Dickins, and H. F. Yin, pp. 146–57. Cambridge University Press.

15 Late Changxingian ammonoids, bivalves, and brachiopods in South Primorye

YURI D. ZAKHAROV, ALEXANDER OLEINIKOV, and GALINA V. KOTLYAR

Late Changxingian (late Dorashamian) ammonoids are known only in three regions of the world: South China (Chao, 1965; Zhao, Liang, and Zheng, 1978; Sheng, 1989), Transcaucasia (Zakharov, 1985, 1986, 1992; Zakharov and Rybalka, 1987; Kotlyar, 1991), and South Primorye (Zakharov and Pavlov, 1986; Zakharov, 1987, 1992). Until recently, their presence in the last two areas had been confirmed only by rare finds.

This chapter comprises excerpts from a lecture presented at an international conference on the Carboniferous–Jurassic strata of Pangaea (Zakharov and Oleinikov, 1993), with some additions. We now have information on two late Changxingian ammonoid complexes discovered in the Primorye region: The first and most representative complex was found on the right bank of the Partizanskaya River (1.0–1.5 km northeast of the railway station of Vodopadnaya) in 1992. The second, comparatively more scanty, was recognized on the left bank of the Artemovka River some years ago (Zakharov, 1987, 1992).

Occurrences of late Changxingian ammonoids in Primorye

In descending order, the sequence of Permian–Triassic sediments in the section at Vodopadnaya is as follows:

Peschanka River Suite (Lower Norian)
8. Dark grey mudstone, coal argillite, and grey sandstone (~5 m): Yields plant remains identified by S. A. Shorokhova and E. B. Volynets as *Ctenis* ex gr. *sulcicaulis* (Phillips) Ward, *Taeniopteris spathulata* Oldham and Morris, *Pseudoctenis* ex gr. *mongugaica* Prynada, *Podozamites* cf. *kiparisovkensis* Srebrodolskaya and Shorokhova, *P.* ex gr. *lanceolata* (Li and Hsu) Braun, *Elatocladus* ex gr. *cephalotaxoides* Florin, *Cycadocarpidium* sp., and *Pachipteris* (*Thinnfeldia*) sp.
7. Grey sandstone and siltstone, with conglomerate at the base (~250 m): Contains the bivalves *Tosapecten subhiemalis* (Kiparisova), *Oxytoma zitteli* (Teller), *O.* cf. *mojsisovicsi* Teller, and *Unionites muensteri* (Wissmann) (E. R. Dorokhovskaya's determinations).
Disconformity

Lyudyanza horizon: Kapreevka siltstone member, *Huananoceras qianjiangense* beds (Upper Dorashamian)
6. Greyish-green and yellowish-green tuff, grey siltstone, and lens of marl (85–90 m): The tuff contains the following: many bivalves (*Posidonia* sp.) and ceratite ammonoids (*Dzhulfoceras* n. sp., *Xenodiscus* aff. *strigatus* Schindewolf, *X.* aff. *jubilaearis* Zakharov, Xenodiscidae n. gen., and *Huananoceras* sp.) 8–16 m from the top (locality numbers K 2653, K 2653/2, K 2653/3); numerous mollusks (*Posidonia* sp., *Xenodiscus* sp., Xenodiscidae n. gen., and *Huananoceras* cf. *perornatum* Chao and Liang) about 25 m from the top (K 2617); many bivalves (*Posidonia* sp.), ceratite ammonoids (*Xenodiscus* aff. *jubilaearis* Zakharov, *Xenodiscus* sp., *Sinoceltites* ex gr. *costatus* Zhao, Liang, and Zheng, and *Huananoceras qianjiangense* Zhao, Liang, and Zheng), and rare nautiloids (*Pseudorthoceras?* sp.) about 45 m from the top (P 2604); numerous bivalves (*Posidonia* sp.), ceratite ammonoids (*Xenodiscus* sp., Xenodiscidae n. gen., and *Sinoceltites* ex gr. *costatus* Zhao, Liang, and Zheng), and rare goniatites (*Changhsingoceras?* sp. indet.) 75 m from the top (P 2504a/2).
5. Thin intercalation of yellowish-green tuff and dark-grey siltstone, with an interbed of acidic tuff (7 cm) in the middle part (~25 m): Contains the following: many bivalves (*Posidonia* sp. and *Chaenomya* sp.), ceratite ammonoids (*Xenodiscus* aff. *strigatus* Schindewolf, *X.* aff. *jubilaearis* Zakharov, Xenodiscidae n. gen., *Mingyuexiaceras* sp. indet., *Huananoceras* cf. *perornatum* Chao and Liang, *H.* cf. *involutum* Chao and Liang, and *H. qianjiangense* Zhao, Liang, and Zheng), rare goniatites (*Changhsingoceras?* sp. indet.), and nautiloids (*Pseudorthoceras?* sp.) 5–10 m above the base (P 2504a); numerous small bivalves (*Posidonia* sp.), ceratites (*Dzhulfoceras* n. sp., *Xenodiscus* aff. *strigatus* Schindewolf, *X.,* aff. *jubilaearis* Zakharov, Xenodiscidae n. gen., *Sinoceltites* ex. gr. *costatus* Zhao, Liang, and Zheng, *Huananoceras* cf. *perornatum* Chao and Liang, *H.* cf. *involutum* Chao and Liang, *H. qianjiangense* Zhao, Liang, and Zheng, Pleuronodo-

ceratidae n. gen. 2, and *Liuchengoceras* cf. *crassi-costatum* Zhao, Liang, and Zheng), and rare nautiloids (*Pseudorthoceras?* sp.), apparently within 5 m of the base (P 2507).

4. Yellowish-green tuff, intercalated with dark-grey silt-stones and fine-grained sandstone (~25 m): Turf-clad interval 5–10 m.

3. Yellowish-green and green acidic tuff, intercalated with dark-grey siltstone and fine-grained calcareous sand-stone (~10 m): Yields many small brachiopods [*Paracrurithyris pygmaea* (Liao), *Crurithyris flabelliformis* Liao, and *Araxathyris minor* Grunt], rare goniatites (Paragastrioceratidae?), and bryozoans (P 2520-3).

2. Intercalation of yellowish-green acidic ashstone and greyish-green siltstone with calcareous nodules (5.5 m): Contains small brachiopods and rare ammonoid jaws (*Anaptychus*) (P 2520-3a).

Yastrebovka Suite, upper part (Lower? Dorashamian)

1. Grey siltstone and mudstone with calcareous nodules, yielding foraminifera, crinoids, and bryozoans (>20 m): The total thickness of the Kapreevka siltstone in the section is about 150 m.

Data on the Kapreevka siltstone at the Vodopadnaya station environs permit us to review the position of the Permian–Triassic boundary in the Artemovka River basin. Until recently, only the units 12–18 (92.5 m) (Zakharov, 1992), overlying the Dzhulfian *Eusanyangites bandoi* beds, were considered to be Dorashamian (Zakharov and Pavlov, 1986; Zakharov, 1987). The lower parts of those sequences (units 12–16) are characterized by am-monoids (*Neogeoceras thaumastum* Ruzhencev, *Eumedlicottia* sp., Cyclolobidae gen. et sp. indet., and *Iranites?* sp. – apparently a new genus), and the upper parts (units 17 and 18) yield bivalves (*Aviculopecten* sp.), gastropods, nautiloids, am-monoids (Pleuronodoceratidae n. gen. 1 and *Liuchengoceras melnikovi* Zakharov), and bryozoans (*Fistulipora*, *Eridopora*, and *Stenodiscus*).

The upper sequences in the section (units 19 and 20) were believed to be Kapreevka siltstone (they were formerly consid-ered lowermost Triassic). The sequences consist of black and greyish-green thin-bedded shale (pelite) and siltstone, containing small nodules of clay and marl, and fine-grained sandstone (39.8 m thick).

Judging from the revision carried out, small ammonoids from the Kapreevka siltstone on the left bank of the Artemovka River seem to be *Xenodiscus* aff. *carbonarius* Waagen, *Xenodiscus* sp., and *Xenodiscus?* sp. indet. The former attribution of some small ammonoid forms to *Dieneroceras*, *Xenoceltites* (Burij and Zharnikova, 1989), and *Ophiceras* (*Lytophiceras*) (Zakharov, 1992) is now believed to be in error. Bivalves from those beds were described by N. K. Zharnikova (Burij and Zharnikova, 1989) as *Posidonia?* and *Claraia?*.

The thickness of the Kapreevka siltstone member in the section is about 40 m. The total thickness of the Dorashamian at the locality is not less than 132 m.

About 300 m downstream, the Lower Olenekian (*Heden-stroemia bosphorensis* Zone) is exposed (Zakharov and Pavlov, 1986). The next Lower Triassic zone, the *Anasibirites nevolini* Zone, was recognized another 400–500 m downstream.

No Induan ammonoids have been discovered in the section. The Permian–Triassic interrelation is not recognized here because of the turf cladding of the corresponding interval.

Analysis of late Dorashamian faunistic complex and some correlation problems

The Kapreevka strata on the left bank of the Partizanskaya River are characterized at the base by Permian-type brachiopods: *Paracrurithyris pygmaea* (Liao), *Crurithyris flabelliformis* Liao, and *Araxathyris minor* Grunt. Representatives of the first two species were found in the Permian–Triassic boundary beds in the Meishan region of South China (Zhao et al., 1978), and representatives of the last species are known from the Dzhulfian and Dorashamian of Transcaucasia (Kotlyar et al., 1983).

Bivalves from the Kapreevka siltstone in South Primorye are abundant, but monotonous in taxonomic aspect. The main body of the bivalve complex is formed by *Posidonia*. The representa-tives of this genus appeared in the Silurian and became extinct in the Jurassic. No *Claraia* have been found in the beds mentioned earlier. *Chaenomya* are found in the upper part of the Kapreevka siltstone. The species of the genus have been recognized only in Palaeozoic (Carboniferous–Permian) sediments.

Ceratites of the family Xenodiscidae (especially *Xenodiscus*) prevail over other groups of ammonoids in the Kapreevka siltstone. Unquestionably, *Xenodiscus* existed during Midian and Dorashamian times. *Xenodiscus* aff. *carbonarius* (Waagen) from the Kapreevka siltstone is closely related to *X. carbonarius* (Waagen) from the Chhidru Formation of the Salt Range (Waagen, 1895), but has narrower lateral saddles of the suture line. Far Eastern *Xenodiscus* aff. *strigatus* Schindewolf are similar to *X. strigatus* Schindewolf from the *Productus* Limestone of the Salt Range (Schindewolf, 1954), but can be distinguished by the higher first lateral saddle. *Xenodiscus* aff. *jubilaearis* Zakharov is very similar to *X. jubilaearis* Zakharov from the Upper Dorashamian of Transcaucasus, but has a more evolute shell without radial folds. Of greater significance for age determination of the Kapreevka siltstone is the discovery of representatives of the Cyclolobidae (goniatites), Araxoceratidae, Huananocera-tidae, Liuchengoceratidae, and Tapashanitidae (ceratites).

Of the araxoceratids, only *Dzhulfoceras* has been found in the Kapreevka siltstone. In the Transcaucasus, representatives of *Dzhulfoceras* were discovered in the upper part of the Dzhulfian (*Vedioceras ventrosulcatum* Zone), but in central Iran they were found in the Dorashamian (member 7 of the Hambast Forma-tion) (Bando, 1979).

The significant representatives of the *Huananoceras* species (Huananoceratidae) in South China are known from the Changxing Formation (Dorashamian), except one species (*Huananoceras involutum* Chao et Liang), which was found in the Wujiaping Formation (Dzhulfian) (Zhao et al., 1978). All

Far Eastern *Huananoceras* were identified as species described from South China.

The representatives of all the other genera determined in the Kapreevka siltstone of South Primorye (*Changhsingoceras?*, *Liuchengoceras*, *Tapashanites*, and *Sinoceltites*) usually are encountered within an interval corresponding to the upper part of the Changxing Formation, except for *Sinoceltites*, which are distributed in both the Upper Dorashamian (Upper Changxingian) and the underlying sediments (the upper part of the *Paratirolites kittli* Zone). In W. W. Nassichuk's opinion (personal communication), *Changhsingoceras?* sp. indet. from Primorye can be identified, on the basis of a fragmentary suture line, as *Paramexicoceras* (Popov, 1970), but the stratigraphic distribution of the genus in the Verkhoyansk region is unknown (the holotype was found in alluvium).

In summary, the Kapreevka siltstone of South Primorye seems to be late Dorashamian (late Changxingian) in age; the lower boundary of the Kapreevka siltstone is believed to be time-transgressive: In the Artemovka River basin it coincides with the top of the Pleuronodoceratidae-*Liuchengoceras* beds, whereas in the Partizanskaya River basin it is situated somewhat lower (at the base of the *Huananoceras qianjiangense* beds, which correspond, apparently, to the base of the Pleuronodoceratidae-*Liuchengoceras* beds of the Artemovka River basin) (Table 15.1).

The following units were originally recognized within the Lyudyanza horizon in South Primorye: (1) *Cyclolobus kiselevae* beds, (2) beds with no ammonoids, (3) *Eusanyangites bandoi* beds, (4) *Iranites?* sp. beds, and (5) Pleuronodoceratidae-*Liuchengoceras* beds. It seems more likely, however, that the Lyudyanza horizon must be accreted to the new member: the Kapreevka siltstone (*Xenodiscus-Posidonia* beds in the Artemovka River basin) or its upper part (upper *Huananoceras qianjiangense* beds in the Partizanskaya River basin). Because in the stratotype of the Lyudyanza horizon the Kapreevka siltstone beds are unknown, we offer the section on the left bank of the Artemovka River (Zakharov and Pavlov, 1986; Zakharov, 1992) as a hypostratotype for this unit.

The Kapreevka siltstone member in the Primorye region, which has its greatest thickness in the Partizanskaya River basin (*Huananoceras qianjiangense* beds), can be correlated with the *Pseudostephanites-Tapashanites* and *Pleuronodoceras-Rotodiscoceras* zones of the uppermost Changxing Formation of South China and the *Pleuronodoceras occidentale* Zone of the Upper Dorashamian in the Dzhulfian Ravine, Transcaucasia (Table 15.2).

The findings from an analysis of the late Dorashamian ammonoid complex in South Primorye show that it is similar (on the species and generic levels) to the late Changxingian of South China. Of 17 species of 10 ammonoid genera from the upper part of the Lyudyanza horizon in South Primorye, 4 species apparently are the same as taxa from the Upper Changxing Formation. Only few taxa (3 species and 3 genera) are new, distributed, obviously, only in the Primorye region.

During the time of the Permian–Triassic boundary, the South

Table 15.1 *Correlation of the Upper Permian of South Primorye*

STAGE	SUBSTAGE	Horizon	ARTEMOVKA		NEIZVESTNAYA	VODOPADNAYA, OREL	
			Suite, Member	B e d s	Suite / B e d s	Suite, Member	B e d s
DORASHAMIAN	UPPER	Lyudyanza	Kapreevka siltstone	Xenodiscus-Posidonia		Kapreevka siltstone	Huananoceras qianjiangense
				Pleuronodo-ceratidae – Liuchengo-ceras	Cotaniella parva		
	LOWER		Iranites?		Iranites?	Yastrebovka	Albaitiella optima (Rudenko and Panasenko, 1990)
DZHULFIAN		Lyudyanza	Eusanyangi-tes bandoi		?		
			?				
MIDIAN (part)			Cyclolobus?		Cyclolobus kiselevae		?

China tropical-sea basins seem to have been more closely connected with the Primorye region than with the western Tethys. To judge from the characteristics of the Permian–Triassic bivalve successions in different regions, one can conclude that the bivalves' occupation of the ecological niche of the extinct brachiopods was a relatively prolonged process. In Primorye, it took place before the Permian–Triassic boundary (late Dorashamian), but in Transcaucasia and some adjoining areas, and apparently in South China, bivalves began to dominate in environments of normal salinity only in the early Induan.

The presence of ash layers and acidic tuff in the Upper Dorashamian of the western Pacific (South Primorye) confirms Yin Hongfu's idea (Yin et al., 1992) about the existence of extensive intermediate-to-acidic volcanism during the time of the Permian–Triassic boundary.

References

Bando, Y. 1979. Upper Permian and Lower Triassic ammonoids from Abadeh, central Iran. *Memoirs of the Faculty of Education, Kagawa University* **29**:103–38.

Burij, I. V., and Zharnikova, N. K. 1989. Paleogeographical and stratigraphical peculiarities of the relationship of the

Table15.2 *Correlation of the Upper Permian of South Primorye with other parts of the world*

STAGE	SUBSTAGE	AKHURA,OGBIN,VEDI		DZHULFA		CHINA		PRIMORYE		JAPAN	
		Suite	Zone (beds)	Suite	Zone (beds)	Formation	Zone (beds)	Suite, member	Beds	Series	Beds
DORASHAMIAN	UPPER	Karabagtyar (part)	Lopingoceras, Haydenella	Karabagtyar (part)	Pleuronodoceras occidentale	Changxing	Rotodiscoceras–Pleuronodoceras	Kapreevka siltstone	Huananoceras qianjiangense	Toioma	?
							Pseudostephan.–Tapashanites				Cotaniella–Palaeofusulina
	LOWER	Akhura	Paratirolites kittli	Dzhulfa (= Dzhulfa)	Paratirolites kittli	Changxing	"Paratirolites"–Shevyrevites	Yastrebovka / Lyudyanza	Iranites?	Toioma	?
			Shevyrevites shevyrevi		Shevyrevites shevyrevi						
			Dzhulfites spinosus		Dzhulfites spinosus						
			Iranites transcaucasius		Iranites transcaucasius						
			Phisonites triangularis		Phisonites triangularis						
DZHULFIAN			Vedioceras ventroplanum	Akhura (= Dzhulfa)	Vedioceras ventrosulcatum	Wujiaping	Sanyangites		Eusanyangites bandoi		Eusanyangites cf. bandoi
			Araxoceras		Araxoceras latissimum (=A.latum)		Araxoceras–Kon. Anders.–Protot.		?		?
MIDIAN (part)		Khatchik (part)	Pseudodumbarulla arpaensis (=Codonofusiella –Araxilevis)		Vescotoceras		Araxoceratidae		Cyclolobus kiselevae		Timorites–Vescotoceras
					Araxilevis intermedius						
					P. arpaensis (=Codonofusiel.)						
		Khatchik				Maokou					

Permian and Triassic in South Primorye (in Russian). In *Paleontologo-Stratigraphicheskie Issledovaniya Phanerozoya Dal'nego Vostoka* [*Paleontological and Stratigraphical Investigations of the Phanerozoic in Far East*], ed. G. I. Buryi and S. V. Tochilina, pp. 11–22. Vladivostok: Dal'nevostochnoye Otdeleniye Akademii Nauk SSSR.

Chao Kingkoo. 1965. The Permian ammonoid-bearing formations of South China. *Scientia Geologica Sinica* 14:1813–26.

Kotlyar, G. V. 1991. Permian–Triassic boundary in Tethys and the Pacific Belt and its correlation. In *Shallow Tethys* 3, ed. T. Kotaka, J. M. Dickins, K. G. McKenzie, K. Mori, K. Ogasawara, and G. D. Stanley, Jr., pp. 387–91. Sendai: Saito Ho-on Kai.

Kotlyar, G. V., Zakharov, Y. D., Koczyrkevicz, B. V., Kropatcheva, G. S., Rostovcev, K. O., Chedija, I. O., Vuks, G. P., and Guseva, E. A. 1983. *Pozdnepermskii Etap Evolutsii Organicheskogo Mira. Dzhulinskii i Dorashamskii Yarusy SSSR.* [*Evolution of the Latest Permian Biota. Dzhulfian and Dorashamian Regional Stages in the USSR*], ed. M. N. Gramm and K. O. Rostovcev. Leningrad: Nauka.

Popov, Y. N. 1970. Ammonoidea (in Russian). *Trudy Nauchno-Issledovatel'skogo Instituta Geologii Arktiki (NIIGA)* 154:113–40.

Schindewolf, O. H. 1954. Ueber die Faunenwende vom Palaeozoicum zum Mesozoicum. *Zeitschrift der Deutschen Geologischen Gesellschaft* 105:153–82.

Sheng, H. 1989. Ammonoidea. Description of new genus and species. In *Study on the Permian–Triassic Biostratigraphy and Event Stratigraphy of Northern Sichuan and Southern Shaanxi*. Geological memoirs, ser. 2, no. 9, pp. 190–204. Beijing: Geological Publishing House.

Waagen, W. 1895. Salt Range fossils. II. Fossils from the Ceratite Formation. *Palaeontologia Indica, ser. 13* 2:1–323.

Yin Hongfu, Huang Siji, Zhang Kexing, Hansen, H. J., Yang Fengqing, Ding Meihua, and Bie Xianmei. 1992. The effects of volcanism on the Permo–Triassic mass extinction in South China. In *Permo–Triassic Events in the Eastern Tethys*, ed. W. O. Sweet, Yang Zunyi, J. M. Dickins, and Yin Hongfu, pp. 146–57. Cambridge University Press.

Zakharov, Y. D. 1985. To the problem of type of the Permian–

Triassic boundary (in Russian). *Bulleten Moskovskogo Obstchestva Ispytatelei Prirody, Otdel Geologicheskii* **60**:59–70.

Zakharov, Y. D. 1986. Type and hypotype of the Permian–Triassic boundary. *Memoire della Società Geologica Italiana* **34**:277–89.

Zakharov, Y. D. 1987. Conclusion (in Russian). In *Problemy Biostratigraphii Permi i Triasa Vostoka SSSR* [*Problems of Permian and Triassic Biostratigraphy of the East USSR*], ed. Y. D. Zakharov and Y. I. Onoprienko, pp. 109–10. Vladivostok: Dalnevostochnyi Nauchnyi Centr Akademii Nauk SSSR.

Zakharov, Y. D. 1992. The Permo–Triassic boundary in the southern and eastern USSR and its international correlation. In *Permo–Triassic Events in the Eastern Tethys*, ed. W. C. Sweet, Yang Zunyi, J. M. Dickins and Yin Hongfu, pp. 46–55. Cambridge University Press.

Zakharov, Y. D., and Oleinikov, A. V. 1993. New data on the problem of the Permian–Triassic boundary in the Far East (abstract). In *Carboniferous to Jurassic Pangea. Program*

and Abstracts, p. 355. Calgary: Canadian Society of Petroleum Geologists.

Zakharov, Y. D., and Pavlov, A. M. 1986. The first find of araxoceratid ammonoid in the Permian of the eastern USSR, (in Russian). In *Permo–Triasovye Sobytiya v Razvitii Organicheskogo Mira Severo-Vostochnoi Azii* [*Permian–Triassic Events During Evolution on the Northeast Asia Biota*], ed. Y. D. Zakharov and Y. I. Onoprienko, pp. 74–85. Vladivostok: Dalnevostochnyi Nauchnyi Centr Akademii Nauk SSSR.

Zakharov, Y. D., and Rybalka, S. V. 1987. A standard for the Permian–Triassic in the Tethys (in Russian). In *Problemy Biostratigraphii Permi i Triasa Vostoka SSSR* [*Problems of Permian and Triassic Biostratigraphy of the East USSR*], ed. Y. D. Zakharov and Y. I. Onoprienko, pp. 6–48. Vladivostok: Dalnevostochnyi Nauchnyi Centr Akademii Nauk SSSR.

Zhao, J., Liang, X., and Sheng, H. 1978. The late Permian cephalopods of South China (in Chinese). *Palaeontologia Sinica, n. ser. B* **1978**(12):1–194.

G. V. Kotlyar considers that the *Cyclolobus kiselevae* beds of the Lyudyanza horizon of Primorye should be correlated with the Dzuhulfian Stage and not the Midian.

16 Radiolaria from Permian–Triassic boundary beds in cherty deposits of Primorye (Sikhote-Alin)

VALERIA S. RUDENKO, EUGENE S. PANASENKO, and SERGEY V. RYBALKA

When considering the problem of determining the Permian–Triassic boundary, one usually uses data on ammonoids, conodonts, bivalves, and, more rarely, brachiopods. Data concerning the character of radiolarian development have not often been used, because representatives of that group are absent from the carbonate-terrigenous deposits of the Permian–Triassic boundary beds in the most representative sections in southeastern China (Yang Zunyi and Li Zishun, 1992), the Transcaucasus (Zakharov, 1992), Iran (Bando, 1979), and the Himalayas (Kapoor, 1992).

Recently, Permian and Triassic Radiolaria from cherty sequences in Japan (Nakaseko and Nishimura, 1979; Caridroit, 1986; Ishiga, 1986, 1990) and Sikhote-Alin (Rudenko and Panasenko, 1990a,b,c; Bragin, 1991; Rudenko, Panasenko, and Rybalka, 1992) have received intensive study. However, a serious obstacle to solving the boundary problem arises because of the absence of radiolarians and conodonts immediately above the Permian–Triassic boundary beds within the discrete cherty plates that are common in the olistostrome terranes of the circum-Pacific.

In Tonishi-type shales in the Ashio Mountains, central Japan, the radiolarian *Neoalbaillella* is found in Lower Dorashamian grey and dark-grey platy cherts, and conodonts are found in Lower Triassic (Smithian) grey cherty argillites (Yamashita, Ishida, and Ishiga, 1992; Sashida, Kamata, and Igo, 1992). A similar picture is seen in Sikhote-Alin.

Recently, we found Permian Radiolaria on the Pantovyi Creek left bank (Taukha Terrane, Tumanovka River basin, 300 m to the west of Zarod Mountain). In an outcrop of red and grey bedded cherts no more than 20 m thick, we found a succession of radiolarian assemblages from six Permian stages: *Pseudoalbaillella scalprata* (Upper Sakmarian), *Spinodeflandrella acutata* (Yakhtashian–Bolorian), *Pseudoalbaillella corniculata* (Kubergandinian), *Pseudoalbaillella globosa* (Murgabian–Lower Midian), *Follicucullus? monacanthus* (Middle Midian), and *Follicucullus porrectus* (Upper Midian). Most of the assemblages are characterized by conodonts.

In a neighbouring section, 100 m to the northeast of the first, radiolarian assemblages from three Upper Permian stages were found in grey bedded cherts no more than 6 m thick: *Follicucullus porrectus* (Upper Midian), *Neoalbaillella optima* (Dzhulfian), and *Neoalbaillella pseudogrypa* (Dorashamian) (Plate 16.1). They also are characterized by conodonts. The two outcrops represent different areas of a single plate. Attention is drawn to the thinness of the cherty beds from the Permian stages (no more than 40 m).

The first assemblage of Dorashamian Radiolaria is represented by *Albaillella excelsa* Ishiga, Kito, and Imoto, *A. levis* Ishiga, Kito, and Imoto, *A.* aff. *levis* Ishiga, Kito, and Imoto, *A. triangularis* Ishiga, Kito, and Imoto, *Albaillella* sp., *Neoalbaillella ornithoformis* Takemura and Nakaseko, *N.* ex gr. *optima* Ishiga and Imoto, *Follicucullus scholasticus* Ormiston and Babcock, *F.* cf. *scholasticus* Ormiston and Babcock, *F.* cf. *dactylinus* Rudenko and Panasenko, *F.* ex gr. *falx* Caridroit and De Wever, *Follicucullus* spp., *Pseudoalbaillella* n. sp., *Pseudoalbaillella* spp., *Copicintra akikawaensis* Sashida and Tonishi, *Copicintra* sp., *Triplanospongos musashiensis* Sashida and Tonishi, *Latentifistula* cf. *asperspinosa* Sashida and Tonishi, and *Ishigaum* sp. A (Sashida and Tonishi, 1986) (Plate 16.2). Together with the Radiolaria of that assemblage there were conodonts: *Gondolella bitteri* Kozur, *G.* ex gr. *orientalis* Barskov and Koroleva, *G. carinata* Clark, *G. subcarinata subcarinata* (Sweet), *G. denticulata* (Clark and Behnken), *G.* ex gr. *latimarginata* (Clark and Wang), and *Hindeodus* sp.

The second assemblage of Dorashamian Radiolaria includes *Neoalbaillella pseudogrypa* Sashida and Tonishi, *N. grypa* Ishiga, Kito, and Imoto, *N.* cf. *gracilis* Takemura and Nakaseko, *Albaillella levis* Ishiga, Kito, and Imoto, *A.* ex gr. *levis* Ishiga, Kito, and Imoto, *Albaillella* sp. *Follicucullus scholasticus* Ormiston and Babcock, *F.* cf. *scholasticus* Ormiston and Babcock, *F.* ex gr. *dactylinus* Rudenko and Panasenko, *F.* ex gr. *falx* Caridroit and De Wever, *Follicucullus* spp. *Pseudoalbaillella* spp., *Copicintra* sp., *Helioentactinia* cf. *nazarovi* Sashida and Tonishi, *Triplanospongos* sp., *Latentifistula* sp., and *Ishigaum* sp. The Radiolaria were found in association with the conodonts *Gondolella carinata* Clark, *Clarkina changxingenis* (Wang and Wang), and *Iranognathus tarazi* Kozur and Mostler.

Thirty kilometres to the southeast of Pantovyi Creek, in the Skalistyj Creek basin, in grey bedded cherts, a late Dzhulfian–early Dorashamian radiolarian assemblage, the *Neoalbaillella*

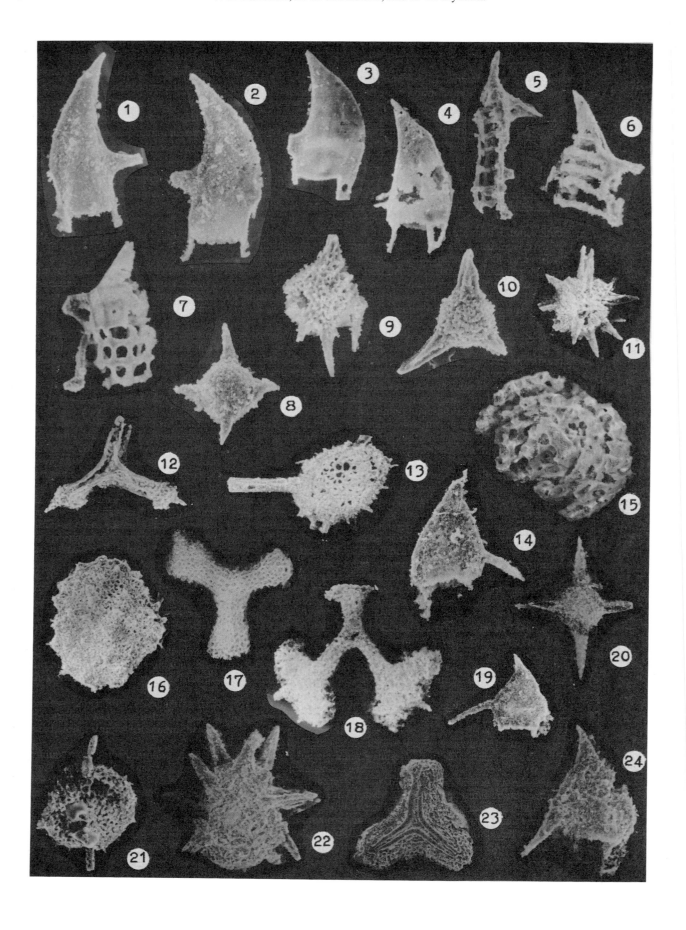

ornithoformis assemblage, was found. The data obtained show that in the section of cherty Permian in Sikhote-Alin there are layers corresponding to the lower Changxing Formation of South China.

The Dorashamian radiolarian assemblages found in the Pantovyi Creek and Skalistyj Creek basins are similar to the late Permian radiolarian associations of Japan (Ishiga, 1986, 1990; Caridroid, 1986). The *Neoalbaillella ornithoformis* assemblage is similar to that from the lower part of the *N. ornithoformis* Zone. The *Albaillella excelsa* assemblage is similar to that from the upper part of the *N. ornithoformis* Zone. The assemblage with *Neoalbaillella pseudogrypa* corresponds in part to that from the upper part of the *N. ornithoformis* Zone and in part to that from a higher stratigraphic level of the Dorashamian, not previously described from the Radiolaria in Japan.

The presence of the conodonts *Gondolella carinata* Clark, *G. subcarinata subcarinata* (Sweet), *G.* ex gr. *orientalis* Barskov and Koroleva, and *Clarkina changxingensis* (Wang and Wang) is testimony to a position near the Permian–Triassic boundary. The predominance of Palaeozoic elements indicates rather a Permian age, not a Triassic age, for those assemblages.

Triassic cherty deposits in Sikhote-Alin have also been described from radiolarians and conodonts (Rybalka, 1987; Buryi, 1989; Bragin, 1991). The least known are their early Triassic representatives.

Until recently, early Triassic conodonts in Sikhote-Alin had been found only in Olenekian (Smithian and Spathian) cherts (Buryi, 1989; Bragin, 1991). Joint occurrences of radiolarians and conodonts were known only in the Dalnegorsk area (Bragin, 1991). "*Stylosphaera*" *fragilis* beds have now been described. In addition to the index species *Follicucullus excelsior* Bragin, *Oertlispongus* sp. and "*Staurosphaera*" sp. are present in the assemblage. A late Olenekian (Spathian) age for the assemblage is supported by the discovery of the conodonts *Neospathodus triangularis* (Bender) and *N. homeri* Bender. Early Triassic Radiolaria usually are poorly preserved, which complicates both their extraction from cherty rocks and their study in thin sections. Recently, we have found early Triassic Radiolaria and conodonts in the cherts of the Taukha Terrane.

Assemblages that apparently are Induan have been recognized in the Fudinov Kamen' Mountain area. Only a few Sphaeroidea were encountered, together with conodont fragments resembling *Hindeodus* ex gr. *parvus* (Kozur). Additional material will be required to define the conodonts with accuracy.

Abundant Sphaeroidea have been found in Lower Triassic cherts in the area of Breevka village. In the bedded chert at the mouth of the Medvedka River (left bank of the Ussuri River) we found *Pseudostylosphaera* and a few Nassellaria and Pylentonemidae. Associated with them is a *Neospathodus* sp. (Wang Cheng-Yuan's determination).

In cherts along the Rudnaya River in the Dalnegorsk area, Sphaeroidea have been found in association with the conodont *Neospathodus* aff. *dieneri* (Sweet). Other radiolarian representatives were not found there. Further study of the late Olenekian (Spathian) Radiolaria of Sikhote-Alin (Pantovyi Creek and the Bolshaya Ussurka River basin), which occur together with the conodonts *N.* cf. *triangularis* (Bender), *N.* aff. *triangularis* (Bender), and *N. homeri* (Bender), has shown abundant Sphaeroidea in those deposits.

Sufficient data have been obtained to indicate the essential changes in radiolarian development at the Permian–Triassic boundary. In Dorashamian time, the leading group comprised representatives of the order Albaillellaria. Sphaeroidea and Latentifistulidea ranked below them in diversity. The Induan and Olenekian stages were characterized by a significant predominance of Sphaeroidea, including *Pseudostylosphaera* representatives found in association with rare Albaillellaria, Nassellaria, and Pylentonemidae. Thus, further investigations of Radiolaria from cherty Permian–Triassic boundary beds will be needed to improve our estimates of the boundary position between the systems.

References

Bando, Y. 1979. Upper Permian and Lower Triassic ammonoids from Abadeh, central Iran. *Memoirs of the Faculty Education, Kagawa University* 29:103–38.

Bragin, N. Y. 1991. *Radiolarii i Nizhnemezozoiskie Tolschi Vostoka SSSR* [*Radiolaria and the Lower Mesozoic Units of the East USSR Regions*]. Moskva: Nauka.

Buryi, G. I. 1989. *Triasovye Konodonty i Stratigaphiya Sikhote-Alinya* [*Triassic Conodonts and Stratigraphy of Sikhote-Alin*]. Vladivostok: Dalnevostochnoye Otdeleniye Akademii Nauk SSSR.

Caridroit, M. 1986. *Contribution a l'Etude Geologique du Japon Sud-Ouest Dans l'Ile de Honshu. 2. Paleontologie de la Faune de Radiolaires Permiens*, pp. 249–500. Orléans: Université d'Orléans U.F.R. des sciences fondamentales et appliques.

Ishiga, H. 1986. Late Carboniferous and Permian radiolarian biostratigraphy of southwest Japan. *Journal of Geosciences, Osaka City University* 29:89–100.

Ishiga, H. 1990. Paleozoic radiolarians. In *Pre-Cretaceous Terranes of Japan. Publications of IGCP Project 224: Pre-Jurassic Evolution of Eastern Asia*, ed. K. Ichikawa, S. Misutani, I. Hara, I. Hada, and A. Yao, pp. 285–95. Osaka: Osaka City University.

Plate 16.1 (opposite). (1–10) *Neoalbaillella ornithoformis* assemblage (Upper Dzhulfian–Lower Dorashamian), sample 703-34, Skalistyj Creek. (11–24) *Albaillella excelsa* assemblage (Lower Dorashamian), sample 49-2, Pantovyi Creek. (1, 2) *Albaillella lauta* Kuwahara (×240). (3, 4) *Albaillella triangularis* Ishiga, Kito, and Imoto (×240). (5–7) *Neoalbaillella ornithoformis* Takemura and Nakaseko (5, ×160; 6 and 7, ×240). (8, 20) Sphaeroidea gen. et sp. indet. (8, ×240; 20, ×160). (9) *Hozmadia?* sp. (×240). (10) *Eptingium?* sp. (×240). (11, 22) *Copicintra?* sp. (11, ×110; 22, ×160). (12, 23) *Triplanospongos musashiensis* Sashida and Tonishi (×110). (13, 16, 21) Discoidea gen. et sp. indet. (×110). (14, 19, 24) *Albaillella* aff. *levis* Ishiga, Kito, and Imoto (14 and 24, ×240; 19, ×160). (15) *Copicintra akikawaensis* Sashida and Tonishi (×240). (17) *Latentifistula* cf. *asperspinosa* Sashida and Tonishi (×110). (18) *Ishigaum* sp. A (Sashida and Tonishi, 1986) (×160).

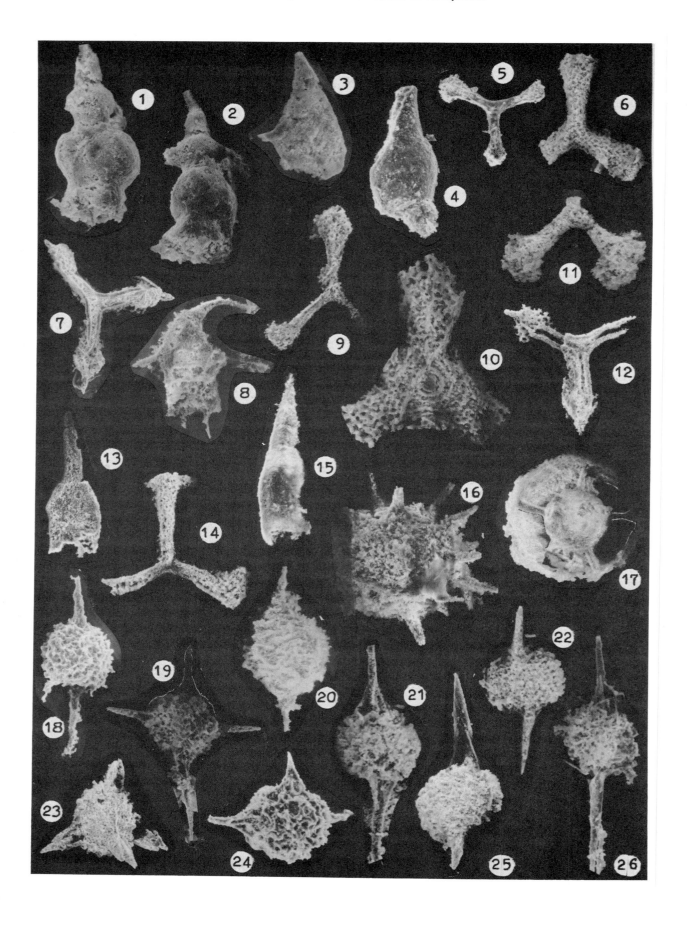

Kapoor, H. 1992. Permo–Triassic boundary of the Indian subcontinent and its intercontinental correlations. In *Permo–Triassic Events in the Eastern Tethys,* ed. W. S. Sweet, Yang Zunyi, J. M. Dickins, and Yin Hongfu, pp. 21–36. Cambridge University Press.

Nakaseko, K., and Nishimura, A. 1979. Upper Triassic radiolaria from southwest Japan. *Science Reports, College of General Education, Osaka University* **28**:61–110.

Rudenko, V. S., and Panasenko, E. S. 1990a. New findings of the Upper Permian radiolarians in Primorye region (in Russian). In *Novye Dannye po Biostratigraphii Paleozoya i Mezozoya Yuga Dal'nego Vostoka [New Data on Palaeozoic and Mesozoic Biostratigraphy of the South Far East],* ed. Y. D. Zakharov, G. V. Belyaeva, and A. P. Nikitina, pp. 117–24. Vladivostok: Dalnevostochnoye Otdeleniye Akademii Nauk SSSR.

Rudenko, V. S., and Panasenko, E. S. 1990b. Permian Albaillellaria (radiolaria) of the Pantovyi Creek sequence in Primorye (in Russian). In *Novye Dannye po Biostratigraphii Paleozoya i Mezozoya Yuga Dal'nego Vostoka [New Data on Palaeozoic and Mesozoic Biostratigraphy of the South Far East],* ed. Y. D. Zakharov, G. V. Belyaeva, and A. P. Nikitina, pp. 181–93. Vladivostok: Dalnevostochnoye Otdeleniye Akademii Nauk SSSR.

Rudenko, V. S., and Panasenko, E. S. 1990c. Permian radiolarian assemblages in siliceous rocks in Primorye (abstract) (in Russian). In *Stratigraphiya Dokembriya i Phanerozoya Zabaikal'ya i Yuga Dal'nego Vostoka. Tezicy Dokladov IV Dal'nevostochnogo Regional'nogo Mezhvedomstvennogo Stratigraphicheskogo Soveschaniya [Precambrian and Phanerozoic Stratigraphy of Trans-Baikal and South Far East. Abstract Book. IV Far-Eastern Stratigraphical Conference],* p. 117. Khabarovsk: Dal'geologiya.

Rudenko, V. S., Panasenko, E. S., and Rybalka, S. V. 1992. Biostratigraphy of Permian deposits in Primorye on radiolaria and conodonts (abstract). In *Late Paleozoic and Early Mesozoic Circum-Pacific Biogeological Events. Abstract Book. International Field Conference on Permian–Triassic Biostratigraphy and Tectonics in Vladivostok,* ed. A. Baud, pp. 27–8. Lausanne: Lausanne University (UNIL).

Rybalka, S. V. 1987. *Konodonty Primor'ya. Sostoyaniye Izuchennosti [Conodonts in Primorye Region. State of Study]* (preprint). Vladivostok: Dal'nevostochnyi Nauchnyi Centr Akademii Nauk SSSR.

Sashida, K., Kamata, Y., and Igo, H. 1992. "Toishi-type shale" in the Ashio Mountains, Central Japan. *Annual Report of the Institut of Geoscience, University of Tsukuba* **18**:59–66

Sashida, K., and Tonishi, K. 1986. Upper Permian sturaxon polycystine radiolaria from Itsukaichi, western part of Tokyo Prefecture. *Science Reports of the Institute of Geoscience, University of Tsukuba, sec. B* **7**:1–13.

Yamashita, M., Ishida, K., and Ishiga, H. 1992. Permian–Triassic boundary in the siliceous rocks of southwest Japan (abstract). In *Abstracts, Vol.* **2/3,** *29 International Geological Congress,* p. 268. Kyoto.

Yang Zunyi and Li Zishun. 1992. Permo–Triassic boundary relations in South China. In *Permo–Triassic Events in the Eastern Tethys,* ed. W. S. Sweet, Yang Zunyi, J. M. Dickins, and Yin Hongfu, pp. 9–20. Cambridge University Press.

Zakharov, Y. D. 1992. The Permo–Triassic boundary in the southern and eastern USSR and its international correlation. In *Permo–Triassic Events in the Eastern Tethys.* ed. W. C. Sweet, Yang Zunyi, J. M. Dickins, and Yin Hongfu, pp. 46–55. Cambridge University Press.

Plate 16.2 (opposite). (1–7) *Albaillella excelsa* assemblage (Lower Dorashamian), sample 64-12. (8–17) *Neoalbaillella pseudogrypa* assemblage (Lower Dorashamian), samples 64-13, 64-14, and 64-15. (18–26) Sphaeroidea assemblage (Olenekian), sample 42-71. All samples from Pantovyi Creek. (1, 2) *Pseudoalbaillella* n. sp. (×240). (3) *Albaillella* sp. (×240). (4, 15) *Follicucullus* sp. (×160). (5, 9, 11, 14) *Ishigaum* sp. (5 and 14, ×110; 9 and 11, ×160). (6, 10) *Triplanospongos* sp. (6, ×110; 10, ×160). (8) *Neoalbaillella pseudogrypus* Sashida and Tonishi (×160). (13) *Follicucullus* aff. *dactylinus* Rudenko and Panasenko (×160). (16) *Copicintra* sp. (×240). (17) *Helioentactinia* cf. *nazarovi* Sashida and Tonishi (×160). (18, 20, 21, 22, 25, 26) *Pseudostylosphaera?* sp. (18, 21, 22, 25 and 26, ×160; 20, ×240). (19) Sphaeroidea gen. et sp. indet. (×160). (23) *Triaenosphaera?* sp. (×110). (24) Pylentonemidae gen. et sp. indet. (×160).

17 Early Mesozoic magmatism in the Russian Far East

VLADIMIR G. SAKHNO

Tectonic activization of the East Asia regions in Triassic and early Jurassic times, when the Pacific Plate began to move to the north, resulted in the crushing of the continental plate into individual blocks, with cores of ancient massifs separated by sutures and faults. Movements along those faults were accompanied by magmatic processes.

In rift troughs, which developed in ancient massifs or along their margins, lavas of basic composition erupted. Those kinds of structures include the Bikin Trough and some small rifts in northern Sikhote-Alin, the Omolon Rift on the Omolon Massif, the Zyryanovsk Graben on the margin of the Kolyma-Omolon block, and the Verkhoyansk fold system, developed on the submerged part of the Siberian Platform.

Rifts and troughs in those areas occur along the zones where the plate and geoblock join and are related to regional faults. They are filled with sedimentary-volcanogenic complexes, in which basic volcanic rocks are represented mainly by alkaline basaltoids, differentiated alkaline gabbro massifs, and tholeiitic basalts.

In the collision zones along the continental margin, island-arc systems were formed. Two groups of island-arc belts can be distinguished: interblock island arcs within the crushed Asian continent and a system of island arcs along its margin. The first group includes the Alazeisk Arc, of latitudinal direction, located between the blocks of the Omolon and Chukotka massifs along the South Anyui Suture. The second group consists of the Uda-Murgal and Samarga-Moneron arcs, located along the margin of the Eurasian continental block at its junction with the Pacific Plate. The second group is characterized by inherited derivatives of earlier (late Palaeozoic) volcanism, which are similar to calc-alkaline series in terms of petrogeochemical parameters (Lychagin et al., 1989). In some structures (Taiganos Zone) (Belyi, 1977, 1978) the Permian–Triassic volcanogenic formations are thicker (3,500–4,000 m).

At the passive margin of the Siberian Platform and the North American Plate in early Triassic time, lavas of basic composition, which can be attributed to trap formation, were injected into terrigenous sediments. Thus, early Triassic and late Triassic–early Jurassic magmatism acted locally on the territory of the northwestern Asian continental block. It coincided in time with the beginning of the Gondwana breakdown and regional extension. An early phase (T_1) was accompanied by trap formation in a stable environment, very similar to the trap formation developed on the Siberian Platform. A late phase (T_3–J_1) corresponded to the beginning of horizontal movements by the entire continent and by its individual blocks, which provided the conditions for formation of island arcs, eruptions of andesite volcanism in the extension zone, formation of a rift trough, and eruption and injections of solutions of basaltic composition.

Rift-trough formation

Basalt/alkaline-basalt and basalt/trachybasalt associations characterize the rifts and rift troughs that arose along the boundaries of different structures or within crystalline massifs. Some of them were collisional, whereas others appeared in fault zones and corresponded to extension zones and deep-trough formation.

The Bikin Rift Trough, a representative of that group, is located on the margin of the Khanka Massif along a major regional fault to the west of the Sikhote-Alin foldbelt (Figures 17.1 and 17.2). The trough is made up of sedimentary volcanogenic Mesozoic formations. In cherty-clayey shales, with tuff and lava sheets of basic composition, Triassic and Jurassic radiolarians and Middle and Upper Triassic conodonts are found (Vrzhosek and Shcheka, 1984). To the southeast and northwest, deep-sea facies are replaced by coastal-sea and coal-bearing facies.

The Lower Mesozoic formations contain basalts, which are subalkaline, according to A. A. Vrzhosek. Their petrochemical compositions and rift-trough structural positions relative to the foldbelt direction suggest their emplacement on a hard basement. In the upper part of the Bikin Trough section there are picrite/alkaline-basalt and picrite/meimechite lavas (Vrzhosek and Shcheka, 1984), typical of the stable areas of the earth's crust.

To the north of the Bikin Trough there are small rift troughs, confined to regional faults, in which volcanogenic and lava formations of basic and moderate compositions are known among the Upper Triassic and Lower Jurassic sediments. They gravitated toward the eastern troughs (Figures 17.1 and 17.2)

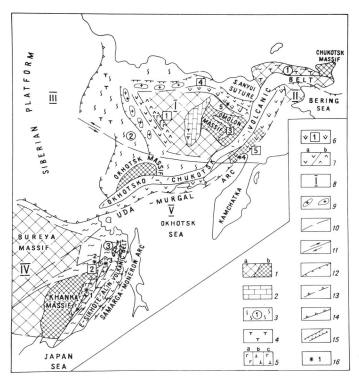

Figure 17.1. Schematic showing the tectonic positions of early Mesozoic magmatic sequences in northeastern Asia: 1, crystalline massifs of geoblocks (a, closed-mantle young sequences; b, massifs with pre-Riphean basement); 2, Palaeozoic rises; 3, Mesozoic fold framing geoblocks (fold system: ① Chukotka, ② Verkhoyansk, ③ Sikhote-Alin); 4–6, early Mesozoic magmatic formations [4, trap formation (Lower Triassic); 5, basaltic formations of rift troughs (a, tholeiitic basalts; b, basalt/alkaline basalt; c, basalt/alkaline basalt and picrite-meimechite); 6, basalt-andesite association of island arcs (T₃–J₁) ① , Zyryanovsk Rift; ② , Bikin Rift; ③ , Omolon Rift; ④ , Alazeisk; ⑤ , Uda-Murgal)]; 7, late Jurassic–early Cretaceous volcanic belts (a, andesite of island-arc type; b, basalt of rift troughs); 8, plates and geoblock (I, Omolon-Kolyma geoblock; II, North American Plate; III, Siberian Platform; IV, Sino-Korean; V, Okhotsk); 9, plutonic belts; 10, faults; 11, wrench-fault zones; 12, boundaries of the passive Mesozoic margins; 13–14, early Mesozoic active margin; 15, rift troughs and rifts; 16, locations of the sections in Figure 17.2.

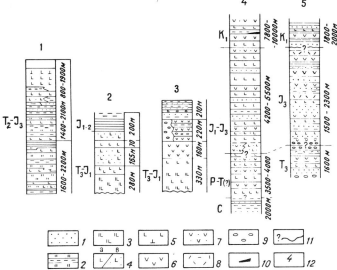

Figure 17.2. Sections of volcanogenic sequences: 1, sandstones, gravelstones; 2, siltstones, clay shales; 3, spilites; 4, lavas (a) and tuffs of basic composition (b); 5, picritic basalt and picrites; 6, andesites; 7, tuffs of andesite; 8, lavas and tuffs of acid composition; 9, conglomerates; 10, coal clasts; 11, boundaries of separation; 12, numbers of sections in Figure 17.1 Rift troughs: 1, Bikin; 2–3, Lower Amur (2, west part; 3, east part). Island arcs: 4, Taigonos Zone of Uda-Murgal Arc; 5, Oloi Zone of Alazeisk Arc.

where a submerged massif (Tastakh) of ancient rocks is suggested to reside (Popeko, Pilatsky, and Kaidalov, 1983).

Geological data and the compositions of petrochemical basalts, along with the spectra of rare-earth elements and large-ion lithophile elements, show differences between the eastern and western parts of the Lower Amur Trough. The basalts of the western part are similar to the basalts of the oceanic floor. Basic volcanic rocks from sections in the eastern parts of the troughs are similar to subalkaline rocks of continental margins. The suggested influence from the basement rocks is supported by the data on strontium-isotope ratios (Popeko et al., 1983). So on a small area of the Lower Amur Rift Trough there are rocks from the oceanic floor (Figure 17.2) and rocks similar in composition to the complexes seen in active margins.

In the structures of northeastern Asia, early Mesozoic rift troughs are known along the southwestern margin of the Omolon-Kolyma block (Zyryanovsk Trough) and within the Omolon crystalline massif (Omolon Rift). In the Zyryanovsk Trough, tuffs, lavas, and dikes of trachybasalts are developed between clay shales, and 400-m-thick limestones contain the remains of a Middle Triassic fauna. Volcanic rocks make up 50-80% of the section and represent high-alumina, high-potassium, and high-barium varieties, which makes them similar to the shoshonites of stable areas (Lychagin et al., 1989).

In the Omolon Rift, sills, dikes, and small massifs and sheets of alkaline basaltoids and differentiated rocks of the gabbro/alkaline-syenite series are common. Among the alkaline basalts there are trachybasalts, trachyandesites, trachytes, phonolites, and other differentiates of the alkaline-basalt series. That series characterized the environments where tectonomagmatic activity took place (Table 17.1).

Formation of island arcs

Geological data on the early Mesozoic formations of northeastern Russia show the formation, in late Triassic–early Jurassic times, of island arcs, accompanied by volcanism; they were formed by movements of the major structures (the Eurasian continent and the Pacific Plate) and of the small geoblocks along their margins. In the late Triassic and early Jurassic, along the

Table 17.1. *Average chemical compositions^a of early Mesozoic magmatic rocks in the Far East*

NN	1	2	3	4	5	6	7	8	9	10
n	1	6	2	21	1	6	6	2	1	1
SiO_2	48.71	49.17	46.40	45.98	47.90	50.93	55.37	61.46	49.54	55.42
TiO_2	2.33	1.87	1.91	1.70	2.13	1.12	0.93	0.49	0.88	0.83
Al_2O_3	13.98	13.63	14.67	14.63	14.69	18.10	17.21	15.67	17.06	17.25
Fe_2O_3	1.95	2.37	2.20	5.35	2.18	3.33	3.26	2.41		
FeO	10.45	10.91	10.44	9.62	10.98	5.67	4.91	4.34	9.84	8.43
MnO	0.22	0.28	0.22	0.56	0.21	0.17	0.13	0.12	0.17	0.16
MgO	6.47	6.91	6.47	4.62	7.72	4.17	3.57	3.40	6.06	3.20
CaO	10.51	8.80	9.96	10.82	7.68	9.21	6.69	5.45	7.71	8.08
Na_2O	2.21	2.43	1.84	1.89	2.50	3.06	3.71	3.10	5.35	3.15
K_2O	0.72	1.02	0.77	0.27	0.60	0.31	0.44	0.72	0.16	1.10
P_2O_5	0.16	0.17	0.18	0.08	0.33	0.19	0.22	0.22	0.14	0.12
H_2O^+	2.32	0.84	2.96	4.55	–	3.31	3.26	2.30	–	–
LOI	1.67	2.18	1.84	–	–	0.46	0.91	0.72	3.43	2.42
Total	101.70	100.58	99.86	100.07	96.92	100.03	100.61	100.40	100.34	100.16
Cr	–	–	–	–	210	60	52	28	33	31
Ni	–	–	–	–	170	76	44	9	46	14
Co	–	–	–	–	43	91	16	7	88	70
V	–	–	–	–	195	170	130	110	–	–
Rb	–	–	–	–	25	42	26	113	3	16
Sr	–	–	–	–	706	370	313	261	253	289
Ba	–	–	–	–	650	120	233	369	95	259
Zr	–	–	–	–	240	163	200	223	–	–
Nb	–	–	–	–	24	15	10	17	–	–
La	–	–	–	–	21	41	45	47	–	–
Ce	–	–	–	–	41	19	30	40	–	–
Nd	–	–	–	–	24	21	31	37	–	–
Y	–	–	–	–	31	41	41	49	–	–

Notes: Trap formation (1–5): 1, diabase, Anyui Zone (Gelman, 1963); 2, diabase, Anyui Zone (Shpetnyi, 1970); 3, diabase Omulevsk Rise (Shpetnyi, 1970); 4, diabase, dike, Bolshoi Anyui (Shpetnyi, 1970); 5, diabase, Verkhoyansk (Shpetnyi, 1970). Basalt-andesite units of island arcs (6–12): 6, basalt, Taiganos Zone of Uda-Murgal Arc (Simonenko and Sakhno, 1984); 7, andesite basalt, Taiganos Zone (Simonenko and Sakhno, 1984); 8, andesite (Simonenko and Sakhno, 1984); 9, Norian basalt, Oloi Zone, Alazeisk Arc (Afitskyi and Lychagin, 1987); 10, Norian andesite-basalts (Afitskyi and Lychagin, 1987);

southeastern margin of northeastern Asia, on the boundary with the Pacific Plate, a basalt-andesite series was formed in the narrow geosynclinal trough. It represents the lower part of a calc-alkaline series of the Uda-Murgal island arc that was active during the Jurassic and early Cretaceous (Parfenov, 1991). V. F. Belyi distinguished it on the Taiganos Peninsula as an andesite geosyncline (Belyi, 1977), the formation of which began in the Palaeozoic and continued in the Mesozoic up to the end of the early Cretaceous (Figure 17.2). The boundary between the Upper Palaeozoic and Triassic has not been established with certainty. The total thickness of the geosynclinal formations, including the Cretaceous sedimentary-volcanogenic formations, is about 20,000 m (Figure 17.2). They are represented by volcanogenic rocks – effusive, pyroclastic, and sedimentary-volcanogenic, among which basalt-andesite associations predominate, and basalts and acidic rocks are subordinate. Basalts and andesites belong to the high-alumina group and contain low amounts of nickel, chromium, zirconium, titanium, barium, and niobium. In terms of petrogeochemical parameters, they are similar to the calc-alkaline series of modern island arcs. Between the volcanogenic rocks of the Uda-Murgal island arc, in the upper part of the section, there are subalkaline basalts with high contents of potassium and barium, making them similar to the shoshonite-latite series seen with mature island arcs. The arc, as a whole, is characterized by lateral petrogeochemical zonation, shown by the change from sodic rocks in the fore-arc part to potassic rocks in the back-arc part. The rocks of the shoshonite-latite series are also found there.

11	12*	13	14	15	16	17	18	19	20
5	1	11	7	2	7	4	10	9	1
50.38	50.40	48.04	49.83	47.35	57.02	58.81	45.34	47.41	47.10
1.02	1.17	1.77	2.25	2.70	0.70	0.08	1.05	1.04	1.31
17.60	18.02	17.16	19.72	17.15	19.96	18.73	14.12	14.78	13.57
	2.96						4.83	4.98	3.09
10.10	4.61	12.19	13.91	12.28	8.03	5.44	8.78	9.53	7.40
–	0.14	–	–	–	–	–	0.18	0.21	0.12
5.61	7.93	8.02	2.43	5.87	0.66	0.24	9.62	8.61	8.28
7.76	8.35	7.69	4.33	9.18	2.04	0.59	9.08	9.73	6.41
4.20	2.43	2.79	5.25	3.76	6.48	9.43	2.33	2.62	1.82
2.54	0.84	1.80	1.49	1.61	4.38	4.88	0.20	0.47	1.17
0.46	0.26	0.37	0.80	0.79	0.24	0.06	0.06	0.05	0.16
–	–	–	–	–	–	–	0.38	0.19	0.86
–	–	–	–	–	–	–	–	–	–
99.67	97.11	99.83	100.01	100.99	99.51	98.26	95.97	99.62	91.29
45	60	97	25	25	24	24	406	344	53
45	76	214	6	9	6	5	–	–	–
113	25	74	17	3	5	3	35	24	21
–	170	–	–	–	–	–	–	–	–
39	50	15	38	60	99	115	12	50	44
813	344	356	391	428	137	97	195	237	200
368	130	1094	816	670	199	60	–	–	–
–	151	–	–	–	–	–	128	138	150
–	12	–	–	–	–	–	–	–	–
–	39	–	–	–	–	–	–	–	–
–	21	–	–	–	–	–	–	–	–
–	23	–	–	–	–	–	–	–	–
–	37	–	–	–	–	–	–	–	–

11–12, basalts (T_3–J_1), east part of Alazeisk Arc (Lychagin et al., 1989). Basaltic units of rift troughs (13–20): 13, Zyryanovsk Rift, trachybasalt (Lychagin et al., 1989); 14–17, Omolon Rift [14, trachybasalt; 15, alkaline gabbro; 16–17, alkaline syenite (Lychagin et al., 1989)]; 18–20, Lower Amur Trough, basalt (Popeko et al., 1983). The author's data are those indicated by an asterisk and also the microelement contents for columns 5–8.

[a] Oxides as percentages by weight; microelements as parts per million.

The Uda-Murgal Arc is a product of long-term evolution. It inherited its structural pattern after the late Palaeozoic. Volcanic evolution in Triassic time shows the signs of succession from a Palaeozoic stage. In the section from the Jurassic, the rocks of andesite composition are widely distributed and alternate with psephite sedimentary-volcanogenic formations developed in coastal-sea subcontinental environments. The composition of the volcanogenic rocks, the content and spectrum of the microelemental lateral zonation, and the structural features allow attribution of the Uda-Murgal Arc to the Andean type (Table 17.1). Triassic volcanites of basic composition are the primary members of a single basalt-andesite association of the calc-alkaline series. They accumulated in narrow troughs that were controlled by a long-term trans-regional fault system.

The Alazeisk island arc is located along the northeastern margin of the Kolyma block, near the South Anyui Suture, as shown by outcrops of melanocratic basement rocks (Parfenov, 1991). The exposures of Triassic volcanites have been dated on the basis of a Norian fauna that was found in the sediments containing basalt and andesite sheets (Afitskyi and Lychagin, 1987). The section of Upper Triassic volcanites is accreted with Upper Jurassic and then Lower Cretaceous volcanogenic and lava sequences developed in sea-basin environments. In the upper part of the section, coastal-sea and continental facies are found. Thus, this structure evolved for a long time and was repeatedly manifested as an island arc in late Triassic and late Jurassic–early Cretaceous times (Natalin, 1984).

Most typical is the eastern fragment of the arc – the Oloi Zone

(Figure 17.1). The petrochemical compositions of the rocks (Table 17.1) and the microelements indicate that the Norian rocks belong to the typical calc-alkaline series characteristic of active margins. The volcanic formations contain high-potassium varieties similar to the shoshonite association in terms of petrogeochemical characteristics (Afitskyi and Lychagin, 1987; Lychagin et al., 1989). They are found in the Oloi Zone and are distributed in the island back-arc portion, whereas sodium varieties are found in the fore-arc part. There are differences in the volcanite compositions of the eastern and western parts of the arc (Table 17.1). The Oloi Arc has lateral and petrochemical zonations. In later epochs (Cretaceous–Palaeogene), along the fault zones on the continental-block margins, volcanite belts of andesite-rhyolite composition were formed (Okhotsk-Chukotka, eastern Sikhote-Alin, etc.).

Trap formation

Rocks of basaltic composition are known in Eurasia in early Mesozoic sedimentary formations. In northeastern Asia they occur between the Lower Triassic sedimentary formations (Figure 17.1). Dikes and intrusive sheets, which occur in the Lower Triassic sedimentary complexes, are not known in Upper Triassic sequences. Effusive members are rare. Data on the section show that volcanogenic rocks are not important in the composition of the Lower Triassic sequences. They occur in greater amounts in intrusive facies in the form of sheets or crosscutting bodies. Intrusions are abundant on rises (Kolyma, Omolon, Kuekvun, Keperveem, etc.). Within rises, intrusions make up numerous sills, dikes, and lens-like lodes that often are oriented subparallel. Intrusion sizes vary widely. The most common intrusions are 10–20 m thick and 1–2 km long. There are isometric intrusives that represent bodies that are patelloid in the section and sometimes large (40–75 km^2).

The compositions of the trap-formation rocks are relatively monotonous. They are predominantly diabases and gabbro-diabases (Table 17.1). The rocks of the trap formation are similar to those of traps in the Tungus Basin on the Siberian Platform in geologic position and composition (Gelman, 1963; Shpetnyi, 1970), as well as the traps of other regions developed on platforms (Cox, 1983).

Conclusion

In this brief space it is impossible to discuss in detail all of the features of the magmatism that affected the vast territory of the eastern margin of the Eurasian continent as it underwent tectonomagmatic activity during an epoch of global reconstruction that spanned the late Palaeozoic–early Mesozoic. Analyses of magmatic formations, using J. A. Pearc's method, and petrochemical data have conferred greater validity on our recognition of geologic and geodynamic regimes of tectono-magmatic activity. We can conclude the following:

1. Early Triassic magmatism in northeastern Asia manifested itself, within limits, in the form of trap eruptions and intrusions into terrigenous rocks on the rises of the Verkhoyansk-Chukotka sedimentary basin generated on the passive margin.
2. In late Triassic time, the region underwent tectonomagmatic activity that resulted in horizontal movements of the Eurasian geoblock as a whole and of its various fragments, as well as extensional and compressional events. The most significant extensions up to the opening of the oceanic crust took place along the faults separating individual blocks. Rift troughs appeared, in which sedimentary-volcanogenic rock masses formed. Within the hard structures, the effects of the magmatism are seen in tholeiitic, alkaline-basalt, and picrite/alkaline-basalt associations. Tholeiitic volcanism, similar to oceanic volcanism, characterized the troughs of the eastern margins, where the extension process (according to petrologic data) was rapid.
3. In the zone of compression, inside the Eurasian continental plate, between its individual blocks and along its eastern margin, island-arc volcanism occurred on the boundary with the Pacific Plate. In early Mesozoic time, volcanism was most intensive within narrow fault zones along the active eastern margin of Eurasia.

References

Afitskyi, A. I., and Lychagin, P. P. 1987. Norian andesite volcanism of the Oloi Zone (in Russian). *Tikhookeanskaya Geologiya* **1987**(3):77–82.

Belyi, V. F. 1977. *Stratigraphiya i Struktury Okhotsko-Chukotskogo Vulkanicheskogo Poyasa [Stratigraphy and Structures of the Okhotsk-Chukotka Volcanic Belt]*. Moskva: Nauka.

Belyi, V. F. 1978. *Formatsii i Tektonika Okhotsko-Chukotskogo Vulkanicheskogo Poyasa [Formations and Tectonics of the Okhotsk-Chukotka Volcanic Belt]*. Moskva: Nauka.

Cox, K. G. 1983. The Karro province of southern Africa: origin of trace element enrichment patterns. In *Continental Basalts and Mantle Xenoliths,* ed. C. J. Hawkesworth and M. J. Norry, pp. 139–57. Cheshire: Shiva Publishing Limited.

Gelman, M. A. 1963. Triassic diabasic formation of the Anyui Zone, Chukotka (in Russian). *Geologiya i Geophizika* **1963**(2):127–34.

Lychagin, P. P., Dylevskyi, E. F., Shpikerman, V. I., and Likman, V. B. 1989. *Magmatizm Central'nykh Raionov Severo-Vostoka SSSR [Magmatism of the Central Regions of the North-East USSR]*. Vladivostok: Dal'nevostochnoye Otdeleniye Akademii Nauk SSSR.

Natalin, B. A. 1984. *Rannemezozoiskaya Eugeosinklinal'naya Sistema Severnoi chasti Tikhookeanskogo Poyasa [Early Mesozoic Eugeosynclinal System of the North Pacific Belt]*. Moskva: Nauka.

Parfenov, L. M. 1991. Tectonics of the Verkhoyansk-Kolyma mesozoids in the context of plate tectonics. *Tectonophysics* **199**:319–42.

Popeko, V. L., Pilatsky, O. O., and Kaidalov, V. L. 1983. Basalts of Upper Triassic–Lower Jurassic complexes in the structure of the North Sikhote-Alin (in Russian). *Tikhookean-skaya Geologiya* **1983**(6):21–31.

Shpetnyi, A. P. 1970. Triassic magmatism (in Russian). In *Geologiya SSSR* [*Geology of the USSR*], vol. **30**, ed. N. E. Drapkin, pp. 61–7. Moskva: Nedra.

Simonenko, V. P., and Sakhno, V. G. 1984. Paleoisland volcanic belts (in Russian). In *Vulkanicheskiye Poyasa Vostoka Azii. Geologiya i Metallogeniya* [*Volcanic Belts of East Asia. Geology and Metallogeny*], ed. A. D. Shcheglov, pp. 146–51. Moskva: Nauka.

Vrzhosek, A. A., and Shcheka, S. A. 1984. Geosynclinal ophiolite belts. Sikhote-Alin Zone (in Russian). In *Vulkanicheskiye Poyasa Vostoka Azii. Geologiya: Metallogeniya* [*Volcanic Belts of East Asia. Geology and Metallogeny*], ed. A. D. Shcheglov, pp. 87–95. Moskva: Nauka.

18 Transgressive conodont faunas of the early Triassic: an opportunity for correlation in the Tethys and the circum-Pacific

RACHEL K. PAULL and RICHARD A. PAULL

Ammonoids have long provided the basis for age assignments and correlations of Lower Triassic (Scythian) marine strata. Although the early Triassic marine faunas were of low diversity and abundance relative to those of the late Permian, ammonoids are widely distributed in Scythian strata. Areas with Lower Triassic fossiliferous marine rocks, including ammonoids, are shown in Figure 18.1. Some 136 ammonoid genera are recognized for the Scythian (Kummel, 1973), and various zonal schemes show detailed subdivisions of that important geologic interval (Tozer, 1967, 1974; Silberling and Tozer, 1968). However, ammonoids are not always present in sufficient numbers for biostratigraphic use.

Coincident with the widespread use of acetic acid for dissolution of conodont-bearing carbonate rocks, the recovery and study of these small phosphatic elements provided a distinctive and cosmopolitan microfossil alternative for Triassic marine biostratigraphy.

Lower Triassic conodonts and evolution of a zonation

Early studies of conodont elements focused primarily on the stratigraphic distribution of diagnostic forms, and the literature burgeoned. By 1950, widespread conodont faunas from Ordovician rocks through Permian rocks had been described in the United States. A full description of a Lower Triassic conodont fauna, however, was not published until the work of Müller (1956) in Nevada, who also suggested the potential for a time-significant biostratigraphic succession. Later, Sweet's prototype Lower Triassic zonation, based on comprehensive collections from West Pakistan, stimulated worldwide study of equivalent faunas (Sweet, 1970). Two new genera and 18 new species were established as a part of that work, and many of those figured prominently in his zonal scheme (Sweet, 1970). The Lower Triassic was divided into nine zones, based on the vertical distribution of conodont species there in the heart of the Tethys, with the oldest spanning the Permian-Triassic boundary.

Since that time, applications of graphic correlation techniques have further refined the zonation by synthesizing data from key Lower Triassic depositional regions (Sweet and Bergström, 1986; Sweet, 1988, 1992). The greater resolution provided by such methods resulted in nine biozones somewhat different from Sweet's original scheme (1970), and the uppermost 1970 biozone is now considered to be earliest Anisian (Middle Triassic) (Nicora, 1977; Sweet, 1988; Orchard, 1994). Most important, however, is the fact that certain time-significant zonal species described nearly 25 years ago now appear to be trans-Tethyan and circum-Pacific in distribution (Paull and Paull, 1992).

Palaeoenvironmental control versus provincialism

Because conodont elements were initially reported from diverse marine lithologies, it was assumed that the animal had been independent of palaeoenvironmental control. The ubiquitous nature of some forms prompted an earlier and somewhat irreverent palaeontologist to remark that "conodonts are like God – they are everywhere" (Lindström, 1976). By the middle 1970s, however, it was apparent that conodonts had had environmental preferences and limitations, as hypothesized in the contrasting models of Seddon and Sweet (1971) and Barnes and Fåhraeus (1975) and in the compilation of palaeoecologic papers edited by Barnes (1976).

Perhaps those key palaeoecologic studies arrived too late on the scene, as many conodont workers were already well along in establishing provincial patterns for the Palaeozoic and Triassic based on apparent regional endemism, primarily attributed to latitudinal differences or isolation of faunas. Huckriede (1958) was one of the first workers to suggest provincialism as a factor in conodont distributions, and he applied that concept to Triassic conodonts. Subsequently, Matsuda (1985) compiled a major study of provincialism for Lower Triassic conodonts by recognizing a twofold Tethys and peri-Gondwanan Tethys. According to his criteria for differentiation of provinces, the western United States belonged in both (Paull, 1988), and the conodont genera distinctive of each biogeographic division strongly suggested palaeoenvironmental control based on near-shore and off-shore conditions, as defined by Carr and Paull (1983). Matsuda's conclusions (1985) eventually were seen as reflecting regional differences in the publication of research results, rather than regional faunal differences (Paull, 1988).

In the western United States, comprehensive Lower Triassic

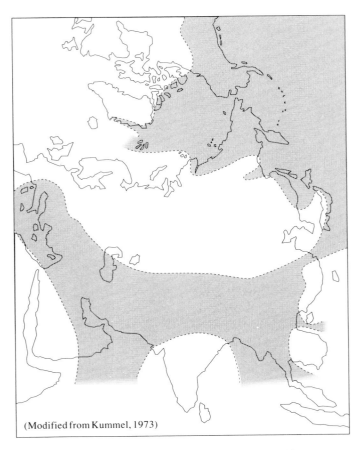

Figure 18.1. Generalized distribution (dark areas) of fossiliferous Lower Triassic marine strata, including the ammonoid genus *Otoceras*. (Adapted from Kummel, 1973.)

(Modified from Kummel, 1973)

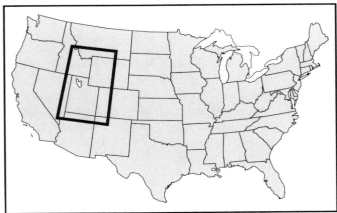

Figure 18.2. The boxed area in the western United States contains marine Lower Triassic miogeoclinal rocks where conodonts have been recovered.

conodont biostratigraphic studies began with Collinson and Hasenmueller (1978), closely followed by Solien (1979), Paull (1982, 1983), Paull and Paull (1983), and Carr and Paull (1983). The Lower Triassic study area of Paull (1982, 1983, 1988), Paull and Paull (1982, 1983, 1984, 1986b, 1987, 1991a, 1993a,b), Paull, Paull, and Anderson (1985), and Paull, Paull, and Kraemer (1989) is shown in Figure 18.2. The studies reported between 1982 and 1994 were biostratigraphic as well as stratigraphic in substance. The study areas of Collinson and Hasenmueller (1978), Solien (1979), and Carr and Paull (1983) also fall within the boundaries outlined on Figure 18.2.

Although exposures of uppermost Permian marine rocks are confined to southern China, Pakistan, Iran, and northern Italy (Ross and Ross, 1988; Sweet, 1992), the transition forms between Upper Permian and Lower Triassic conodonts suggest that the Permian–Triassic extinction event had little effect on the surviving conodont lineages. There is no evidence for provincialism among Lower Triassic conodonts from the western United States, and their distribution patterns indicate interchange throughout the Tethys-Pacific. Global occurrence of conodont species in the early Triassic was endorsed in a previous biofacies

study by Budurov et al. (1985). Examination of late Permian neogondolellids from South China revealed that those species were also cosmopolitan (Clark and Wang, 1988).

Eustasy and faunal change

Although climate is a major influence on the biogeographic distributions of faunas today, in more equable times in the geologic past its role was of lesser importance, and expansions and contractions of epicontinental seas were the major factors controlling the degrees of faunal endemism. When sea levels were high, increased communication between marine faunas resulted in the spread of cosmopolitan forms because of an increase in palaeoecologic niches in near-shore and shelf environments.

Lateral shifts in biofacies and lithofacies have provided a wealth of data for determining changes in relative sea levels within the Phanerozoic. Chamberlin (1909) introduced the hypothesis that transgressions allowed the spread of new marine faunas. Subsequently, Newell (1967) explicitly related marine faunal radiations and extinctions to eustasy, based on a correlation between the area of available habitat and the biotic diversity. More recently, other palaeontologists and biostratigraphers have observed that radiations commonly resulted from an expansion of shallow-shelf environments during normally oxygenated marine transgressive events and eustatic highstands (Ross and Ross, 1988; Hallam, 1992). Fåhraeus (1976) and Hirsch (1994) applied that concept specifically to conodont faunas.

Conversely, extinctions resulted from habitat reductions caused by relative lowerings of the sea level. Temperature and salinity variations in very shallow marine settings accompanied the other deteriorating conditions, to the detriment of stenotopic faunas. Hallam (1992) noted that the risk for extinction increases with time in a stressful environment, even without diminution of habitat area. The lowered sea levels associated with most Phanerozoic transgressive–regressive sequences may have been all the more devastating because of their long durations. As a result, the slow and modest decreases in the levels of the broad, extremely shallow epicontinental seas had major adverse effects on the marine biota.

Shoreward successions of shallow-marine environments during a transgressive event controlled the timing for the development and radiation of different conodont genera and species, as well as other marine faunas. Littoral forms developed during the initial stages of sea-level rise, and rapid transgression allowed them to spread more widely than could a coeval off-shore assemblage (Hallam, 1992). Brett and others, cited by Sageman (1992), described the shifting of benthic communities in response to migrations of marine environments as "faunal tracking." By characterizing and mapping a series of recurrent assemblages representing a relative bathymetric gradient, one can observe shoaling and deepening trends. In turn, that information can be critical for differentiating sequence-stratigraphic systems and tracts. Recently, Lower Ordovician conodont biostratigraphic correlations were integrated with analyses of stratigraphic sequences to identify and trace eustatic changes across the North American continent (Smith, Byers, and Dott, 1993).

No discussion of faunal migration and implied synchroneity of events would be complete without a disclaimer in regard to homotaxy. Similarities in sequences of conodont faunas are evident worldwide, but absolute isochroneity is unlikely. Klapper (1991) cited a study documenting a 0.5-m.y. migration time for a late Neogene planktonic-foraminiferan species in the Pacific based on the magnetic-polarity scale. Although strontium-isotope studies of conodonts may provide finely calibrated correlations in the future, a travel lag of 0.5 m.y. is negligible when compared with geohistory as a whole.

Transgressive–regressive sequences in the early Triassic

The basal Lower Triassic marine rocks around the world are transgressive deposits that resulted from a global rise in sea level after the Permian extinction event. The early Triassic rise in sea level has been estimated as somewhat greater than 200 m (Forney, 1975), and three transgressive–regressive (T–R) cycles in the western United States and most other Tethyan and circum-Pacific depositional basins were superimposed on that general increase in sea level. A T–R cycle, in the sense of Johnson, Klapper, and Sandberg (1985, p. 568), consists of "sedimentary rocks deposited during the time between the beginning of one deepening event and the beginning of the next, following one of the same scale." Comparable eustatic changes (100–200 m) were reported by Ross and Ross (1985, 1988) for 60 T–R depositional sequences in the Carboniferous and Permian. Although the pace and pattern of earlier Palaeozoic transgressive episodes were attenuated by comparison with those of the late Palaeozoic (Ross and Ross, 1985, 1988) and early Triassic, the vertical sea-level changes were similar.

Embry (1988) reported nine T–R cycles for the entire Triassic period in the Canadian Arctic, and three of those occurred in the

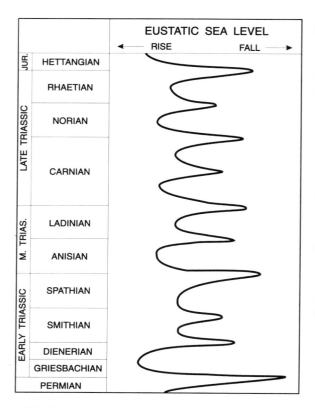

Figure 18.3. Eustatic sea-level curve for the Triassic based on T–R cycles in the Canadian Arctic described by Embry (1988).

early Triassic (Figure 18.3). Those cycles are especially well documented in sandstone-shale sedimentary packages, with biostratigraphic control provided primarily by ammonoids. Brandner (1984) illustrated a similar, closely synchronous, series of Triassic events in the alpine regions of Europe (Figure 18.4). Other synchronous T–R events elsewhere have been documented using both conodonts and ammonoids, although the reports of the magnitudes and total number of events have varied (Figure 18.5). Such differences reflect the influences of regional tectonics.

Particularly good comparisons can be made between the transgressions in the Canadian Arctic islands and those in the western United States, Svalbard, and Italy. Those episodes, of probable tectono-eustatic origin, reflected changes in the volume of the oceans that produced nearly synchronous sedimentary sequences in many Triassic depositional basins (Embry, 1988). Abrupt changes in genera and species, related to expansions into newly created shallow-marine environments, were associated with those transgressions.

Studies of early Triassic sedimentary patterns in the western United States (Paull and Paull, 1984, 1986a, 1987, 1991b) and Arctic Canada (Embry, 1984, 1988) have supported the idea of rapid or "geologically instantaneous" transgressions of relatively short duration. Embry (1988) favoured a symmetrical eustatic sea-level curve for Triassic depositional sequences characterized by a rapid rise that slowed to a stillstand. That would have been

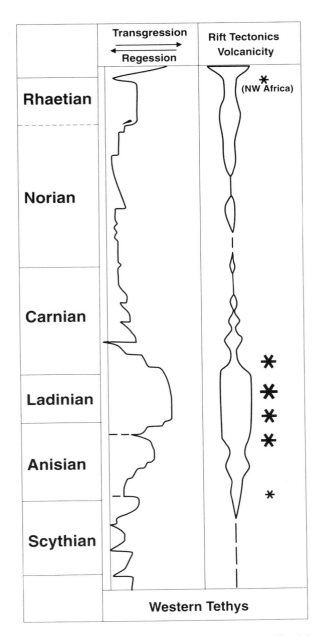

Figure 18.4. Triassic sea-level changes, rift tectonics, and volcanicity in the alpine region of the western Tethys. Asterisks signify episodes of volcanicity. (Adapted from Hallam, 1992, and Brandner, 1984.)

followed by a slow fall in sea level that would have become progressively faster. The three upward-shallowing early Triassic sequences in the western United States originated in the Griesbachian, Smithian, and earliest Spathian to middle Spathian (Collinson and Hasenmueller, 1978; Carr and Paull, 1983; Paull and Paull, 1991b, 1992), and they correspond in scale to third-order events, as described by Vail et al. (1977). The periods of time required for those T–R sequences appear to average about 1.7 m.y., based on the geologic time scale of Harland et al. (1990). That figure falls at the midpoint of the 1.2–

4-m.y. range for sequence durations reported for the late Palaeozoic by Ross and Ross (1988).

Transgressive conodont faunas of the early Triassic

The conodont faunas associated with the three early Triassic (Scythian) transgressive events were distinctive and had broad distributions in Tethyan and circum-Pacific areas where marine sedimentation was close to continuous. Data from some 200 stratigraphic sections in the western United States provide the basis for evaluating the depositional sequences in that region in terms of extent and timing.

First early Triassic transgression in the western United States

The initial Mesozoic advance in the early–middle Griesbachian was accompanied by *Hindeodus typicalis, Isarcicella isarcica,* and *Hindeodus parvus* (*Isarcicella? parva* of Sweet, 1992). In environments farther from shore, *Neogondolella carinata* was also present. Those conodonts of biozone 1 (Carr and Paull, 1983) provide a widespread correlation horizon encompassing some 194,000 km^2 in the western United States. They are also known from localities in Asia, Europe, the Canadian Cordillera and Canadian Arctic, and eastern Greenland; see the Appendix to this chapter.

Second early Triassic transgression in the western United States

The second transgressive episode began in early Smithian time. That flood was somewhat different in pattern and extent than the initial event, with a major southward incursion that was half again as extensive (310,000 km^2) as the Griesbachian advance. The conodonts of biozone 4 (Carr and Paull, 1983) provide an excellent widespread correlation horizon across several major depositional settings. The most time-significant off-shore form was *Neospathodus waageni*, often accompanied by *N. pakistanensis*. On the shallow shelf to the south there was a less diverse near-shore fauna composed of *Parachirognathus ethingtoni* and robust ellisonids (*Pachycladina* and *Hadrodontina*) (Paull and Paull, 1993b). Those conodonts are also known from localities in Asia, Europe, the Canadian Cordillera and Canadian Arctic, Alaska, Svalbard, Australia, and Indonesia (Appendix). In the western United States, that interval is equivalent to ammonoid faunas of the *Meekoceras gracilitatus* Zone and the overlying *Anasibirites* beds (Silberling and Tozer, 1968).

Third early Triassic transgression in the western United States

The third transgression took place in the early–middle Spathian Stage. In the western United States, its areal extent was comparable to that of the second inundation, but the general pattern shows a spread westward and a farther southern

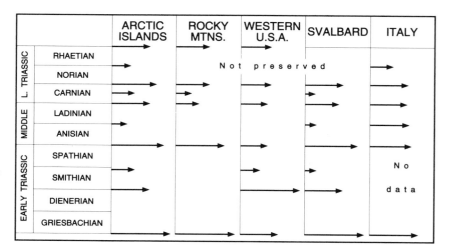

Figure 18.5. Regional comparison of Triassic transgressive sequences in five regions, indicating relative magnitude and number of events. The "Rocky Mountains" designation refers to localities in western Canada, many of which are allochthonous terranes. (Adapted from Embry, 1988.)

extension. The conodont assemblage associated with that episode was also distinctive and widely distributed, consisting of *Neospathodus homeri, Neogondolella jubata,* and, less commonly, *Neospathodus triangularis.* That fauna has also been reported from various localities in Asia, Europe, the Canadian Cordillera, Alaska, Svalbard, Australia, New Zealand, and Indonesia (Appendix). Those forms were associated with biozones 5 through 7 (Carr and Paull, 1983) and were contemporaneous with ammonoids of the *Columbites-Tirolites* beds and the overlying *Prohungarites-Subcolumbites* beds (Silberling and Tozer, 1968) in the western United States.

Nammal Gorge section, West Pakistan

The classic Tethyan section at Nammal Gorge in the Salt Range of West Pakistan provides a palaeontologic and sea-level record of the latest Permian and earliest Triassic. Conodonts from that area provided the basis for the initial zonal scheme proposed by Sweet (1970).

Haq, Hardenbol, and Vail (1988) combined the outcrop studies of Nakazawa et al. (1985) in Nammal Gorge with the conodont zonation of Matsuda (1985) to produce the modified sequence-stratigraphic chart shown in Figure 18.6. Ammonoids, ostracods, and dinoflagellates were also used to tie the major depositional features to a global stratigraphic framework. That allowed accurate dating of transgressive surfaces (TS), surfaces of maximum flooding (MFS) (downlap surfaces, DLS, in the subsurface record), and sequence boundaries (SB), the three systems-tracts boundaries that can be recognized in surface exposures (Haq et al., 1987).

The three global early Triassic T–R sequences are illustrated in Figure 18.6, with the transgressive conodont faunas, in a general way, occupying the interval between each TS and its associated DLS. In the western United States, however. *Neospathodus waageni* and *N. pakistanensis* do not occur in separate zones or subdivisions of a single zone, but they may represent condensed faunas. In addition, recent work suggests

that *Neospathodus timorensis* is lowest Anisian (middle Triassic) in age, and it would properly be associated with the overlying transgressive sequence in regions of continuing marine deposition (Nicora, 1977; Sweet, 1988; Orchard, 1994).

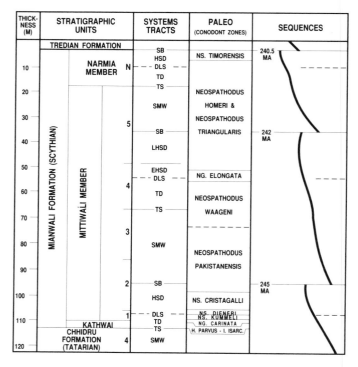

Figure 18.6. Stratigraphic sequences and palaeontologic data from the Lower Triassic (Scythian) section at Nammal Gorge, Salt Range, West Pakistan. The conodont zonation is adapted from Matsuda (1985). The three types of depositional surfaces recognized in the outcrops include transgressive sequences (TS), maximum flooding or downlap surfaces (DLS), and sequence boundaries (SB). The diagram is adapted from Haq et al. (1988), who defined sequences by shallowing events, rather than deepening events. That approach is in contrast to the proposal by Johnson et al. (1985) used in this chapter.

To correlate or to partition?

A biostratigraphic methodology for development and application of biozones was introduced in 1856 by Albert Oppel during his study of Jurassic strata. Oppel visualized an "ideal profile," based on the vertical ranges of some 10–30 species in widely separated localities with diverse sedimentary environments, ignoring any local palaeontologic or lithic successions (Hancock, 1977). His approach, by repetitively homogenizing the facies-controlled aspects of faunas, somewhat resembles the assembly of a composite section by graphic correlation, but does not include the incremental stadial units. Oppel's concept involved the development of *unifying* criteria, rather than mutually exclusive criteria, and distinguished useful biozones across a range of environments. With the application of those principles, parochial differences were "massaged" away, resulting in a cosmopolitan palaeontologic sequence.

Unfortunately, few attempts have been made to utilize Oppel's comparative technique, which focused on faunal similarities. An emphasis on regional *differences* (provincial or palaeoenvironmental), particularly in poorly fossiliferous, restricted, or very shallow marine settings, has led to the development and application of "local" or biofacies conodont zonations, with limited value for widespread regional correlation (Clark and Carr, 1984; Perri, 1991). Integration of locally unique shallow-water forms within the off-shore zonal framework would emphasize inter-regional similarities.

Oppel's "ideal profile" was facilitated by the radiation of species during transgressions. Hence, Lower Triassic conodont faunas related to T–R sequences and the eustatic sea-level curve present an opportunity for global correlation.

Conclusions

Early Triassic conodonts were more widespread and are stratigraphically more significant than was first thought. Since the development of Sweet's prototype zonation (Sweet, 1970), regional studies have established the commonality of many marine invertebrate faunal elements, including conodonts, in Eurasia and North America (Tollmann and Kristan-Tollmann, 1985; Budurov et al., 1985; Paull, 1988; Paull and Paull, 1992). Time-significant conodont species described from the Tethys nearly 25 years ago now appear to be trans-Tethyan and circum-Pacific in distribution. Conversely, early Triassic species first described outside of the Tethys seaway are also cosmopolitan. As a result, there is no suggestion of provincialism among early Triassic conodonts, and their distribution patterns indicate interchange throughout the Tethys-Pacific.

Three periods of relative sea-level rise in the early Triassic are well documented in many Permian–Triassic depositional regions around the world. The accompanying radiations and expansions of cosmopolitan conodont biofacies during those transgressive episodes provide a unique opportunity for global correlation.

References

Barnes, C. R. (ed.) 1976. *Conodont Paleoecology.* Special paper 15. Geological Association of Canada.

Barnes, C. R., and Fåhraeus, L. E. 1975. Provinces, communities, and the proposed nektobenthic habit of Ordovician conodontophorids. *Lethaia* 8:133–49.

Brandner, R. 1984. Meeresspiegelschwankungen und Tektonik in der Trias der NW-Tethys. *Jahrbuch der Geologisches Bundesanstalt* 126:435–75.

Budurov, K. J., Gupta, V. J., Sudar, M. N., and Buryi, G. I. 1985. Conodont zonation, biofacies and provinces in the Triassic. *Geological Society of India, Journal* 26:84–94.

Carr, T. R., and Paull, R. K. 1983. Early Triassic stratigraphy and paleogeography of the Cordilleran miogeocline. In *Mesozoic Paleogeography of the West-Central United States,* ed. E. D. Dolly, M. W. Reynolds, and D. R. Spearing, pp. 39–55. Rocky Mountain Paleogeography Symposium 2, Denver. Society of Economic Paleontologists and Mineralogists.

Chamberlin, T. C. 1909. Diastrophism as the ultimate basis of correlation. *Journal of Geology* 17:689–93.

Clark, D. L., and Carr, T. R. 1984. Conodont biofacies and biostratigraphic schemes in western North America: a model. In *Conodont Biofacies and Provinicalism,* ed. D. L. Clark, pp. 1–9. Special paper 196. Boulder: Geological Society of America.

Clark, D. L., and Wang Cheng-Yuan. 1988. Permian neogondolellids from South China: significance for evolution of the *serrata* and *carinata* groups in North America. *Journal of Paleontology* 62:132–8.

Collinson, J. W., and Hasenmueller, W. A. 1978. Early Triassic paleogeography and biostratigraphy of the Cordilleran miogeosyncline. In *Mesozoic Paleogeography of the Western United States,* ed. D. G. Howell and K. A. McDougall, pp. 175–87. Pacific Coast Paleogeography Symposium 2, Los Angeles. Society of Economic Paleontologists and Mineralogists.

Embry, A. F. 1984. Triassic eustatic sea level changes – evidence from the Canadian Arctic Archipelago. *Geological Society of America, Abstracts* 16:501.

Embry, A. F. 1988. Triassic sea-level changes: evidence from the Canadian Arctic Archipelago. In *Sea-Level Changes: An Integrated Approach,* ed. C. K. Wilgus et al., pp. 249–59. Special publication 42. Society of Economic Paleontologists and Mineralogists.

Fåhraeus, L. E. 1976. Conodontophorid ecology and evolution related to global tectonics. In *Conodont Paleoecology,* ed. C. R. Barnes, pp. 11–26. Special paper 15. Geological Association of Canada.

Forney, G. G. 1975. Permo–Triassic sea level change. *Journal of Geology* 83:773–9.

Hallam, A. 1992. *Phanerozoic Sea-Level Changes.* New York: Columbia University Press.

Hancock, J. M. 1977. The historic development of concepts of biostratigraphic correlation. In *Concepts and Methods of Biostratigraphy,* ed. E. G. Kaufmann and J. E. Hazel, pp. 3–22. Stroudsburg, PA: Dowden, Hutchinson & Ross, Inc.

Haq, B. U., Hardenbol, J., and Vail, P. R. 1987. Chronology of fluctuating sea levels since the Triassic. *Science* 235:1156–67.

Haq, B. U., Hardenbol, J., and Vail, P. R. 1988. Mesozoic and Cenozoic chronostratigraphy and eustatic cycles. In *Sea-Level Changes: An Integrated Approach,* ed. C. K. Wilgus et al., pp. 71–108. Special publication 42. Society of Economic Paleontologists and Mineralogists.

Harland, W. B., Armstrong, R. L., Cox, A. V., Craig, L. E., Smith, A. G., and Smith, D. G. 1990. *A Geologic Time Scale 1989*. Cambridge University Press.

Hirsch, F. 1994. Triassic multielement conodonts versus eustatic cycles. In *Recent Developments on Triassic Stratigraphy*, ed. J. Guex and A. Baud, pp. 35–52. Mémoire de Géologie 22. Lausanne.

Huckriede, R. 1958. Die Conodonten der Mediterranean Trias und ihr stratigraphischer Wert. *Paläontologische Zeitschrift* 32:141–75.

Johnson, J. G., Klapper, G., and Sandberg, C. A. 1985. Devonian eustatic fluctuations in Euramerica. *Geological Society of America, Bulletin* 96:567–87.

Klapper, G. 1991. Accuracy of biostratigraphic zones. *Palaios* 6:1.

Kummel, B. 1973. Lower Triassic (Scythian) molluscs. In *Atlas of Paleobiogeography*, ed. A. Hallam, pp. 225–33. Amsterdam: Elsevier.

Lindström, M. 1976. Conodont provincialism and paleoecology – a few concepts. In *Conodont Paleoecology*, ed. C. R. Barnes, pp. 3–9. Special paper 15. Geological Association of Canada.

Matsuda, T. 1985. Late Permian to early Triassic conodont paleobiogeography in the "Tethys Realm." In *The Tethys*, ed. K. Nakazawa and J. M. Dickins. pp. 157–70. Tokyo: Tokai University Press.

Müller, K. 1956. Triassic conodonts from Nevada. *Journal of Paleontology* 30:818–30.

Nakazawa, K., et al. 1985. Permian and Triassic systems in the Salt Range and Surghar Range, Pakistan. In *The Tethys*, ed. K. Nakazawa and J. M. Dickins, pp. 221–312. Tokyo: Tokai University Press.

Newell, N. D. 1967. *Revolutions in the History of Life*. Special paper 89. Boulder: Geological Society of America.

Nicora, A. 1977. Lower Anisian platform-conodonts from the Tethys and Nevada: taxonomic and stratigraphic revision. *Palaeontographica* A157:88–107.

Orchard, M. J. 1994. Conodont biochronology around the early–middle Triassic boundary: new data from North America, Oman and Timor. In *Recent Developments on Triassic Stratigraphy*, ed. J. Guex and A. Baud, pp. 105–14. Mémoire de Géologie 22. Lausanne.

Paull, R. K. 1982. Conodont biostratigraphy of Lower Triassic rocks, Terrace Mountains, northwestern Utah. In *The Overthrust Belt of Utah*, ed. D. L. Nielson, pp. 235–49. Publication 10. Utah Geological Association.

Paull, R. K. 1983. Definition and stratigraphic significance of the Lower Triassic (Smithian) conodont *Gladigondolella meeki* n. sp. in the western United States. *Journal of Paleontology* 57:188–92.

Paull, R. K. 1988. Distribution pattern of Lower Triassic (Scythian) conodonts in the western United States: documentation of the Pakistan connection. *Palaios* 3:598–605.

Paull, R. K., and Paull, R. A. 1982. Permian–Triassic unconformity in the Terrace Mountains, northwestern Utah. *Geology* 10:582–7.

Paull, R. K., and Paull, R. A. 1983. Revision of type Lower Triassic Dinwoody Formation, Wyoming, and designation of principal reference section. *Contributions to Geology* (*University of Wyoming*) 22:83–90.

Paull, R. K., and Paull, R. A. 1984. Early Triassic marine transgressions over southwest Montana – a lesson in planar hydrodynamics (abstract). In *Geology, Tectonics, and*

Mineral Resources of Western and Southern Idaho, ed. P. C. Beaver, p. 18. Dillon, MT: Tobacco Root Geological Society.

Paull, R. K., and Paull, R. A. 1986a. Epilogue for the Permian in the western Cordillera – a retrospective view from the Triassic. *Contributions to Geology* (*University of Wyoming*) 24:243–52.

Paull, R. A., and Paull, R. K. 1986b. Depositional history of Lower Triassic Dinwoody Formation, Bighorn Basin, Wyoming and Montana. In *Geology of the Beartooth Uplift and Adjacent Basins*, ed. P. B. Garrison, pp. 13–25. Yellowstone-Bighorn Research Association Guidebook. Montana Geological Society.

Paull, R. A., and Paull, R. K. 1987. Lower Triassic rocks within and adjacent to the thrust belt of Wyoming, Idaho, and Utah. In *The Thrustbelt Revisited*, ed. R. Miller, pp. 149–61. 38th field conference guidebook. Wyoming Geological Association.

Paull, R. A., and Paull, R. K. 1991a. Allochthonous rocks from the western part of the early Triassic miogeocline: Hawley Creek area, east-central Idaho. *Contributions to Geology* (*University of Wyoming*) 28:145–54.

Paull, R. K., and Paull, R. A. 1991b. Early Triassic sea-level changes along the western margin of the North American craton. *Geological Society of America, Abstracts* 23:348–9.

Paull, R. K. and Paull, R. A. 1992. Worldwide distribution of "Tethyan" Lower Triassic conodont faunas and their role in the correlation of global transgressive sequences. In *Abstracts of the 29th International Geological Congress*, Kyoto.

Paull, R. A., and Paull, R. K. 1993a. Stratigraphy of the Lower Triassic Dinwoody Formation in the Wind River Basin area, Wyoming. In *Oil and Gas and Other Resources of the Wind River Basin*, ed. W. R. Keefer, W. J. Metzger, and L. H. Godwin, pp. 31–47. Wyoming Geological Association.

Paull, R. A., and Paull, R. K. 1993b. Interpretation of early Triassic nonmarine–marine relations, Utah, U.S.A. In *Nonmarine Triassic Symposium*, ed. S. G. Lucas and M. Morales, pp. 403–9. Bulletin 3. Albuquerque: New Mexico Museum of Natural History & Science.

Paull, R. K., Paull, R. A., and Anderson, A. L. 1985. Conodont biostratigraphy and depositional history of the Lower Triassic Dinwoody Formation in the Meade Plate, southeastern Idaho. In *Orogenic Patterns and Stratigraphy of North-Central Utah and Southeastern Idaho*, ed. G. J. Kerns and R. L. Kerns, pp. 55–56. Publication 14. Utah Geological Association.

Paull, R. A., Paull, R. K., and Kraemer, B. R. 1989. Depositional history of Lower Triassic rocks in southwestern Montana and adjacent parts of Wyoming and Idaho. In *Montana Centennial Guidebook*, ed. D. E. French, pp. 69–90. Montana Geological Society.

Perri, M. C. 1991. Conodont biostratigraphy of the Werfen Formation (Lower Triassic), Southern Alps, Italy. *Bollettino della Società Paleonotologica Italiana* 30:23–46.

Ross, C. A., and Ross, J. R. P. 1985. Late Paleozoic depositional sequences are synchronous and worldwide. *Geology* 13:194–7.

Ross, C. A., and Ross, J. R. P. 1988. Late Paleozoic transgressive-regressive deposition. In *Sea-Level Changes: An Integrated Approach*, ed. C. K. Wilgus et al., pp. 227–47. Special publication 42. Society of Economic Paleontologists and Mineralogists.

Sageman, B. 1992. Paleoecology – a cure for sequence syndrome? *Palaios* 7:485–6.

Seddon, G., and Sweet, W. C. 1971. An ecologic model for conodonts. *Journal of Paleontology* 45:869–80.

Silberling, N. J., and Tozer, E. T. 1968. *Biostratigraphic Classification of the Marine Triassic in North America*. Special Paper 110. Boulder: Geological Society of America.

Smith, G. L., Byers, C. W., and Dott, R. H., Jr. 1993. Sequence stratigraphy of the Lower Ordovician Prairie du Chien Group on the Wisconsin arch and in the Michigan basin. *American Association of Petroleum Geologists, Bulletin* 77:49–67.

Solien, M. A. 1979. Conodont biostratigraphy of the Lower Triassic Thaynes Formation, Utah. *Journal of Paleontology* 53:276–306.

Sweet, W. C. 1970. Uppermost Permian and Lower Triassic conodonts of the Salt Range and Trans-Indus Ranges, West Pakistan. In *Stratigraphic Boundary Problems: Permian and Triassic of West Pakistan*, ed. B. Kummel and C. Teichert, pp. 207–75. Special paper 4. Department of Geology, University of Kansas.

Sweet, W. C. 1988. A quantitative conodont biostratigraphy for the Lower Triassic. *Senckenbergiana Lethaea* 69:253–73.

Sweet, W. C. 1992. A conodont-based high-resolution biostratigraphy for the Permo–Triassic boundary interval. In *Permo–Triassic Events in the Eastern Tethys*, ed. W. C. Sweet, Yang, Z., Dickins, J. M., and Yin, H., pp. 120–33. Cambridge University Press.

Sweet, W. C., and Bergström, S. M. 1986. Conodonts and biostratigraphic correlation. *Annual Review of Earth and Planetary Sciences* 14:85–112.

Tollmann, A., and Kristan-Tollmann, E. 1985. Paleogeography of the European Tethys from Paleozoic to Mesozoic and the Triassic relations of the eastern part of the Tethys and Panthalassa. In *The Tethys*, ed. K. Nakazawa and J. M. Dickins, pp. 3–22. Tokyo: Tokai University Press.

Tozer, E. T. 1967. *A Standard for Triassic Time*. Bulletin 156. Geological Survey of Canada.

Tozer, E. T. 1974. Definitions and limits of Triassic stages and substages: suggestions prompted by comparisons between North America and the Alpine-Mediterranean region. In *Schriftenreihe Erdwissenschaftlichen Kommission, Österreichische Akademie Wissenschaften*, vol. 2, pp. 195–206. Berlin: Springer-Verlag.

Vail, P. R., Mitchum, R. M., Jr., Todd, R. G., Widmier, J. M., Thompson, S., III, Sangree, J. B., Bubb, J. N., and Hatlelid, W. G. 1977. *Seismic Stratigraphy and Global Changes of Sea Level*. Memoir 26. American Association of Petroleum Geologists.

Appendix

Table 18.1 shows the distributions of 20 stratigraphically significant conodont species over 12 geographic regions. The three early Triassic transgressive sequences are indicated on the left side. Lower Triassic conodont species are from Sweet (1988, 1992). The geographic regions listed in Table 18.1 and the sources of data are as follows:

A. Himalayas: Matsuda (1981, 1982, 1983, 1984), Nicora (1991)

B. Eastern Asia (including China and Primorye): Buryi (1979), Ding (1992), Yang and Li (1992)

C. European Alps: Perri (1991)

D. Eastern Europe: Kolar-Jurkovšek (1990), Sudar (1986), Uroševic and Sudar (1980)

Table 18.1. *Conodont distributions during early Triassic transgressive sequences*

Transgression sequence	Stratigraphically significant conodont species	Geographic region											
		A	B	C	D	E	F	G	H	I	J	K	L
I	*Ng. timorensis*	×	×			×	×	×	×	×	×	×	×
III	*Platyvillosus asper.*												×
	Neospathodus collinsoni		×			×	×	×					×
	Neospathodus homeri	×	×		×	×		×	×	×		×	×
	Neospathodus triangularis	×	×	×	×	×	×	×		×		×	×
	Neogondolella jubata	×	×					×	×	×	×		×
II	*Neogondolella milleri*	×	×					×				×	×
	Neospathodus? conservativus	×	×			×	×	×	×				×
	Parachirognathus ethingtoni		×		×			×		×			×
	Furnishius tiserratus		×		×			×					×
	Neospathodus waageni	×	×					×	×	×	×	×	×
	Platyvillosus costata	×	×					×		×		×	×
	Neospathodus pakistanensis	×	×					×	×	×	×	×	×
I	*Neospathodus cristagalli*	×	×							×	×	×	
	Neospathodus kummeli	×	×									×	×
	Neospathodus dieneri	×	×	×				×	×	×	×	×	×
	Isarcicella isarcica	×	×	×		×							×
	Isarcicella? parva	×	×	×		×					×	×	×
	Neogondolella carinata	×	×			×				×	×	×	×
	Hindeodus typ. or sp.	×	×	×		×					×	×	×

E. Mediterranean: Gaetani et al. (1992), Mietto, Fratoni, and Perri (1991)

F. Middle East and Transcaucasia: Gedik (1975), Kapoor (1992), Orchard (1992, 1994)

G. Western and southwestern Pacific (including Japan): Berry, Burrett, and Banks (1984), Koike (1981, 1982), Koike, Kobayashi, and Ozawa (1985)

H. South Pacific (Australia and New Zealand): McTavish (1973), Paull (unpublished data)

I. Northern Asia (Siberia): Dagys (1984)

J. North Atlantic and arctic North America: Dagys and Korchinskaya (1987), Hatleberg and Clark (1984), Sweet (1976), Wardlaw and Jones (1980)

K. Canadian Cordillera and terranes: Beyers and Orchard (1991), Mosher (1973), Orchard (1991)

L. Western United States: Carr (1981), Collinson and Hasenmueller (1978), Paull (1982, 1988)

The following sources were used to prepare Table 18.1; it is not meant to be all-inclusive.

References

Berry, R., Burrett, C., and Banks, M. 1984. New Triassic faunas from East Timor and their tectonic significance. *Geologica et Palaeontologica* 18:127–37.

Beyers, J. M., and Orchard, M. J. 1991. Upper Permian and Triassic conodont faunas from the type area of the Cache Creek Complex, south-central British Columbia, Canada. In *Ordovician to Triassic Conodont Paleontology of the Canadian Cordillera*, ed. M. J. Orchard and A. D. McCracken, pp. 269–97. Bulletin 417. Geological Survey of Canada.

Buryi, G. I. 1979. *Lower Triassic Conodonts, South Primor'ye* (in Russian). Moskva: Nauka.

Carr, T. R. 1981. Paleogeography, depositional history and conodont paleoecology of the Lower Triassic Thaynes Formation in the Cordilleran miogeosyncline. Unpublished Ph.D. dissertation, University of Wisconsin, Madison.

Collinson, J. W., and Hasenmueller, W. A. 1978. Early Triassic paleogeography and biostratigraphy of the Cordilleran miogeosyncline. In *Mesozoic Paleogeography of the Western United States*, ed. D. G. Howell and K. A. McDougall, pp. 175–87. Pacific Coast Paleogeography Symposium 2, Los Angeles. Society of Economic Paleontologists and Mineralogists.

Dagys, A. A. 1984. *Early Triassic Conodonts of Northern Middle Siberia* (in Russian). Moskva: Nauka.

Dagys, A. A., and Korchinskaya, M. V. 1987. First conodonts from *Otoceras* beds of Svalbard (in Russian). In *Boreal Triassic*, ed. A. S. Dagys, pp. 110–13. Moscow: Nauka.

Ding, M. 1992. Conodont sequences in the Upper Permian and Lower Triassic of South China and the nature of conodont faunal changes at the systemic boundary. In *Permo–Triassic Events in the Eastern Tethys*, ed. W. C. Sweet, Yang Zunyi, J. M. Dickins, and Yin Hongfu, pp. 109–19. Cambridge University Press.

Gaetani, M., Jacobshagen, V., Nicora, A., Kauffmann, G., Tselepidis, V., Fantini Sestini, N., Mertmann, D., and Skourtsis-Coroneou, V. 1992. The early–middle Triassic

boundary at Chios (Greece). *Rivista Italiana di Paleontologia e Stratigrafia* 98:181–204.

Gedik, I. 1975. Die Conodonten der Trias auf der Kocaeli-Halbinsel (Turkei). *Palaeontographica, Abt. A (Stuttgart)* 150:99–160.

Hatleberg, E. W., and Clark, D. L. 1984. Lower Triassic conodonts and biofacies interpretations: Nepal and Svalbard. *Geologica et Palaeontologica* 18:101–25.

Kapoor, H. M. 1992. Permo–Triassic boundary of the Indian subcontinent and its intercontinental correlation. In *Permo–Triassic Events in the Eastern Tethys*, ed. W. C. Sweet, Yang Zunyi, J. M. Dickins, and Yin Hongfu, pp. 21–36. Cambridge University Press.

Koike, T. 1981. Biostratigraphy of Triassic conodonts in Japan. *Science Reports of Yokohama National University, sec. II* 28:25–42.

Koike, T. 1982. Triassic conodont biostratigraphy in Kedah, West Malaysia. In *Geology and Palaeontology of Southeast Asia*, vol. 23, ed. T. Kobayashi, R. Toriyama, and W. Hashimoto, pp. 9–51. University of Tokyo Press.

Koike, T., Kobayashi, F., and Ozawa, T. 1985. Smithian (Lower Triassic) conodonts from Iwai, Hinode-machi, Nishitama-gun, Tokyo-to, Japan. *Science Reports of Yokohama National University, sec. II* 32:45–56.

Kolar-Jurkovšek, T. 1990. Smithian (Lower Triassic) conodonts from Slovenia, NW Yugoslavia. *Neues Jahrbuch für Geologie und Paläontologie, Mh.* 9:536–46.

McTavish, R. A. 1973. Triassic conodont faunas from Western Australia. *Neues Jahrbuch für Paläontologie und Geologie, Abhandlungen* 143:275–303.

Matsuda, T. 1981. Early Triassic conodonts from Kashmir, India. Part 1: *Hindeodus* and *Isarcicella*. *Journal of Geosciences, Osaka City University* 24:75–108.

Matsuda, T. 1982. Early Triassic conodonts from Kashmir, India. Part 2: *Neospathodus* 1. *Journal of Geosciences, Osaka City University* 25:87–102.

Matsuda, T. 1983. Early Triassic conodonts from Kashmir, India. Part 3: *Neospathodus* 2. *Journal of Geosciences, Osaka City University* 26:87–110.

Matsuda, T. 1984. Early Triassic conodonts from Kashmir, India. Part 4: *Gondolella* and *Platyvillosus*. *Journal of Geosciences, Osaka City University* 27:119–41.

Mietto, P., Fratoni, R. P., and Perri, M. C. 1991. Spathian and Aegean conodonts from the Capelluzzo calcarenites of the Monte Facito Group (Lagonegro Sequence, Southern Apennines). *Memorie di Scienze Geologiche (Padova)* 43:305–17.

Mosher, L. C. 1973. *Triassic Conodonts from British Columbia and the Northern Arctic Islands*. Contributions to Canadian Paleontology, Bulletin 222. Geological Survey of Canada.

Nicora, A 1991. Conodonts from the Lower Triassic sequence of central Dolpo, Nepal. *Rivista Italiana di Paleontologia e Stratigrafia* 97:239–68.

Orchard, M. J. 1991. Conodonts, time and terranes: an overview of the biostratigraphic record in the western Canadian Cordillera. In *Ordovician to Triassic Conodont Paleontology of the Canadian Cordillera*, ed. M. J. Orchard and A. D. McCracken, pp. 1–25. Bulletin 417. Geological Survey of Canada.

Orchard, M. J. 1992. Lower Triassic conodonts from Oman. *Geological Society of America, Abstracts* 24:57–8.

Orchard, M. J. 1994. Conodont biochronology around the early–middle Triassic boundary: new data from North America, Oman and Timor. In *Recent Developments on Triassic*

Stratigraphy, ed. J. Guex and A. Baud, pp. 105–14. Mémoire de Géologie 22. Lausanne.

Paull, R. K. 1982. Conodont biostratigraphy of Lower Triassic rocks, Terrace Mountains, northwestern Utah. In *The Overthrust Belt of Utah,* ed. D. L. Nielson, pp. 235–49. Publication 10. Utah Geological Association.

Paull, R. K. 1988. Distribution pattern of Lower Triassic (Scythian) conodonts in the western United States: documentation of the Pakistan connection. *Palaios* 3:598–605.

Perri, M. C. 1991. Conodont biostratigraphy of the Werfen Formation (Lower Triassic), Southern Alps, Italy. *Bolletino della Società Paleontologica Italiana* 30:23–46.

Sudar, M. 1986. Triassic microfossils and biostratigraphy of the Inner Dinarides between Gucevo and Ljubišnja mts., Yugoslavia. *Ann. Géol. Péninsule Balkanique (Beograd)* 50:151–394.

Sweet, W. C. 1976. Conodonts from the Permian–Triassic boundary beds at Kap Stosch, East Greenland. *Meddelelser om Grønland* 197:51–3.

Sweet, W. C. 1988. A quantitative conodont biostratigraphy for the Lower Triassic. *Senckenbergiana Lethaea* 69:253–73.

Uroševic, D., and Sudar, M. 1980. Triassic microfossils from the area of the Mountain Gucevo. In *Symposium de Géologie Régionale et Paleontologie,* pp. 491–507. Belgrade: Geologie Universite.

Wardlaw, B. R., and Jones, D. L. 1980. Triassic conodonts from eugeosynclinal rocks of western North America and their tectonic significance. *Riv. It. Paleont. Strat.* 80:895–908.

Yang, Z., and Li, Z. 1992. Permo–Triassic boundary relations in South China. In *Permo–Triassic Events in the Eastern Tethys,* ed. W. C. Sweet, Yang Zunyi, J. M. Dickins, and Yin Hongfu, pp. 9–20. Cambridge University Press.

19 Triassic biostratigraphy and palaeobiogeography of East Asia

YIN HONGFU

East Asia occupies the key position in the Triassic correlation of the circum-Pacific region, Tethys, and marginal Gondwana, because it includes important areas from all of those realms. This chapter presents biostratigraphic correlations for the major Triassic groups of organisms (Tables 19.1–19.3). Detailed discussion of how these charts were worked out would be lengthy and will be omitted here.

Early Triassic biogeography of East Asia

The East Asian area of the early Triassic has been subdivided into the following realms and regions: the northern temperate North Laurasian Realm, including the Boreal Sea Region (northeastern Asia, I_1) and the North Eurasian Region (I_2); the northern warm-temperate Central Laurasian Realm, including the Central Eurasian Region (II_2) and its Marginal Seas Region (II_1); the tropical–subtropical Laurasian Tethys Realm, including the Cathaysian Tethys Region (III_1) and its Southern Margin Region (III_2); the southern temperate Gondwanan Realm, including the Indian Subcontinent Region (V); and the Gondwanan Tethys Realm, including the Himalaya–Banda Loop Region (IV) (Figure 19.1)

Boreal Sea Region of the North Laurasian Realm (northeastern Asia, I_1)

This region covers the Verkhojan (Verchojan)–Chukote (Chukchi) and Mongol–Okhotsk areas, where carbonate sediments are extremely rare. The biota was dominated by ammonoids and bivalves (Dagys, Arkhipov, and Bychkov, 1979): early Griesbachian ammonoids *Otoceras, Tompophiceras, Hypophiceras, Anotoceras*; late Griesbachian Boreal ophiceratids (*Ophiceras boreale, Ophiceras indigirense, Ophiceras commune*); a Spathian *Keyserlingites* assemblage, including *Olenikites, Svalbardiceras, Sibirites*, and *Nordophiceras*; there were no Smithian *Owenites* and Spathian *Columbites* and *Tirolites* [*Tirolites morpheus* (Popov) from Verkhojan has been renamed *Tompoites*]. The abundant tropical bivalve *Eumorphotis-Claraia* assemblage was represented there by only two cosmopolitan species: *Eumorphotis multiformis* and *Claraia stachei* (Yakutsk and Olenek rivers). Instead, the forms that flourished there were *Atomodesma, Promyalina, Nuculana*, and the *Posidonia* group, such as the Smithian *Posidonia mimer* and the Spathian *Posidonia aranea*. *Atomodesma* was a relict of the Boreal Permian and showed typical bipolarity during that period (Boreal Sea and Gondwana Tethys). The previously mentioned *Posidonia* species were typical of the Boreal early Triassic and also inhabited Arctic Canada. In general, the Boreal biota was characterized by low diversity and high cosmopolitanism. There was no ice sheet in the Arctic during the Triassic. The associated plants found in the terrestrial interbeds are northern temperate ferns of the *Pseudoaraucarites* assemblage, and the Triassic temperature measured from oxygen isotopes in ammonoid shells was 14°C (Zakharov, 1973). Such findings indicate that those areas belonged to the northern temperate zone.

North Eurasian Region of the North Laurasian Realm (I_2)

This region is characterized by the Maltzev fauna and the Korvonchan flora. The former includes the *Palaeanodonta-Palaeomutella-Microdontella* bivalve assemblage, the *Pseudoestheria tomiensis–Glyptoasmussia wetlugensis* conchostracan assemblage, and the *Darwinula* ostracod assemblage. The *Palaeanodonta* assemblage has been found at Tunguss, Kuznetsk, Taohaiyinzi in Inner Mongolia, and Jiutai in Changchun, Jilin province, and the estheriid *Cornia* has been found in the Pechora Basin (53 in Figure 19.1), Jiutai, and Junggar. The Korvonchan flora, dominated by temperate ferns, was widespread in the Tunguss Basin, but was replaced by the *Pseudoaraucarites* assemblage in the northern Siberian marginal depressions and Vilui Basin at its eastern margin. The foregoing flora and fauna are both found at Kuznetsk.

Central Eurasian Region of the Central Laurasian Realm (II_2)

This region is distinguished by a lystrosaur-labyrinthodont fauna and the *Voltzia-Pleuromeia* flora. Lystrosaurs have been found at Jimsar in Xinjiang and along the lower Dvina River on the Russian Platform; labyrinthodonts have been found at Fugu and Wubao in Shaanxi province. *Pleuromeia* are widespread in this region and are distributed westward through the Pamir, Darwaz, and pre-Caspian area to western Europe. They are also found in the Middle Triassic Ermaying Formation at Wubao in Shaanxi and at Jungar Qi in Inner Mongolia. In southern Fergana, no *Pleuromeia* have been found, but the related forms *Pleuromeiopsis* and *Pseudovoltzia* are found instead. *Voltzia* occurs in Kuznetsk and Shanxi, and *Voltzia heterophylla* in Junggar-Aratau and Seme'itau. The latter area also yields flora of the northern Eurasian type, thus representing a mixed area of central and north Eurasian biogeographic regions.

The lystrosaurs show a bipolar distribution and thus lived in nontropical, most probably warm-temperate, areas. *Pleuromeia* and *Voltzia* have been used to indicate a tropical–subtropical arid climate, but they occur with relatively abundant nontropical elements such as *Tersiella*, *Madygenia*, and *Schizoneura* in the central Eurasian region, and with mixed Boreal and Tethyan marine faunas in Japan and Russian Primorye, both belonging to the Marginal Seas Region of the Central Laurasian Realm. In Yushe, Shanxi, they have been reported together with the glossopteridales *Neoglossopteris*, *Gangamopteris(?)*, and *Gondwanidium*. Thus, the occurrence of *Pleuromeia* in deposits of the Central Eurasian Region does not necessarily mean that the region was tropical. In general, it would seem that it should have been a warm-temperate or warm-temperate–subtropical region.

Marginal Seas Region of the Central Laurasian Realm (II₁).

The marginal seas covering the areas of Sikhote-Alin, Japan, the southern margin of the North China Platform (southernmost Ordos, the southern Qilian Mountains, the Burhan Buda Mountains), the northern belt of the southern Qinling Mountains, and probably Darwaz in Pamir were distinguished by a mixed temperate (Boreal) and tropical (Tethyan) biota. Ammonoids and bivalves are the main components of the fossil fauna. The former comprises both the Boreal *Keyserlingites* assemblage and the Tethyan *Tirolites-Columbites* assemblage. To the south of Primorye, *Keyserlingites* appears only in Anisian deposits and thus is absent from the Lower Triassic deposits of North China's southern margin, but other Boreal elements such as *Nordophiceras*, *Svalbardiceras*, and *Gurleyites* are found there (Wang, 1985). Moreover, temperate ammonoids, including *Otoceras(?)* (Changxing), *Hypophiceras* (Changxing, Guangyuan, Daye). *Anotoceras*, *Wordieoceras*, and *?Dunedinites* (Madoi), have been found along a belt following the northern margin of the Cathaysian Tethys, which may have been due to invasion of Boreal water through a seaway between the Yangtze and North China (Yin, 1991). Bivalves constitute the bulk of the relatively rich *Eumorphotis-Claraia* fauna, which differs from that of the Eurasian Tethys in composition (Yin, 1990).

Cathaysian Tethys Region of the Laurasian Tethys Realm (III₁)

This is a typical tropical-sea area (Dang, Dagys, and Kiparisova, 1965; Yin, 1988). The prolific fauna consists mainly of ammonoids, bivalves, brachiopods, and conodonts. The bipolar *Otoceras* is not among the Griesbachian ammonoids found there, and its group of *Ophiceras* species differs from the Boreal group. The widespread Dienerian and Smithian ammonoids are similar to those of other regions, except for the Boreal Region. The early Spathian ammonoids are represented by Dinaritidae and Columbitidae. *Tirolites*, of the former family, lived mainly in epicontinental seas like the Yangtze, whereas *Columbites*, of the latter family, lived mainly in shelf areas and basins. Well-developed *Tirolites* assemblages have also been found in Idaho and Utah in the western United States and in southern Primorye in the Russian Far East, revealing their distribution farther into the middle palaeolatitudes of circum-Pacific areas. *Tirolites* did not flourish in the Gondwanan Tethys Realm and the Boreal Region, although some ill-preserved and age-doubtful specimens have been reported from the Indian Himalayas and southern Tibet. Columbitidae also were distributed mainly in the Eurasian Tethys and in the circum-Pacific middle palaeolatitudes (Japan, Primorye, western United States), but they had wider ranges than *Tirolites* and also appeared in the Gondwanan Tethys Realm. Spathian Keyserlingitidae and Sibiritidae, typical of the Boreal Sea Region, are not found in the Cathaysian Tethys Region.

The bivalves of the tropical Laurasian Tethys Realm can be divided into a western group centering on the Mediterranean and an eastern group centering on East Asia (Kobayashi and Tamura, 1983; Yin, 1990). In East Asian areas, the early Triassic *Eumorphotis-Claraia* assemblage was extremely prolific, each genus consisting of several dozens of species. Contrasting with the Boreal *Lingula borealis*, the brachiopods of this region are typified by *Lingula tenuissima* and a few articulates. The thriving conodonts differed from those of both the Boreal and the peri-Gondwanan seas; *Hindeodus parvus* and *Isarcicella isarcica* have been found in more than 20 localities in southern China alone.

This region connects westward with the pre-Caspian, Caucasus, and Alps along a palaeolatitude of approximately 20°N and is typified by widespread red clastics and evaporites, indicating a hot, arid climate.

Southern Margin Region of the Laurasian Tethys Realm (III₂).

This region differs from the Cathaysian Tethys Region in that its Lower Triassic deposits are not well developed. In Sibumasu and East Malaya, *sensu* Metcalfe (1988), the scarce marine findings include *Claraia intermedia* from Nampang in Thailand and *C. concentrica* and *C. intermedia multistriata* from Gua Musang in Malaysia, both areas having similarities to the Gondwanan Tethys Realm (Tamura, 1968). The Lhasa (or Gangdise) block should be connected with Sibumasu, but separated from

Table 19.1. *Correlation chart of the Lower and Middle Triassic of East Asia based on ammonoids and bivalves*

	Tethys (Krystyn, 1983)	Canada Tozer, 1974, 1981)	Qomolongma (Wang & He, 1976)	Spiti (Bhalla, 1983)
Ladinian Longobard	*Frankites? regoledanus* *Protrachyceras archelaus,* *Protrachyceras gredleri*	*Frankites sutherlandi* *M. maclearni, M. meginae,* *Protrachyceras poseidon*	*Protrachyceras, Joannites*	*Daonella indica*
Fassan	*Eoprotrachyceras curionii* *Nevadites*	*Eoprotrachyerras subasperum*		*Daonella lommeli*
Anisian Illyr	*Parakellnerites (Tecinites polymorphus, A. avisianum, P. reitzi)* *Paraceratites trinodosus*	*Frechites chischa* *Frechites deleeni*	*Ptychites*	*Ptychites rugifer*
Pelson	*Balatonites balatonicus*			*Keyserlingites dieneri,* *Sibirites* prahlada
Bithyn	*Anagymnotoceras ismidicum, Nicomedites osmani*	*Anagymototceras varium*	*Anacrochordiceras nodosus*	
Aegeum	*"Aegeiceras ugra"*	*Lenotropites caurus*	*Japonites magnus*	
Olenekian Spathian		*Keyserlingites subrobustus*		*Rhynchonella griesbachi*
	Tozericeras pakistanum, *Tirolites cassianus*	*Keyserlingites pilaticus*	*Procarnites, Anasibirites*	
Smithian	*Wasatchites spiniger, A. pluriformis, A. prahiada*	*Wasatchites tardus*		*Hedenstroemia mojsisovicsi*
	Meekoceras gracilitatis	*Euflemingites romunderi*	*Owenites*	*Meekoceras varaha*
Induan Dienerian Griesbachian	*Flemingites rahilla* *Gyronites frequens* *Ophiceras connectens,* *Ophiceras tibeticum*	*Vavilovites sverdrupi* *Proptychites candidus* *Proptychites strigatus,* *Ophiceras commune*	*Gyronites psilogyrus* *Lytophiceras sakuntala*	*Lytophiceras sakuntala*
	Otoceras woodwardi	*Otoceras boreale, Otoceras concavum*	*Otoceras latilobatum*	*Otoceras woodwardi*

Note: We are not able to fit Dagys's important proposal (1991) into the scheme, pending more detailed information.

Qiangtang (northern Tibet), because there was a vast branch of the Meso-Tethys along Nujiang in between, and it was not until the end of the Jurassic that the brackish *Peregrinoconcha* fauna bypassed Nujiang. In accordance with the scheme of Audley-Charles (1988), these areas are called the Southern Margin Region. The fresh-water *Lystrosaurus* from Luang Prabang, in Laos, although doubted by Metcalfe (1988), suggests a relationship between Gondwana and Indochina, which was echoed again in the late Triassic, as discussed later. Thus, Indochina is also tentatively set within this region.

Gondwanan Tethys Realm (IV)

The Gondwanan Tethys Realm embraces Kashmir, the Himalayas, and Banda Loop and, like the Boreal Region, is dominated by ammonoids and bivalves. The Griesbachian ammonoids *Otoceras* and *Anotoceras* have been reported from the Himalayas, Kashmir, and Timor, showing the same nontropical bipolarity as the Boreal Region. In the Dienerian and Smithian, the ammonoids, such as *Gyronites, Owenites,* and *Anasibirites,* were similar to those of the Eurasian Tethys. The Spathian featured the Boreal *Svalbardiceras* (the Himalayas, Kashmir, Salt Range, and southern Tibet), *Kingites* (Kumaon), *Arctomeekoceras* (Salt Range), and *Sibirites* (Byans), whereas *Tirolites* was uncommon.

The bivalve diversity was much lower than in the Laurasian Tethys Realm (Yin, 1990). The lower part of the Lower Triassic is characterized by *Claraia griesbachi,* the *C. intermedia* group (including *C. perthensis*), and the *C. concentrica* group (including *C. hupensis*), whereas such Laurasian Tethys elements as *C.*

Verchojan (Rostovtsev & Zhamoida, 1984)	Japan (synthesized)	Primoria (Zakharov, 1978; Oleinikov & Paevskaya, 1978)	SW China (Yang et al., 1982)
Nathorstites mcconnelli			
Indigirites krugi, A. omolonjensis *Longobardites oleshkoi*	*Protrachyceras* cf. *archelaus*	*Daonella*	*Protrachyceras deprati*
Frechites humboldtensis	*Protrachyceras reitzi*	*Ptychites oppeli*	*Protrachyceras prinum*
Gymnotoceras rotelliforme	*Paraceratites* cf. *trinodosus*	*Paraceratites trinodosus* *Acrochordiceras kiparisovae*	*Paraceratites trinodosus* *Paraceratites binodosus*
Arctohungarites kharaulakhensis	*Hollandites*		*Nicomedites yohi*
Czekanowskites decipiens *L. tardus, G. taimyrensis*	*Leipohy.* cf. *pseudopradyumna*	*Leiophyllistes pradyumna*	*Parapopanoceras nanum*
Olenekites spiniplicatus, K. subprobustus,	*Subcolumbites*	*Subcolumbites multiformis*	*Procarnites oxynostus*
P. grambergi, Dieneroceras demokidovi	*Columbites parsianus*	*Neocolumbites insignis*	*Columbites costatus*
Wasatchites tardus	*Anasibirites*	*Anasibirites nevolini*	*Pseudowenites oxynostus*
Hedenstroemia hedenstroemi *Vavilovvites compressus, V. turgious*	*Owenites* *Entolium, Eumorphotis*	*Hedenstroimia bosphorense* *Gyronites subdharmus*	*Owenites costatus* *Koninkites lingyunensis, Proptychites kwangsiensis*
Glyptochiceras nielseni	*Glyptophiceras*		*Vishnuites marginalis, Ophiceras sinensis*
O. boreale, Otoceras indigirense, Otoceras concavum			

aurita, C. clarai, and *C. wangi* are missing. *Eumorphotis* is rarely seen. Represented in the upper part of the Lower Triassic are *C. decidens, Leptochondria minima,* and the endemic Himalayan genus *Zandaia* (*"Pseudomonotis" himaica*). *Leptochondria minima* and *Eumorphotis huancangensis* are found in the Marginal Seas Region of the Central Laurasia and in southern Ngari, Qomolangma Mountain, and Kashmir of the Gondwanan Tethys Realm, again showing bipolarity and suggesting a similar palaeoclimatic situation.

Indian Subcontinent Region of the Gondwanan Realm (V).

This region yields the latest *Glossopteris* flora, including *Schizoneura gondwanensis, Glossopteris communis, G. indica, G. ampla,* and *?Dicroidium,* as well as the *Lystrosaurus-Chasmatosaurus* vertebrate fauna. They are related to those from South Africa and Australia and lived in the southern warm-temperate belt.

Middle Triassic biogeography of East Asia

The East Asian provincialization of the Middle Triassic was basically the same as that of the early Triassic, except that the Southern Margin Region of the Laurasian Tethys Realm disappeared (Figure 19.2).

Boreal Sea Region of the North Laurasian Realm (northeastern Asia, I_1)

As was the case for the Lower Triassic, the deposits of this region are dominated by clastic sediments, though of sandier

Table 19.2. *Correlation chart of the Upper Triassic of East Asia based on ammonoids and bivalves*

	Tethys (Krystyn, 1983)		Canada (Tozer, 1974, 1981)	Qomolongma Mt. (Wang & He, 1976)

Norian
Rhaetian
	Choristoceras marshi	*Choristoceras marshi,*
		  Vandaites
		   stuerzenbaumi
			Ch. crickmayi	

Sevatian	*Rhabdoceras suessi*	*Sagenites reticulatus,*	*C. amoenum*	

Alaunian	*Halorites macar*	*Sagenites*
		  quinquepunctatus
		  "catenate Haloriten,"
		  Amarussites
		   semiplicatus
			G. cordilleranus
			
			H. columbianus	

	Himavatites hagarti	*Himavatites hagarti,* *Himavatites watsoni*		*Himavatites columblanus*
Lac	*Cyrtopleurites bicrenatus* *Juavvites magnus* *Malayites paulckei*	*Malayites paulckei,* *Malayites tingriensis*	*D. rutherfordi* *J. magnus* *M. dawsoni*	*Cyrtopleurites socius* *Indojuvarites angulatus* *Griesbachites, Goniontites*

Carnian
Tuval
	Guembalites jandianus
	
	Anatropites
	
	Tropites subbullatus
	
	Tropites dilleri
		Dimorphites selectus,
		  Dimorphites n. sp. 1
		Gonionotites italicus,
		  Discotropites plinii
		Tropites subbullatus,
		  Projuvavites
		   crassaplicatus
			S. kerri
			
			"*K. macrolobatus*"
			"Upper *T. wellleri*"
			
			"Lower *T. welleri*"
			
			Tropites dillleri
				Nodotibetites nodosus
				
				Parahauerites acutus
				
				Hoplotropites
				
				Indoesites dieneri

Jul	*Austrotrachyceras*
	  austriacum
	Trachyceras aonoides
		"*Neoprotrachyceras oedi
		  pus,*" *Austrotrachyceras*
		  triadicum
		Trachyceras anoides,
		  Trachyceras aon
			Austrotrachyceras
			  obseum
			Trachyceras aonoides	

composition, and by ammonoid-bivalve faunas (Bychkov et al., 1976; Dagys et al., 1979). Typical Anisian ammonoids are the Popanoceratidae (*Amphipopanoceras, Stenopopanoceras*), Longobarditidae (*Czekanowskites, Longobardites, Arctohungarites*), and Groenlanditinae (*Lenotropites, Pearylandites, Tienzunites*). Popanoceratidae were also widespread in the Austral Realm, such as in New Zealand and New Caledonia, thus showing bipolarity. Groenlanditinae have also been found in Arctic Canada. Ladinian ammonoids include nathorstitids, again typically Boreal. Tethyan families such as Paraceratinae, Japonitidae, and Protrachyceratidae are not well represented. Of the bivalves, *Janopecten* is typically Boreal, and the daonellid group here has species different from those of the Tethys. The Ladinian saw the appearance of the *Pennospiriferina-Fletcherithyroides* brachiopod assemblage, endemic for northeastern Asia.

North Eurasian Region of the North Laurasian Realm (I_2)

The terrestrial biota of this region is represented by the flora of the Tunguss Basin. However, uncertainty remains in regard to the stratigraphic correlation. Dobrushkina (1982) maintained that this third assemblage of the Korvonchan flora is from the Anisian, on the ground that unlike the first two assemblages it no longer contains Palaeozoic-type palynomorphs and that *Quadrocladus* and *Lutuginia* are abundant. Kiparisova, Radchenko, and Gorskii (1973), however, on the basis of ostracods and estheriids, would have set the same strata mostly in the Lower Triassic; therefore, in Figure 19.2 the flora localities are numbered with question marks. The flora is dominated by ferns and is different from the *Pleuromeia-Voltzia* flora in the Central Eurasian and Cathaysian Tethys regions. During the middle Triassic, the early

Spiti (Bhalla, 1983)	Verchojan (Dagys et al., 1979)	Primoria (Zakharov, 1978; Oleinikov & Paevskaya, 1978)	Japan (synthesized)	SW China (Yang et al., 1982)
Megalodon ladakhensis, Dicerocardium himalayense	Tosapecten efimovae			
	Entomonotis ochotica	E. ochotica, Ambin Fm.	E. zabaicallica?, E. ochotica, E. densistriata	
Spirigera maniensis Monotis Spiriferina griesbachi and corals Indojuvavites angulatus	Entomonotis scutiformis	E. scutiformis	E. typica	Yunnanophorus – Indosinion
	Otapiria ussuriensis	Otapiria ussuriensis	Otapiria dubia Juvaviites cf. kellyi [Takaguchi Fm.]	
	Pinacoceras verchojanicum	Pterosirenites kiparisovae		
Dielasma julicum				
		Sadgorod Fm	Halobia, Tosapecten, Oxytoma (L&M Mine Gr.) and Sandlingites aff. oribasus, Sirenites cf. nanseni (Nakijin Fm.)	
Tropites subullatus	Sirenites yakutensis			Thisbites-Discotropites
	Neosirenites pentastichus	Kiparisova Fm		
	Neoprotrachyceras seimkanense			Trachyceras, Protrachyceras
Halobia cf. comata, Joannites thanamensis	Protrachyceras omkutchanikum Nathorstites tenuis	Halobia talajaensis		

Indosinian Orogeny uplifted areas south of the Tunguss, including Hinggan, Mongolia, Sayan, and Altay, into highlands that then underwent erosion, and their southern margins composed the boundary between the Central Laurasian Realm and the North Laurasian Realm, as in the early Triassic.

Central Eurasian Region of the Central Laurasian Realm (II$_2$)

The kannemeyeriids (*Parakannemeyeria, Sinokannemeyeria, Shanbeikannemeyeria*) are the most remarkable faunal forms in this region. As to flora, *Annalepis* occurs in Shanxi (Tongchuan Formation), Gansu (Dingjiayao Formation), the lower Yangtze area (Huangmaqing Formation), and Hubei (Badong Formation); *Pleuromeia* occurs in southern Junggar (Lower Kalamay Formation), Shanxi (Ermaying Formation); *Tongchuanophyllum* occurs in the Ordos Basin (Ermaying Formation) and Gansu (Dingjiayao Formation). For the North China part of this region, Zhou and Zhou (1983) established a lower *Voltzia–Aipteris wuziwanensis* flora and an upper *Annalepis-Tongchuanophyllum* flora, distinguishable from the synchronous *Annalepis–Neocalamites meriani* flora of the Yangtze.

Given their common occurrences in Gondwana, it seems that the kannemeyeriids were bipolarly distributed and thus non-tropical. Besides, pioneers of the warm-temperate *Danaeopsis-Symopteris* flora and fresh-water *Shaanxiconcha-Sibireconcha* assemblage had already appeared in several localities. Sedimentation patterns show a less humid climate than that in the Cathaysian Tethys. In general, this region belonged to the warm- temperate belt. Along its margin, the Aso Formation in Japan yields Ladinian dipteridales, denoting hot and humid weather.

Table 19.3. *Correlation chart of the Triassic of East Asia based on conodonts*

	International & Nevada (Sweet et al., 1971, 1986)	Bulgaria (Budurov, 1976; Budurov et al., 1985)	Lena River mouth (Dagys et al., 1979)	Sikhote-Alin (Buryi, 1989)	Qinling (Lai, 1992)	South Jiangsu (Duan, 1987)
Rhaetian		*Miskella posthernsteini*		*Miskella posthernsteini*		
Norian						
Sevat	*Epigondolella biodentata*	*E. bidentata*		*E. bidentata*		
Alaun	*E. multidentata*	*E. postera*		*E. postera*		
Lac	*E. abneptis*	*E. abneptis*		*E. abneptis*		
Carnian						
Tuval	*Parag. poly- gnathiformis*	*E. nodosa*		*Metapolygnathus nodosus*		
Jul	*Neospathodus newpassensis*	*Parag. polygnathiformis*		*Parag. polygnathiformis*		
Ladinian						
Longobard	*Carinella mungoensis*	*Parag. foliata, Car. mungoensis*		*Parag. foliata*		
Fassan	*Neogondolella mombergensis*	*Ng. mombergensis*			*Ng. mombergensis*	
Anisian						
Illyr	*Ng. constricta*	*Ng. cornuta, Parag. excelsa*		*Parag. excelsa*	*Ng. constricta*	
Pelson		*N. kockeli*		*N. kockeli*		
Bithyn + Aegeum	*Ng. regale*	*Ng. regali*			*Ng. regale*	
Olenekian						
Spathian	*N. timorensis* *Ng. jubata* *N. collinsoni* *N. triangularis*	*N. timorensis* *N. homer* *N. triangularis*	*N. timorensis* *Ng. jubata* *Prionodina*	*N. timorensis* *N. homeri*	*N. hungaricus, N. homeri* *N. triangularis* *Pachycladina – Parachirognathus*	*N. anhui. + Gladio.* *N. anhuiensis, N. homeri* *N. collinsoni, N. triangularis*
Smithian	*Ng. milleri* *N. waageni*		*N. waageni*	*N. cf. waageni*		*N. waageni*
Induan Dienerian	*N. pakistanensis* *N. cristagalli* *N. kummeli, N. dieneri*		*N. pakistanensis* *N. dieneri*		*N. pakistanensis* *N. dieneri*	*N. cristagalli* *N. dieneri* *N. kummeli*
Griesbachian	*Isarcicella isarcica* *H. typicalis* *H. parvus*				*Hindeodus parvus*	*H. parvus*

Marginal Seas Region of the Central Laurasian Realm (II₁)

As in the early Triassic, this region was characterized by mixed faunas of temperate (Boreal) and tropical (Tethyan) aspects. Its Japan–Sikhote-Alin section differs from the section embracing the southern Qilian, western Qinling, and Burhan Buda Mountains in that the former bears features of the circum-Pacific (e.g., its temperate-water components are mainly popanoceratid and arctohungaritid ammonoids and the *Daonella densisulcata* bivalve assemblage). The latter section contains both temperate keyserlingitid ammonoids and the tropical *Leptochondria illyrica*

Guangxi (Yang et al., 1986)	Tibet, Hubei (Wang & Wang, 1976; Wang & Cao, 1981; Wang et al., 1981)	Salt Range (Pakistani – Japanese Group, 1981)	Kashmir (Matsuda, 1981, 1985)	West Malaysia (Igo & Koike, 1975)	West Australia (McTavish, 1973)
	E. multidentata *E. abneptis* *Parag. polygnathiformis* *E. diebeli*			*Parag. polygnathiformis* + *G. malavensis* + *C. mungoensis*	
	Parag. excelsa – *Ng. mombergensis*				
	Ng. constricta *N. germanicus* – *N. kockeli* *Ng. regale*			*Ng. aegaea* + *Parag. excelsa* + *N. timorensis* + *N. germanicus*	
N. homeri, *N. triangularis*	*N. timorensis* *Ng. N. homeri jubata Pachycladina*	*N. timorensis* *N. homeri* – *N. triangularis*	*N. timorensis* *N. homeri* – *N. triangularis*		*N. timorensis*
N. waageni *Platy. costatus* *N. cristagalli* *N. dieneri*	*Platyvillosus* *N. waageni* *N. pakistanensis* *N. cristagalli* *N. dieneri* *N. kummeli*	*Ng. jubata* *N. waageni* *N. pakistanensis* *N. cristagalli* *N. dieneri* *N. kummeli*	*Ng. elongata* + *Ng. milleri Ng. waageni* *N. pakistanensis* *N. cristagalli* *N. dieneri* *N. kummeli*	*N. conservativus* +*N. bicuspidatus*	*Ng. jubata* *N. waageni* *N. dieneri*
Ng. carinata *Isarcicella isarcica* *H. parvus*	*Ng. carinata* *Is. isarcica* *H. parvus*	*Ng. carinata* *H. parvus*	*Ng. carinata* *I. isarcica* *H. parvus*		

bivalve assemblage. The Anisian brachiopod assemblage (*Pseudospiriferina, Paranptychina, Septalirhynchia*) discovered in the southern Qilian, western Qinling, and Burhan Buda mountains in this section is closely related to the Qingyan assemblage in Guizhou, thus bearing Tethyan features (Yang et al., 1983; Yin et al., 1992). In general, this region belonged to a subtropical–warm-temperate belt.

Cathaysian Tethys Region of the Laurasian Tethys Realm (III)

This region yields tropical fauna, including the Anisian ammonoids *Japonites, Balatonites, Danubites,* and *Beyrichites* and Ladinian Protrachyceratidae, bivalves of the *Costatoria goldfussi* group (*chegarperahensis, pahangensis, singapurensis, submul-*

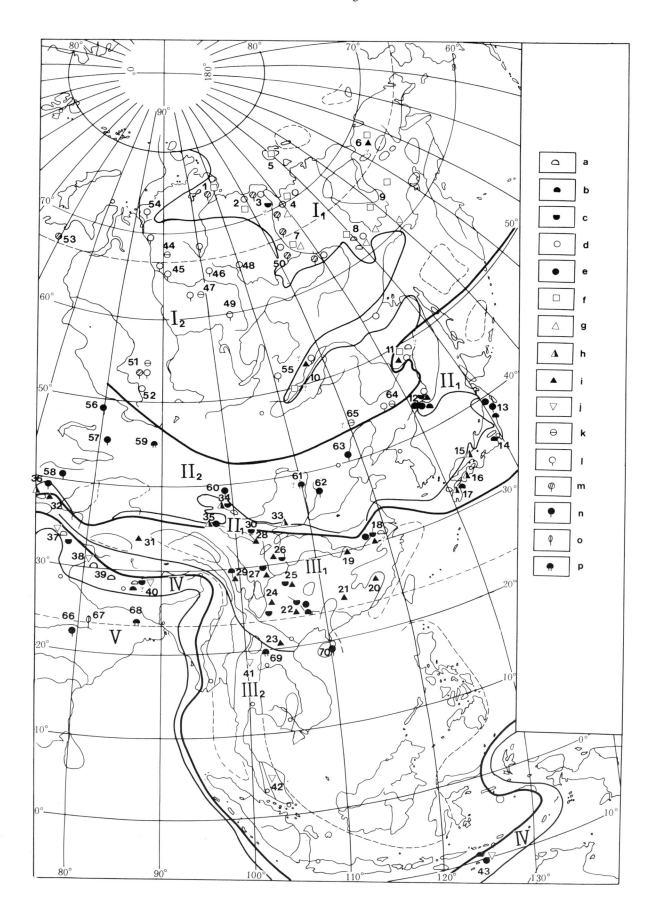

tistriata, radiata, mansuyi), Anisian longidaonellids, and the Ladinian *Daonella lommeli–D. indica* assemblage. The fauna shows high degrees of diversity and endemism. For example, 372 species from 84 genera of early and middle Triassic bivalves have been reported in South China alone, including approximately 40% endemic species (Yin, 1991). Trigoniids characteristic of the circum-Pacific are not found here. The middle Triassic plants were similar to contemporaneous European floras, but different from those of the Gondwanan Realm and North Laurasian Realm.

Himalaya–Banda Loop Region of the Gondwanan Tethys Realm (IV)

This region is marked by a mixed tropical and temperate fauna. The ammonoids found in exotic blocks in Timor and Chitichun in the Himalayas have not been counted. Comparison of early Anisian ammonoids from Spiti, Shalshal, and Bambanag in the Himalayas with those from the Eurasian Tethys shows lower diversity in this region. Boreal *Keyserlingites* has been found in Lilang, Po, Shalshal, and Bambanag in the Himalayas and in Timor, whereas Eurasian elements like Danubitidae, *Paracrochordiceras,* and *Japonites* are uncommon here. Lhasa and Qomolangma Mountain are now close to each other, but palaeobiogeographically they belonged to Laurasian Tethys Realm and the Gondwanan Tethys Realm, respectively. Among 16 late Anisian ammonoid species described from Lhasa, 15 are identical with species found in the Alps, whereas only 4 are identical with species from the Himalayas, and none match those from Qomolangma (Gu, He, and Wang, 1980); the Ladinian ammonoids of Qomolangma are closely related to those from other areas of the Himalayas, but there is no species in common with Guizhou, although both areas are dominated by *Protrachyceras* (Wang and He, 1976). The same situation holds for brachiopods. The *Holcorhynchella, Tetractinella,* and *Puncto-*

spirella group of Pamir (Eurasian Tethys) and the *Spirigerellina, Coenothyris, Schwagerispira,* and *Piarorhynchella* group of the Himalayas and Kashmir (Gondwanan Tethys) are mutually exclusive, although the two areas are now close neighbours (Talent and Mawson, 1979). *Tulongospirifer stracheyi* is found in this region and in Japan and northeastern Asia, but not in the Eurasian Tethys, thus again showing bipolarity.

Indian Subcontinent Region of the Gondwanan Realm (V)

A kannemeyeriid *Wadiasaurus* has been reported from the Yerapalli Formation in Godavali Valley, together with *Rechnisaurus, Trirachodon,* and Prestosuchidae (Colbert, 1984), showing affinities with South Africa and South America. The *Dicroidium* flora from Pasora, regarded as early–middle Triassic (Dobrushkina, 1982) or late Triassic (Shah, Singh, and Sastry, 1971), denotes temperate arid to subarid weather.

Late Triassic biogeography of East Asia

The early Indosinian Orogeny, which occurred between the middle and late Triassic, caused the incorporation and uplift of many eastern Asian blocks and the formation of the Palaeo-Pacific Tectonic Realm. That, plus an expansion of the belt of hot, humid climate, greatly modified the late Triassic biogeography of East Asia. The main changes were that central Asia and northern Asia were incorporated into one terrestrial biogeographic unit, areas of the circum-Pacific formed one marine biogeographic unit, and the Cathaysian Tethys and Gondwanan Tethys were unified into the tropical Tethys. Consequently, the late Triassic biogeography of East Asia was tripartite, consisting of Palaeo-Asia, the Circum-Pacific Realm, and the Tethys (Figure 19.3).

Figure 19.1 (opposite). Biogeographic map of East Asia in the early Triassic: I$_1$, Boreal Sea Region (northeastern Asia) of the North Laurasian Realm; I$_2$, North Eurasian Region of the North Laurasian Realm; II$_1$, Marginal Seas Region of the Central Laurasian Realm; II$_2$, Central Eurasian Region of the Central Laurasian Realm; III$_1$, Cathaysian Tethys Region of the Laurasian Tethys Realm; III$_2$, Southern Margin Region of the Laurasian Tethys Realm; IV, Himalaya–Banda Loop Region of the Gondwana Tethys Realm; V, Indian Subcontinent Region of the Gondwanan Realm; 1, eastern Taymir; 2, Olenek River mouth; 3, lower Lena River; 4, Kular area of Jana River; 5, Novosibirski Island; 6, Small Aniuyi River, Chukchi (Chukote); 7, western Verchojan; 8, Lekeer River, eastern Verchojan; 9, Omolon; 10, Onon River of Transbaikalia and Ondorchaan; 11, lower Amur River; 12, South Primorye; 13, southern Kitakami; 14, Kwanto; 15, Maizuru; 16, Kochi, in Shikoku; 17, southern Kyushu; 18, lower Yangtze; 19, eastern Hubei; 20, Fujian; 21, Hunan and Guangdong; 22, south Guizhou and north Guangxi; 23, northern Vietnam and northern Laos; 24, western Guizhou and eastern Yunnan; 25, central and northern Guizhou; 26, Sichuan Basin; 27, Longmenshan and Emeishan; 28, Songpan and Zoige; 29, western Sichuan and northern Tibet; 30, southern Qinling; 31, Shuanghu, in northern Tibet; 32, central and southeastern Pamir; 33, southern margin of Ordos; 34, southern Qilian; 35, Burhan Buda; 36, Darwaz; 37, Spiti; 38, Kumaon and southern Ngari; 39, Nepal; 40, Qomolangma Mt.; 41, southern Thailand; 42, Gua Musang, Malaya; 43, Nifoekoko, Timor; 44, Kurei, northern Lechen; 45, Tutonchan River; 46, Korvonchan River; 47, lower Tungusk River; 48, upper Vilui River; 49, upper Ilimpen River; 50, Vilui Basin; 51, Kuznetsk; 52, Rudnayi Altay; 53, Pechora Basin; 54, Obi River mouth; 55, Selenga-Baikal area; 56, Seme'itau Mts.; 57, Junggar Alatau; 58, Fergana; 59, Jimsar, southern margin of Junggar; 60, northern Qilian; 61, Wubao and Fugu, Shanxi; 62, Jiaochen, Pingyao, and Yushe, in Shanxi; 63, Chende, Hebei; 64, Jiutai, Jilin; 65, Taohaiyinzi, Inner Mongolia; 66, southern Rewa; 67, Nidpur; 68, Asansol (Panchet Formation); 69, Luang Prabang; 70, Qionghai, Hainan; a, *Otoceras;* b, *Owenites;* c, *Tirolites* assemblage; d, *Keyserlingites* assemblage; e, *Columbites* group; f, *Posidonia (mimer, aranea);* g, *Atomodesma* assemblage; h, middle-latitude *Eumorphotis* assemblage; i, tropical *Claraia-Eumorphotis* assemblage; j, Gondwanan *Claraia* assemblage; k, *Palaeanodonta* assemblage; l, Korvonchan flora; m, *Pseudoaraucarites* flora; n, *Pleuromeia-Voltzia* flora; o, late *Glossopteris* flora; p, *Lystrosaurus-Chasmatosaurus* assemblage.

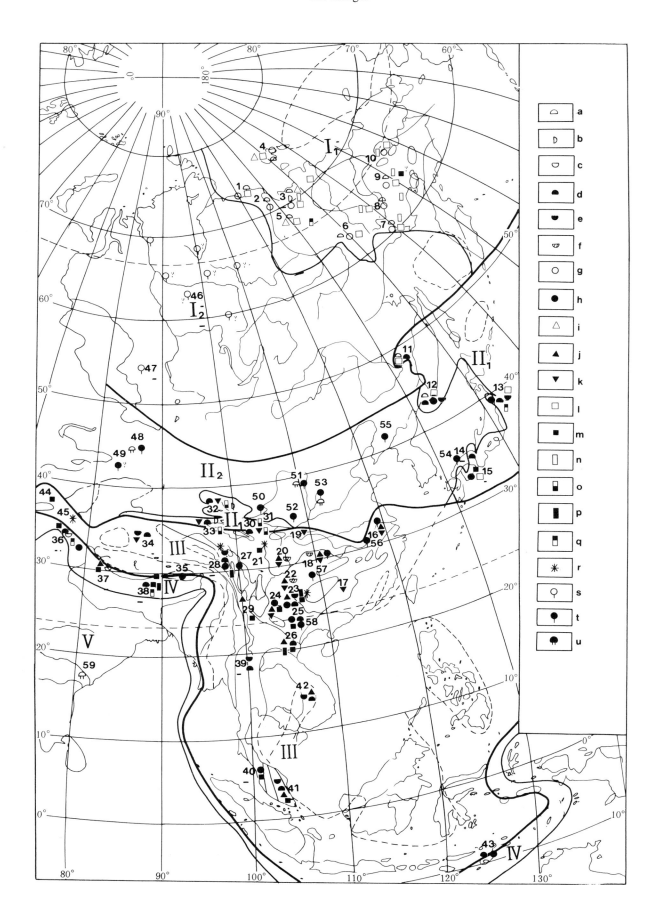

Circum-Pacific Realm (I)

This realm became distinct after the late Triassic and was characterized by entomonotiids, trigoniids, and a few other endemic genera. In East Asia, or the northwestern section of the Circum-Pacific Realm, typical groups were the bivalves *Entomonotis* and *Eomonotis*, the *Tosapecten-Oxytoma-Otapiria* assemblage (also *Chlamys mojsisovics, Palaeopharus, Ochotomya,* and *Bureimya*), the *Halobia aotii* assemblage, and certain genera and species of brachiopods (Spiriferinidae) and ammonoids (Sirenitidae). Norian ammonoids were quite few, as compared with the abundant Tethyan fauna (Bychkov et al., 1976).

The northwestern section of the Circum-Pacific Realm can be subdivided into the Northeast Asian Region, the Japan–South Primorye Region, and the Southeast China Region. The Northeast Asian fauna bears the typical features mentioned earlier and basically shows no intermingling of tropical elements. In eastern Taymir, eastern Verkhoyan, and the northeastern depression of Mongolia they coexist with or neighbour with temperate floras (Dobrushkina, 1987), thus representing a temperate marine fauna.

The Japan–South Primorye Region is peculiar for its transitional biota (Kiparisova, 1972; Hayami, 1975): The marine fauna combined abundant temperate taxa such as entomonotiids and certain halobiid groups (18 species in Japan and 4 in Primorye), together with tropical corals, megalodontids, and ammonoids (Tamura, 1983); the land was dominated by the rain-forest *Dictyophyllum-Clathropteris* flora. Palaeogeographically, this region corresponded to the southern transitional belt of the Asian continent and belonged to a warm-temperate–subtropical belt.

In the Southeast China Region, the land flora was represented by the *Dictyophyllum-Clathropteris* assemblage, whereas the marine fauna belonged to the Circum-Pacific Realm. This region yields plentiful endemic estuarine–lagoon bivalves (*Jiangxiella, Guangdongella, Nanglingella*), gastropods (*Fascivalvata, Peri-*

carinata), ostracods (*Oncocythere, Minicythere*), and estheriids (*Anyuanestheria*) (Ichang Institute, 1977; Chen, 1983). Ammonoids and entomonotiids have not yet been reported. In many localities the *Dictyophyllum-Clathropteris* flora is mixed with taxa from the *Danaeopsis-Symopteris* flora. In general, the biota of this region belonged to a tropical–subtropical belt. If only the palaeoclimatic situation were considered, this region and the Japan–South Primorye Region could be incorporated into the Tropical Tethys Realm. However, there were land-sea barriers between them, for after the early Indosinian the Palaeo-Pacific was separated from the Tethys along a western Hunan–western Guangdong–Indochina line, and that caused the disappearance of the Tethyan ammonoids and the *Indopecten* and *Burmesia* assemblages in these regions. Citing the marine biota as the main basis for provincialization, these two regions are here attributed to the Circum-Pacific Realm.

Laurasian Realm (II)

The central and northern parts of Laurasia of the early and middle Triassic merged into one realm during the late Triassic, symbolized by the *Danaeopsis-Symopteris* flora (Wu, 1983). Also, the ostracod *Darwinula-Lutkevichinella-Tongchuania* assemblage may serve as another symbol, because most of its components were distributed within this realm from western Europe to eastern Asia (Geological Institute, 1980). Its late Triassic floras have been divided into eastern Asian, central Asian, and Europe sections (Dobrushkina, 1982, 1987), and thus far the important fresh-water bivalve *Shaanxiconcha-Sibireconcha* assemblage also appears to have been restricted to eastern and central Asia. Hence the Asian sections – Palaeo-Asia – can be distinguished from the European part. Palaeo-Asia can be further subdivided into an Asian Region (Asia proper) and its Southern Margin Mixed Region. Biotas of the Asian Region displayed typical aspects of Palaeo-Asia and were representative of a temperate belt.

Figure 19.2 (opposite). Biogeographic map of East Asia in the middle Triassic: I$_1$, Boreal Sea Region (northeastern Asia) of the North Laurasian Realm; I$_2$, North Eurasian Region of the North Laurasian Realm; II$_1$, Marginal Seas Region of the Central Laurasian Realm; II$_2$, Central Eurasian Region of the Central Laurasian Realm; III, Cathaysian Tethys Region of the Laurasian Tethys Realm; IV, Himalaya–Banda Loop Region of the Gondwanan Tethys Realm; V, Indian Subcontinent Region of the Gondwanan Realm; 1, Olenek River mouth; 2, lower Lena River; 3, Kular area of Jana River; 4, Novosibirski Island; 5, northern Bakay; 6, Lekeer River, eastern Verchojan; 7, Jana-Okhotsk River, southern Bakay; 8, Omulev area; 9, Omolon Massif; 10, upper Great Aniuyi River; 11, lower Amur River; 12, South Primorye; 13, southern Kitakami; 14, Maizuru; 15, Shikoku; 16, lower Yangtze; 17, Hunan and Jiangxi; 18, western Hubei; 19, Zhen'an, eastern Qinling; 20, Sichuan Basin; 21, Songpan and Zoige; 22, southern Sichuan and northern Guizhou; 23, Qingyan, central Guizhou; 24, western Guizhou and eastern Yunnan; 25, southern Guizhou and northern Guangxi; 26, northern Vietnam and northern Laos; 27, Yidun; 28, Changdu and northern Three Rivers area; 29, western Yunnan; 30, western Qinling; 31, Qing'an and Tongwei; 32, southern Qilian; 33, Burhan Buda; 34, northern Tibet; 35, Lhasa; 36, Spiti; 37, Kumaon and southern Ngari; 38, Qomolangma Mt.; 39, southern Thailand; 40, Songkhla, Taiping; 41, Pahang and Singapore; 42, western Cambodia; 43, Timor; 44, southeastern Pamir; 45, Heweitan, Karakorum; 46, Tunguss; 47, Kuznetsk (Sosnov Formation); 48, Jimsar-Manas, southern Junggar; 49, Kuqa; 50, Baiying, Gansu (Dingjiayao Formation); 51, northern Ordos (Fugu, Lishi, Xinxian, Shenmu); 52, Tongchuan, southern Ordos; 53, central Shanxi (Ermaying Formation); 54, Atsu, Chikoku Belt; 55, Benxi, Liaoning; 56, Huaining, Anhui; 57, Sangzhi, Hunan; 58, Luodian and Bianyang, Southern Guizhou; 59, Godavali Valley (Yerapalli Formation); a, *Popanoceras* group and *Arctohungarites* group; b, Groenlanditinae; c, *Keyserlingites*; d, *Japonites, Balatonites*; e, *Paraceratites*; f, *Progonoceratites*; g, *Nathorstites*; h, *Protrachyceras*; i, *Janopecten*; j, *Costatoria goldfussi* group; k, *Leptochondria illyrica* assemblage; l, *Daonella dubia* assemblage; m, *D. lomelli* and *D. indica*; n, *Pennospiriferina* assemblage; o, *Pseudospiriferina-Paranptychina* assemblage; p, Qingyan-type brachiopods; q, *Tulongospirifer stracheyi*; r, hexacorals, diploporids; s, late Korvonchan flora; t, *Annalepis, Pleuromeia*; u, *Wadiasaurus*.

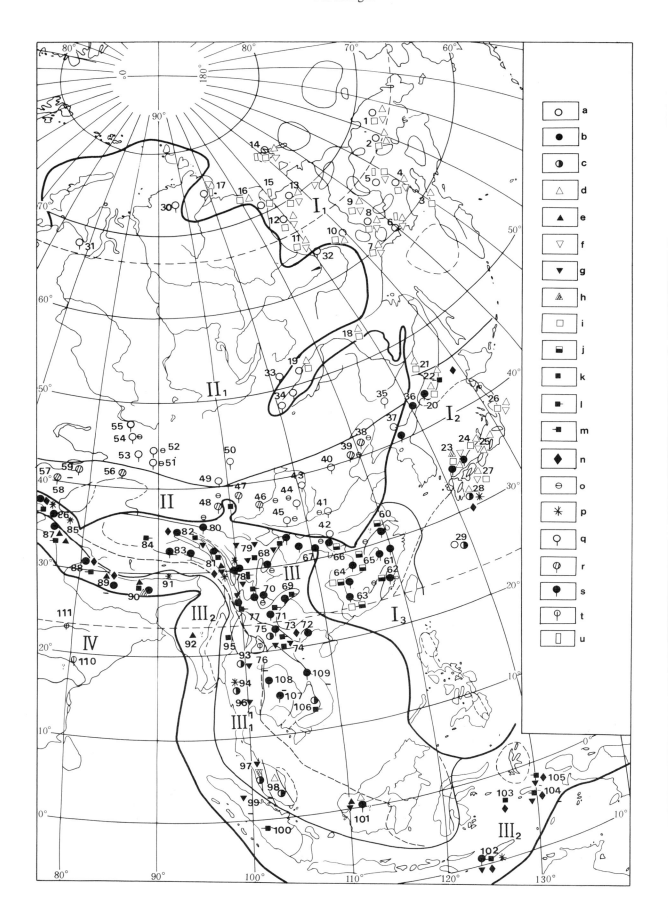

The Southern Margin Mixed Region, consisting of North China, Tarim, and the southern part of Central Asia, shows the following features, which are different from those of Asia proper: (1) There was infiltration of *Dictyophyllum-Clathropteris* components. *Dictyophyllum, Clathropteris,* and *Hausmannia* have been reported from Linyuan, Caoyang, and Beipiao in western Liaoning, Funing and Pingchuan in Hebei, Jingyuan and Wuwei in Gansu, Kuqa in Xinjiang, and Issak-kul and Fergana in Central Asia. The Yanchang flora of the Ordos Basin also yields taxa of that flora. (2) The estheriid *Euestheria minuta* assemblage of southern China penetrated into Jiyuan, Tongchuan, Hanchen, Linwu, and Jingyuan in this region, but not farther northward. (3) The bivalve *Shaanxiconcha-Sibireconcha* assemblage originated from this region and expanded southward into southern China, probably through western Henan and northwestern Hubei, showing connection with the Tethys. (4) A tropical Tethyan *Burmesia-Yunnanophorus* faunule was reported at Datong, in the southern Qilian Mountains, one of the transgressed areas of this region. Its eastern neighbour, the Japan–South Primorye Region of the Circum-Pacific Realm, is also marked by mixed temperate and tropical biotas. Therefore, there was a mixed belt, both on land and at sea. To sum up, the southern margin should be separated as an independent Southern Margin Mixed Region, representing a warm-temperate climatic belt, slightly cooler than the Japan–South Primorye Region, which was strongly influenced by Palaeo-Pacific monsoons and warm currents.

Tropical Tethys Realm (III)

Compared with its extent in the early and middle Triassic, the scope of this realm was enlarged in the late Triassic to cover Cathaysian Tethys and what was formerly Gondwanan Tethys. Typical taxa include heteromorphic and other tropical–subtropical ammonoids (Juvavitidae, Thisbitidae, Trachyceratidae, Tropitidae, etc.), the *Halobia comata* assemblage (*austriaca, superba, talauna, yunnanensis, fallax, superbescens,* etc.), megalodontids, and colonial corals. This realm is subdivided into eastern and western sections roughly along Jordan and Transcaucasia. The eastern or Asian section is symbolized by tibetitid ammonoids, the *Indopecten* and *Burmesia* bivalve assemblages, and flora (Dobrushkina, 1982) and bryozoans (Hu, 1984) different from those of the western or Mediterranean section.

Gondwanan Tethys Region (III₂)

Within the eastern section of the Tethyan Realm, which had become tropical–subtropical as a whole, were the Gondwanan Tethys Region and the Cathaysian Tethys Region. The Gondwanan Tethys Region was adjacent to the temperate Gondwana continent, where *Dicroidium* flora flourished, and was latitudinally and climatically more subtropical than tropical, based on the following features: (1) Tropical biotas were not fully developed; heteromorphic ammonoids, ostreids and *Ostrea*-form bivalves,

---

Figure 19.3 (opposite). Biogeographic map of East Asia in the late Triassic: I_1, Northeast Asian Region; I_2, Japan–South Primorye Region; I_3, Southeast China Region; II, Laurasian Realm, Paleo-Asia; II_1, Asian (proper) Region; II_2, Southern Margin Mixed Region; III, Tropical Tethys Realm, eastern section; III_1, Cathaysian Tethys Region; III'_1, Cimmerides Subregion; III_2, Gondwanan Tethys Region, eastern part; IV, Indian Subcontinent Region, Gondwanan Realm; 1, Chukote (Chukchi); 2, Great Aniuyi River; 3, eastern Penzhen Bay, Koriak; 4, Konni-Taigono area; 5, Omolon Massif; 6, Jana-Okhotsk River; 7, Okhotsk; 8, southeastern Bakay; 9, Omulev; 10, eastern Verchojan; 11, central western Verchojan; 12, northwestern Bakay; 13, Kular area of Jana River; 14, Novosibirski Island; 15, Lena River mouth; 16, Olenek River mouth; 17, eastern Taymir; 18, western Preokhotsk; 19, southern Cita (Chita); 20, South Primorye; 21, Nadanhada; 22, Tetiukhe, coastal Sikhote-Alin; 23, Mine and Nariwa; 24, Maizuru (Nabae Formation); 25, Tambo and Joetsu; 26, southern Kitakami (Saragai Formation); 27, Shikoku (Kochigatani Formation); 28, southern Kyushu; 29, Okinawa; 30, Fajiu-Kuda area; 31, Pechora Basin; 32, Tumar River; 33, Serenga-Baikal area; 34, Dzargalantuin Gol, Dashibalbar, and southern Uldz River, northeastern Mongolia; 35, Shuangyang, Jilin; 36, Wangqing and Dongning, Jilin; 37, Xiaohekou, Hunjiang, Jilin; 38, Lingyuan, Chaoyang, and Beipiao, western Liaoning; 39, Pingchuan and Funing, Hebei; 40, Western Hills, Beijing; 41, Jiyuan, Yima, and Mianchi, northern Henan; 42, Lushi and Nanzhao, southwestern Henan; 43, Shenmu, northern Shanxi; 44, Yanchang, Yan'an, and Zichang, central Ordos; 45, Tongchuen and Hanchen, southern Ordos; 46, Jinyuan and Jintai, Gansu; 47, Wuwei, Gansu Corridor; 48, Muli, southern Qilian; 49, Beishan Mts.; 50, southern Mongolia; 51, Turpan; 52, Xiaochuangou, southern Junggar; 53, Manas River; 54, Kalamay; 55, Saur, Kazakhstan; 56, Kuqa, southern Tienshan; 57, Fergana; 58, Pamir; 59, Isaac-Kul; 60, lower Yangtze; 61, Loping and Quxian in Jiangxi; 62, Fujian; 63, Xiaoping, northern Guangdong; 64, southern Hunan; 65, Liling and Anyuan, Hunan–Jiangxi border; 66, Jingmeng-Dangyang Basin; 67, Xiangxi, western Hubei; 68, Longmenshan and Emeishan; 69, Guiyang and Anlong, central Guizhou; 70, Yongren and Yipinglang, central Yunnan; 71, southwestern Guizhou and eastern Yunnan (Huobachong); 72, Hongay, northern Vietnam; 73, Tonkin, northern Vietnam; 74, Ninh-binh and Song Da; 75, Xam Nua (Sam Neua) and Dienbienphu; 76, Phong Saly, Laos; 77, western Yunnan; 78, Yidun, western Sichuan; 79, Yajiang and Markam, western Sichuan; 80, Burhan Buda; 81, Qamdo, eastern Tibet; 82, Gyiza and Tumenggla, southeastern Qinghai; 83, Dingqing and Ando, Qinghai; 84, Shuanghu, northern Tibet; 85, Heweitan, Karakorum; 86, southeastern Pamir; 87, Spiti; 88, northern Kumaon and southern Ngari; 89, northeastern Kumaon, Nepal; 90, Qomolangma Mt. and Lhaze; 91, Lhasa; 92, Chin Mts.; 93, Napeng, Burma; 94, Nampang and Mae Sot, Thailand; 95, Kamawkala, Thai–Burma border; 96, Kanchanaburi, southwestern Thailand; 97, Songkhla, Taiping; 98, Kelantan, Singapore; 99, Toba Lake, northern Sumatra; 100, Padang, central Sumatra; 101, Kuching, western Sarawak; 102, Timor, Roti, and Sawu Is.; 103, Seran I.; 104, Buru I.; 105, Misol I.; 106, Ban Don and Nha Trang, southern Vietnam; 107, Korat Plateau; 108, northeastern Thailand (Huai Hin Lat Formation); 109, Kaomin, southern Vietnam; 110, Maleri; 111, Pasora; a, Boreal ammonoids (Sirenitidae); b, Tethyan ammonoids with Tibetitidae; c, Tethyan ammonoids, Tibetitidae not reported; d, *Entomonotis, Eomonotis;* e, *Monotis;* f, *Halobia aotii* assemblage; g, *H. comata* assemblage; h, *Minetrigonia;* i, *Tosapecten* assemblage; j, *Jiangxiella* assemblage; k, *Indopecten* assemblage and *Burmesia* assemblage; l, *Burmesia* assemblage; m, *Indopecten* assemblage; n, Megalodontidae; o, *Shaanxiconcha* assemblage; p, colonial corals; q, *Danaeopsis-Symopteris* flora; r, *Danaeopsi-Symopteris* flora, with *Dictyophyllum, Clathropteris,* etc.; s, *Dictyophyllum-Clathropteris* flora; t, *Dicroidium* flora; u, Boreal brachiopods (Spiriferinidae).

chaetitids, diploporid dasycladales, and sphinctozoans were uncommon. (2) The fresh-water *Shaanxiconcha-Sibireconcha* bivalve assemblage and the *Euestheria minuta* conchostracan assemblage, typical of Palaeo-Asia, did not extend to this region. (3) The brachiopods *Hagabirhynchia, Spiriferina abbichi,* and *Misolia* and the bivalves *Serania* and *Nucula misolensis* show endemism of this region. (4) The thriving state of the Tibetitidae and the *Indopecten* assemblage (*Palaeocardita burruca, Serania, Pergamidia*), as well as the relative rarity of the *Burmesia* assemblage (*Yunnanophorus, Costatoria napengensis*), reflects the open-sea aspect of this region, in contrast to the more or less restricted environments widespread in the Cathaysian Tethys.

Cathaysian Tethys Region (III₁)

This region featured a typically tropical, highly diversified, and strongly endemic marine biota (Yin, 1991). Among its taxa were hexacorals (65 species of 29 genera), megalodontids (20 species of 5 genera), heteromorphic ammonoids, ostreids and *Ostrea*-form bivalves, halorellid brachiopods, sphinctozoans, chaetitids, diploporids, and other forms. The land flora was dominated by the highly diversified *Dictyophyllum-Clathropteris* assemblage, representing a tropical rain forest. In southern China and Indochina the *Burmesia* assemblage flourished, denoting an epicontinental environment.

Cimmerides Subregion (III'₁)

The intermediate blocks, including Sibumasu, Kalimantan, and probably Lhasa, roughly corresponding to the Cimmerides (Sengör, 1984), show the following features that distinguish this subregion from the Cathaysian Tethys Region: (1) The *Shaanxiconcha- Sibireconcha* and *Euestheria minuta* assemblages did not reach these blocks. (2) Tibetitidae have not yet been reported. (3) A Gondwanan flora (*Glossopteris indica, Schizoneura gondwanensis, Pecopteris tonkinensis*) was reported in Phong Saly, Laos (Fontaine and Workman, 1978), whereas the *Dictyophyllum-Clathropteris* flora has also been reported from western Sarawak in Kalimantan. (4) It is of special interest that western Sarawak has yielded monotids typical of all three important biotas (i.e., *Monotis salinaria* of the Tethys, *Entomonotis ochotica* of the northwestern Circum-Pacific, and *E. subcircularis* of the southwestern and eastern Circum-Pacific) (Kobayashi and Tamura, 1983); the Japanese taxa *Halobia aotii* and *Minetrigonia* have also been reported in Malaya (Tamura et al., 1975). Points 3 and 4 seem to indicate that these blocks were at the junction of the tropical Tethys, Circum-Pacific, and Gondwana. This is in accordance with the tectonic viewpoint of a late Triassic collision of these blocks with Eurasia (Mitchell, 1981).

Indian Subcontinent Region of the Gondwanan Realm (IV)

Late Triassic Gondwanan vertebrates and plants have been found in the Maleri Formation (Colbert, 1984), and a *Dicroidium* flora

has been reported in the Pasora Formation (Bhalla, 1983). These denote typical Gondwanan provincialization.

Implications of the Triassic biogeography of East Asia for plate tectonics

1. A realm boundary reflects the double control of a temperature–latitude barrier and a continent–ocean barrier. The former is implicitly in the east–west direction. In East Asia during the early and middle Triassic, the latter was also mainly east–west (e.g., Central Asia–Mongolian Sea, Palaeo-Tethys, and Meso-Tethys). Therefore, during the early and middle Triassic, realm boundaries were in the east–west direction. The superficial curvature along the Yarlung Zangbo–Timor line apparently is due to a later northward push of the Indian subcontinent.

2. Palaeogeography and biogeography show that there was no oceanic barrier in East Asia east of 100°E. The Central Asia–Mongolian Sea had closed by the end of the Permian; the Palaeo-Tethys section east of the Qinling closed after the middle Triassic. In the process of a plate collision, leading to mountain building and leaving shallow seas, with biotic separations and alterations, there is always some time lag, and therefore biogeographic boundaries will last beyond plate incorporation. It would be oversimplified to demonstrate the existence of land-ocean barriers between plates only by means of biogeographic boundaries. In such cases, special attention should be paid to whether or not there existed biotas reflecting an intermediate climatic belt. Because north–south climatic changes are gradual, a climatic saltation between the northern and southern sides of a biogeographic boundary implies the subduction of a large area – probably a vast ocean. Instead, the existence of transitional climatic belts, such as a warm-temperate belt in between temperate and tropical belts, indicates the nonexistence of any vast land-ocean barriers. Palaeomagnetic and tectonic specialists (e.g., Sengör, 1984; Lin, Fuller, and Zhang, 1985) have suggested an oceanic barrier between North China and the Yangtze Platform in the Triassic that was closed during the Indosinian Orogeny, but that is discordant with the transitional climatic situation in the Qinling area. Detailed research has shown that there was a deep sea yielding radiolarians and even a few nannofossils in western Qinling, but no ocean (Yin et al., 1992). The Qinling Sea may have opened westward to a palaeo-ocean west of 100°E, just as the Wharton Sea opened westward into the Indian Ocean, although that is not certain. On the other hand, along the Ganze-Litang zone and the Changning-Menglian zone of the Three Rivers Region there were small oceans, as revealed by ophiolites and radiolarites (Liu, Feng, and Fang, 1991). The area between the Yarlung Zangbo–Banda Loop line and the Kunlun–Qinling line probably was a vast Triassic water basin, with archipelagos, displaying a pattern of interspersed subcontinents and small oceans or seas, as is the case in modern Southeast Asia.

3. The Yarlung Zangbo–Banda Loop line existed through the entire Triassic as a realm boundary, representing a Meso-Tethys barrier, which is in accordance with palaeomagnetic and ophiolite records. Thus far, the best-developed radiolarites of Triassic

East Asia have also been distributed along that line: southeastern Pamir (nearly 40 species, mainly Carnian), the Himalayas (Gyirong Formation, 15 species), Sumatra (Tuhur Formation of the Padang Group), Timor, and adjacent islands. During the early and middle Triassic, south of that line there existed a temperate sea along the northern margin of Gondwana – the Gondwanan Tethys (IV in Figures 19.1 and 19.2) – which was added to the Tropical Tethys Realm (III in Figure 19.3) in the late Triassic owing to southward expansions of tropical–subtropical biotas into the Gondwanan Tethys. The causes of those expansions may have been a geographic expansion of hot weather, because simultaneously there was an expansion of warmer weather into the southern margin of Laurasia (II in Figure 19.3); alternatively, the cause may have been northward migration of the Indian Plate, with its northern margin reaching the subtropical belt. East (probably north during the Permian and Triassic) of the Yarlung Zangbo–Banda Loop line, Fang (1991) recognized a transitional Permian Sibumasu biotic province. Such a transitional biotic province also existed in the early Triassic (Southern Margin Region) and late Triassic (Sibumasu and Kalimantan). These findings indicate that all of the aforementioned blocks did not fully merge with southern China and Indochina during the Triassic. That assumption is strengthened by the findings of ophiolites and radiolarites along the Ganze-Litang and Changning- Menglian zones.

4. The late Triassic circum-Pacific provincialization indicates the existence of the Palaeo-Pacific Tectonic Realm. The Mesozoic Neocathaysian pattern of uplifted and depressed relief in a north-northeast direction and the influences of monsoons and currents in a northeast direction (comparable to the modern Kuro and Ashio currents), all evidently generated and controlled by that tectonic pattern, combined to form a nearly longitudinally distributed northwestern circum-Pacific biogeographic unit. Farther east, outside of that unit, along the Tetiukhe (Dalnegorsk) zone in Sikhote-Alin and the Japanese Sambosan zone, from Kyushu to Okinawa, Triassic oceanic ophiolites and radiolarites developed, comparable to those in ocean basins outside of the modern northwestern Pacific island arcs (Kimura, Hayami, and Yoshida, 1991; Golozubov et al., 1992).

References

Audley-Charles, M. G. 1988. Evolution of the southern margin of Tethys (North Australian regime) from early Permian to late Cretaceous. *Geological Society, Special Publications* **37**:80–100.
Bhalla, S. N. 1983. India. In *The Phanerozoic Geology of the World II, The Mesozoic, B,* ed. M. Moullade and A. E. M. Nairn, pp. 305–50. Amsterdam: Elsevier.
Budurov, K. J. 1976. Die triassischen Conodontendes Ostbalkans. *Geologica Balcanica* **6**:95–104.
Budurov, K. J., Gupta, V. J., Sudar, M. N., and Buryi, G. I. 1985. Conodont zonation, biofacies and provinces in the Triassic. *Journal of the Geological Society of India* **26**:84–94.
Buryi, G. I. 1989. *Triassic Conodonts and Stratigraphy of Sikhote-Alin.* Vladivostok: Academy of Science, USSR, Far Eastern Branch, Far East Geological Institute.

Bychkov, Y. M., Dagys, A. S., Efimova, A. F., and Polubotko, I. V. 1976. *Atlas of Triassic Fauna and Flora of Northeastern USSR* (in Russian). Moscow: Nedra.
Chen Jinhua. 1983. Late Triassic and early Jurassic bivalve assemblages and palaeogeography of southeastern China, with a bivalve-biogeographic division in China. In *Palaeobiogeographic Provinces of China,* ed. Palaeontological Society of China, pp. 100–20. Beijing: Science Press.
Colbert, E. H. 1984. Mesozoic reptiles, India and Gondwanaland. *Indian Journal of Earth Sciences* **11**:25–37.
Dagys, A. S., Arkhipov, Y. V., and Bychkov, Y. M. 1979. *Stratigraphy of the Triassic System of Northeastern Asia* (in Russian). No. 447. Moskva: Trudy IGG SO, Akademia Nauk, USSR.
Dang Vu Khuc, Dagys, A. S., and Kiparisova, L. D. 1965. *Les fossiles caracteristiques du Trias au Nord Viet-Nam* (in Vietnamese). Hanoi: Directoire Generale de Geologie de la Vietnam.
Dobrushkina, I. A. 1982. *Triassic Flora of Eurasia* (in Russian). No. 365. Moskva: Trudy Geologiskii Instituta Akademia Nauk, USSR.
Dobrushkina, I. A. 1987. Phytogeography of Eurasia during the early Triassic. *Palaeogeography, Palaeoclimatology, Palaeoecology* **58**:75–86.
Duan Jinying. 1987. Permian–Triassic conodonts from southern Jiangsu and adjacent areas, with index of their colour alteration. *Acta Micropaleontologica Sinica* **4**:351–68.
Fang Zhonjie. 1991. Sibumasu biotic province and its position in Paleo-Tethys. *Acta Palaeontologica Sinica* **30**:511–32.
Fontaine, H., and Workman, D. R. 1978. Review of the geology and mineral resources of Kampuchea, Laos and Vietnam. In *Proceedings of the 3rd Regional Conference on Geology and Mineral Resources of Southeast Asia,* ed. P. Nutalaya, pp. 541–603. Bangkok.
Geological Institute, Chinese Academy of Geosciences. 1980. *Mesozoic Stratigraphy and Paleontology of the Shanxi-Gansu-Ningxia Basin.* Beijing: Geological Publishing House.
Golozubov, V. V., Khanchuk, A. I., Kemkin, I. V., Panchenko, I. V., and Simanenko, V. P. 1992. *Taukha and Zhuravlevka Terranes, South Sikhote-Alin.* Moscow: Russian Academy of Sciences, Far East Branch.
Gu Qingge, He Guoxiong, and Wang Yigang. 1980. Discovery of the late Anisian *Paraceratites trinodosus* fauna (Ammonoidae) from Doilungdeqen, Tibet, and its significance. *Acta Palaeontologica Sinica* **11**:343–56.
Hayami, I. 1975. *A Systematic Survey of the Mesozoic Bivalvia from Japan.* Bulletin 10. University Museum, University of Tokyo.
Hou Liwei, Luo Daixi, Fu Deming, Hu Shihua, and Li Kaiyuan. 1991. *Triassic Sedimentary-Tectonic Evolution in Western Sichuan and Eastern Xizang Region.* Geological Memoirs, ser. 3, no. 13. Ministry GMR of China.
Hu Zhaoxun. 1984. Triassic Bryozoa from Xizang (Tibet) with reference to their biogeographical provincialism in the world. *Acta Palaeontologica Sinica* **23**:568–77.
Ichang Institute of Geology and Mineral Resources, Chinese Academy of Geosciences. 1977. *Palaeontological Atlas of Central-Southern China,* vol. 3. Beijing: Geological Publishing House.
Igo, H., and Koike, T. 1975. Triassic conodonts from West Malaysia. In *Triassic System of Malaysia, Thailand and Some Adjacent Areas,* ed. M. Tamura, pp. 135–42. *Geology and Palaeontology of Southeast Asia,* vol. **15**. University of Tokyo Press.

Kimura, T., Hayami, I., and Yoshida, S. 1991. *Geology of Japan.* Tokyo: University of Tokyo Press.

Kiparisova, L. D. 1972. *Paleontologicheskoe obosnovanie stratigraphii triasovaikh otlozhenii Primorskovo Kraya. Ch. 2. Pozdnetriasovaie dvustvopchataie molliuski i obschaya stratigraphia* [Paleontological Characteristics of the Stratigraphy of Triassic Deposits of Primoria Border. Part 2. Late Triassic Bivalve Molluscs and General Stratigraphy]. Moscow: Nedra.

Kiparisova, L. D., Radchenko, G. P., and Gorskii, V. P. 1973. *Stratigraphy of the Triassic System of USSR,* vol. 7. Moscow: Nedra.

Kobayashi, T., and Tamura, M. 1983. The Arcto-Pacific realm and Trigoniidae in the Triassic period. *Proceedings of Japan Academy, ser. B* 59:207–10.

Kozur, H. 1980. The main events in Upper Permian and Triassic conodont evolution and their bearing on the Permo-Triassic stratigraphy. *Rivista Italiana di Paleontologia e Stratigrafia* 85:741–66.

Krystyn, L. 1973. Probleme der biostratigraphischen Gliederung der Alpen-Mediterranean Obertrias. In *Die Stratigraphie der Alpine – Mediterranen Trias,* ed. H. Zapfe, pp. 137–44. Schriftenreich der Erdwissenschaftlichen Komitee 2. Wien: Oesterreiche Akademie der Wissenschaft.

Lai Xulong. 1992. Triassic marine faunas and fossil zonations: conodonts. In *The Triassic of Qinling Mountains and Neighboring Areas,* ed. Yin Hongfu, Yang Fengqing, Huang Qisheng, Yang Hengshu, and Lai Xulong, pp. 66–8. Wuhan: China University of Geosciences Press.

Lin Jinlu, Fuller, M., and Zhang Wenyou. 1985. Preliminary Phanerozoic polar wander paths for the North and South China blocks. *Nature* 313:444–9.

Liu Benpei, Feng Qinglai, and Fang Nianqiao. 1991. Tectonic evolution of the Paleo-Tethys in Changning-Menglian Belt and adjacent regions, western Yunnan. *Journal of China University of Geosciences* (*English issue*) 2:18–28.

McTavish, R. A. 1973. Triassic conodont faunas from Western Australia. *Neues Jahrbuch für Paläontologie und Geologie, Abhandlungen* 143:275–303.

Matsuda, T. 1981. Early Triassic conodonts from Kashmir, India. Pt. 1: *Hindeodus* and *Isarcicella*. *Journal of Geosciences, Osaka City University* 24:75–108.

Matsuda, T. 1985. Late Permian to early Triassic conodont paleobiogeography in the Tethys realm. In *The Tethys, Her Paleogeography and Paleobiogeography from Paleozoic to Mesozoic,* ed. K. Nakazawa and J. M. Dickins, pp. 57–70. Tokai University Press.

Metcalfe, I. 1988. Origin and assembly of south-east Asian continental terranes. In *Gondwana and Tethys,* ed. M. G. Audley-Charles and A. Hallam, pp. 101–18. Geological Society, Special Publication 37.

Mitchell, A. H. G. 1981. Phanerozoic plate boundaries in mainland SE Asia, the Himalayas and Tibet. *Journal of the Geological Society* 138:109–22.

Oleinikov, A. N., and Paevskaya, E. B. 1978. Stratigraphy of the Upper Triassic deposits of Primorye region (in Russian). *Soviet Geology* 1978(2):31–47.

Pakistani-Japanese Research Group. 1981. *Stratigraphy and Correlation of the Marine Permian–Lower Triassic in the Surghar Range and Salt Range, Pakistan.* Kyoto: Kyoto University Press.

Rostovtsev, K. O., and Zhamoida, A. I. 1984. *A Standard for Triassic Time in the USSR.* St. Petersberg: VSEGEI.

Sengör, A. M. C. 1984. *The Cimmeride Orogenic System and Tectonics of Eurasia.* Special paper 195. Boulder: Geological Society of America.

Shah, S. C., Singh, G., and Sastry, M. V. A. 1971. Biostratigraphic classification of Indian Gondwana. In *Proceedings of the Symposium on the Gondwana System, Aligarh Muslim,* pp. 306–26. Aligarh: Muslim University.

Sweet, W. C., and Bergström, S. M. 1986. Conodonts and biostratigraphic correlation. *Annual Review of Earth and Planetary Sciences* 14:85–112.

Sweet, W. C., Mosher, L. C., Clark, D. L., Collinson, J. H., and Hasenmueller, W. A. 1971. Conodont biostratigraphy. *Geological Society of America, Memoirs* 127:411–65.

Talent, J. A., and Mawson, R. 1979. Palaeozoic–Mesozoic biostratigraphy of Pakistan in relation to biogeography and the coalescense of Asia. In *Geodynamics of Pakistan,* ed. A. Forab and K. A. DeJong, pp. 81–102. Geological Survey of Pakistan.

Tamura, M. 1968. *Claraia* form north Malaya, with a note on the distribution of *Claraia* in SE Asia. *Geology and Palaeontology of Southeast Asia,* 5:78–87.

Tamura, M. 1983. Megalodonts and megalodont limestones in Japan. *Memoirs of the Faculty of Education, Kumamoto University* 32:7–28.

Tamura, M., Hashimoto, W., Igo, H., Ishibashi, T., Iwai, J., Kobayashi, T., Koike, T., Pitakraivan, K., Sato, T., and Yin, E. H. 1975. The Triassic system of Malaysia, Thailand and some adjacent areas. *Geology and Palaeontology of Southeast Asia* 41:103–49.

Tozer, E. T. 1974. Definitions and limits of Triassic stages: suggestions prompted by comparisons between North America and the Alpine-Mediterranean regions. In *Die Stratigraphie der Alpine–Mediterranean Trias,* ed. H. Zapfe, pp. 195–206. Schriftenreich der Erdwissenschaftlichen Komitee 2. Wien: Oesterreiche Akademie der Wissenschaft.

Tozer, E. T. 1981. Triassic Ammonoidea. In *The Ammonoidea,* pp. 66–100, 394–431. Systematics Association special volume 18. London: Academic Press.

Wang Chenyuan and Wang Zhihao. 1976. Triassic conodonts in the Qomolangma Peak region. In *A Report of the Scientific Expedition in the Qomolangma Peak Region, Paleontology,* fas. 3:387–425. Beijing: Science Press.

Wang Yigang. 1985. Remarks on the Scythian–Anisian boundary. *Rivista Italiana di Paleontologia e Stratigrafia* 90:515–44.

Wang Yigang, Chen Chuchen, He Guoxiong, and Chen Jinhua. 1981. *An Outline of the Marine Triassic in China.* Publication 7. International Union of Geological Sciences.

Wang Yigang and He Guoxiong. 1976. Triassic ammonites in the Qomolangma Peak region. In *A Report of the Scientific Expedition in the Qomolangma Peak Region, Paleontology,* Fas. 3, pp. 223–502. Beijing: Science Press.

Wang Zhihao and Cao Yanyu. 1981. Early Triassic conodonts from Lichuan, western Hubei. *Acta Palaeontologica Sinica* 20:363–75.

Wu Shunqing. 1983. On the Late Triassic, Lower and Middle Jurassic floras and phytogeographic provinces of China. In *Palaeobiogeographic Provinces of China,* ed. Palaeontological Society of China, pp. 121–30. Beijing: Science Press.

Yang Shouren, Wang Xinping, and Huo Weichen. 1986. Early and middle conodont sequence of western Guangxi. *Acta Scientiarum Naturalium Universitatis Pekinensis* 4:90–102.

Yang Zunyi, Li Zishun, and Qu Lifan. 1982. The Triassic system of China. *Acta Geologica Sinica* 20:1–21.

Yang Zunyi, Yin Hongfu, Xu Guirong, Wu Shunbao, He Yuanliang, Liu Guancai, and Yin Jiarun. 1983. *Triassic of* the South Qilian Mountains. Beijing: Geological Publishing House.

Yin Hongfu (ed.) 1988. *Paleobiogeography of China.* Wuhan: China University of Geosciences Press.

Yin Hongfu. 1990. Palaeogeographical distribution and stratigraphical range of the Lower Triassic *Claraia* and *Eumorphotis* (Bivalvia). *Journal of China University of Geosciences* (*English issue*) **1**:98–110.

Yin Hongfu. 1991. Triassic paleobiogeography of China. In *Shallow Tethys 3,* ed. T. Kotaka et al., pp. 403–21. Special publication 3. Sendai: Saito Ho-on Kai.

Yin Hongfu, Yang Fengqing, Huang Qisheng, Yang Hengshu, and Lai Xulong. 1992. *The Triassic of Qinling Mountains and Neighboring Areas.* Wuhan: China University of Geosciences Press.

Yin Hongfu, Yang Fengqing, Lai Xuelong, and Yang Hengshu. 1988. Triassic belts and Indosinian development of the Qinling Mountains. *Geoscience, Journal of the Graduate School, China University of Geosciences* **2**:364–72.

Zakharov, E. T. 1973. The importance of the paleobiogeographical data for the solution of the problem on the Lower Triassic division. In *Die Stratigraphie der Alpine–Mediterranen Trias,* ed. F. Zapfe, pp. 101–20. Schriftenreich der Erdwissenschaftlichen Komitee 2. Wien: Oesterreiche Akademie der Wissenschaft.

Zakharov Y. D. 1978. *Early Triassic Ammonoidea of the Far East of USSR.* Moscow: Nauka.

Zapfe, H. 1983. *Das Forschungsprojekt "Triassic of the Tethys Realm" (IGCP Project 4) (Abschlussbericht).* Schriftenreich Erdwissenschaftlichen Komitee 5. Wien: Oesterreiche Akademie der Wissenschaft.

Zhou Tongshun and Zhou Huiqing. 1983. Triassic non-marine strata and flora of China. *Bulletin of the Chinese Academy of Geological Sciences* **5**:95–107.

20 Classification and correlation of Triassic limestones in Sikhote-Alin on the basis of corals

TATIANA A. PUNINA

Triassic biogenic limestones in Sikhote-Alin are not ubiquitous, but are found only in the central part of the Dalnegorsk region (Rudnaya River basin), where they compose individual massifs within a Lower Cretaceous olistostrome. The massifs range from 6–8 m to 3,500 m across. In the Lower Cretaceous terrigenous matrix there are smaller fragments of limestone (side by side with bodies of cherty composition). In the Dalnegorsk region there are, in all, about 10 limestone massifs. Their characteristic limestones show that they were formed in middle and late Triassic time. Recently, I proposed a stratigraphic scheme on the basis of the corals in the Dalnegorsk region. The following beds have been distinguished: (1) *Coryphyllia moisseevi* beds (Ladinian–Lower Carnian), (2) *Volzeia badiotica* beds (Upper Carnian), (3) *Margarosmilia charlyana* beds (Lower Norian), (4) *Gablonzeria kiparisovae* beds (Middle Norian), (5) *Meandrostylis tener* beds (Upper Norian), and (6) *Retiophyllia buonamici* beds (Rhaetian) (T. A. Punina, unpublished data).

As a type for the lowest subdivision, I offer the section in trench no. 1558 (southwest slope of Bolnichnaya Mountain). Other beds with corals are recognized in a single section on Sakharnaya Mountain, which is proposed as a type for those subdivisions.

The characteristics of the main limestone massifs in the Dalnegorsk region are described in the following sections. This should help to correlate the Middle and Upper Triassic strata of the region and to compare the Triassic coral assemblages of Sikhote-Alin with those of the circum-Pacific region.

Besides myself, B. Briner, G. I. Buryi, I. V. Burij, T. G. Ilyina, L. N. Khetchikov, L. D. Kiparisova, Z. I. Myasnikova, V. P. Parnjakov, G. P. Volarovich, and N. K. Zharnikova have sampled the fossiliferous limestones in the Dalnegorsk region at various times.

Triassic limestone massifs of the Dalnegorsk region

Sakharnaya Mountain

The limestone massif of Sakharnaya Mountain is on the Rudnaya River right bank 1 km from its mouth, near the entrance to Dalnegorsk from Kavalerovo (Figure 20.1). The massif, composing the upper part of the mountain, is 1,500 m × 700 m × 600 m.

In descending order, the sequence of Upper Triassic beds in this section is as follows:

Retiophyllia buonamici beds (Rhaetian)

10. Grey framework limestone, replaced by organic-slime and organic-detrital varieties (facies of reef core) (15 m): In the limestone, calcareous algae, hydroids, and corals [*Retiophyllia buonamici* (Stoppani), *R. cyathophylloides* (Frech), *Toechastraea* sp., *Heterastraea profunda* (Reuss), *Pamiroseris rectilamellosa* (Winkler)] are abundant.

9. Light-grey and grey framework limestone, often oolitic (20 m): Contains calcareous algae, foraminiferans, sponges, hydroids, and corals [*Astraeomorpha confusa* (Winkler), *Retiophyllia wanneri* (Vinassa de Regny), and *R. norica* (Frech)].

8. Light-grey, strongly recrystallized limestone, with caverns filled by calcite (bioherm facies) (26 m): Organic remains are represented by calcareous algae, hydroids, sponges, sphinctozoans, and corals [*Pamiroseris* sp., *Retiophyllia minima* (Melnikova), *R. clathrata* (Emmrich), *Primorodendron* sp.].

Meandrostylis tener beds (Upper Norian)

7. Light-grey framework limestone (bioherm facies) (80 m): Yields calcareous algae, sponges, sphinctozoans, hydroids, and corals [*Meandrostylis tener* (Punina), *Pamiroseris rectilamellosa* (Winkler), *Astraeomorpha crassiseptata* (Reuss), and *Palaeastraea alnigmata* (Punina)].

Gablonzeria kiparisovae beds (Middle Norian)

6. Light-grey and grey framework limestone, locally oolite and clay (biostrome facies) (180–185 m): Contains calcareous algae, hydroids, sponges, and corals [*Gablonzeria kiparisovae* Punina, *G. singularis* (Punina), *Margarosmilia multigranulata* (Melnikova), *Distichomeandra tenuiseptata* Punina].

5. Light-grey framework limestone (biostrome facies) (220 m): Yields calcareous algae, sponges, sphinctozoans (*Sollasia* sp.), hydroids (*Bauneia* sp., *Stro-*

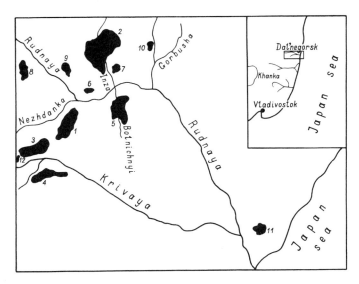

Figure 20.1. Locations of the main Triassic carbonate massifs in Sikhote-Alin: 1, Sakharnaya (Sakharnaya Golova) Mountain; 2, Verkhnij Rudnik; 3, Bolnichnaya Mountain; 4, Kamennye Vorota; 5, Partizansky Massif; 6, Podstantsiya Massif; 7, Vokzalnaya Mountain; 8, Gorelaya Mountain; 9, Pimorsky Massif; 10, Karyernaya Mountain; 11, Trench no. 1558.

matomorpha sp.), corals [*Margarosmilia charlyana* (Frech), *M. melnikovae* (Punina), *Gablonzeria reussi* Cuif, *Protoheterastraea konosenis* (Kanmera)] (Punina, 1990), bivalves (*Chlamys* cf. *hinnitiformis* Gemmellaro and Di Blasi, *Pteria* cf. *tofaniea* Bittner) (Burij and Zharnikova, 1980, 1981), and conodonts [*Epigondolella abneptis* (Huckriede), *E, postera* (Kozur and Mostler)] (Buryi, 1989).

Volzeia badiotica beds (Upper Carnian)

4. Grey, dense, framework limestone (100 m): Yields abundant calcareous algae (*Solenopora* and *Diplopora*), hydroids, sponges (*Molengraffia* sp.), and corals [*Protoheterastraea* sp., *Volzeia subdichotoma* (Muenster), and *Pachysolenia primorica* (Ilyina)].

3. Grey and dark-grey limestone (12 m): Contains the remains of small bivalves, gastropods, and, more rarely, calcareous algae, hydroids, and sphinctozoans.

2. Dark-grey pelitoidal limestone, yielding bivalves (Megalodontidae) and gastropods (6 m).

1. Dark-grey pelitoidal limestone occurring on basalts (3 m). The total thickness of the limestones is 670 m.

Verkhnij Rudnik

The Verkhnij Rudnik massif is located along the middle course of the Inza River (Rudnaya River left bank). The massif is broken into several blocks, dislocated relative each other by 10–20 m. The massif is 3,500 m long, 500–1,000 m wide and more than 600 m high. The section is represented by the following deposits (in descending order):

Retiophyllia buonamici beds (Rhaetian)

12. Dark limestone with lenses of black limestone (reef-core facies) (5 m): Yields calcareous algae, foraminiferans, hydroids (*Blastochaetetes*, *Circopora*, *Stromatomorpha*), and corals [*Retiophyllia buonamici* (Stoppani), *R. cyathophylloides* (Frech), *Heterastraea profunda* (Reuss)].

11. Dark-grey framework limestones, with interbeds of pelitoidal and oolitic varieties (lagoon facies) (10 m): Contains calcareous algae, foraminiferans, sponges, and corals [*Retiophyllia buonamici* (Stoppani), *R. norica* (Frech), and *Blastodendron* sp.].

Meandrostylis tener beds (Upper Norian)

10. Light-grey framework limestone (reef-core facies) (100 m): Yields calcareous algae (*Solenopora*, Dasicladaceae), foraminiferans, hydroids (*Blastochaetetes*, *Balatonia*, *Stromatomorpha*), sponges (*Molengraffia*), sphinctozoans, corals [*Gablonzeria singularis* (Punina), *Pamiroseris rectilamellosa* (Winkler), *Astraeomorpha crassiseptata* Reuss, *Palaeastraea alnigmata* (Punina), *Retiophyllia minima* (Melnikova), *R. buonamici* (Stoppani), *Meandrostylis* sp.], bivalves (including isolated Megalodontidae), and gastropods.

9. Light-grey oolite-oncholite limestone (lagoon facies) (50 m): Contains calcareous algae, foraminiferans, crinoids, rare sponges, and corals [*Stylophyllopsis bortepensis* (Melnikova)].

8. Light-grey biogenic-detrital limestone, yielding abundant echinoid spicules, coral fragments, and other organisms (slope facies) (2 m).

7. Light-grey clay limestone (8 m).

Gablonzeria kiparisovae beds (Middle Norian)

6. Light-grey framework limestone (bioherm facies) (100 m): Yields calcareous algae (*Solenopora*), hydroids (*Bauneia*, *Blastochaetetes*), sponges, sphinctozoans (*Celyphia*, *Parauvanella*, *Colospongia*) (Boiko, Belyaeva, and Zhuravleva, 1991), and corals (*Gablonzeria reussi* Cuif, *G. kiparisovae* Punina, *Distichomeandra primorica* Punina, *D. tenuiseptata* Punina, *Margarosmilia melnikovae* Punina, *M. multigranulata* Melnikova) (Pezhenina and Punina, 1989).

Margarosmilia charlyana beds (Lower Norian)

5. Light-grey framework limestone, locally with distinct algal hummocky bedding (biostrome facies) (100 m): Contains abundant remains of foraminiferans, sponges, hydroids, sphinctozoans (*Celyphia*) (Boiko et al., 1991), corals [*Volzeia badiotica* (Volz), *Volzeia* sp., *Craspedophyllia* sp., *Margarosmilia cultus* Melnikova, *Distichomeandra attenuata* Punina], and bivalves [*Parallelodon curioni* (Bittner)] (Burij, Zharnikova, and Buryi, 1986).

Volzeia badiotica beds (Upper Carnian)

4. Light-grey framework limestone (biostrome facies) (40 m): Yields abundant remains of calcareous algae, sponges, sphinctozoans (*Sollasia* and *Uvanella*) (Boiko et al., 1991), and poorly branched corals (*Volzeia badio-*

tica Volz, *Volzeia* sp., *Pachysolenia primorica* Iljina) (Punina, 1990). Sporadic accumulations of preserved megalodontid valves are found (in 0.5-m interbeds).

3. Grey pelitoidal limestone (50 m): Contains large Megalodontidae, *Parallelodon* cf. *dieneri* Boehm, *P.* cf. *imbricarius* Bittner (Burij et al., 1986), and conodonts (*Metapolygnathus vialovi* Buryi, *Ancyrogondolella triangularis* Budurov) (Buryi, 1989).

2. Marbled limestone without traces of organisms (30 m).

1. Dark-grey, dense, pelitoidal limestone occurring on basalts (120 m): Yields foraminiferans, bivalves (*Parallelodon curonii* Bittner, *P.* cf. *imbricarius* Bittner) (Burij et al., 1986), gastropods, and conodonts (*Hindeolella triassica* Müller, *Hindeodella* sp.).

The total thickness of this section is 600–610 m.

Bolnichnaya Mountain

The limestone massif of Bolnichnaya Mountain is located 3.5 km to the southwest of Sakharnaya Mountain. Its height is about 300 m. length about 4 km, and width 500 m. In the section, the following deposits are observed, in descending order:

Gablonzeria kiparisovae beds (Middle Norian)

3. Light-grey massive limestone (150 m): Yields foraminiferans, bryozoans, sponges, hydroids, corals [*Retiophyllia fenestrata* (Reuss), *Distichomeandra* sp., *Gablonzeria krasnovi* Punina, *G. kiparisovae* Punina] (Punina, 1990), bivalves, gastropods, and conodonts (*Epigondolella multidentata* Mosher) (Buryi, 1989).

Margarosmilia charlyana beds (Lower Norian)

2. Light-grey framework limestone (120 m): Yields calcareous algae, foraminiferans, sponges, hydroids, corals [*Retiophyllia weberi* (Vinassa de Regny), *Gablonzeria krasnovi* Punina, *Distichomeandra* sp., *Margarosmilia charlyana* Frech, *M. melnikovae* Punina)], bivalves, and gastropods.

Carnian (with no corals)

1. Light-grey massive limestone (30 m): Characterized by foraminiferans, sponges, hydroids, bivalves [*Chlamys* cf. *hinnitiformis* Gemmellaro and Di Blasi, *Neoschisodus laevigatus* (Ziethen), *Palaeocardita crenata* Muenster] (Kiparisova, 1972), and gastropods.

The total thickness of the limestones in this section is 200 m.

Kamennye Vorota

The Kamennye Vorota massif, in the Nezhdanka River and Krivaya River interfluve, is 1,000 m long, about 200 m wide, and 806 m high.

The section is represented by the following deposits (in descending order):

?*Meandrostylis tener* beds (Upper Norian)

6. Light-grey massive limestone (bioherm facies) (350 m):

The lower part contains calcareous algae, sponges, hydroids, and corals [*Retiophyllia fenestrata* (Reuss), *R. weberi* (Vinassa de Regny), Ceriopheterastraea, and *Meandrostylis* sp.].

Gablonzeria kiparisovae beds (Middle Norian)

5. Light-grey massive limestone (bioherm facies) (150 m): Yields calcareous algae, foraminiferans, sponges, hydroids, corals [*Retiophyllia wanneri* (Vinassa de Regny), *Margarosmilia cultus* (Melnikova), *Stylophyllopsis* sp., *Gablonzeria kiparisovae* Punina (Punina, 1989), *Gablonzeria* sp.], bivalves (*Halobia* cf. *austriaca* Mojsisovics) (Burij et al., 1986), and conodonts (*Epigondolella* cf. *multidentata* Mosher) (Buryi, 1989).

Margarosmilia charlyana beds (Lower Norian)

4. Grey and dark-grey massive limestone (biostrome facies) (120 m): Contains calcareous algae, foraminiferans, sponges, hydroids, corals [*Margarosmilia charlyana* (Frech), *M. melnikovae* Punina, *M. multigranulata* (Melnikova), *Distichomeandra* sp., *Palaeastraea* sp.] (Punina, 1990), bivalves (*Chlamys* cf. *subliliemensis* Kiparisova) (Burij and Zharnikova, 1984), gastropods, and conodonts [*Metapolygnathus primitia* (Mosher) and *M. vialovi* Buryi] (Buryi, 1989).

Volzeia badiotica beds (Upper Carnian)

3. Grey and light-grey massive limestone (biostrome facies) (60 m): Yields calcareous algae, foraminiferans, corals (*Protoheterastraea konosenis* Kanmera, *Ceriopheterastraea* sp., *Volzeia badiotica* Volz) (Punina, 1990), bivalves (*Chlamys* sp.) (Burij and Zharnikova, 1984), and gastropods.

2. Dark-grey and black pelitoidal limestone (lagoon facies) (20 m): Yields calcareous algae, sponges, small corals (*Volzeia* sp., *Margarophyllia* sp.), bivalves, and gastropods.

1. Black calcareous marl occurring on basalts (7–10 m): Contains bivalves (*Prospondylus* cf. *crassus* Broili, *Newaagia kinzuchensis* Kiparisova) (Burij and Zharnikova, 1984), and conodonts (*Metapolygnathus vialovi* Buryi, *Ancyrogondolella triangularis* Budurov) (Buryi, 1989).

The total thickness of the limestones and marl is 710 m.

Partizansky Massif

The Partizansky Massif is located on the Rudnaya River right bank in the center of the town of Dalnegorsk. The massif is separated into two large blocks by Bolnichnyi Creek. The north block (right bank of Bolnichnyi Creek), 160 m × 800 m × 1,000 m in size, is upside down. The section is represented by the following deposits (in descending order):

Meandrostylis tener beds (Upper Norian)

6. Light-grey limestone, with lenses of black limestone (65 m): Yields calcareous algae, hydroids, and corals [*Meandrostylis tener* Punina, *Palaeastraea alnigmata*

Punina, *Retiophyllia norica* (Frech), *Primorodendron asher* Punina].

?Middle Norian (beds with corals have not been recognized)

5. Grey pelitoidal, partially oolitic limestone (lagoon facies) (30 m): Contains calcareous algae, sponges, hydroids, and corals (*Retiophyllia* sp., *Volzeia* sp.).
4. Grey and dark-grey pelitoidal limestone, with rare remains of reef-builders (algae, sponges) (40 m).
3. Dark-grey, dense limestone, with lenses of black, thin-bedded limestone (13 m): Yields Megalodontidae (small accumulations) and gastropods.
2. Dark-grey, dense limestone, yielding the remains of foraminiferans, bivalves (Megalodontidae), and gastropods (15 m).
1. Dark-grey, dense limestone, yielding foraminiferans and gastropods (15 m).

The total thickness of limestones in the section of the north block is 200 m.

The south block of limestones (left bank of Bolnichnyi Creek) is 100 m × 600 m × 800 m in size. It is partially wooded. The section is represented by the following deposits (in descending order):

?Middle Norian (beds with corals have not been recognized)

5. Grey, coarse-bedded, detrital-biogenic limestone, locally conglomerate-like (20 m).
4. Coarse-bedded limestone of different tinges (4 m).
3. Light- and dark-grey, massive, biogenic and biogenic-detrital limestone (30 m): Contains foraminiferans, corals (*Retiophyllia* sp., *Stylophyllopsis* sp.), abundant bivalves (Megalodontidae), and gastropods.

Gablonzeria kiparisovae beds (Middle Norian)

2. Light- and dark-grey massive limestone (80 m): Yields calcareous algae, sponges, hydroids, and corals (*Distichomeandra primorica* Punina, *Distichomeandra* sp., *Gablonzeria* sp., *Volzeia* sp., *Retiophyllia* sp.).

Meandrostylis tener beds (Upper Norian)

1. Dark-grey dense limestone (6 m): Contains calcareous algae, sponges, hydroids, and corals [*Retiophyllia norica* (Frech), *Primorodendron* sp., *Meandrostylis* sp.].

The total thickness of limestones in the section of the south block is 140 m.

Massif in the region of Podstantsiya

A limestone outcrop (35 m × >50 m × 120 m) is located on the left bank of the Rudnaya River, 1.5 km to the north of Sakharnaya Mountain. The limestone is light grey and has been metamorphosed (about 50 m thick). On the weathered surface one can see distinct dendroid colonies of Late Carnian corals (*Volzeia* sp.), algae, sponges, and bivalves. At the base of the limestone body,

conglomerate breccias that contain basalts and limestones, with rare remains of gastropods, are found. The limestone is tentatively considered comparable to the *Volzeia badiotica* beds of Sakharnaya Mountain.

Vokzalnaya Mountain

Vokzalnaya Mountain is on the bank of the Rudnaya River (Partizanskaya Pad), 1.2 km to the south of the Verkhnij Rudnik massif. Limestone outcrops in the upper part of the mountain are 50 m × 70 m × 100 m in size. The section is represented by the following deposits (in descending order):

***Retiophyllia badiotica* beds (Rhaetian)**

5. Light-grey marbled limestone (3 m).
4. Light-grey framework limestone (15 m): Yields calcareous algae (*Solenopora* and Dasycladaceae), sponges, hydroids (*Bauneia* and *Stromatomorpha*), and corals [*Heterastraea profunda* (Reuss), *Margarosmilia multigranulata* Melnikova, *Retiophyllia norica* (Frech), *Astraeomorpha* sp.].

***Gablonzeria kiparisovae* beds (Upper Norian)**

3. Grey framework limestone, with lenses of dark-grey limestone (25 m): Contains calcareous algae, foraminiferans, sponges, and corals [*Gablonzeria singularis* (Punina), *Gablonzeria* sp., *Margarosmilia* sp., *Retiophyllia* sp.), bivalves, and gastropods.
2. Grey and dark-grey limestones, yielding megalodontid bivalves (3 m).
1. Grey and dark-grey pelitoidal limestone, yielding calcareous algae, sponges, hydroids, large megalodontid bivalves, and gastropods (26 m).

The total thickness of the limestones is more than 70 m.

Gorelaya Mountain

The limestone massif of Gorelaya Mountain is located on the Rudnaya River right bank, 4 km to the west of Sakharnaya Mountain. The height of the massif is about 100 m, and it is 150 m long. Limestone crops out on the northwest slope of the mountain. The following deposits represent the section (in descending order):

?Norian (with no corals)

3. Light-grey, fine-grained limestone, yielding foraminiferans, bryozoans, sponges, coral (*Retiophyllia* sp.), and bivalves. Bryozoans described as *Buria improvisa* Morozova and Zharnikova (Morozova and Zharnikova, 1984) appear to have been reported from this part of the section.
2. Light-grey oolitic limestone (3 m).
1. Light-grey, recrystallized, framework limestones (60 m): Yields calcareous algae, foraminiferans, sponges, and poorly preserved corals (*Retiophyllia* sp., *Gablonzeria* sp., *Margarosmilia* sp.).

The total thickness of the limestones in the section is more than 100 m.

Primorsky Massif

The limestone massif is located on the Rudnaya River left bank, 6 km to northwest of Sakharnaya Mountain. Its size is 50 m × >50 m × 75 m. Reliably established corals (*Heterastraea* sp.) of Late Norian age have been found only in the upper part of the massif, represented by light-grey recrystallized limestone. Other areas of the massif are composed of grey and light-grey recrystallized limestone that contains the remains of hydroids, corals, and poorly preserved mollusks. In the east part of the massif, the limestone contains an admixture of clay material, and in the west, interbeds of oolitic varieties. The total thickness is about 50 m.

Nikolaevskaya Mountain

The massif of Nikolaevskaya Mountain is located in the northern part of the region on the Rudnaya River left bank, 3 km to the north of the Verkhnij Rudnik massif. The massif is dome-shaped and distinct in relief; its height is 150 m, length 120 m, and width 40 m.

The Nikolaevsky massif represents a bioherm of late Carnian–early Norian age, composed of sponges, bryozoans, hydroids, and various reef-lovers: foraminiferans, echinoids, and mollusks. A specific feature is the absence of corals, sphinctozoans, and megalodontid bivalves. In contrast, bryozoans are more abundant than in other massifs. Bryozoans are rock-forming organisms that live at depths of several tens of meters. The Nikolaevsky massif may have been formed at a greater depth than other massifs in the Dalnegorsk region. The total thickness of the Nikolaevsky massif limestones is more than 150 m.

Karyernaya Mountain

The Karyernaya limestone massif is on the Rudnaya River left bank, 25 km to the east of Dalnegorsk (Monomakhovo village). The massif is 150 m high and makes up the upper part of the mountain. The section is represented by the following sequence (in descending order):

Upper Carnian–Middle Norian (with no corals)

3. Dark-grey pelitoidal limestone that, in its central part, is oolitic-oncholitic (30 m): Yields foraminiferans (*Trocholina* sp., *Nodosaria* sp.) (A. P. Nikitina's determination), bivalves [*Parallelodon imbricarius* (Bitner)], megalodontids (Burij et al., 1986), and gastropods.

2. Grey, micritic, dolomitized limestone, not distinctly bedded, yielding foraminiferans, megalodontids, and gastropods (40 m).

1. Dark-grey micritic limestone with megalodontids (15 m).

The total thickness of the limestones is about 95 m.

Trench no. 1558

Trench no. 1558 was made along the southwest slope of Bolnichnaya Mountain, 450 m from its top. In the trench, dark-grey, almost black, marls (*Coryphyllia tenuiseptata* beds) 8 m thick were recovered. The trench is interesting because the rocks contain late Ladinian–early Carnian coral fossils that are not found in the limestones of the massifs described earlier. The findings include the corals *Margarophyllia* cf. *capitata* (Muenster), *M. inculata* Deng and Kong, *Coryphyllia tenuiseptata* (Melnikova), and *C. moisseevi* Punina and Melnikova (Punina, 1990), the bivalves *Nucula* cf. *goldfussi* (Alberti), *Leda* cf. *polaris* Kiparisova, and *Parallelodon* cf. *beyrichi* Stromberg (Zharnikova, 1984), and gastropods.

Discussion

In the opinion of some researchers (Baud, 1993; Boni et al., 1993), following the Permian–Triassic crisis the communities of the Tethyan carbonate platforms became capable of producing abundant carbonate material only in the middle Triassic (late Anisian). Data from Sikhote-Alin support that idea. Regarding Middle Triassic carbonate deposits, which are very limited here, only Ladinian marl has been identified. Most of the Triassic carbonate massif in Sikhote-Alin is composed of Carnian–Norian limestones. Beds with corals, recognized in the type section at Sakharnaya Mountain, can be traced in many massifs of the Dalnegorsk region. A certain trend is seen in the evolution of biogenic construction: from isolated banks (Ladinian–early Carnian), through biostromes (late Carnian–early Norian) and bioherms (middle Norian), to typical reefs (late Norian–Rhaetian), which have been distinguished in the massifs of Sakharnaya and Verkhnij Rudnik, for instance. It seems that the main Triassic carbonate massifs in Sikhote-Alin originally represented separated biogenic bodies. Some of the data cited earlier, and the fact that the coeval carbonate deposits of the massifs differ in facies and thickness, testify to that supposition. The best-preserved biogenic constructions are the massifs of Sakharnaya and Verkhnij Rudnik, where the main recognized subdivisions are observed and the main reef elements (core, slope, and lagoon) are established. Small massifs (Partizansky, Primorsky, Vokzalnaya, and Gorelaya) represent the individual fragments of primary biogenic construction. Rarely, in some massifs, portions of the basement rocks (basaltoids) have been preserved.

Correlation of the carbonate Triassic of the circum-Pacific and adjacent areas on the basis of corals

The late Ladinian and early Carnian coral complexes in Sikhote-Alin, like those in other regions of the world, are not diverse, which makes it difficult to correlate the enclosing deposits. In the circum-Pacific, the most typical genera of those stages are *Coryphyllia*, *Margarophyllia*, *Volzeia*, and *Graspedophyllia*. The

Coryphyllia moisseevi beds (Ladinian–Lower Carnian) can be correlated with the upper part of the Ohse Formation, Kyushu Island (Kanmera and Furukawa, 1964), the lower part of the limestone of the Yangliujing Formation, Guizhou province, South China (Deng and Kong, 1984), and the Augusta Mountain Formation, Nevada (Stanley, 1986). In addition to material from the Upper Carnian–Middle Norian (*Volzeia badiotica* beds, *Margarosmilia charlyana* beds, and *Gablonzeria kiparisovae* beds) in Sikhote-Alin, I was able to study a collection of corals from the Verkhnenutekinskian member (limestone, tuff-siltstone, and sandstone) of the Nutevingenkyveem River basin, Kankaren Ridge, Koryak Upland, that was given to me by E. V. Krasnov (sampling by Y. M. Bytchkov). Hermotypic colonial corals were associated there with megalodontid bivalves and early–middle Norian ammonoids of Tethyan type (Melnikova and Bytchkov, 1986). Some genera (*Retiophyllia, Astraeomorpha, Pamiroseris, Palaestraea*) occurring in the limestones of Kankaren Ridge are common in Norian complexes of Sikhote-Alin. At the same time, representatives of *Stuorezia, Thamnotropis, Rhaetiastraea*, and *Benekastraea* were found there, which are typical for the Upper Carnian and Lower Norian of South China (but they have not been found in Sikhote-Alin). A significant part of the Verkhnenutekinskian member at Kankaren Ridge containing coral remains apparently was accumulated during the early–middle Norian boundary time. So I propose to correlate the *Margarosmilia charlyana* beds (Lower Norian) and *Gablonzeria kiparisovae* beds (Middle Norian) with the Verkhnenutekinskian member. The *Meandrostylis tener* beds and *Retiophyllia buonamici* beds (late Norian–Rhaetian) in Sikhote-Alin appear to correspond to the Koguchi Formation, Kyushu Island (Okuda and Yamagiwa, 1978), to limestones in Yunnan province, South China (Xia and Liao, 1986), to the upper part of the Martin Bridge Formation, Idaho, and to the upper part of the Kamishek Series, Alaska (Stanley, 1989).

References

Baud, A. 1993. Tethys just after the end of the Permian mass extinction (abstract). In *Carboniferous to Jurassic Pangea. Program and Abstracts,* p. 15. Calgary: Canadian Society of Petroleum Geologists.

Boiko, E. V., Belyaeva, G. V., and Zhuravleva, I. T. 1991. *Sphinctozoa Phanerozoya Territorii SSSR [Phanerozoic Sphinctozoa of the USSR Territory]*. Moskva: Nauka.

Boni, M., Jannace, A., Torre, M., and Zamparelli, V. 1993. A Ladino-Carnian reefal facies assemblage in southern Italy: its significance in the western Tethys paleogeography (abstract). In *Carboniferous to Jurassic Pangea. Program and Abstracts,* p. 39. Calgary: Canadian Society of Petroleum Geologists.

Burij, I. V., and Zharnikova, N. K. 1980. Flora-bearing layers of the Ladinian Stage (Middle Triassic) in South Primorye (in Russian). *Bulleten Moskovskogo Obstchestva, Ispytatelei Prirody, Otdel Geologicheskii* **55**:45–53.

Burij, I. V., and Zharnikova, N. K. 1981. The age of carbonate sequences of the Tetyucha Suite in Dalnegorsk region (in Russian). *Sovetskaya Geologiya* **1981**(3):75–80.

Burij, I. V., and Zharnikova, N. K. 1984. The stratigraphy of

Triassic deposits of the Pribrezhnaya Zone, Sikhote-Alin area (in Russian). In *Novye Dannye po Geologii Dal'negorskogo Rudnogo Raiona [New Data in Geology of Dalnegorsk Ore District]*, ed. E. V. Krasnov and G. I. Buryi, pp. 19–36. Vladivostok: Dal'nevostochnyi Nauchnyi Centr Akademii Nauk SSSR.

Burij, I. V., Zharnikova, N. K., and Buryi, G. I. 1986. Triassic deposits of the Nezhdanka River right bank (Primorye) (in Russian). *Sovetskaya Geologiya* **1986**(7):50–8.

Buryi, G. I. 1989. *Konodonty i Stratigraphiya Triasovykh Otlozhenii Sikhote-Alinya [Triassic Conodonts and Stratigraphy of Sikhote-Alin]*. Vladivostok: Dalnevostochnoye Otdeleniye Akademii Nauk SSSR.

Deng Zhan-qin and Kong Lei. 1984. Middle Triassic corals and sponges from southern Giizhou and Eastern Yunnan. *Acta Palaeontologica Sinica* **23**:489–502.

Kanmera, K., and Furukawa, H. 1964. Triassic coral faunas from the Konose Group in Kyushi. *Memoirs, Faculty of Science, Kyushu University, ser. D* **15**:117–47.

Kiparisova, L. D. 1972. *Paleontologicheskoye Obosnovaniye Stratigraphii Triasovykh Otlozhenii Primorskogo Kraya [Palaeontological Basis of Triassic Stratigraphy of Primorye region]*, vol. 2. Trudy Vsesoyuznogo Nauchno-Issledovatel'skogo Geologicheskogo Instituta (VSEGEI), n. ser. **181**. Leningrad: Nedra.

Melnikova, G. K., and Bytchkov, Y. M. 1986. The Upper Triassic Scleractinia of the Kankaren Ridge (the Koryak Upland) (in Russian). In *Korrelyatsiya Permo-Triasovykh Otlozhenii Vostoka SSSR [Correlation of Permo-Triassic Sediments of the East USSR]*, ed. Y. D. Zakharov and Y. I. Onoprienko, pp. 63–81. Vladivostok: Dal'nevostochnyi Nauchnyi Centr Akademii Nauk SSSR.

Morozova, I. P., and Zharnikova, N. K. 1984. On new Triassic bryozoans (in Russian). *Paleontologicheskii Zhurnal* **1984**(4):73–9.

Okuda, H., and Yamagiwa, N. 1978. Triassic corals from M. T. Daifugen, Nara Prefecture, Southwest Japan. *Transactions and Proceedings of the Palaeontological Society of Japan, n.s.* 110:297–305.

Pezhenina, L. A., and Punina, T. A. 1989. Carbonate constructions of the Dalnegorsk region (in Russian). *Litogenez i Rudoobrazovaniye v Drevnikh i Sovremennykh Morskikh Basseinakh Dal'nego Vostoka [Lithogenesis and Ore Formation in Ancient and Modern Sea Basins of the Far East]*, ed. O. V. Chudaev, pp. 99–111. Vladivostok: Dal'nevostochnoye Otdeleniye Akademii Nauk SSSR.

Punina, T. A. 1990. Ladinian–Carnian and Norian coral assemblage of Sikhote-Alin (in Russian). In *Novye Dannye po Biostratigraphii Paleozoya i Mezozoya Yuga Dal'nego Vostoka [New Data on Palaeozoic and Mesozoic Biostratigraphy of the South Far East]*, eds. Y. D. Zakharov, G. V. Belyaeva, and A. P. Nikitina, pp. 137–40. Vladivostok: Dal'nevostochnoye Otdeleniye Akademii Nauk SSSR.

Punina, T. A. 1992. Stages of organic buildup development in Triassic of Sikhote-Alin (abstract). In *Late Paleozoic and Early Mesozoic Circum-Pacific Biogeological Events, Abstract Book, International Field Conference on Permian–Triassic Biostratigraphy and Tectonics in Vladivostok*, ed. A. Baud, p. 22. Lausanne: Lausanne University (UNIL).

Stanley, E. D. 1986. Upper Triassic Dachstein-type, reef limestone from the Wallowa Mountains, Oregon: first reported occurrence in the United States. *Palaios* **1986**(1): 172–7.

Stanley, E. D. 1989. Triassic corals and spongiomorphs from Hell's Canyon, Wallowa terrane, Oregon. *Paleontology* **63**:800–19.

Xia Jin-bao and Liao Wei-hua. 1986. Some scleractinian corals of Procyclolitidae from Lhasa. *Acta Palaeontologica Sinica* **25**:46–53.

Zharnikova, N. K. 1984. Ladinian bivalves in the Dalnegorsk region (in Russian). *Novoe v Geologii Dal'negorskogo Rudnogo Raiona* [*News in Geology of the Dalnegorsk Ore Region*], ed. E. V. Krasnov and G. I. Buryi. pp. 48–70. Vladivostok: Dal'nevostochnoye Otdeleniye Akademii Nauk SSSR.

21 Evolution of the platform elements of the conodont genus *Metapolygnathus* and their distribution in the Upper Triassic of Sikhote-Alin

GALINA I. BURYI

The late Triassic was the time interval of greatest morphologic diversity for the conodont genera. Those short-ranging taxa, with distinctive shapes, often were represented by only a few species, and they have proved particularly useful, serving as excellent biostratigraphic keys (e.g., *Paragondolella, Metapolygnathus, Epigondolella,* and *Misikella*). The genus *Metapolygnathus* is regarded as characteristic of the Tuvalian, Lacian, and Alaunian. Its morphology has not been adequately investigated, and the species composition has been recognized only to a first approximation. The literature data on this problem are very contradictory (Hayashi, 1968; Kozur, 1972; Mosher, 1973; Budurov, 1977; Krystyn, 1980; Orchard, 1991).

I have recognized six species of *Metapolygnathus* in the Upper Triassic of Sikhote-Alin. This genus is defined in the paper by S. Hayashi's (1968) original diagnosis, aided by K. Budurov (1977). Those sources provided the basis for this chapter.

Phylogenetic relationships of the *Metapolygnathus* platform elements

The reconstructed organisms described as *Metapolygnathus* represent conodont elements that, along with the platform elements discussed here, contain such elements as cypridodelliform, hindeodelliform, chirodelliform, diplododelliform, prioniodelliform, and enantiognathiform structures (Buryi, 1985). *Metapolygnathus* was erected by Hayashi in 1968, with *M. communisti* as the type species, collected from cherts of the Adoyama Formation in the Ashio Mountains, central Japan. In the generic diagnosis, he emphasized how the genus *Metapolygnathus* differed in the mode of its lower surface from other similar genera (*Polygnathus, Gondolella,* and *Gladigondolella*) known at that time: "Lower surface bears stout, wide, flat and sharply edged keel, posterior margin always sharply squared and frequently fork off into branches" (Hayashi, 1968, p. 72).

Thus, Hayashi separated the platform elements with a bifurcated loop from *Polygnathus* (the same year that Mosher renamed the genus *Epigondolella*) and included them in his new genus *Metapolygnathus*. Later, Kozur (1972) assigned all middle and late Triassic platform elements, including the elements with a bifurcated loop (i.e., all elements having a platform and free

blade) to *Metapolygnathus* and thereby significantly extended the generic diagnosis. A year later, Mosher (1973) changed the *Metapolygnathus* diagnosis by the incorporation of *Paragondolella* platform elements in the genus. At the same time, he attributed the platform elements with a bifurcated loop to *Epigondolella*. Budurov (1976, 1977), who had looked more closely at the biostructures of the platform elements (structures of the platform edges, denticles of the carina, structures and forms of the basal field and pit) using scanning electron microscopy (SEM), has clearly justified the relationship of the basal-field biostructures to the evolution of the platform elements, which allowed their coherent classification. It was concluded that the following conodont genera existed in the late Triassic: *Metapolygnathus, Ancyrogondolella, Epigondolella,* and *Paragondolella*. Following Hayashi, Budurov demonstrated that *Metapolygnathus* is characterized by a bifurcated loop, and he supposed that the genus originated from *Paragondolella*. The elements of *Epigondolella* are similar in structure to the platform upper surface and differ in having a pointed elliptical loop. According to Budurov's classification, *Metapolygnathus* consists of *M. communisti* Hayashi, *M. linguiformis* Hayashi, *M. nodosus* (Hayashi), *M. permicus* (Hayashi), and *M. spatulatus* (Hayashi) (Budurov, 1977).

Recently, Krystyn (1980) and Orchard (1991) have confirmed the differences between *Metapolygnathus* and *Epigondolella*. It should be noted that Orchard recognized only the two genera *Metapolygnathus* (Carnian) and *Epigondolella* (Norian) in the Upper Triassic Kunga Group of the Queen Charlotte Islands, British Columbia.

I have found some species of *Metapolygnathus* [*M. primitia* (Mosher), *M. linguiformis* Hayashi, *M. nodosus* (Hayashi), *M. permicus* (Hayashi), and *M. spatulatus* (Hayashi)] in Upper Triassic cherts and biogenic limestones of Sikhote-Alin (Figures 21.1 and 21.2). In addition, in the Sikhote-Alin collection there are specimens similar in appearance to *Epigondolella abneptis* (Huckriede), but having a bifurcated loop. They are characterized by platforms that are rather wide and rounded or have rounded-off right angles on the upper surface, slightly narrowed at the lateral sides at one-third of the distance from the front end, and gradually widening to the back end. Large sharp denticles are developed on the front half of the platform. All

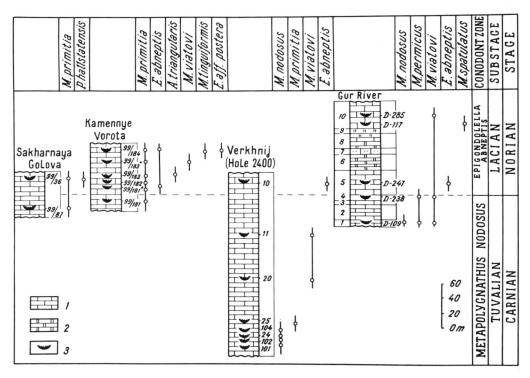

Figure 21.1. Stratigraphic distribution of platform elements of the genus *Metapolygnathus* in Upper Triassic carbonaceous sequences of Sikhote-Alin (Taucha and Khungary terranes); 1, limestone; 2, chert; 3, conodonts.

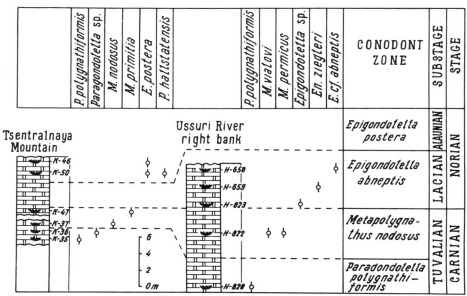

Figure 21.2. Stratigraphic distribution of *Metapolygnathus* platform elements in the Upper Triassic cherts of Sikhote-Alin (Taucha and Samarka terranes). Designations as in Figure 21.1.

Plate 21.1 (opposite). (1–3) *Metapolygnathus linguiformis* Hayashi, DVGI-1B/7-99/184, Primorye, Kamennye Vorota, Norian, *Epigondolella abneptis* Zone: 1, ×330, upper view; 2, ×1,130, lower view; 3, ×330, lower view. (4–7) *Metapolygnathus vialovi* Buryi: 4, holotype, DVGI-1B/15-99/183a, Primorye, Kamennye Vorota, Norian, *Epigondolella abneptis* Zone, ×180, upper view; 5, DVGI-1B/10-99/183a, same locality, ×190, upper view; 6 and 7, DVGI-1B/11-99/169, Primorye, Verkhnij, Carnian, *Metapolygnathus nodosus* Zone; 6, ×550; 7, ×1,300, lower view. (8, 9) *Metapolygnathus primitia* (Mosher), DVGI-2B/40-K-47, Primorye, Tsentralnaya Mountain quarry, Carnian,

Metapolygnathus nodosus Zone; 8, ×250; 9, ×350, lower view. (10–14) *Metapolygnathus spatulatus* (Hayashi): 10 and 11, DVGI-1B-21-K-585-1, Primorye, Katen River basin, Norian, *Epigondolella abneptis* Zone; 10, ×400; 11, ×240, upper view; 12–14, DVGI-1B-20, same locality; 12, ×180; 13, ×1,000, upper view; 14, ×440, lower view. (15–18) *Metapolygnathus nodosus* (Hayashi): 15, DVGI-1B-22-101, Primorye, Verkhnij, hole no. 2400, Carnian, *Metapolygnathus nodosus* Zone; 16, DVGI-1B-23-101, same locality, ×150, upper view; 17 and 18, DVGI-1B-14-D-109, Primorye, Gur River basin, Carnian, *Metapolygnathus nodosus* Zone; 17, ×200; 18, ×480, upper view.

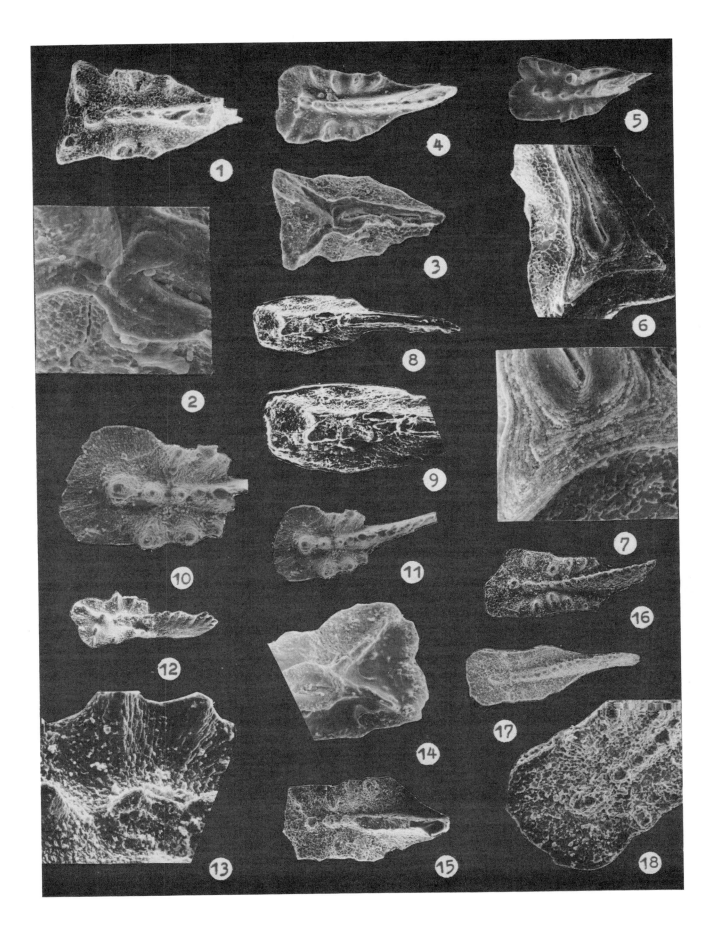

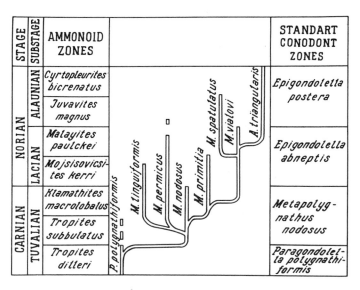

Figure 21.3. Phylogenetic lineage of the *Metapolygnathus* species

these specimens have the bifurcate loop on the lower surface and must be attributed to the genus *Metapolygnathus* [Plate 21.1 (4–7)]. At the same time, they differ from all known species of this genus: from *M. nodosus* (Hayashi) [Plate 21.1 (15–18)] in the widened back end and a much more bifurcated loop on the lower surface, from *M. spatulatus* (Hayashi) [Plate 21.1 (10–14)] in having a much longer platform. The fissures on the lateral sides of *M. permicus* (Hayashi) (Hayashi, 1968, pl. 2, figs. 3a–c) occur on the back end, not on the front one-third part of the more narrow platform. They do not look like *M. linguiformis* (Hayashi) [Plate 21.1 (1–3), as the platform of the latter is not wide, but gradually narrows from the back right-angled end to the front end, and it has shallow hollows on the lateral sides near the back end. I have attributed these specimens to *M. vialovi* Buryi (1989).

Metapolygnathus evolved from *Paragondolella*, apparently in the early Tuvalian (*Tropites dilleri* Zone) (Figure 21.3). At that time, the most progressive changes took place in the structure of the basal field and loop. The basal field of late *Paragondolella* (*P. polygnathiformis* Budurov and Stefanov) is rather wide, with rounded right angles on the back end, and it protrudes outward a little and in some adult specimens is slightly bifurcated. That insignificant loop bifurcation that the ancestors had became a distinct generic feature for early representatives of *Metapolygnathus* [*M. nodosus* (Hayashi)]. This species (especially juvenile individuals) shows similarities with the ancestral form in the outline of the platform back edge, the absence of denticles on it, and the presence of a short free blade.

In the middle Tuvalian (*Tropites subulatus* Zone), several morphologic varieties of *M. nodosus* (Hayashi) appeared. *M. permicus* (Hayashi) was characterized by a strong constriction on the back half of the platform (Hayashi, 1968, pl. 2, fig. 3). *M. linguiformis* Hayashi showed a narrowing of the platform edges,

in the form of small hollows near the back edge (the back third) (Buryi, 1989, pl. 9, fig. 1). Specimens with the platform slightly widened to the back edge are usually attributed to *M. primitia* (Mosher) (Mosher, 1970, pl. 110, figs. 7–13, 16, and 17).

Further development of *Metapolygnathus* resulted in larger bifurcation of the loop, as well as in widening and complication of the back edge of the platform. At the end of the Tuvalian (*Klamathites macrolobatus* Zone), the platform of *Metapolygnathus* became rather wide. It gradually widened from the forward edge to the back edge. At one-third of that distance, on the platform sides, a small narrowing is observed; sometimes it is larger on one side than on the other. Juvenile individuals have the smooth platform on the back half, and more mature individuals have the node-like marginal denticles. The loop is strongly bifurcated. I have attributed such representatives of *Metapolygnathus* to *M. vialovi* Buryi. The last representative of *Metapolygnathus* – *M. spatulatus* (Hayashi) – developed a much more complicated platform. *M. spatulatus* (Hayashi) appears to have originated from *M. vialovi* Buryi in the middle Lacian. It is of regular rounded form and resembles a teaspoon. The radial ribbing is regular. The ribs develop along the platform edges and just reach or almost reach the center of the platform, where the posterior denticle of the carina is located.

Synthesis

Thus, the following tendencies are observed in the development of *Metapolygnathus*:

1. gradual bifurcation of the loop on the lower surface: from faintly visible, which *Metapolygnathus nodosus* (Hayashi) had in the middle Carnian, to the appearance of two branches, as in *M. vialovi* Buryi (late Carnian);
2. complication of the morphology of the platform's posterior end: the appearance of strong narrowings [*M. permicus* (Hayashi)] and hollows (*M. linguiformis* Hayashi) at the first stage (middle Carnian), and significant widening of the platform edges (*M. vialovi* Buryi) (late Carnian), with increased ornamentation [*M. spatulatus* (Hayashi)], at the second stage (Norian, middle Lacian).

These tendencies are extremely pronounced in the specimens that Budurov (1972) attributed to *Ancyrogondolella triangularis* Budurov (Buryi, 1989, pl. 7, figs. 9–11). I believe that *A. triangularis* Budurov developed from *Metapolygnathus vialovi* Buryi in the late Lacian (*Malayites paulckei* Zone).

References

Budurov, K. J. 1972. *Ancyrogondolella triangularis* gen. et sp. n. (Conodonta). *Mitteilungen der Gesellschaft der Geologischen Bergbaustud* **21**:853–60.

Budurov, K. J. 1976. Structures, evolution and taxonomy of the Triassic platform conodonts. *Geologica Balcanica* **6**:13–20.

Budurov, K. J. 1977. Revision of the late Triassic platform conodonts. *Geologica Balcanica* **7**:31–48.

Buryi, G. I. 1985. Upper Triassic conodont associations of Sikhote-Alin (abstract). In *Fourth European Conodont Symposium* (*ECOS IV*), abstracts vol., ed. R. J. Aldridge, R. L. Austin, and M. P. Smith, pp. 6–7. Nottingen: University of Southampton.

Buryi, G. I. 1989. *Konodonty i Stratigraphiya Triasovykh Otlozhenii Sikhote-Alinya* [*Triassic Conodonts and Stratigraphy of Sikhote-Alin*]. Vladivostok: Dal'nevostochnoye Otdeleniye Akademii Nauk SSSR.

Hayashi, S. 1968. The Permian conodonts in chert of the Adoyama Formation, Ashio Mountains, Central Japan. *Earth Science* **22**:63–77.

Kozur, H. 1972. Die Conodontengattung *Metapolygnatus* Hayashi, 1968 und ihr stratigraphisher Wert. *Geologischen-Palaeontologischen Mitteilungen* **2**:1–37.

Krystyn, L. 1980. Triassic conodont localities of the Salzkammergut region (Northern Calcareous Alps). In *Guidebook and Abstracts of ECOS II,* ed. H. P. Schonlaub, pp. 61–98. Abhandlungen der Geologischen Bundesanstalt **35**.

Mosher, L. C. 1970. New conodont species as Triassic guide fossils. *Journal of Paleontology* **44**:737–42.

Mosher, L. C. 1973. Triassic conodonts from British Columbia and the Northern Arctic Islands. *Contributions to Canadian Paleontology* **222**:141–93.

Orchard, M. J. 1991. Late Triassic conodont biochronology and biostratigraphy of the Kunga Group, Queen Charlotte Islands, British Columbia. In *Evolution and Hydrocarbon Potential of the Queen Charlotte Basin, British Columbia,* pp. 173–93. Paper **90–10**. Geological Survey of Canada.

22 Late Triassic North American halobiid bivalves: diversity trends and circum-Pacific correlations

CHRISTOPHER A. McROBERTS

By virtue of its facies independence, widespread distribution, and high species turnover, the late Triassic pteriid bivalve *Halobia* has been of considerable importance for correlating marine strata throughout much of the Upper Triassic. The age significance and taxonomy of *Halobia* and its thin-shelled cousin *Daonella* were established largely by the classic monographs of Mojsisovics (1874), Gemmellaro (1882), and Kittl (1912) describing material from the Hallstatt facies of the Austrian Salzkammergut and from Sicily. Today, *Halobia* from the Alpine-Mediterranean region continues to play a major role in the construction of a biochronology of the Upper Triassic, as seen in recent works by Gruber (1976), Cafiero and De Capoa Bonardi (1982), and De Capoa Bonardi (1984).

The early success in establishing the age significance of *Halobia* in Europe was quickly repeated along the Pacific margins following the discovery of *Halobia* in New Zealand (Hochstetter, 1863; Trechmann, 1918) and North America (Gabb, 1864). After the initial discovery of *Halobia* in the Pacific margin, significant occurrences were documented from Timor (Krumbeck, 1924) and North America (Smith, 1927). A large number of studies revealing the extent and diversity of circum-Pacific *Halobia* appeared in the middle part of the twentieth century because of the efforts of T. Kobayashi (Japan and Southeast Asia), Chen-Chu Chen (China), and I. Polubotko (Russia) and their colleagues.

Throughout the circum-Pacific, *Halobia* continues to play an important role in correlating Carnian–middle Norian marine strata. With the exception of the South American Cordillera, *Halobia* is well represented in the marine Triassic around the circum-Pacific and is now known from the North American Cordillera, northeastern Russia, China, Japan, Indochina, islands in the western Pacific, and New Zealand (Figure 22.1A). Because of post-Triassic movements of the tectonostratigraphic terranes in which many of those halobiid faunas lived (e.g., Tozer, 1982; Hallam, 1986), the present distribution of *Halobia* in circum-Pacific deposits probably differs greatly from its original late Triassic biogeography (Figure 22.1B). Unfortunately, the post-Triassic tectonism obscured many of the stratigraphic and palaeogeographic relationships among halobiid-bearing strata of the circum-

Pacific, thus often preventing accurate sequencing of halobiid faunas.

Halobiid species had high origination and extinction rates; among the marine macrofauna, those rates were surpassed only by the ammonoids. Because taxa with short-lived species seem to be susceptible to higher rates of extinction (Stanley, 1979, 1990), *Halobia*, in spite of its widespread geographic range and facies tolerance, may provide an excellent and independent means to test the geographic and taxonomic extent of extinctions during Carnian–middle Norian time.

One drawback of using an analysis of halobiid diversity is the lack of a stable specific and supraspecific taxonomy for the group. Although it may be desirable to subdivide such a large taxonomic group into genera and subgenera, such divisions should be based on phylogenetic criteria, rather than on subjective usefulness, in order to provide a more accurate measure of biologic diversity within the group. Accordingly, the new generic and subgeneric terminologies proposed by Gruber (1976), Polubotko (1984, 1988), Polubotko, Alabushev, and Bychov (1990), and Campbell (1985) will, for the moment, be held in abeyance until the validity and monophyly of the supraspecific characters used in those halobiid taxonomies can be assessed.

This chapter will illustrate the usefulness of halobiid bivalves for establishing a biochronology for the Carnian–middle Norian rocks of North America. The data presented here should provide a new measure with which to correlate late Triassic biozones within North America and across the circum-Pacific, as well as to assess the extent and nature of local changes in species diversity.

Palaeoecologic controls on distribution

The cornerstone of *Halobia*'s value for biochronology and correlation undoubtedly lies in its palaeoautoecology. Although aspects of the life habits of halobiid species remain a mystery, for lack of a certain *in situ* fossil occurrence, several hypotheses have been proposed to explain their widespread distribution and varied facies occurrences in shallow-water carbonates, oxygen-poor black shales, and, rarely, even cherts. Of the several

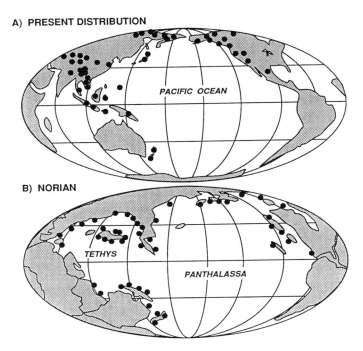

A) PRESENT DISTRIBUTION

PACIFIC OCEAN

B) NORIAN

TETHYS

PANTHALASSA

Figure 22.1. (A) Present-day distribution of circum-Pacific *Halobia* localities (filled circles). (B) Distribution based on Norian palaeogeographic reconstruction. (Adapted from Tozer, 1984.)

competing hypotheses, a pseudoplanktonic or epibenthic life mode adapted to an oxygen-poor environment would seem to offer the most plausible scenario. A complete account of halobiid autoecologic arguments is beyond the scope of this work; the reader is referred to the more comprehensive discussions by Krumbeck (1924), Hayami (1969), Gruber (1976), and Campbell (1985) and the sources cited therein.

Although undoubtedly rare in modern oceans, let alone the fossil record, a pseudoplanktonic life habit for *Halobia* should not be discounted. Several adaptations, such as a thin shell and a flat-valve morphology for streamlining, together may have aided in a pseudoplanktonic existence. Halobiids need not have been obligate on their planktonic hosts (*sensu* Wignall and Simms, 1990), but may have been opportunistic, attaching themselves to any available firm substrate, whether planktonic or not. Analogous associations of fossil bivalves, similar in size and shape to *Halobia*, with wood or algae attest to the possibility of that life mode, such as the Carboniferous pteriomorphs *Actinopteria* and *Caneyella* illustrated by McRoberts and Stanley (1989) and Wignall and Simms (1990).

Wignall (1993) prefers to see *Halobia* and other "flat clams" as being soft-substrate-dwellers, adopting to the soft mud by attaching to one another and together building mutually supporting mats. He suggests a less critical role for oxygen concentrations in determining their facies occurrences, because many halobiids are known from pelagic oozes that were deposited in waters that probably had normal oxygen contents. On the other hand, the adult *Halobia* that often are found in thick shell beds

numbering thousands of individuals per square metre may represent *in situ* occurrences, reflecting an opportunistic response to an oxygenation event in an otherwise relatively oxygen-poor bottom. Contrary to that idea is the observation that most of the deposits of that type that I have seen, as well as those recorded in the literature (e.g., Gruber, 1975), do not appear to have been deposited *in situ* and are composed of disarticulated single valves, suggesting some post-mortem transport. In spite of Wignall's assertion that the substrate is a major control on life habit, there may have been some morphologic adaptations to counter the effects of low oxygen levels. One possible adaptation would include a flat-valve morphology which would optimize the surface-to-volume ratio, thus aiding in the diffusion of oxygen through a very thin shell (e.g., Kauffman, 1988; Oschmann, 1993).

Despite some adaptations that enabled *Halobia* to live in oxygen-poor environments, there is only circumstantial evidence for a chemosymbiotic association with sulphur-reducing algae or bacteria. On functional morphologic grounds, the "sulphur pump" of Seilacher (1990) is perhaps better interpreted as a tube from which the byssus emerged. Additional evidence supporting a role for that structure as a byssal tube comes from the possible fossilized byssal threads on the tube interior of an exceptionally well preserved halobiid (Campbell, 1985). Furthermore, modern chemosymbiotic bivalves and other taxa are known to live primarily in tidal mud flats, hydrothermal vents, or cold-seep environments (Vetter, Powell, and Somero, 1991), and yet halobiid occurrences have not been documented from any such environments. One of the more convincing arguments against a chemosymbiotic life habit in an anoxic environment comes from analogy to known chemosymbiotic bivalves, in which basic biochemical reactions dictate a metabolism employing glycolysis and a citric acid cycle, which require oxygen (Oschmann, 1993).

If the halobiids were either benthic or planktic, then an extended planktotrophic larval stage may indeed have aided in long-distance dispersal throughout the Triassic seas. Such holopelagic adaptations are rare among modern bivalves, but have been suggested for several fossil taxa, including some "flat clams" (Oschmann, 1993). A holopelagic (or telepelagic) stage might allow for a planktonic life mode of more than a year, enough time for dispersal across even large oceans (e.g., Sheltema, 1977). Of course, such a dispersal mechanism in itself could not account for adult individuals that presumably had lived in or had fallen into deep-water environments and, occasionally, oxygen-poor environments.

North American succession

Halobiid bivalves in North America occur in sequences along the palaeomargins of cratonic North America and also in great abundance in tectonostratigraphic terranes accreted to the craton during later Mesozoic time (Figure 22.2). A summary of the North American halobiid succession, derived from numerous sequences throughout the North American Cordillera, is discussed later and is illustrated in Figures 22.3–22.5.

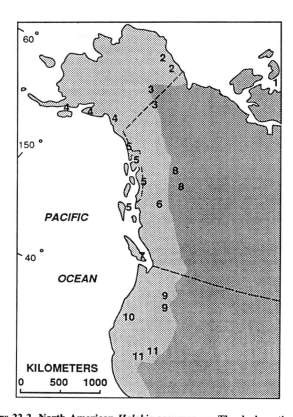

Figure 22.2. North American *Halobia* occurrences. The darker stippled area represents cratonic North America, and the lightly stippled area represents Mesozoic accreted terranes; data from Jones et al. (1987), Monger and Berg (1987), and Silberling et al. (1987). Numbers refer to halobiid-bearing regions. See the Appendix for primary sources and other locality information: 1, Arctic Canada; 2, Arctic Alaska; 3, east-central Alaska and western Yukon Territory; 4, southern and peninsular Alaska; 5, southeastern Alaska and Queen Charlotte Islands, British Columbia; 6, west-central British Columbia; 7, northeastern British Columbia; 8, Vancouver Island; 9, northeastern Oregon; 10, northern California; 11, west-central Nevada.

The earliest *Halobia* in North America, *Halobia zitteli,* occurs in the Lower Carnian deposits in the Arctic regions of Alaska and Canada. In Arctic Canada, this species occurs with the ammonoid *Sirenites nanseni* of the *Obesum* Zone (Tozer, 1967). *Halobia zitteli* has wide distributions only in the high palaeo-latitudes of North America, Siberia, and Spitsbergen. It is notably absent throughout the low-palaeolatitude terranes of North America, as well as most of the circum-Pacific realm. As such, *H. zitteli* is perhaps the only halobiid species in North America, and perhaps elsewhere, that shows evidence of latitudinal controls on its distribution. Less abundant in North America is *Halobia rugosa* Gümbel, a name given to Lower Carnian forms lacking plicae in the pre-growth-stop field. This species is found in craton-bound strata in the *Obesum* Zone of west-central Nevada and possibly in northeastern British Columbia and Arctic Canada (Tozer, 1967; Kristan-Tollmann and Tollmann, 1983). In the cordilleran terranes, this species occurs below typical tropitid ammonoid fauna and has been associated with the Lower Carnian ammonoid *Trachyceras* from the Shasta region of California and possibly from northeastern Oregon (Smith, 1927; Kristan-Tollmann and Tollmann, 1983) and Kupreanof Island off southeastern Alaska (Muffler, 1967).

The most abundant and best represented halobiid group in North America is that of *Halobia superba.* Two subspecies are recognized in North America: *H. superba superba* Mojsisovics (for the form with a simple break in its ribs) and *H. superba ornatissima* Smith (for the form with multiple breaks in its ribs). Although both forms are relegated primarily to Carnian rocks occurring with several tropitid ammonoid species, *H. superba superba* is known from several undisputed Lower Norian localities, such as northeastern British Columbia [Tozer, 1967; unpublished collections of the Geological Survey of Canada (GSC)], whereas *H. superba ornatissima* is known only from Upper Carnian rocks in California, Oregon, British Columbia, and Alaska; see the discussion by McRoberts (1993).

Figure 22.3. Stratigraphic distribution of North American *Halobia* species. (Adapted from McRoberts, 1993.)

AGE		AMMONOID ZONES (after Tozer, 1984)	NORTH AMERICAN *HALOBIA*
NORIAN (PART)	MID	*H. columbianus*	
		D. rutherfordi	
	EARLY	*J. magnus*	
		M. dawsoni	
		S. kerri	
CARNIAN	LATE	*K. macrolobatus*	
		T. welleri	
		T. dilleri	
	EARLY	*A. obesum*	
		T. desatoyense	

Figure 22.4. Representative North American *Halobia* species.
(1) *Halobia zitteli* Lindström, Lower Carnian, Blaa Mountain Formation, Axel Heiberg Island, Arctic Canada, GSC 85851 (×1). (2) *Halobia rugosa* Gümbel, Lower Carnian, poorly preserved from the Pit Shale, Shasta region, California, USNM 316112 (×2.5). (3) *Halobia superba superba* Mojsisovics, Lower Norian, Pardonet Formation, Mount McLearn, British Columbia, CGS 85852 (×2). (4) *Halobia superba ornatissima* Smith, Upper Carnian, Pardonet Formation, Alaska Highway, British Columbia, CGS 85853 (×1). (5) *Halobia radiata radiata* Gemmellaro, Upper Carnian, Burnt River Conglomerate, Admiralty Island, southeastern Alaska, USNM 7118 (×2.5). (6) *Halobia cordillerana* Smith, Lower Norian, Otuk Formation, Otuk Creek, Arctic Alaska, D 11100 (×1). (7) *Halobia beyrichi* (Mojsisovics), holotype of *Halobia alaskana* Smith, Lower Norian, Gravina Island, southeastern Alaska, USNM 74192 (×1.25). (8) *Halobia fallax* Mojsisovics, Middle Norian, Yukon River region near Nation, Alaska, USNM 74179 (×3). (9) *Halobia halorica* Mojsisovics, USNM 316113 (×2). (10) *Halobia* cf. *lineata* (Münster), Middle Norian, Martin Bridge Formation, Wallowa Mountains, Oregon, UMIP 16060 (×2).

NORTH AMERICAN HALOBIA Norian ■ Carnian-Norian ▲ Carnian ●	west-central Nevada	northern California	northeast Oregon	Vancouver Island, B.C.	western British Columbia	southeast Alaska & Queen Charlotte Islands, B.C.	southern & peninsular Alaska	northeast British columbia	east-central Alaska & Yukon	Arctic Alaska & Canada
Halobia halorica	■		■			■	■		■	■
Halobia lineata			□			■	■		■	■
Halobia salinarum			□							■
Halobia plicosa						■				■
Halobia fallax	■		■			■		□	■	■
Halobia cordillerana	■	■	□			■		□	■	■
Halobia beyrichi	■		■	■	□	■	□			
Halobia austriaca		●	●			▲	▲			
Halobia radiata radiata			●			●	▲	△		
Halobia superba superba		●	●		●	▲	●	▲	●	●
Halobia superba ornatissima		●	●			●		●	●	
Halobia rugosa		●	○			●				○
Halobia zitteli										●

Figure 22.5. Distribution of *Halobia* species throughout North American regions (see Figure 22.1 and the Appendix for localities). Closed symbols represent certain occurrences; open symbols represent uncertain occurrences.

Another species that spans the Carnian–Norian boundary is *Halobia radiata* Gemmellaro, including forms described as *Halobia dalliana* Smith, *Halobia septentrionalis* Smith, and *Halobia symmetrica* Smith. Although there are at least two subspecies of this taxon, as interpreted by Gruber (cited by Kristan-Tollmann and Tollmann, 1983) and McRoberts (1993) – *H. radiata radiata* for forms with the beak centrally located, and *H. radiata hyatti* Kittl for forms with the beak anteriorly situated – only *H. radiata radiata* is known from North America. It occurs in several Upper Carnian localities (e.g., northeastern Oregon and western British Columbia), and yet it is known from both Carnian and lowest Norian localities in southeastern Alaska and craton-bound northeastern British Columbia (McRoberts, 1993).

Halobia austriaca Mojsisovics is known from several North American localities, in craton-bound strata as well as in terranes. *Halobia brooksi*, a name used by Smith (1927) for *H. austriaca* with a growth-stop, fits within the concept of *H. austriaca* and is treated as a junior synonym. In North America, that species has been reported from Zacatecas in central Mexico (Kittl 1912), from an undescribed Lower Norian collection from the Wallowa Terrane of Oregon, and from Upper Carnian and Lower Norian rocks in the Wrangell Mountains of southeastern Alaska, as well as from Carnian localities in southeastern Alaska, where it co-occurs with the ammonoids *Tropites* and *Anatropites* [Silberling, 1959; unpublished collections of the U.S. Geological Survey (USGS)].

In North America, the lowermost Norian *Kerri* Zone is recognized by the first appearance of *Halobia beyrichi* (Moj-

sisovics). This species (for which *H. alaskana* Smith is considered a junior synonym) is known to occur throughout the cordilleran terranes from Alaska to Nevada. It is well represented in Vancouver and the Queen Charlotte Islands of British Columbia (Tozer, 1967; Carter, Orchard, and Tozer, 1989), in southeastern Alaska (Muffler, 1967), in the Wallowa Terrane in Oregon (McRoberts, 1993), and in the Shoshone Mountains of Nevada, where it occurs with the ammonoid *Stikinoceras kerri* (Silberling, 1959; Silberling and Tozer, 1968; Kristan-Tollmann and Tollmann, 1983). In North America, *H. beyrichi* takes the place of *H. styriaca* (Mojsisovics) in serving as a key zonal fossil for the lowermost Norian of the Tethys realm and the western Pacific regions, for *H. styriaca* is conspicuously absent from North America.

The remainder of the Lower Norian is characterized by *Halobia cordillerana* Smith, a form similar to, and probably descendant from, the *H. superba* group, but which can be distinguished by subtleties in the plicae. *Halobia cordillerana* is known throughout the North American Cordillera, with significant occurrences in Nevada (*H. hochstetteri*, listed by Kristan-Tollmann and Tollmann from the Luning Formation, is here treated as a synonym for *H. cordillerana*), California, and possibly northeastern Oregon and southern, eastern, and Arctic Alaska. In British Columbia, this species (or a similar species) is known as *Halobia parcalis* McLearn, from the Peace River foothills of northeastern British Columbia.

At the end of the Lower Norian, five species can be recognized that continue into the Middle Norian, where they coexisted with, and later were replaced by, the pectinacean bivalve *Monotis*. Those five are *H. fallax* Smith, *H. plicosa* Mojsisovics, *H. salinarum* (Bronn), *H. lineata* Münster, and *H. halorica* Mojsisovics. Although only two of those species, *H. fallax* and *H. halorica*, occur in sufficient numbers for adequate characterization of their temporal ranges, a complete sequence of all five species is known (without ammonoids) from the Brooks Range in northern Alaska (undescribed USGS collection). In addition to the Brooks Range specimens, *Halobia fallax* occurs in the *Dawsoni* Zone of northeastern British Columbia (Tozer, 1967), as well as in the Yukon Valley area of eastern Alaska (undescribed USGS collections). *Halobia halorica* (a senior synonym of *H. dilitata* Kittl) is perhaps the best represented, with occurrences in the Yukon Valley and southeastern Alaska, the Queen Charlotte Islands, northeastern Oregon, and west-central Nevada (Smith, 1927; Muffler, 1967; Tozer, 1967; McRoberts, 1993).

Circum-Pacific correlations

Surprisingly few correlations based on *Halobia* have been made between North American sequences and other circum-Pacific sequences. It remains unclear whether that is due to endemism among circum-Pacific *Halobia* or is due to failures in collection, or perhaps that disparity reflects superficial differences in nomenclature among conspecific forms. As for the third point, many of the circum-Pacific *Halobia* described or named before

Age		North American Ammonoid Zones [1]	North American Halobia	Northeast Russia [2]		China [3]	Japan [4]	Southeast Asia [5]	New Zealand [6]
NORIAN (PART)	MID	H. columbianus	Halobia halorica	Halobia obruchevi		Halobia norica	Halobia longissima		Halobia zealandica / Halobia plicosa / Halobia hoernesi
		D. rutherfordi							
	EARLY	J. magnus	Halobia cordillerana	H. indigirensis	H. aotii	Halobia superbescens & Halobia subrugosa	Halobia obsoleta / Halobia aotii	Halobia aotii	Halobia styriaca / Halobia austriaca
		M. dawsoni							
		S. kerri	Halobia beyrichi		H. kawadai		Halobia kawadai	Halobia styriaca	
CARNIAN	LATE	K. macrolobatus	Halobia superba	H.asperella	H. kudleyi	Halobia pluriradiata & Halobia superba	Halobia talauana	Halobia talauana	
		T. welleri			H.omkutchanica				
		T. dilleri		Halobia subfallax					
	EARLY	A. obesum	Halobia zitteli	Halobia popowi / Halobia talajensis / Halobia zitteli		Halobia comata & Halobia rugosoides	Halobia comata	Halobia comata	
		T. desatoyense							

Figure 22.6. Selected halobiid zones of the circum-Pacific compared to an informal "taxon range" zonation of North American Halobia: 1, ammonoid zonation of Tozer (1984); 2, Polubotko (1984) (also aided in 3– 5); 3, Chen (1964, 1976), Chen and Ba (1986); 4, Kobayashi and Aoti (1943), Kobayashi et al. (1967); 5, Kobayashi et al. (1967), Kobayashi and Tamura (1984); 6, Campbell (1985).

the late 1970s need to be reconsidered. Specific criteria have shifted, since the 1970s and 1980s, away from highly interspecifically variable shell shape and ribbing characteristics and toward the use of details in ontogenetic series, as well as the nature of the anterior byssal tube and auricle. It remains a distinct possibility that after further study, many circum-Pacific halobiid species will fall into synonymy.

It is particularly significant that many of the earlier age determinations for western Pacific halobiid sequences may be in need of revision. For example, *Halobia styriaca,* formerly believed to be a Carnian species, is now considered an earliest Norian index fossil from the Alpine-Mediterranean region; see De Wever et al. (1979) and De Capoa Bonardi (1984) and the discussion by Campbell (1985). The question whether the western Pacific *H. styriaca* should be considered Norian (as treated herein) or Carnian must await further age data in the form of co-occurrences with ammonoids and conodonts. Similar shifts in temporal interpretations may therefore be necessary for other species that are more restricted in geographic distribution, such as *Halobia kawadai* and *Halobia aotii.* An overview of halobiid successions throughout the circum-Pacific regions and their relationships to the North American sequences is presented next, as summarized in Figure 22.6.

Northeastern Russia

Northeastern Russia, particularly eastern Siberia and the Korayak Mountains, is an especially important region for halobiid sequences and has been studied in detail for at least 50 years. Since Kiparisova's initial treatment (Kiparisova, 1937, 1938), and the major bivalve treatise of Kiparisova, Bychkov, and Polubotko (1966), the halobiids of northeastern Russia have been described in a variety of papers by Polubotko and her colleagues (e.g., Bychkov and Yefimova, 1968; Polubotko, 1980, 1984, 1986, 1988; Polubotko et al., 1990). In 1984, Polubotko proposed a halobiid zonation for all of northeastern Russia, as

well as a new supraspecific taxonomy that was later clarified and expanded (Polubotko, 1988; Polubotko et al., 1990).

Similarities between halobiids in North America and northeastern Russia are evident from both Carnian and Norian sequences. Direct correlation of the earliest Carnian can be made on the basis of simultaneous occurrences of *Halobia zitteli* in numerous localities in both Russian Far East and Arctic North America. *Halobia omkuchanica* Polubotko and, to a lesser extent, *H. nutekinensis* (Polubotko and Alabushev), known from the Upper Carnian of Siberia and the Korayak Mountains (e.g., Polubotko, 1984; Polubotko et al., 1990), appear to converge in morphology with *H. superba ornatissima* from the North American Cordillera, and that may serve to provide additional correlation levels.

Correlation of the Carnian–Norian intervals in Russia and North America can be made on the basis of *Halobia kenkerensis* (Polubotko) (Polubotko et al., 1990), which, in terms of morphology and in its Upper Carnian–Lower Norian stratigraphic range, is similar to, and possibly synonymous with, *H. austriaca.* The remainder of the Lower Norian sequence in northeastern Russia does not permit direct correlation to North America and is characterized by the *Halobia indigirensis* Zone, which can be further divided into the *H. kawadai* and *H. aotii* subzones. The Middle Norian can be recognized by the *H. obruchevi* Zone, which contains less abundant species, including *Halobia fallax* and *Halobia lineata,* both of which allow for direct correlation to sequences in Alaska and elsewhere in North America.

China

General discussions of the distributions of Chinese *Halobia* have been published by Reed (1927), Chen (1976), and Chen and Ba (1986). In general, similarities to the halobiid faunas of the eastern and central Tethyan regions are to be expected, because of geographic proximity, and significant co-occurrences of several species have provided limited correlations to North American strata.

At the western limit of the circum-Pacific realm in central China, Chen and Ba (1986) identified the lowermost Carnian *Halobia rugosoides* and *H. subcomata* assemblages as being characteristic zonal indices. It is not until the Middle and Upper Carnian, characterized by *H. superba* and *H. pluriradiata* (which occur throughout southern China), as well as *H. radiata* [listed as *H. septentrionalis* by Chen and Ba (1986)], that direct correlations with the uppermost Carnian or lowermost Norian North American sequences are possible. Although *Halobia styriaca* occurs in Tibet and other areas of central and eastern China (Chen and Ba, 1986), providing taxonomic links to much of the circum-Pacific and the Tethyan realm, the early Norian is more abundantly represented by *H. superbescens* and *H. subrugosa*. Clear biogeographic ties to the well-studied western Tethyan regions in the middle Norian can be made with *H. norica*, which is well represented throughout much of China. Additionally, assuming correct species assignments, ties to North America can be based on *H. lineata, H. halorica, H. plicosa*, and *H. fallax*, listed by Chen and Ba (1986) as occurring in the Lower Norian of central and western China, rather than the Middle Norian as in North America (e.g., Figure 22.3). It is also noteworthy, albeit unverified with ammonoids, that Chen and Ba (1986) listed *H. superba ornatissima* within the Lower Norian, rather than the typical assignment to the middle part of the Carnian, where it is placed in North America.

Japan

Significant occurrences of *Halobia* in Japan have been described by Kobayashi and colleagues (e.g., Kobayashi and Aoti, 1943; Kobayashi and Ichikawa, 1949; Kobayashi and Tokuyama, 1959; Nakazawa, 1964; Nakazawa and Nogami, 1967; Hayami, 1975). A generalized zonal scheme compiled from those works, as well as from the work of Polubotko (1984), is shown in Figure 22.6. The halobiid sequence in Japan shows closer correspondences to the Russian and Southeast Asian sequences than to the occurrences in China and areas farther west in the Tethyan belt.

Any direct halobiid correlations between the Carnian sequences of North America and Japan are as yet unclear. The sequence begins with *Halobia comata* of the Lower Carnian, which appears to be replaced vertically by the Carnian species *H. talauana* in the western hills of Kyoto and other localities (Hayami, 1975; Polubotko, 1984).

As in northeastern Russia, the lowermost Norian is characterized by *Halobia kawadai*, followed by *H. aotii*, both of which occur in the Mine region and other areas (Hayami, 1975). There is one reference to *Halobia styriaca* from the Carnian of the Motobu area (Kobayashi and Ishibashi, 1970), but that occurrence may possibly be Lower Norian, considering the age revisions for the species (Gruber, 1975, 1976). The upper part of the Lower Norian is marked by *H. obsoleta*, which in turn is overlain by the Middle Norian *H. longissima*. Although less abundant than *H. longissima, H.* cf. *obruchevi* and *H.* cf. *fallax*

were reported by Nakazawa (1964) among other species from the latest part of the Middle Norian, permitting correlations to both Russia and western North America.

Indonesia and Southeast Asia

Some of the earliest circum-Pacific work concerned the faunas of Timor and Roti (Krumbeck, 1924); more recent studies have focused on the Thai-Malay Peninsula (Kobayashi and Tamura, 1968, 1984; Kristan-Tollmann, Barkham, and Gruber, 1987; Chonglakmani and Grant-Mackie, 1993). Many halobiids from those regions are known, but few occur in clearly defined sequences with age-diagnostic ammonoids or conodonts. Thus a tentative sequence, as shown in Figure 22.6, is here considered typical of the region. The species content of the halobiid fauna of Southeast Asia resembles those of China and Japan, in addition to having specific New Zealand affinities.

As in Japan, the Carnian of Southeast Asia is first represented by *Halobia comata*. That species is later replaced by *H. talauana* and the earliest Norian species, *H. styriaca*, followed by *H. aotii*; see Kobayashi and Tamura (1984) for a more extensive treatment and for Asian comparisons. Although halobiids are poorly represented or unknown in the Middle Norian sequences in most of Southeast Asia and Indonesia, Chonglakmani and Grant-Mackie (1993) recently described the halobiid succession of Thailand, including some forms that may be Middle Norian. Of particular importance are the occurrences of *Halobia styriaca, H. charlyana*, and *H. distincta*, allowing correlation to the well-dated Norian sequences in New Zealand. Similarly, *H. halorica* (listed as *H. dilitata*) and *H. fallax* allow correlation to the Lower and Middle Norian of North America.

New Zealand

Following Trechmann (1918) and Marwick (1953), the halobiid succession of New Zealand has been made known largely by the efforts of Campbell (1985). Campbell's work allows correlation to North America, as well as to the classic Alpine-Mediterranean region. Although there are few occurrences of Carnian *Halobia* (mostly from tectonically disturbed localities of the Torlesse Terrane), the most reliable and abundant occurrences are from the Norian of the stratigraphically complex Murihiku Terrane, in both North Island and South Island.

As in most other regions of the circum-Pacific, exclusive of North America, *H. styriaca* occurs in the lowest Norian (*Kerri Zone*) of the Murihiku Terrane, providing excellent correlation to the classic western Tethyan sequences of the Austrian Alps. The Lower Norian of New Zealand contains *Halobia hochstetteri* (a possible synonym of *H. cordillerana*, as used in North America) in both the Murihiku and Torlesse terranes, *H. hoernesi* in the Murihiku Terrane, and *H. superba superba* and *H. austriaca* (possibly including *H. lilliei* Marwick), possibly in the earliest Norian of the Torlesse Terrane. The latter two, *H. superba* and *H. austriaca*, allow correlations to well-dated sequences in North America and Europe. Likewise, the Lower–

Middle Norian *Halobia plicosa* from the Murihiku Terrane provides an additional correlation with North America. Other forms representative of New Zealand *Halobia,* such as the Lower–Middle Norian *Halobia zealandica,* are poorly represented elsewhere.

North American diversity and the Carnian–Norian "extinction"

The halobiid speciation and extinction rates exceeded those for other late Triassic macrofauna, with the possible exception of ammonoids. The average duration for North American halobiid species was less than three ammonoid zones (Figure 22.3) and has been estimated, using the time scale of Harland et al. (1990), as being around 2.5 m.y. Because of such rapid species turnover, it follows that *Halobia,* like other short-lived species, may be sensitive to extinction (e.g., Stanley, 1979, 1990), and can serve as a useful marker for examination of bioevents throughout Carnian–middle Norian time.

Our halobiid diversity metrics (Figure 22.7) are controlled by varying speciation and extinction rates and are modified by phylogenetic constraints and taxonomic concepts. Because a satisfactory phylogenetic hypothesis for halobiid species or species groups is not available, the fundamental unit considered here is the nominal taxonomic species. Once lineages and chronospecies are identified, the diversity metrics presented here undoubtedly will change. Although any such changes in themselves will not alter our understanding of the standing diversity for any one time, they will consistently reduce our understanding of origination and extinction rates. Such diversity stasis, however, is not contingent on any future systematic revision, which unquestionably would alter the standing diversity. Given these uncertainties, the general statements concerning the diversity of North American *Halobia* presented here await further substantiation by additional taxonomic and biostratigraphic data.

The standing diversity of North American halobiid species increased from two species in the early Carnian to five in the late Carnian and reached a peak of eight species in the early Norian. By the middle Norian, the standing diversity declined to five species, and it dropped to zero at the close of the substage.

A more comprehensive characterization of the diversity metrics must take into account the number of additions (origins) and deletions (extinctions) of taxa throughout *Halobia*'s range. Like the diversity curve, the number of extinctions increased throughout the Carnian and reached a peak at the middle Norian. The Carnian and early Norian extinctions eliminated some 40–60% of halobiid species; extinctions increased to 100% at the close of the middle Norian. Conversely, the number of new species increased from two in the early Carnian to five in the early Norian, but decreased to two by the middle Norian. Those originations accounted for ever-decreasing proportions of the standing diversity throughout *Halobia*'s range in North America.

It is particularly important that across the Carnian–Norian boundary and into the early Norian, extinctions kept pace with originations, but diversity increased because of Carnian holdover

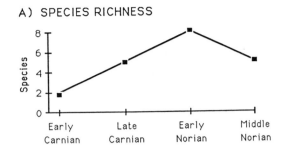

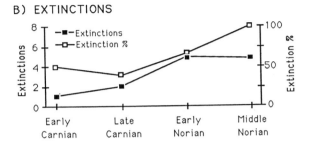

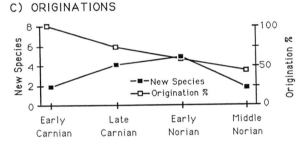

Figure 22.7. Halobiid diversity metrics. (A) Species richness. (B) Extinctions. (C) Originations. Note that both extinctions and originations are given as species counts and as percentages of taxonomic diversity.

taxa. That suggests, as shown in Figure 22.7A, that there was no major extinction event at the Carnian–Norian boundary (Figure 22.3). The absence of a halobiid mass extinction can be illustrated in cordilleran sections, where the Carnian–Norian boundary is marked by the first appearance of *Halobia beyrichi,* rather than by extinction (Carter, Orchard, and Tozer, 1989; McRoberts, 1993). That pattern of complex turnover is even more pronounced in European halobiid sequences (Gruber, 1976); however, it has not been assessed in other circum-Pacific regions.

The trend of increasing *Halobia* diversity across the Carnian–Norian is in contrast to the trends seen in some other fossil taxa, including some non-halobiid bivalves and other invertebrate and vertebrate groups (e.g., Benton, 1986, 1991; Stanley, 1988; Johnson and Simms, 1989). The reasons for that discrepancy may include differences in scale and sample size among the studies cited. The patterns of Carnian–Norian extinctions documented for European bivalves and crinoids (Johnson and Simms, 1989) were culled primarily from temporally and geographically disparate horizons, rather than from continuous marine sections spanning the Carnian–Norian interval. Given that only few localities worldwide preserve such a boundary (the stratotype in the Northern Alps is composed of condensed and temporally mixed fossil assemblages), accurate documentation of the timing of the faunal turnover must await more extensive sampling and a more systematic treatment of fossil groups. This is particularly true for many circum-Pacific regions, where confirmation of the diversity patterns for halobiid and non-halobiid bivalves will require further testing.

Summary and conclusions

A composite sequence for North American *Halobia* offers nearly complete coverage, from the earliest Carnian through the middle Norian. Although the majority of those occurrences are in tectonostratigraphic terranes from Alaska to Nevada, a few occur in craton-bound strata in northeastern British Columbia, Canada. Several key *Halobia* species occur in sufficient abundance in those regions for interregional, circum-Pacific, and even global correlations. The value of halobiid biochronology may be due in part to its unique adult life strategy of mud-sticking or pseudoplanktonism, as well as an extended planktotrophic larval stage that aided in long-distance dispersal.

The first halobiid species in North America and elsewhere in the circum-Pacific is *Halobia zitteli,* which makes its appearance in the lowest Carnian. The Upper Carnian, on the other hand, is more diverse in species, but can confidently be characterized by the highly ornamented forms attributed to subspecies of *H. superba* and the straight and finely ribbed forms regarded as *H. radiata radiata.* The lowest Norian is almost everywhere characterized by the first occurrences of *H. styriaca* or *H. beyrichi* and related forms with an undivided anterior byssal tube. Several species occur in the Middle Norian of North America, but *H. halorica* and *H. fallax* are the most abundant.

The North American sequence also serves as an important test for the validity of the idea of a global bioevent at the Carnian–Norian boundary. Complete sequences in British Columbia and Oregon suggest that the boundary may represent a period of diversification and complex turnover, rather than of major extinction, a pattern observed for European halobiid sequences. The periods of accelerated halobiid extinction, assigned to later in the early Norian, differ from those observed for other taxa, including bivalves, and tend to suggest that such extinction scenarios need further collaboration in detailed sequences before their global nature is confirmed.

Acknowledgments

I thank J. M. Dickins for inviting this contribution, which is adapted from an original presentation at the Fifth North American Paleontological Convention, Chicago. I am deeply grateful to N. J. Silberling and E. T. Tozer, who, after years spent getting most of this story together, have allowed access to their collections and libraries and have provided invaluable advice and comments on the marine Triassic. Thanks also to M. Bizzaro and J. T. Scallan, whose comments improved earlier drafts of the manuscript, and to Yin Hongfu and H. J. Campbell for thoughtful reviews.

References

Benton, M. J. 1986. More than one event in the late Triassic mass extinction. *Nature* 321:857–61.
Benton, M. J. 1991. What really happened in the Late Triassic? *Historical Biology* 5:263–78.
Blome, C. D., Reed, K. M., and Tailleur, I. L. 1988. *Radiolarian Biostratigraphy of the Otuk Formation in and near the National Petroleum Reserve in Alaska,* pp. 725–76. Professional paper 1399. U. S. Geological Survey.
Bychkov, Y. M., and Yefimova, A. F. 1968. New late Triassic *Monotis* and *Halobia* of the northeastern USSR (in Russian). In *New Species of Early Plants and Invertebrates of the USSR,* no. 2, part 1, pp. 186–8. Moscow: Nedra.
Cafiero, B., and De Capoa Bonardi, P. 1980. Stratigraphy of the pelagic Triassic in the Budva-Kotor area (Crna-Gora, Montenegro, Yugoslavia). *Bollettino della Società Paleontologica Italiana* 21:179–204.
Cafiero, B., and De Capoa Bonardi, P. 1982. Biostratigrafia del Trias pelagico della Sicilia. *Bollettino della Società Paleontologica Italiana* 21:35–71.
Campbell, H. J. 1985. Stratigraphic significance of the bivalves *Daonella* and *Halobia* in New Zealand and New Caledonia, Unpublished doctoral dissertation, Cambridge University.
Carter, E. S., Orchard, M. J., and Tozer, E. T. 1989. *Integrated Ammonoid-Conodont-Radiolarian Biostratigraphy, Late Triassic Kunga Group, Queen Charlotte Islands, British Columbia,* pp. 23–30. Paper 89-1H. Geological Survey of Canada.
Chen C.-c. 1964. On the occurrence of *Halobia* fauna from the Ganzi region, western Sichuan (Szechuan) and its significance. *Acta Palaeontologica Sinica* 2:75–8.
Chen C.-c. 1976. *Lamellibranch Fossils of China.* Beijing: Science Press.
Chen J.-h. and Ba D.-z. 1986. *Halobia* fauna from Zedang of South Xizang with a discussion on the *Halobia* assemblages in China. *Acta Palaeontologica Sinica* 25:1–9.
Chonglakmani, C., and Grant-Mackie, J. A. 1993. Biostratigraphy and facies variation of the marine Triassic sequences in Thailand. In *International Symposium on Biostratigraphy of Mainland Southeast Asia: Facies and Paleontology,* vol. 1; pp. 97–123.
De Capoa Bonardi, P. 1984. *Halobia* zones in the pelagic late Triassic sequences of the central Mediterranean area. *Bollettino della Società Paleontologica Italiana* 23:91–102.
De Wever, P., Sanfilippo, A., Riedel, W., and Gruber, B. 1979. Triassic radiolarians from Greece, Sicily and Turkey. *Micropaleontology* 25:75–110.
Gabb, W. M. 1864. Paleontology of California. *Geological Survey of California, Palaeontology* 1:17–35.
Gemmellaro, G. 1882. Sul Trias regione occidentale della Sicilia. *Accedemia Nazionale dei Lincei, Memorie* 12:451–73.
Gruber, B. 1975. Unternorische Halobien (Bivalvia) aus Bosnien Jugoslavian. *Sitzungsberichte der Oesterreichische Akademie Wissenschaften, Math-naturwiss.* (Abt. 1) 183:119–30.
Gruber, B. 1976. Neue Ergebnisse auf dem Gebiete der Oekologie, Stratigraphie und Phylogenie der Halobien (Bivalvia). *Mitteilungen der Gesellschaft der Geologie und Bergbaustudented in Oesterreich* 23:181–98.

Hallam, A. 1986. Evidence of displaced terranes from Permian to Jurassic faunas around the Pacific margins. *Journal of the Geological Society (London)* 143:209–16.

Harland, W. B., Cox, A. V., Llewellyn, P. G., Pickton, C. A. G., Smith, A. G., and Walters, R. 1990. *A Geologic Time Scale, 1989.* Cambridge University Press.

Hayami, I. 1969. Notes on Mesozoic "planktonic" bivalves. *Journal of the Geological Society of Japan* 75: 375–83.

Hayami, I. 1975. A systematic survey of the Mesozoic Bivalvia from Japan. *Bulletin of the University Museum, University of Tokyo* 10:1–203.

Hochstetter, F. von. 1863. *Neu-Seeland.* Stuttgart.

Jeletzky, J. A. 1950. *Stratigraphy of the West Coast of Vancouver Island between Kyuquot Sound and Esperanza Inlet, British Columbia.* Paper 50-57. Geological Survey of Canada.

Johnson, A., and Simms, M. J. 1989. The timing and cause of late Triassic marine invertebrate extinctions: evidence from scallops and crinoids. In *Mass Extinctions: Processes and Evidence*, ed. S. K. Donovan, pp. 174–94. London: Bellhaven.

Jones, D. L., Silberling, N. J., Coney, P. J., and Plafker, G. 1987. *Lithotectonic Terrane Map of Alaska (West of the 141st Meridian).* Miscellaneous field studies map MF-1874-A. U.S. Geological Survey.

Kauffman, E. G. 1988. The case of the missing community: low-oxygen adapted Paleozoic and Mesozoic bivalves ("flat clams") and bacterial symbiosis in typical Phanerozoic seas. *Geological Society of America, Abstracts with Programs* 20(7):A48.

Kiparisova, L. 1936. Upper Triassic pelecypods from the Kolyma-Indigirka Land (in Russian). *Transactions of the Arctic Institute* 30:71–136.

Kiparisova, L. 1937. Fauna of the Triassic deposits of the Arctic regions of the Soviet Union. *Transactions of the Arctic Institute* 31:135–256.

Kiparisova, L. D. 1938. *Pelecypoda of the Triassic System of the USSR.* Leningrad: Central Geological and Prospecting Institute.

Kiparisova, L. D., Bychkov, Y. M., and Polubotko, I. V. 1966. *Upper Triassic Bivalve Molluscs from the Northeast USSR* (in Russian). Magadan: Vsesoyuznyy Nauchno-issledovatel'skii Instituta (VSEGEI).

Kittl, E. von. 1912. *Materialen zu einer Monographie der Halobiidae und Monotidae der Trias, Resultate der Wissenschaftlichen Erforschung des Balatonsees, I Band, I Teil, Paläontologie Anhang.* Wien.

Kobayashi, T. 1963. *Halobia and some other fossils from Kedah, northwest Malaya. Japanese Journal of Geology and Geography* 34:113–28.

Kobayashi, T., and Aoti, K. 1943. Halobiae in Nippon. *Journal Shigenkagaku Kenkyusho* 1:241–55.

Kobayashi, T., Burton, C. K., Tokuyama, A., and Yin, E. H. 1967. The *Daonella* and *Halobia* facies of the Thai Malay Peninsula compared with those of Japan. *Geology and Paleontology of Southeast Asia* 3:96–122.

Kobayashi, T., and Ichikawa, K. 1949. *Myophoria* and other Upper Triassic pelecypods from the Sakawa Basin in Shikoku, Japan. *Japanese Journal of Geology and Geography* 21:177–92.

Kobayashi, T., and Ishibashi, T. 1970. *Halobia styriaca*, Upper Triassic pelecypod discovered in Okinawa-Jima, the Ryuku Islands. *Proceedings of the Paleontological Society of Japan*, n.s. 77:243–8.

Kobayashi, T., and Masatani, K. 1968. Upper Triassic *Halobia* (Pelecypoda) from North Sumatra with a note on the

Halobia facies in Indonesia. *Japanese Journal of Geology and Geography* 39:113–23.

Kobayashi, T., and Tamura, M. 1968. Upper Triassic pelecypods from Singapore. *Geology and Paleontology of Southeast Asia* 5:135–50.

Kobayashi, T., and Tamura, M. 1984. The Triassic Bivalvia of Malaysia, Thailand and adjacent areas. *Geology and Paleontology of Southeast Asia* 25:201–27.

Kobayashi, T., and Tokuyama, A. 1959. The Halobiidae from Thailand. *Journal of the Faculty of Science, University of Tokyo* 12:27–30.

Kristan-Tollmann, E., Barkham, S., and Gruber, B. 1987. Potschenschichten, Zlambachmergel (Hallstatter Obertrias) und Liasfleckenmergel in Zentraltimor, nebst ihren Faunenelementen. *Mitteilungen der Gesellschaft der Geologie und Bergbaustudenten in Oesterreich* 80:229–85.

Kristan-Tollmann, E., and Tollmann, A. 1983. Tethys-Faunenelemente in der Trias der USA. *Mitteilungen der Gesellschaft der Geologie und Bergbaustudenten in Oesterreich* 76:213–32.

Krumbeck, L. 1924. Die Brachiopoden, Lamellibranchiaten und Gastropoden der Trias von Timor. In *Paläontologie von Timor. Vol. 13: Lieferung*, pp. 1–275. Stuttgart: Schweizerbart.

McRoberts, C. A. 1993. Systematics and biostratigraphy of halobiid bivalves from the Martin Bridge Formation (Upper Triassic), northeast Oregon. *Journal of Paleontology* 67:198–210.

McRoberts, C. A., and Stanley, G. D., Jr. 1989. A unique bivalve-algal life association from the Bear Gulch Limestone (Upper Mississippian) of central Montana. *Journal of Paleontology* 63:578–81.

Marwick, J. 1953. *Divisions and Faunas of the Hokonui System (Triassic–Jurassic).* Palaeontological bulletin 21. New Zealand Geological Survey.

Mojsisovics, E. von. 1874. Ueber die Triadischen Pelecypoden-Gattungen *Daonella* und *Halobia. Abhandlungen der k.k. Geologischen Reichsanstalt* 7:1–35.

Monger, J. W. H., and Berg, H. C. 1987. *Lithotectonic Terrane Map of Western Canada and Southeastern Alaska. Miscellaneous field studies map MF-1874-B.* U.S. Geological Survey.

Muffler, L. J. P. 1967. *Stratigraphy of the Keku Islets and Neighboring Parts of Kuiu and Kupreanof Islands, Southeastern Alaska.* Bulletin 1241-C. U.S. Geological Survey.

Nakazawa, K. 1964. On the *Monotis typica* zone in Japan. *Memoirs of the Faculty of Science, University of Kyoto* 30:249–79.

Nakazawa, K., and Nogami, Y. 1967. Problematic occurrences of the Upper Triassic fossils from the western hills of Kyoto. *Memoirs of the Faculty of Science, University of Kyoto* 34:9–22.

Oschmann, W. 1993. Environmental fluctuations and the adaptive response of marine benthic organisms. *Journal of the Geological Society (London)* 150:187–91.

Polubotko, I. V. 1980. Early Carnian Halobiidae of northeast Asia. *Paleontological Journal* 1:34–41.

Polubotko, I. V. 1984. Zonal and correlation significance of Late Triassic halobiids (in Russian). *Sovetskaya Geologiya* 6:40–51.

Polubotko, I. V. 1986. Zonal complexes of late Triassic halobiids in the northeastern USSR (in Russian). In *Biostratigrafiya mezozoya Sibiri i Dal'nego Vostoka*, ed. A. L. Yanshin and A. S. Dagys, pp. 118–26. Novosibirisk: Trudy Instituta Geologii i Geofiziki.

Polubotko, I. V. 1988. On the morphology and systematics of the Late Triassic Halobiidae (bivalve mollusks) (in Russian). *Annual of the All-Union Palaeontological Society* 31:90–103.

Polubotko, I. V., Alabushev, A. I., and Bychkov, Y. M. 1990. Late Triassic halobiids (bivalve mollusks) from the Kenkeren Range (northeast USSR) (in Russian). *Annual of the All-Union Palaeontological Society* 33:122–39.

Reed, F. R. C. 1927. Palaezoic and Mesozoic fossils from Yun-Nan. *Palaeontologia Indica* 10:1–327.

Seilacher, A. 1990. Aberrations in bivalve evolution related to photo- and chemosymbiosis. *Historical Biology* 3:289–311.

Sheltema, R. S. 1977. Dispersal of marine invertebrate organisms: paleobiogeographic and biostratigraphic implications. In *Concepts and Methods of Biostratigraphy*, ed. E. G. Kauffman and J. E. Hazel, pp. 73–108. Stroudsburg, PA: Hutchinson & Ross.

Silberling, N. J. 1959. *Pre-Tertiary Stratigraphy and Upper Triassic Paleontology of the Union District, Shoshone Mountains, Nevada.* Professional Paper 322. U.S. Geological Survey.

Silberling, N. J., Jones, D. L., Blake, M. C. J., and Howell, D. G. 1987. *Lithotectonic Terranes of the Western Conterminous United States.* Miscellaneous field studies map MF-1874-C. U.S. Geological Survey.

Silberling, N. J., and Tozer, E. T. 1968. *Biostratigraphic Classification of the marine Triassic in North America.* Special paper 110. Boulder: Geological Society of America.

Smith, J. P. 1927. *Upper Triassic Marine Invertebrate Faunas of North America.* Professional paper 141. U.S. Geological Survey.

Stanley, G. D., Jr. 1988. The history of early Mesozoic reef communities: a three-step process. *Palaios* 3:170–83.

Stanley, S. M. 1979. *Macroevolution: Pattern and Process.* San Francisco: Freeman.

Stanley, S. M. 1990. The general correlation between rate of speciation and rate of extinction: fortuitous causal linkages. In *Causes of Evolution*, ed. R. M. Ross and W. D. Allmon, pp. 103–27. University of Chicago Press.

Tozer, E. T. 1967. *A Standard for Triassic Time.* Bulletin 156. Geological Survey of Canada.

Tozer, E. T. 1982. Marine Triassic faunas of North America; their significance for assessing plate and terrane movements. *Geologische Rundschau* 7:1077–104.

Tozer, E. T. 1984. *The Trias and Its Ammonoids: The Evolution of a Time Scale.* Miscellaneous report 35. Geological Survey of Canada.

Trechmann, C. T. 1918. The Trias of New Zealand. *Geological Society Quarterly (London)* 73:165–245.

Vetter, R. D., Powell, M. A., and Somero, G. N. 1991. Metozoan adaptations to hydrogen sulphide. In *Metozoan Life Without Oxygen*, ed. C. Bryant, London: Chapman & Hall.

Wignall, P. B. 1993. Distinguishing between oxygen and substrate control in fossil benthic assemblages. *Journal of the Geological Society (London)* 150:193–6.

Wignall, P. B., and Simms, M. J. 1990. Pseudoplankton. *Palaeontology* 33:359–78.

Appendix

North American locality information (see Figures 22.2 and 22.5):

1. Arctic Canada: Localities from the Canadian Arctic Archipelago representing cratonic North America; fossil localities described by Tozer (1967).

2. Arctic Alaska: Presumably allochthonous, as the North Slope Terrane, following Jones et al. (1987); fossil localities described by Blome et al. (1988), as well as many from undescribed USGS collections.

3. East-central Alaska and Yukon Territory: Presumably cratonic North America; fossil localities and descriptions from Smith (1927) and from undescribed USGS collections.

4. Southern and peninsular Alaska: Allochthonous, as part of the Alexander and Peninsular terranes (Jones et al., 1987); fossil localities mentioned by Smith (1927), as well as from undescribed USGS collections and undescribed personal collections.

5. Southeastern Alaska, Queen Charlotte Islands, British Columbia: Allochthonous, as part of the Alexander, Peninsular, and Wrangellian terranes (Monger and Berg, 1987); fossil localities mentioned by Muffler (1967), Tozer (1967), and Carter et al. (1989), as well as from undescribed USGS collections.

6. West-central British Columbia: Allochthonous, part of the Stikine Terrane (Monger and Berg, 1987); scattered localities mentioned by Tozer (1967).

7. Vancouver Island: Presumably a southern extension of Wrangellian Terrane (Monger and Berg, 1987); fossil localities from Tozer (1967) and Jeletzky (1950).

8. Northeastern British Columbia: Represents the western margin of cratonic North America (Tozer, 1982; Monger and Berg, 1987); fossil localities listed by Tozer (1967), as well as from undescribed CGS collections.

9. Northeastern Oregon: Wallowa Mountains and Hell's Canyon region of the Wallowa Terrane (Silberling et al., 1987); fossil localities mentioned by Smith (1927) and McRoberts (1993).

10. Northern California: Shasta region, part of Eastern Klamath Terrane (Silberling et al., 1987); collections and fossil localities described by Smith (1927), as well as from undescribed USGS collections and personal collections.

11. West-central Nevada: Probably allochthonous to North America as part of the Walker Lake Terrane, but presumably close to the craton during late Triassic time (N. J. Silberling, personal communication, 1992); fossil localities mentioned by Kristan-Tollmann and Tollmann (1983) and Silberling and Tozer (1968).

23 Upper Triassic Chinle Group, western United States: a nonmarine standard for late Triassic time

SPENCER G. LUCAS

Nonmarine Triassic strata have wide distributions on all the world's continents. Their precise correlation is an obvious step towards better elucidating the Triassic history of physical and biological events on land. At present, Triassic chronology has been developed almost exclusively on the basis of marine rocks and fossils. To that end, standard sections of strata and stratotype sections have been identified for the Triassic portion of the Standard Global Chronostratigraphic Scale (SGCS) (Tozer, 1984). Although much work remains to be done on the Triassic portion of the SGCS, a workable Triassic biochronology, based principally on ammonoids and conodonts, is already in place, rooted in standard and stratotype sections of marine strata.

In regard to the nonmarine stratigraphic terranes of the Triassic, quite a different situation exists. No such standard or stratotype rock sections have been identified for nonmarine Triassic strata, and their correlations are made either by direct (though usually imprecise) reference to the marine SGCS or by nonmarine biochronologic constructs such as the *Lystrosaurus* "Zone." Recently, however, Lucas (1992) and Lozovsky (1993) have suggested that standard sections of nonmarine Triassic strata be identified for use in correlation. These would not be intended to be the stratotypes of stages, but rather reference sections to aid in the correlation of nonmarine Triassic strata. Here, I follow their lead, arguing that the Upper Triassic nonmarine Chinle Group of the western United States (Figure 23.1) provides an excellent standard section for Upper Triassic nonmarine chronology. To develop this argument, I shall first briefly review the Upper Triassic portion of the SGCS to make clear my use of terms such as Carnian, Norian, and Rhaetian. I shall also discuss briefly the geochronometric and magnetostratigraphic calibration of the Upper Triassic SGCS. I shall then review the Chinle Group lithostratigraphy and biostratigraphy/biochronology, with special emphasis on the biochronologic utility of Chinle Group fossils. Finally, I shall argue that the Chinle Group, despite some failings, meets the criteria by which any standard section of rocks to be used as a basis for correlation should be judged.

Late Triassic SGCS

Tozer (1984) has reviewed the historical development of the Upper Triassic SGCS in some detail. With the exception of the Rhaetian Stage, my use of late Triassic stage and substage definitions and terminology follows his recommendations (Figure 23.2).

A long and sometimes acrimonious debate about the validity of the Rhaetian Stage was recently ended by a nearly unanimous vote of the IUGS Subcommission on Triassic Stratigraphy to recognize the Rhaetian as the youngest Triassic stage (Visscher, 1992). Nevertheless, there is as yet no agreement on a definition for the base of the Rhaetian. Here, my definition of the Rhaetian is an operational one. It refers to the time represented by the post-Knollenmergel Triassic strata of the German Keuper, essentially the Contorta and Triletes beds of northern Mecklenburg and their correlatives (Kozur, 1993, table 4b). Only this definition has any biostratigraphic/biochronologic utility when discussing nonmarine correlations, especially those based on tetrapods.

Late Triassic geochronometry and magnetostratigraphy

One of the largest problems of the Triassic time scale is the general dearth of reliable radiometric age determinations that can be unambiguously related to precise biochronology. The result of this has been wide variance in the numerical ages that have been assigned to Triassic, especially late Triassic, stage boundaries (Figure 23.3). Forster and Warrington (1985) provided the most exacting review of the available radiometric age constraints for the Triassic time scale. They identified only seven reliable numerical age determinations relevant to the boundaries of the late Triassic stages, and fewer than 10 ages relevant to numerical calibration of the Triassic–Jurassic boundary. Their boundary determinations for the Ladinian–Carnian (230 ± 5 Ma), Carnian–Norian (220 ± 5 Ma), Norian–Rhaetian (210 ± 5 Ma), and Rhaetian–Jurassic (205 ± 5 Ma) are close to the boundary determinations of Harland et al. (1990) used here.

Late Triassic magnetostratigraphy has been studied in both

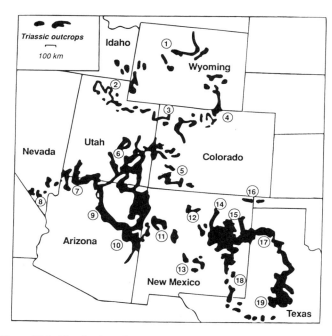

Figure 23.1. Distribution of Upper Triassic nonmarine strata of the Chinle Group in the western United States. Numbers serve to locate the stratigraphic columns shown in Figure 23.5.

SERIES	STAGES	SUBSTAGES	AMMONITE ZONES/ BIOCHRONS
LATE TRIASSIC	RHAETIAN		*Choristoceras marshi*
	NORIAN	Upper (Sevatian)	*Cochloceras amoenum*
			Gnomohalorites cordilleranus
		Middle (Alaunian)	*Himavavites columbianus*
			Drepanites rutherfordi
		Lower	*Juvavites magnus*
			Malayites dawsoni
			Stikinoceras kerri
	CARNIAN	Upper (Tuvalian)	*Klamathites macrolobatus*
			Tropites welleri
			Tropites dilleri
		Lower (Julian)	*Austrotrachyceras obesum*
			Trachyceras desatoyense

Figure 23.2. The late Triassic SGCS used in this chapter. (Adapted from Tozer, 1984.)

marine and nonmarine strata, especially in North America. Molina-Garza et al. (1991, 1993) have reviewed the publications on late Triassic magnetostratigraphy (Figure 23.4). Their review reveals reasonable consistency between the magnetization of the Chinle Group and those of other strata. Witte, Kent, and Olsen (1991) and Kent, Witte, and Olsen (1993) reported briefly on the magnetostratigraphy of the Upper Triassic–Lower Jurassic strata of the Newark Supergroup in the Newark Basin in Pennsylvania and New Jersey (USA). When the magnetostratigraphy of the

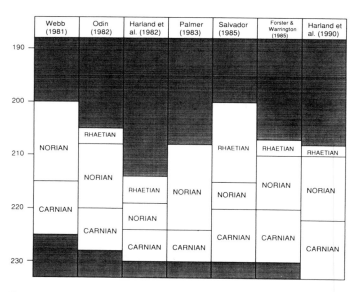

Figure 23.3. Comparison of recently proposed numerical time scales for the late Triassic.

5.5-km-thick core recovered from the Newark Basin is published in detail, it should provide a strong basis for magnetostratigraphic characterization of most of the Upper Triassic.

Chinle Group lithostratigraphy

Upper Triassic nonmarine strata are exposed in the western United States from northern Wyoming to West Texas and from southeastern Nevada to northwestern Oklahoma (Figure 23.1). Those strata, exposed over an area of about 2.3 million km^2, were deposited in a single depositional basin (Chinle Basin), with a palaeoslope down to the west-northwest (Lucas, 1993). They are assigned to the Chinle Group of late Carnian–Rhaetian age. The Chinle Group strata are mostly redbeds, though some portions are variegated blue, purple, olive, yellow, and grey. The sandstones are mostly fluvial-channel deposits that range from mature quartz arenites to very immature litharenites and greywackes. The conglomerate clasts can be extrabasinal (silica pebbles and Palaeozoic limestone pebbles) or intrabasinal (mostly nodular calcrete rip-ups), or a mixture of both. Most of the mudstones are bentonitic, except in the youngest Triassic strata. The extensive lacustrine deposits encompass analcimolite and pisolitic limestone. Within that variety one can see overall sandstone immaturity, red colouration, textures and sedimentary structures of fluvial origin, and a general abundance of volcanic detritus, lending the Chinle Group strata a lithologic character that facilitates their ready identification.

Lucas (1993) reviewed in detail the lithostratigraphy of the Chinle Group and presented a comprehensive correlation of all Chinle strata based on lithostratigraphy and biostratigraphy (Figure 23.5). About 50 lithostratigraphic terms are presently applied to strata of the Chinle Group. To simplify discussion in this chapter, I identify five regionally extensive stratigraphic

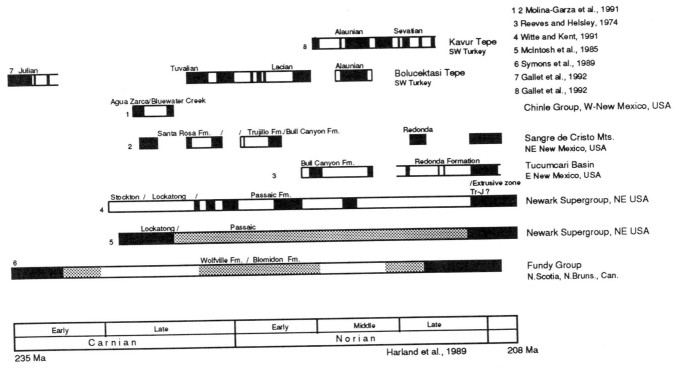

Figure 23.4. Global magnetostratigraphic correlation of the late Triassic. Normal-polarity intervals are black; grey indicates intervals of mixed polarity. (From Molina-Garza et al., 1993, with permission.)

intervals in the Chinle Group and label them A, B, C', C, and D (Figure 23.5). In terms of the classic Chinle Group stratigraphy of the Colorado Plateau, interval A refers to the Shinarump Formation and its correlatives, B to the Monitor Butte Formation plus the Blue Mesa member of the Petrified Forest Formation and their correlatives, C' to the Sonsela member of the Petrified Forest Formation and its correlatives, C to the Painted Desert member of the Petrified Forest Formation plus the Owl Rock Formation and their correlatives, and D to the Rock Point Formation and its correlatives.

Chinle Group sequence stratigraphy

Lithostratigraphic and biostratigraphic correlation of Chinle Group strata has identified two intra–Chinle Group unconformities that delimit three depositional sequences (Figure 23.5). The oldest Chinle Group sequence is the Upper Carnian Shinarump–Blue Mesa sequence. It begins with sandstones and silica-pebble conglomerates (stratigraphic interval A) that rest unconformably on older Triassic or Palaeozoic strata. Above the sandstones and conglomerates are variegated mudrock, sandstone, and minor carbonate (lower part of stratigraphic interval B). Those strata are overlain by mudrock-dominated lithofacies (upper part of stratigraphic interval B) that show extensive pedogenic modification (Lucas, 1993).

The Lower–Middle Norian second sequence of the Chinle

Group is the Moss Back–Owl Rock sequence. It begins with pervasive intrabasinal conglomeratic sandsheets (stratigraphic interval C') that rest disconformably on rocks at the top of stratigraphic interval B or older Chinle Group strata. Above stratigraphic interval C' are fluvial- and floodplain-deposited redbeds (majority of stratigraphic interval C). These redbeds are gradationally overlain by carbonate- siltstone strata of the Owl Rock Formation, which is restricted to the Colorado Plateau region of the Chinle Group outcrop area (Dubiel, 1989; Lucas, 1993).

The upper Chinle Group sequence is the Rhaetian Rock Point sequence (stratigraphic interval D). The base of the Rock Point sequence is everywhere defined by an unconformity that truncates various formations of the underlying sequences. The Rock Point lithofacies are varied, but consist mostly of repetitive, laterally persistent beds of siltstone, litharenite, and minor carbonate. The Rock Point sequence is unconformably overlain by formations of the Lower Jurassic Glen Canyon Group or younger strata.

On the Colorado Plateau, the basal Chinle Group locally consists of the late(?) Carnian Spring Mountains Formation (southeastern Nevada) and Temple Mountain Formation (east-central Utah) and a palaeo-weathering zone informally termed "mottled strata," which is variably present in other parts of the Chinle Basin (Stewart, Poole, and Wilson, 1972; Lucas, 1991a, 1993; Lucas and Marzolf, 1993). Those strata are as much as 35 m thick and may represent a depositional sequence older than

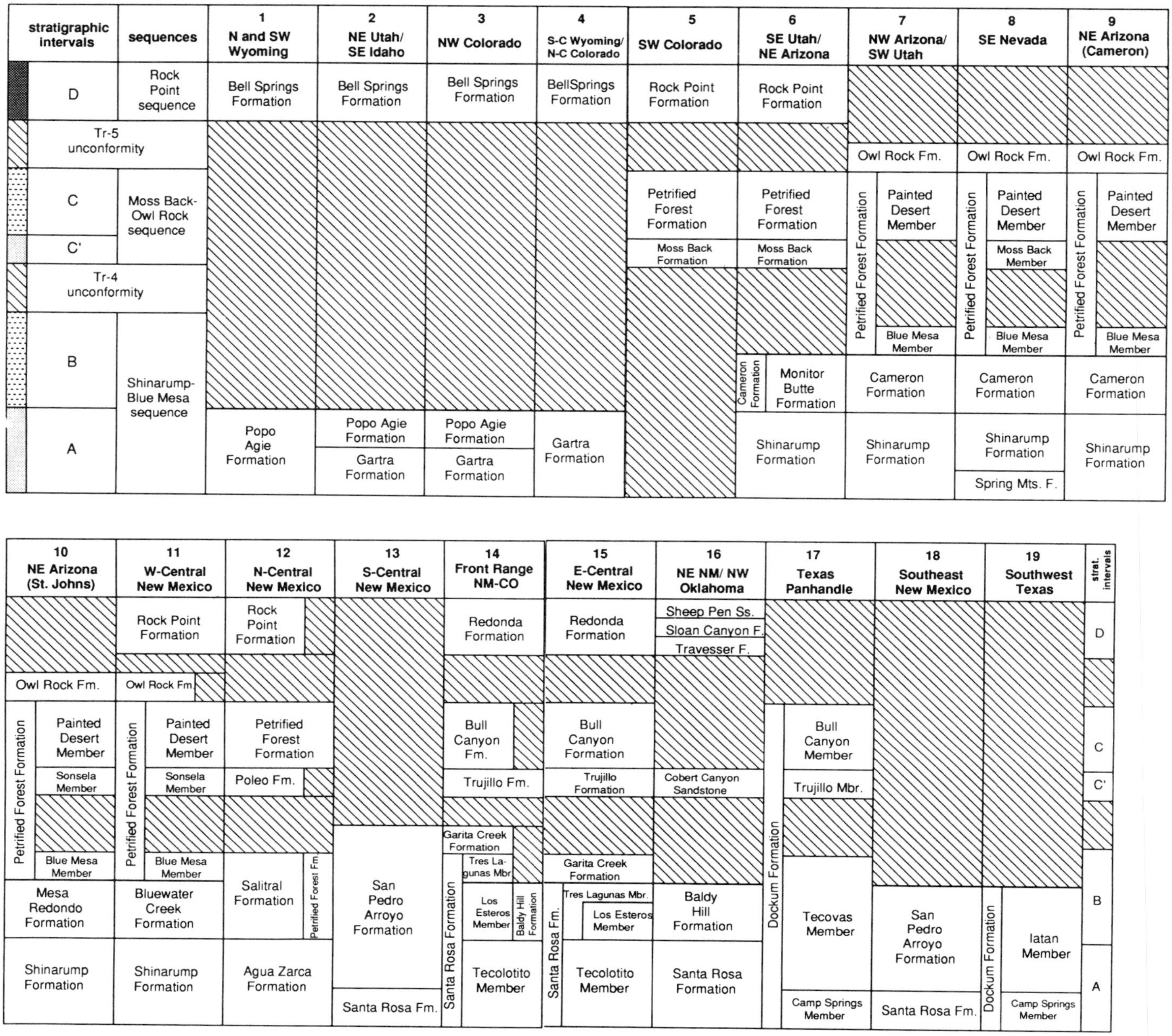
stratigraphic intervals
sequences
1 N and SW Wyoming
2 NE Utah/ SE Idaho
3 NW Colorado
4 S-C Wyoming/ N-C Colorado
5 SW Colorado
6 SE Utah/ NE Arizona
7 NW Arizona/ SW Utah
8 SE Nevada
9 NE Arizona (Cameron)
D
Rock Point sequence
Bell Springs Formation
Bell Springs Formation
Bell Springs Formation
BellSprings Formation
Rock Point Formation
Rock Point Formation
Tr-5 unconformity
Owl Rock Fm.
Owl Rock Fm.
Owl Rock Fm.
C
Moss Back-Owl Rock sequence
Petrified Forest Formation
Petrified Forest Formation
Painted Desert Member
Painted Desert Member
Painted Desert Member
C'
Moss Back Formation
Moss Back Formation
Moss Back Member
Tr-4 unconformity
Petrified Forest Formation
Petrified Forest Formation
Petrified Forest Formation
Blue Mesa Member
Blue Mesa Member
Blue Mesa Member
B
Shinarump-Blue Mesa sequence
Cameron Formation
Monitor Butte Formation
Cameron Formation
Cameron Formation
Cameron Formation
A
Popo Agie Formation
Popo Agie Formation
Popo Agie Formation
Gartra Formation
Gartra Formation
Gartra Formation
Shinarump Formation
Shinarump Formation
Shinarump Formation
Spring Mts. F.
Shinarump Formation
10 NE Arizona (St. Johns)
11 W-Central New Mexico
12 N-Central New Mexico
13 S-Central New Mexico
14 Front Range NM-CO
15 E-Central New Mexico
16 NE NM/ NW Oklahoma
17 Texas Panhandle
18 Southeast New Mexico
19 Southwest Texas
strat. intervals
Rock Point Formation
Rock Point Formation
Redonda Formation
Redonda Formation
Sheep Pen Ss.
Sloan Canyon F.
Travesser F.
D
Owl Rock Fm.
Owl Rock Fm.
Painted Desert Member
Painted Desert Member
Petrified Forest Formation
Bull Canyon Fm.
Bull Canyon Formation
Bull Canyon Member
C
Sonsela Member
Sonsela Member
Poleo Fm.
Trujillo Fm.
Trujillo Formation
Cobert Canyon Sandstone
Trujillo Mbr.
C'
Petrified Forest Formation
Petrified Forest Formation
Blue Mesa Member
Blue Mesa Member
Garita Creek Formation
Tres Lagunas Mbr
Garita Creek Formation
Tres Lagunas Mbr.
Dockum Formation
Mesa Redondo Formation
Bluewater Creek Formation
Salitral Formation
Petrified Forest Fm.
San Pedro Arroyo Formation
Los Esteros Member
Baldy Hill Formation
Los Esteros Member
Baldy Hill Formation
Tecovas Member
San Pedro Arroyo Formation
Iatan Member
B
Santa Rosa Formation
Santa Rosa Fm.
Dockum Formation
Shinarump Formation
Shinarump Formation
Agua Zarca Formation
Santa Rosa Fm.
Tecolotito Member
Tecolotito Member
Santa Rosa Formation
Camp Springs Member
Santa Rosa Fm.
Camp Springs Member
A

Figure 23.5. Correlation of nonmarine Upper Triassic strata of the Chinle Group. See Figure 23.1 for locations of numbered columns. Five informal stratigraphic intervals (A, B, C′, C, and D) identified here are used in the text and in subsequent figures.

and disconformably overlain by the Shinarump–Blue Mesa sequence (Marzolf, 1993). However, the detailed stratigraphic relationships of those oldest Chinle strata and weathering horizons have not been well studied. At present, I consider them to represent early, incised valley fills of the Shinarump–Blue Mesa sequence.

In the Mesozoic marine province of northwestern Nevada, shelf and basinal rocks are juxtaposed along the trace of the late Mesozoic Fencemaker thrust fault (Speed, 1978a,b; Oldow, 1984; Oldow, Bartel, and Gelber, 1990). Lucas and Marzolf (1993; cf. Lupe and Silberling, 1985) considered the Cane Spring Formation of the Star Peak Group and overlying strata of the Auld Lang Syne Group to be correlative and genetically related to Chinle Group strata (Figure 23.6).

In the northeastern part of the Star Peak outcrop area, the base of the Cane Spring Formation contains chert-pebble conglomerate, cobble conglomerate, and planar-crossbedded conglomeratic sandstone, up to 100 m thick (Nichols and Silberling, 1977), containing lenses of deeply weathered clastic rocks (Nichols, 1972). Those basal clastics closely resemble the

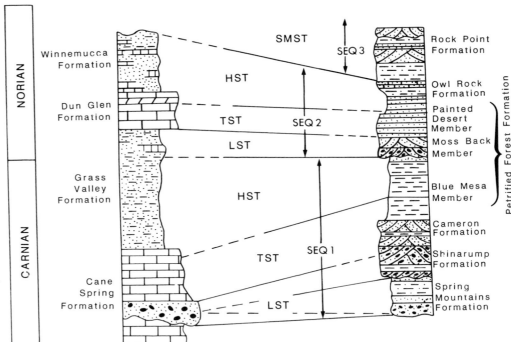

Figure 23.6. Sequence stratigraphic correlation of marine/nonmarine Upper Triassic strata in northwestern Nevada (left) with Chinle Group strata (right). LST, lowstand-system tract; HST, highstand-system tract; SMST, shelf-margin-system tract; TST, transgressive-system tract.

Shinarump Formation of the Chinle Group and were deposited on a subaerially eroded, channelized and karsted surface developed on the underlying Middle Triassic Smelser Pass member of the Augusta Mountain Formation.

The basal coarse clastics are overlain by bioclastic wackestone up to 300–400 m thick (Nichols and Silberling, 1977). In the western part of the Star Peak outcrop area, those carbonates have been informally divided into a lower, brownish-weathering, evenly bedded silty and argillaceous limestone and an upper, more massive and thickly bedded grey limestone.

The latter is overlain by the Grass Valley Formation or the equivalent Osobb Formation. Those two formations represent a voluminous influx of siliciclastic sediment, interpreted by Silberling and Wallace (1969) as a deltaic system. Palaeocurrent indicators and a westward increase in mud-to-sand ratio indicate that distributaries transported sand from delta plains in the east to delta fronts and prodeltas in the west. Wood fragments and logs are locally abundant in the fine-to-coarse sandstones of the eastern sections.

The deltaic sediments of the Grass Valley/Osobb formations are conformably overlain by massive, thick-bedded dolostone and limestone of the Dun Glen Formation. The Dun Glen is uniform in composition and thickness across its outcrop area. Its fossils suggest a shallow-water depositional environment. The Dun Glen is gradationally overlain by mixed siliciclastic and carbonate sediments of the Winnemucca Formation. The Winnemucca contains a much higher proportion of carbonate, compared with sandstone and clay, than do the deltaic sediments of the Grass Valley Formation and is the stratigraphically highest unit of the shelf sequence.

Ammonoids provide a reasonably precise biochronology for the shelfal strata of the Cane Spring Formation and Auld Lang Syne Group (Silberling, 1961; Silberling and Tozer, 1968; Silberling and Wallace, 1969; Nichols, 1972; Burke and Silberling, 1973; Nichols and Silberling, 1977). The lower Cane Spring Formation clastics approximate the *dilleri* Zone, and the Lower Norian *kerri* Zone is found in basal calcareous beds of the Osobb Formation. *Magnus* Zone ammonites are present in the uppermost Grass Valley Formation and the Dun Glen Formation, and the Winnemucca Formation probably is as young as the *columbianus* Zone.

Recent refinements of Chinle Group stratigraphy and biochronology prompted Lucas (1991b; Lucas and Marzolf, 1993) and Marzolf (1993) to reexamine the Lupe and Silberling (1985) proposal for a possible genetic relationship between deposition of Chinle Group strata and upper Star Peak–Auld Lang Syne Group strata (Figure 23.6). As noted earlier, the Chinle Group is composed of three third-order cycles that are bounded by unconformities. The criteria that define the regional extents of those unconformities are (1) evidence of extensive, subaerial weathering and channeling at the base of each depositional sequence, (2) major shifts in dominant lithologies (and facies) at the base of each sequence, (3) correlative rocks immediately above each unconformity that overlie rocks of different ages in different regions, and (4) each unconformity corresponding to a significant reorganization of the biota (Lucas, 1991b, 1993).

The conglomeratic sandsheets at the bases of the Shinarump–Blue Mesa and Moss Back–Owl Rock sequences were deposited in a broad alluvial basin characterized by extensive palaeovalley incision and prolonged subaerial exposure during periods of nondeposition (e.g., Blakey and Gubitosa, 1983; Lucas, 1991a; Lucas and Anderson, 1993). In each sequence, the basal sandsheets are overlain by fluvial and/or lacustrine facies throughout the Chinle depositional basin. Each sequence is capped by paludal carbonate and siltstone that show evidence of channeling and subaerial weathering prior to deposition of the overlying sequence.

Lucas (1991a,b, 1993; Lucas and Marzolf, 1993) interpreted the basal sandsheets as lowstand-system tracts (LSTs) whose deposition occurred in response to initial coastal onlap at the onset of a transgressive–regressive cycle. The overlying fluvial and fluviolacustrine siliciclastics represented transgressive-system tracts (TSTs), and the highstand-system tracts (HSTs) were defined as aggradational deposits of paludal-lacustrine siltstone and carbonate.

In Nevada, the Shinarump equivalent is the Cane Spring conglomerate (LST), which is overlain by shelfal, dolomitized carbonate that represents the TST. Overlying basal clastics of the Grass Valley Formation are identified as representing the HST. The lowstand surface of the next sequence is an unconformity in the Grass Valley Formation. Because the Osobb Formation contains basal Norian ammonoids (*kerri* Zone) and thus straddles the Carnian–Norian boundary, I suggest that it and its correlative, the Grass Valley Formation, contain an unconformity that reflects a basinward strand-line shift that accompanied the regression–transgression cycle that defines the Carnian–Norian boundary on the Colorado Plateau.

The Dun Glen Formation is a platform carbonate that was interpreted to represent the TST, where the transgressing base level entrapped sediment landward of the deepening shelf, and it is correlated to most of the upper Petrified Forest Formation. The Winnemucca and Owl Rock thus represent homotaxial highstand deposits. The shelf sequence, however, does not preserve age-correlative strata of the Rock Point sequence.

The importance of the sequence stratigraphic correlations just outlined lies not only in their suggestion that eustasy was a driving force in Chinle Group sedimentation but also in the fact that they provide a rationale for correlating selected Upper Triassic ammonoid zones to Chinle Group strata (Lucas, 1991b; Lucas and Luo, 1993). Those correlations are consistent with the palynologic and tetrapod-based correlations of the Chinle Group outlined later. They identify the base of the Chinle Group as approximately equivalent to the Upper Carnian (Tuvalian) *dilleri* Zone. The Carnian–Norian boundary (base of the *kerri* Zone) is about at the base of stratigraphic interval C'. Stratigraphic interval C is no younger than the Middle Norian *columbianus* Zone. Chinle Group stratigraphic interval D has no equivalent in the Nevada shelfal terrane.

Chinle Group biostratigraphy and biochronology

The first fossils from the Chinle Group brought to scientific attention were petrified logs reported by Simpson (1850). In the more than 140 years since the first discoveries by Simpson, fossils

Chinle strata		Palynomorph Zones	Megafossil plant zones
	D	Zone III	Sanmiguelia Zone
	C		
	C'		Dinophyton Zone
	B	Zone II	
	A	Zone I	Eoginkgoites Zone

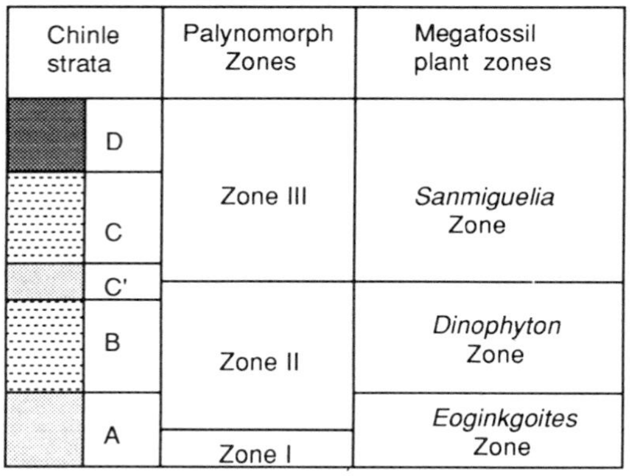

Figure 23.7. Comparison of the three palynomorph zones of Litwin et al. (1991) with the megafossil plant zones defined by Ash (1980, 1987).

of enormous diversity and abundance have been collected from Chinle Group strata across its outcrop belt. The volumes edited by Lucas and Hunt (1989) and Lucas and Morales (1993) review much of that record and the massive literature that it has prompted, obviating the need for any historical review of those discoveries here. Instead, my focus is to evaluate the biostratigraphic and biochronologic utility of each taxonomic group represented by Chinle Group fossils.

Palynology

Litwin, Traverse, and Ash (1991) and Cornet (1993) have reviewed Chinle Group palynostratigraphy in some detail. Palynomorphs are abundant and well preserved throughout the Chinle Group and have been studied for at least 30 years. Litwin et al. (1991) defined three palynomorph zones that nearly parallel the megafossil plant zones of Ash (1980, 1987), discussed later (Figure 23.7). Zone I, from the Temple Mountain Formation, is characterized by two taeniate bisaccate taxa, *Lunatisporites* aff. *L. noviaulensis* and *Infernopollenites claustratus* (the latter is also found in the Shinarump Formation). Zone II is widely distributed and is characterized by more than 100 taxa. Key taxa are *Brodispora striata*, *Michrocachrydites doubingeri*, *Lagenella martinii*, *Samaropollenites speciosus*, *Plicatisaccus badius*, *Camerosporites secatus*, and *Infernopollenites claustratus*. Zone II includes a large number of FADs (first-appearance datums) and LADs (last-appearance datums). This zone is of Tuvalian age, based on correlation to European palynomorph zones and cross-correlation to Tuvalian ammonite-bearing strata with palynomorphs (Dunay and Fisher, 1974, 1979). The youngest zone II assemblage is at or near the base of stratigraphic interval C'.

The overlying zone III assemblage encompasses all of the upper Chinle Group (stratigraphic intervals C', C, and D). This zone lacks many common-to-cosmopolitan late Carnian palynomorphs, and the FADs of several taxa – *Foveolatitriletes potoniei*, *Kyrtomisporis speciosus*, *K. laevigatus*, and *Camerosporites verucosus* – indicate a Norian age (Litwin et al., 1991).

Litwin et al. (1991) claimed that the presence of *Pseudenzonalasporites summus* indicates an early Norian age for all of zone III, citing Visscher and Brugman (1981) as authority for an early Norian age of *P. summus*. However, Visscher and Brugman (1981) indicate that *P. summus* extends into the late Norian. Indeed, *P. summus* is known from the youngest Triassic (Rhaetian) strata of the Newark Supergroup in eastern North America (upper Passaic Formation) (Cornet, 1993; Huber, Lucas, and Hunt, 1993a), so it cannot be considered indicative of only an early Norian age. Litwin et al. (1991, p. 280) also claimed that the absence of *Corollina* (= *Classopollis*), *Triancoraesporites ancorae*, *Rhaetipollis germanicus*, *Ricciisporites tuberculatus*, and *Heliosporites reissingeri* in zone III "precludes a younger age assignment for the Chinle because these palynomorphs occur commonly in late Norian (i.e., 'Rhaetian') strata in Europe, the North Atlantic (Greenland) and the Arctic." Nevertheless, *Classopollis*, *R. tuberculatus*, *H. reissingeri*, and other supposed Rhaetian index palynomorphs are known from ammonoid-bearing early Norian strata in Svalbard, calling into question their validity as Rhaetian index taxa (Smith, 1982). Furthermore, I do not consider the absence of taxa to be as strong an indicator of age as the presence of taxa, so the absence of a few so-called Rhaetian index palynomorphs from zone III is of doubtful biochronologic significance. I thus conclude that the zone III palynomorph assemblage of Litwin et al. (1991) is post-late Carnian Triassic, but I do not believe that it can provide a more precise correlation within the Norian–Rhaetian interval.

Cornet (1993) recently presented an extensive data set of palynomorphs from two wells drilled in Chinle Group strata in West Texas (stratigraphic intervals A and B). His data largely reinforce the conclusion of Litwin et al. (1991) that zones I and II are late Carnian.

Palynomorphs thus provide an important means by which Chinle Group strata are correlated. Particularly significant is the potential that palynomorphs may provide for direct linkage to the marine SGCS, thus allowing precise assignment of the Chinle Group strata to the late Carnian, the Carnian–Norian boundary, and the post–late Carnian Triassic. Clearly, the frontier for Chinle Group palynostratigraphy is in the upper part of the group, the zone III assemblage of Litwin et al. (1991). This assemblage needs more extensive documentation to subdivide it and/or arrive at a more precise, palynomorph-based correlation of the upper Chinle Group.

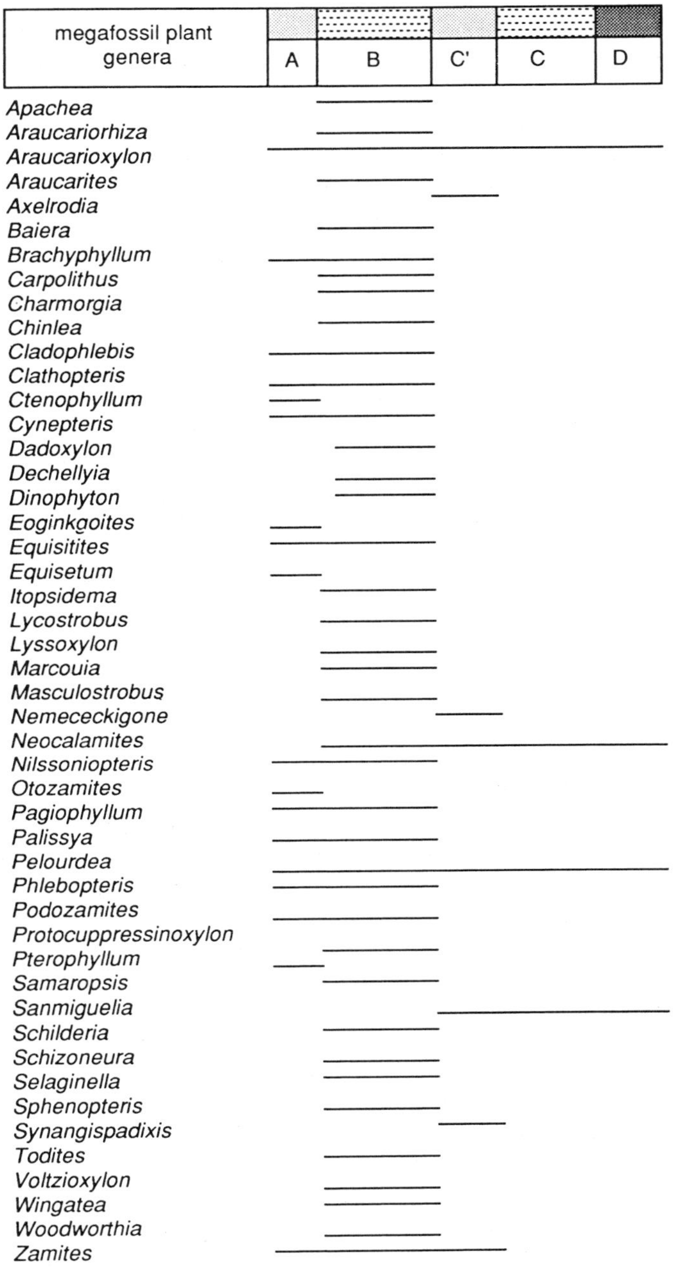

Megafossil plants

Study of Chinle Group fossil plants extends back to 1850, but the works of Daugherty (1941) and Ash (1989) and the sources they cite provide most of our knowledge of Chinle Group megafossil plants. Ash (1980, 1987) proposed that three floral zones can be recognized in Chinle Group strata: (1) the *Eoginkgoites* Zone, from stratigraphic interval A, (2) the *Dinophyton* Zone, from stratigraphic interval B, and (3) the *Sanmiguelia* Zone, from stratigraphic intervals C', C, and D.

When the stratigraphic ranges of all Chinle megafossil plant

Figure 23.8. Stratigraphic ranges of megafossil plant genera in the Chinle Group, based on the work of Ash (1989) and the sources cited therein.

genera are plotted (Figure 23.8), some clear patterns emerge: (1) the majority of genera (26 of 49, or 53%) are confined to stratigraphic interval B; (2) very few genera (8 of 49, or 16%) are found in stratigraphic intervals C', C, and D; and (3) a minority of genera (18 of 49, or 37%) are found in interval A, of which only 5 are restricted to that interval. *Eoginkgoites* is restricted to interval A, but the other genera in that interval either are rare or are known only from one locality.

The *Dinophyton* Zone of Ash (1980) is confined to stratigraphic interval B. *Dinophyton* and several other genera are restricted to that zone. Clearly, that accounts for the bulk of the Chinle megafossil flora, probably because of preservational biases. Ash's *Sanmiguelia* Zone corresponds to stratigraphic intervals C′, C, and D (Ash, 1987). *Nemececkigone* is a possible seed of *Sanmiguelia,* and *Synangispadixis* is its possible pollen-bearing organ (Cornet, 1986); so these two taxa in interval C′ are redundant of *Sanmiguelia.* Clearly, the *Sanmiguelia* Zone cannot be characterized except for the presence of *Sanmiguelia,* which is known from about half a dozen localities and is endemic to the Chinle Group.

The two older Chinle Group megafossil plant zones do allow internal correlations of Chinle Group strata that reinforce tetrapod-based correlations (i.e., the *Eoginkgoites* Zone is of Otischalkian age, and the *Dinophyton* Zone is of Adamanian age, as discussed later). Furthermore, those two zones can be correlated to strata in several of the Newark Supergroup basins, correlations that are consistent with tetrapod-based correlations (Ash, 1980; Axesmith and Kroehler, 1988; Lucas and Huber, 1993; Huber et al., 1993b). I conclude that Chinle Group late Carnian plants provide a strong basis for correlation, but that the Norian–Rhaetian megaflora of the Chinle Group needs further collection and study before it can be of much biostratigraphic/biochronologic utility.

Charophytes

Charophytes are well preserved and locally abundant in Chinle Group strata, but have been little studied. Kietzke (1987) first described Chinle Group charophytes – specimens of *Stellatochara* and *Altochara.* Lucas and Kietzke (1993) described *Porochara abjecta* Saidakovsky from the Petrified Forest National Park, Arizona. There are other reports of Chinle Group charophytes in the literature, but these are the only two documented occurrences. At present, these identified Chinle Group charophytes are from the upper part (Norian–Rhaetian) of the group: *Altochara* and *Stellatochara* from the Bull Canyon Formation, *Porochara abjecta* from the Painted Desert member of the Petrified Forest Formation, and *Stellatochara* from the Sloan Canyon Formation. Those three taxa had relatively long temporal ranges, and it is presently uncertain how useful charophytes will be in the biostratigraphy/biochronology of the Chinle Group.

Invertebrate trace fossils

Tracks, trails, burrows, and other trace fossils of invertebrates are common in most Chinle Group strata. Limuloid trackways (*Kouphichnium*) and crayfish burrows (*Camborygma*) have been studied in some detail (Caster, 1938, 1944; Hunt et al., 1993a,b; Hasiotis and Mitchell, 1993; Hasiotis and Dubiel, 1993a,b). Other, less well studied trace fossils represent a typical nonmarine ichnofacies dominated by *Scoyenia, Skolithos,* and other unornamented burrows (e.g., Lucas, Hunt, and Hayden, 1987; Hester, 1988; Hasiotis and Dubiel, 1993a). Much work needs to be undertaken on Chinle Group invertebrate trace

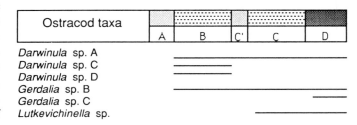

Figure 23.9. Stratigraphic distribution of ostracods in the Chinle Group (data primarily from Kietzke, 1989).

fossils as an aid to sedimentologic and palaeoecologic interpretations. Whether these fossils will be of any biostratigraphic/biochronologic utility remains to be seen.

Ostracods

Like charophytes and conchostracans, Chinle Group ostracods are widely distributed and locally abundant, but relatively little studied. Thus far, Kietzke (1987, 1989), Kietzke and Lucas (1991a), and Lucas and Kietzke (1993) have provided all of the published descriptions of Chinle Group ostracods. These ostracods are darwinulaceans, except for the cytheracean *Lutkevichinella* (Figure 23.9).

Permian–Recent darwinulaceans are unornamented nonmarine ostracods of limited biochronologic utility, because they essentially lack structural diversity beyond basic shape characteristics. Crushing and plastic deformation often are difficult to detect in these ostracods and frequently confuse proper identification. Vast taxonomic diversities among Triassic darwinulaceans have been recognized in China (e.g., Pang, 1993) and Russia (e.g., Belousova, 1961) and have been employed in biostratigraphy/biochronology, though I am skeptical of much of the alpha taxonomy on which that is based.

Instead, Kietzke and I have taken a very conservative approach to Chinle Group ostracod taxonomy, recognizing three genera encompassing six unnamed species. Two ostracod zones can be recognized in the Chinle Group: (1) the lower Chinle zone (stratigraphic intervals A and B), characterized by a large *Darwinula* associated with *Gerdalia* and a small *Darwinula,* and (2) the upper Chinle zone (stratigraphic intervals C and D), characterized by small *Darwinula* similar to *D. liulingchuanensis, Gerdalia,* and *Lutkevichinella.* This biozonation divides the Chinle Group into late Carnian and post–late Carnian portions, but the ostracod distribution is too patchy to be of great biostratigraphic/biochronologic significance. Broad correlations with, for example, Chinese late Triassic ostracods – the *Tongchuania-Darwinula-Lutkevichinella* assemblage of Xu (1988) and Pang (1993) – are obvious, but no more precise correlations are possible.

Conchostracans

Conchostracans are known from a variety of Chinle Group strata: the Tecovas member of the Dockum Formation in West

Texas, the Monitor Butte Formation in southeastern Utah, the Bluewater Creek Formation in west-central New Mexico, the Blue Mesa and Sonsela members of the Petrified Forest Formation in northeastern Arizona, the Redonda Formation in east-central New Mexico, and the Bell Springs Formation in northwestern Colorado. However, only those from the Bluewater Creek Formation have been described and illustrated. Tasch (1978) named those conchostracans "*Cyzicus (Lioestheria) wingatella.*" This is a late Carnian record, but the lack of a late Triassic conchostracan biostratigraphy/biochronology makes it impossible to use Chinle Group conchostracans for correlation. Further work on Chinle Group conchostracans is needed, as is work on the extensive conchostracan record from the Newark Supergroup of eastern North America, to develop a useful conchostracan biostratigraphy/biochronology for the late Triassic.

Insects

Insect fossils are present in the Bluewater Creek Formation at Fort Wingate, west-central New Mexico, and in the Sonsela member of the Petrified Forest Formation in the Petrified Forest National Park, Arizona, but no specimens have been described or illustrated. Insect-body fossils appear to be so rare and so poorly preserved in Chinle Group strata that I expect them to be of little biostratigraphic/biochronologic utility.

Decapod crustaceans

Crayfish-body fossils are known from Chinle Group strata in Utah and Arizona (Miller and Ash, 1988; Hasiotis and Mitchell, 1989). Miller and Ash (1988) identified a specimen from the Blue Mesa member of the Petrified Forest Formation in Arizona as *Enoploclytia porteri.* At present, Chinle Group crayfish-body fossils are too rare to be of biostratigraphic/biochronologic utility.

Bivalves and gastropods

Nonmarine mollusks (unionid bivalves and prosobranch mesogastropods) are widespread in the Chinle Group and were among the first Chinle Group fossils described (Good, 1989, 1993a,b). As Lucas (1991b, 1993) and Good (1993a,b) indicated, these fossils are much more abundant in the upper Chinle Group (stratigraphic intervals C and D) than in the lower. That probably is because of differences in favourable living habitats and preferential preservation in the more oxidized upper Chinle Group sediments. Kietzke (1987, 1989) reported "spirorbids" from the Chinle Group, but more likely they were vermiform gastropods (Weedon, 1990; Kietzke and Lucas, 1991b).

Good (1993a,b) recognized two "molluskan faunas" based on Chinle Group nonmarine mollusks that I shall refer to as "zones" (Figure 23.10): (1) a lower zone (stratigraphic interval B) characterized by two unionid taxa, *Uniomerus(?) hanleyi* and

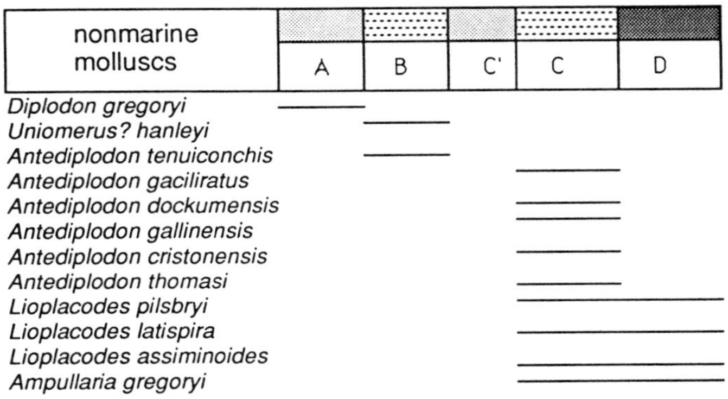

Figure 23.10. Stratigraphic distribution of nonmarine mollusks in the Chinle Group. (Adapted from Good, 1993b.)

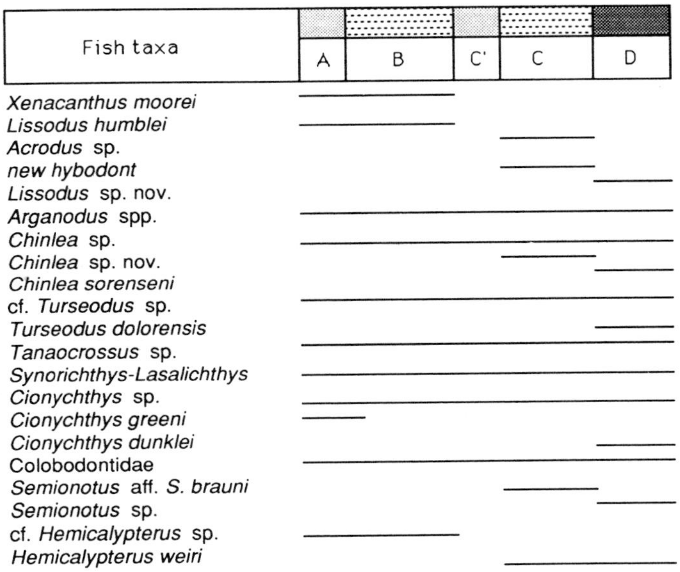

Figure 23.11. Stratigraphic distribution of fishes in the Chinle Group (data from Huber et al., 1993c).

Antediplodon hanleyi, and (2) an upper zone (stratigraphic intervals C', C, and D) of various species of *Antediplodon,* and with gastropods of the genera *Lioplacodes* and *Ampullaria.* *Diplodon gregoryi* is known from one problematic specimen from the Shinarump Formation (Reeside, 1927) and is of no biostratigraphic/biochronologic utility. The gastropod-dominated interval without unionids in uppermost Chinle Group strata (Figure 23.11) reflects more arid facies and may prove to be of biostratigraphic/biochronologic utility.

Unionids and gastropods provide a robust internal correlation of Chinle Group strata into two time intervals. They are more abundant than ostracods and are more useful in Chinle Group correlations. However, no effort has been made to compare Chinle Group nonmarine mollusks from other Upper Triassic nonmarine strata; so their utility in broader correlations remains to be tested.

Vertebrate coprolites

The Chinle Group has a prolific record of vertebrate coprolites, some of which have been described (e.g., Case, 1922; Ash, 1978; Lucas, Oakes, and Froehlich, 1985). This type of record is not uncommon in nonmarine Triassic redbeds Pangaea-wide (e.g., Rusconi, 1949; Ochev, 1974; Jain, 1983). The principal problem with interpretation of Chinle Group vertebrate coprolites, as with all vertebrate coprolites, is identification of their perpetrators.

Despite this, two coprolite morphologies are restricted to the late Carnian portion of the Chinle Group section (stratigraphic intervals A and B) (Hunt, 1992): (1) heteropolar spiral coprolites that usually are less than 3 cm long, taper to blunt points, and have four to six spirals at the wider end (Case, 1922, fig. 33A-B; Ash, 1978, fig. 2h; Hunt, 1992, fig. 4A-B); (2) broad, large (6–9-cm long) coprolites with no external features (Hunt, 1992, fig. 4C). Other morphologies of vertebrate coprolites occur throughout the Chinle Group section. Thus, a crude biostratigraphy of two zones can be based on Chinle Group vertebrate coprolites.

Fishes

Chinle Group fossil fishes range from isolated scales to complete articulated skeletons and are found at a wide variety of outcrops throughout the stratigraphic range of the Chinle Group. Huber et al. (1993c) provided a comprehensive review of Chinle Group fishes and identified three assemblages (Figure 23.11): (1) a late Carnian (stratigraphic intervals A and B) assemblage with cf. *Turseodus*, *Tanaocrossus* sp., *Cionychthys greeni*, representatives of the *Synorichthys-Lasalichthys* complex, indeterminate colobodontids, cf. *Hemicalypterus*, *Chinlea* sp. *Arganodus* sp., *Xenacanthus moorei* and *Lissodus humblei;* (2) an early–middle Norian assemblage with cf. *Turseodus*, *Tanaocrossus* sp., indeterminate redfieldiids and colobodontids, *Semionotis* cf. *S. brauni*, *Chinlea* n. sp. and *Chinlea* sp., *Arganodus*, and *Acrodus;* (3) a Rhaetian assemblage with *Turseodus dolorensis*, *Tanaocrossus kalliokoski*, *Cionychthys dunklei*, *Synorichthys stewarti*, *Lasalichthys hillsi*, indeterminate colobodontids, *Semionotis* sp., *Hemicalypterus weiri*, *Chinlea sorenseni*, *Arganodus* sp., and *Lissodus* n. sp. Most of those taxa either were long-ranging or unique to a particular assemblage; so they are of little biostratigraphic/ biochronologic utility. I do not expect that outlook to change with further collecting and study, although much work remains to be done on Chinle Group fossil fishes.

Tetrapods

Tetrapod biochronology. Tetrapod vertebrates (amphibians and reptiles) provide one of the strongest and most refined means for correlating Upper Triassic nonmarine strata. The Chinle Group has an extensive tetrapod-fossil record that has long played a key role in late Triassic correlations. Lucas and Hunt (1993b) recently organized Chinle Group tetrapod stratigraphic ranges to define four land-vertebrate faunachrons (lvfs) of late Triassic age (Figure 23.12). These lvfs rely heavily on the distributions of four groups of abundant, widespread late Triassic tetrapods: metopo-

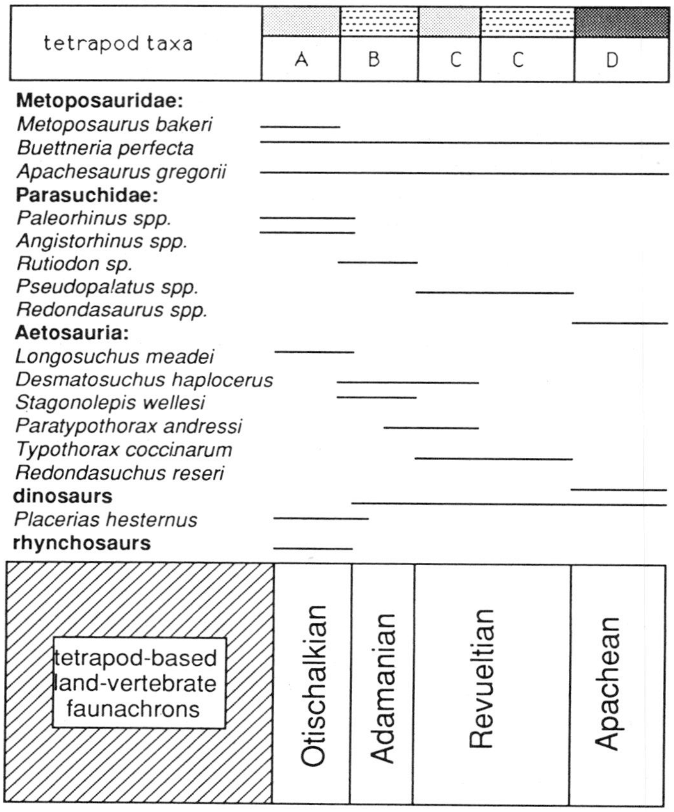

Figure 23.12. Stratigraphic distribution of biostratigraphically/ biochronologically significant tetrapods in the Chinle Group and their relationships to the land-vertebrate faunachrons of Lucas and Hunt (1993b).

saurs, phytosaurs, aetosaurs, and dicynodonts. Their biostratigraphy/biochronology is reviewed here, as is that of Chinle Group tetrapod footprints. At present, other Chinle Group tetrapods are less useful biostratigraphically/biochronologically because of inadequate sampling and/or confused taxonomy, in dire need of revision. Fraser (1993) has well emphasized the need to document better the distribution and taxonomy of small tetrapods of late Triassic age, especially sphenodontians, as an aid to correlation. This work is well under way by several palaeontologists and promises further reinforcement and refinement of Chinle Group tetrapod biochronology.

Metoposauridae. All Chinle Group temnospondyl amphibians are metoposaurs. Hunt (1993b) revised the metoposaurids and identified three biochronologically useful Chinle Group taxa (Figure 23.13): (1) *Metoposaurus bakeri*, known only from Otischalkian-age strata in West Texas; (2) *Buettneria perfecta*, known mostly from Otischalkian–Adamanian-age strata, though it occurs less frequently in Revueltian–Apachean-age strata; and (3) *Apachesaurus gregorii*, most common in Revueltian–Apachean-age strata, but also present less frequently in Otischalkian–Adamanian-age strata. Thus, the Otischalkian–Adamanian is an acme zone for *B. perfecta*, whereas the

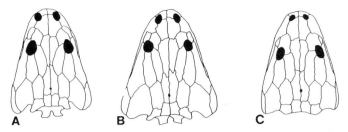

Figure 23.13. Dorsal skull roofs of Chinle Group metoposaurids: A, *Metoposaurus bakeri* Case; B, *Buettneria perfecta* (Case); C, *Apachesaurus gregorii* Hunt. Not drawn to scale. (From Hunt, 1993b, with permission.)

Revueltian–Apachean is an acme zone for *A. gregorii* (Hunt and Lucas, 1993a).

Phytosauria. The use of phytosaurs in Chinle Group biostratigraphy/biochronology has a long tradition (e.g., Camp, 1930; Gregory, 1957; Colbert and Gregory, 1957), and their fossils are abundant. Phytosaurs had a broad distribution across late Triassic Pangaea. Ballew (1989) most recently revised the taxonomy of the phytosaurs, and based on her revision five biochrons (Figure 23.12) can be defined using Chinle Group phytosaurs; for details, see Hunt (1991) and Hunt and Lucas (1991a, 1993a):

1. *Paleorhinus* biochron: *Paleorhinus* is the most primitive phytosaur. All Chinle Group occurrences of *Paleorhinus,* except its youngest occurrence in eastern Arizona,

are of Otischalkian age. *Paleorhinus* occurs in marine Tuvalian strata in Austria (Opponitzer Schichten), and its other occurrences (Figure 23.14) are generally considered to be of late Carnian age (Hunt and Lucas, 1991a). It provides important evidence of the Tuvalian age of the base of the Chinle Group and an important cross-correlation between Chinle Group nonmarine biochronology and the marine late Triassic SGCS. *Angistorhinus* co-occurs with *Paleorhinus* in the Chinle Group.

2. Overlap biochron of *Paleorhinus, Angistorhinus,* and *Rutiodon:* The oldest Chinle Group localities of Adamanian age in eastern Arizona and northern New Mexico produce rare *Paleorhinus* and *Angistorhinus* and more common *Rutiodon.*

3. The remainder of the Adamanian has produced only one phytosaur genus, *Rutiodon* (*sensu* Ballew, 1989).

4. Revueltian-age strata of the Chinle Group have produced only one phytosaur genus, *Pseudopalatus* (*sensu* Ballew, 1989). The German Stubensandstein of early–middle Norian age produces phytosaurs that Ballew (1989) identified as *Belodon, Mystriosuchus,* and *Nicrosaurus.* Some of those specimens appear to be congeneric with North American specimens she termed *Pseudopalatus.* This supports a Revueltian–Stubensandstein correlation and assignment of an early–middle Norian age to the Revueltian.

5. The youngest Chinle Group phytosaur, of Apachean age, is *Redondasaurus* (Hunt and Lucas, 1993b). This endemic taxon is the most evolutionarily advanced

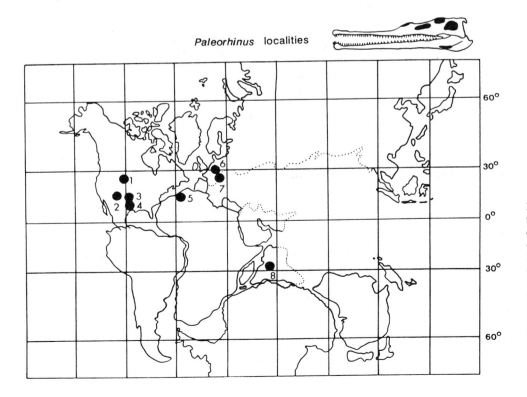

Paleorhinus localities

Figure 23.14. *Paleorhinus* localities of the late Triassic Pangaean supercontinent: 1, Popo Agie Formation, central Wyoming, USA; 2, lowermost Blue Mesa member of Petrified Forest Formation, east-central Arizona, USA; 3, Camp Springs member of Dockum Formation, West Texas, USA; 4, Iatan member of Dockum Formation, West Texas, USA; 5, Argana Formation, Morocco; 6, Blasensandstein, Germany; 7, Opponitzer Schichten, Austria; 8, Maleri and Tiki formations, India.

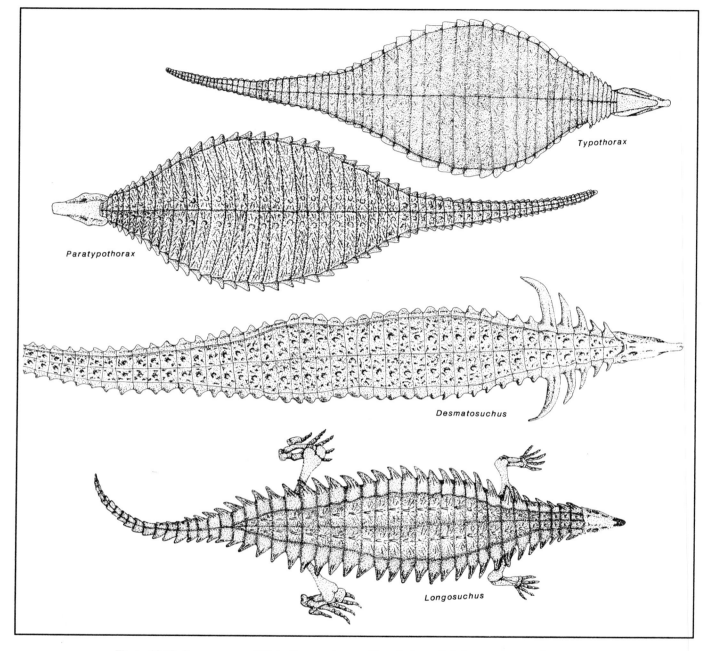

Figure 23.15. Some common Chinle Group aetosaurs (dorsal views of skeletons, not to scale). (Drawing by Randy Pence.)

phytosaur and thus suggests that the Apachean is of late Norian or Rhaetian age.

Hunt (1993a) has recently completed, but not yet published, a revision for the phytosaurs that somewhat alters the taxonomy of Ballew (1989). However, his taxonomy does not change the phytosaur-based biochronology and the correlations outlined here. The problems posed by phytosaurs as index fossils reside in the need to have a nearly complete phytosaur skull to arrive at

precise identification, thus rendering the vast majority of phytosaur fossils, which are isolated bones, teeth, and skull fragments, useless for correlation. Despite that, sufficient numbers of phytosaur skulls are known from the Chinle Group and elsewhere to continue their long-standing use in late Triassic tetrapod biochronology.

Aetosauria. Aetosaur fossils (Figure 23.15) are at least as abundant as phytosaur fossils in strata of the Chinle Group.

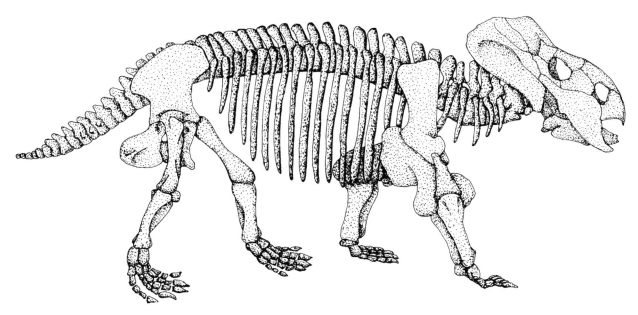

Figure 23.16. Skeletal reconstruction of the 2-m-long dicynodont *Placerias*. (Drawing by Randy Pence.)

Furthermore, a genus-level identification of an aetosaur can be made from an isolated armor plate or fragment of a plate. For example, the syntype specimens of *Typothorax coccinarum* Cope, a common upper Chinle Group aetosaur, are only fragments of paramedian plates, but they are diagnostic (Lucas and Hunt, 1992). Skulls are not needed; so aetosaurs provide index fossils that are much easier to identify than those of phytosaurs. Aetosaurs also had a broad distribution across late Triassic Pangaea, and some aetosaur genera found in the Chinle Group (i.e., *Longosuchus, Stagonolepis, Desmatosuchus,* and *Paratypothorax*) are also known from the Newark Supergroup and/or western Europe. Aetosaurs thus provide an important basis for correlating Chinle Group strata and other nonmarine late Triassic strata.

Six aetosaur biochrons can be identified in the Chinle Group (Figure 23.12):

1. *Longosuchus* biochron: *Longosuchus* is of Otischalkian age and co-occurs in the lowermost Chinle Group with *Desmatosuchus,* a co-occurrence also documented in the late Carnian Pekin Formation of the Newark Supergroup (Hunt and Lucas, 1990).

2. *Stagonolepis* (= *Calyptosuchus*) biochron: *Stagonolepis* is confined to strata of Adamanian age in the Chinle Group. It is also known from the late Carnian Lossiemouth Sandstone of Scotland (Hunt and Lucas, 1991b).

3. *Paratypothorax* biochron: *Paratypothorax* in the Chinle Group ranges in age from Adamanian to Revueltian (Hunt and Lucas, 1992a). In Germany it has a shorter temporal range, being known only from the early Norian Lower Stubensandstein (Long and Ballew, 1985).

4. *Desmatosuchus* biochron: *Desmatosuchus* ranges in age from Otischalkian to early Revueltian in the Chinle Group.

5. *Typothorax* biochron: This endemic Chinle Group aetosaur is of Revueltian age.

6. *Redondasuchus* biochron: This endemic Chinle Group taxon (Hunt and Lucas, 1991c) is of Apachean age.

Dicynodonts. Non-archosauromorph reptiles and mammals are rare in the Chinle Group, with the exception of the dicynodont *Placerias* (Figure 23.16). The Chinle Group dicynodonts are *Placerias hesternus* (= *P. gigas*) and cf. *Ischigualastia* sp. *Placerias* is known from Otischalkian–Adamanian strata in Arizona and Wyoming, whereas the possible *Ischigualastia* is known from earliest Adamanian strata in New Mexico (Lucas and Hunt, 1993a). *Placerias* (= *Mohgreberia*) is also known from the Pekin Formation of North Carolina and the Argana Formation of Morocco. Its occurrences in Wyoming, Arizona, North Carolina, and Morocco define a *Placerias* biochron of late Carnian age. The possible *Ischigualastia* in the Chinle Group suggests a possible direct correlation to Argentinian and Brazilian strata of late Carnian age that contain this large dicynodont (Cox, 1965; Araujo and Gonzaga, 1980; Rogers et al., 1993).

Tetrapod footprints. Tetrapod footprints are abundant in the uppermost strata of the Chinle Group, stratigraphic interval D (Hunt and Lucas, 1992b). Only a handful of tetrapod footprints are known from older Chinle Group strata (Hunt et al., 1993a); so they are of no biostratigraphic/biochronologic significance.

The tetrapod ichnofauna of stratigraphic interval D is dominated by the ichnotaxa *Brachychirotherium, Grallator, Pseudotet-*

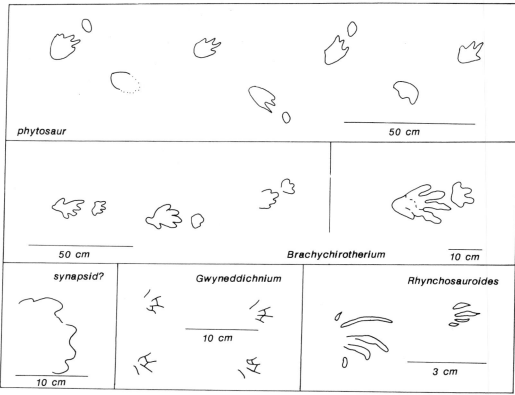

Figure 23.17. Characteristic Chinle Group tetrapod footprints from stratigraphic interval D.

rasauropus, Tetrasauropus, and *Gwyneddichnium* (e.g., Lockley et al., 1992; Hunt et al., 1989, 1993a,b) (Figure 23.17). These footprints provide a basis for intra-Chinle correlation of strata of interval D in Wyoming, Utah, Colorado, New Mexico, and Oklahoma. They also indicate a late Triassic age, by comparison with tetrapod-footprint assemblages in eastern North America, Europe, and South Africa. Most significant is the prosauropod footprint *Tetrasauropus*. Prosauropod footprints are also known from the lower Elliot Formation (lower Stormberg Group) of South Africa (Ellenberger, 1970; Olsen and Galton, 1984) and marginal marine strata of Rhaetian age in Switzerland (Furrer, 1993). Their distribution may define a *Tetrasauropus* biochron of Rhaetian age recognizable across much of Pangaea.

Chinle Group geochronometry and magnetostratigraphy

Large numbers of radiometric age determinations have been reported from Chinle Group rocks, mostly from uranium ores on the Colorado Plateau (e.g., Miller and Kulp, 1963; Young, 1964; Stewart et al., 1986; Ash, 1992). Those ages are either too young or too old (well outside the 230–200-Ma time span within which Chinle Group deposition falls by any numerical time scale). Most of the Chinle Group contains mudstones and sandy mudstones rich in volcanic detritus, consisting of rounded grains of altered tuff, lava fragments, euhedral biotite, sanidine, and plagioclase

(Allen, 1930; Waters and Granger, 1953; Cadigan, 1963; Schultz, 1963). Basal conglomerates of the Chinle Group also contain local deposits of a few pebbles and cobbles of volcanic rocks (Stewart et al., 1972; Dodge, 1973). Little effort has been made to obtain numerical ages from this volcanic detritus. It may hold promise for direct numerical calibration of Chinle Group deposition.

Molina-Garza et al. (1991, 1993) summarized the published Chinle Group magnetostratigraphy, which has largely been undertaken on outcrops in New Mexico. Their cumulative magnetostratigraphy for the Chinle Group (Figure 23.18) will be evaluated further by ongoing and unpublished research by D. Bazard, D. Kent, M. Steiner, and others. Rocks of the Chinle Group have characteristic well-defined high-temperature magnetizations carried by hematite. These are early acquired chemical-remnant magnetizations that identify Chinle Group strata as reliable sources of late Triassic magnetic-polarity history (Bazard and Butler, 1991; Molina-Garza et al., 1991, 1993).

A nonmarine standard

As a first step toward developing more precise correlations of nonmarine Triassic strata, we need to consider those stratigraphic sections that can serve as standards from which to correlate other sections, and then ultimately correlate those standards to the marine SGCS. Such standards can then serve as

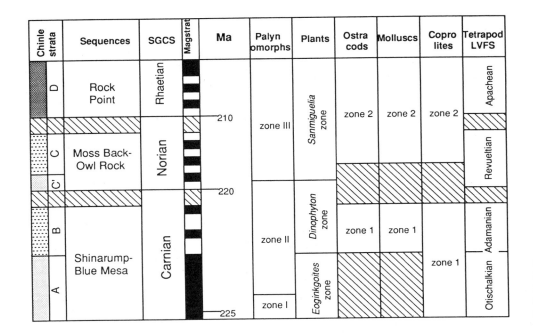

Figure 23.18. Summary of Chinle Group lithostratigraphy, biostratigraphy/biochronology, and magnetostratigraphy (numerical time scale from Harland et al., 1990).

the hubs of correlation networks for nonmarine Triassic strata. Ultimately, when it comes time to establish GSSPs (global stratotype sections and points) for the important nonmarine intervals of Triassic time, they will most likely be in the standard sections.

To establish standard sections, we must have strata that meet the following criteria: (1) extensive outcrops, (2) ready accessibility, (3) abundantly and diversely fossiliferous, (4) amenable to dating methods other than biochronology (i.e., geochronometry, magnetostratigraphy), and (5) complete temporal representation. The Chinle Group certainly has an extensive outcrop area (Figure 23.1). Most Chinle Group outcrops are on lands owned by the U.S. government, ensuring continuous access. Such access is by paved roads or all-weather unpaved roads to most outcrops. The foregoing summary indicates the abundance, diversity, and biostratigraphic/biochronologic utility of Chinle Group fossils (Figure 23.18). The chemical-remnant magnetizations of Chinle Group rocks faithfully record part of late Triassic magnetic-polarity history, and the abundant volcanic detritus in much of the Chinle Group may represent unrealized potential for radiometric dating.

Despite these great advantages as a standard, the Chinle Group does not represent all of late Triassic time. That is its greatest drawback as a standard. No early Carnian time is represented, and the two intragroup unconformities (Figures 23.5 and 23.18) are definite hiatuses of uncertain duration. Furthermore, most of the Chinle deposition was by fluvial processes, which produced strata riddled with numerous hiatuses of varying duration. Such incompleteness is typical of most Upper Triassic nonmarine sequences, and the Chinle Group has superior temporal representation when compared with them. An exception to this is the Newark Supergroup, in eastern North America, which contains thick sequences of lacustrine strata that record more of late Triassic time than do Chinle Group strata. Yet, in regard to most other criteria (extent of outcrop, accessibility, fossil record), the Newark Supergroup is vastly inferior to the Chinle Group. Thus, in most respects the Chinle Group is an ideal nonmarine standard for late Triassic time, although it does not record all of the late Triassic.

Summary

Nonmarine Upper Triassic strata of the Chinle Group are exposed over a 2.3-million km^2 outcrop area in the western United States (Idaho, Wyoming, Utah, Nevada, Colorado, Oklahoma, Texas, New Mexico, and Arizona). Chinle Group strata were deposited during most of the late Triassic (late Carnian–Rhaetian) in a single depositional basin (Chinle Basin) with a regional palaeoslope down to the west-northwest. The Chinle Group is as much as 600 m thick and consists of fluvial, lacustrine, and minor eolian facies of siliciclastic sediments that are mostly redbeds. Chinle Group fossils have been collected for more than 140 years and include palynomorphs, megafossil plants, charophytes, invertebrate trace fossils, ostracods, conchostracans, insects, decapod crustaceans, bivalves, gastropods, tetrapod footprints, fishes, and tetrapod-body fossils. Magnetostratigraphic characterization of many Chinle Group intervals has been undertaken, but Chinle volcanic detritus has not provided reliable syn-depositional numerical ages.

Lithostratigraphy supported by diverse biostratigraphy provides a precise correlation of Chinle Group strata across their outcrop belt. This correlation identifies three intragroup depositional sequences bounded by basinwide unconformities. The two older sequences have a genetic relationship to late Carnian–

middle Norian marine shelfal rocks in northwestern Nevada. This relationship allows selected late Triassic ammonite zones to be correlated to Chinle Group strata. Nonmarine biochronology, especially of palynomorphs, megafossil plants, and tetrapod vertebrates, and magnetochronology indicate that the oldest Chinle strata are of late Carnian (Tuvalian) age. Younger Chinle Group strata are of early Norian–Rhaetian age. The Chinle Group has an extensive outcrop area, is relatively thick, and is very accessible. However, it does not represent all of late Triassic time – no early Carnian (Julian) strata are present, and the two intragroup unconformities represent hiatuses of unknown duration. Nevertheless, its extensive fossil record, replicatable magnetostratigraphy, and potential for numerical age dating make the Chinle Group an excellent standard for correlation of the nonmarine Upper Triassic strata of Pangaea.

Acknowledgments

Adrian Hunt, Phillip Huber, John Marzolf, and Orin Anderson have been instrumental in developing much of the Chinle Group biostratigraphy/biochronology and lithostratigraphy presented here and elsewhere. The New Mexico Museum of Natural History and Science, the National Geographic Society, the Petrified Forest Museum Association, and the New Mexico Bureau of Mines and Mineral Resources have supported this research. Numerous landowners and land-management agencies have granted access to land, and numerous museum curators and collection managers have allowed study of specimens under their care. My thank to P. Huber and A. Hunt, whose comments on an earlier draft of this manuscript improved its content.

References

Allen, V. T. 1930. Triassic bentonite of the Painted Desert. *American Journal of Science* 19:283–8.

Araujo, D. C., and Gonzaga, T. D. 1980. Uma nova especie de *Jachaleria* (Therapsida, Dicynodonta) do Triassico de Brasil. In *Actas II Congreso de Paleontologia y Bioestratigrafia y I Congreso Latinoamericano de Paleontologia Buenos Aires,* vol. 1, pp. 159–74.

Ash, S. R. 1978. Coprolites. In *Geology, Paleontology, and Paleoecology of a Late Triassic Lake, Western New Mexico,* ed. S. R. Ash, pp. 69–73. Geology studies 25. Brigham Young University.

Ash, S. R. 1980. Upper Triassic floral zones of North America. In *Biostratigraphy of Fossil Plants,* ed. D. L. Dilcher and T. N. Taylor, pp. 153–70. Stroudsburg, PA: Dowden, Hutchinson & Ross.

Ash, S. R. 1987. The Upper Triassic red bed flora of the Colorado Plateau, western United States. *Journal of the Arizona-Nevada Academy of Science* 22:95–105.

Ash, S. R. 1989. A catalog of Upper Triassic plant megafossils of the western United States through 1988. In *Dawn of the Age of Dinosaurs in the American Southwest,* ed. S. G. Lucas and A. P. Hunt, pp. 189–222. Albuquerque: New Mexico Museum of Natural History.

Ash, S. R. 1992. The Black Forest Bed, a distinctive unit in the Upper Triassic Chinle Formation, northeastern Arizona. *Journal of the Arizona-Nevada Academy of Science* 24–25:59–73.

Axesmith, B. J., and Kroehler, P. A. 1988. Upper Triassic *Dinophyton* zone plant fossils from the Stockton Formation in southeastern Pennsylvania. *The Mosasaur* 4:45–7.

Ballew, K. L. 1989. A phylogenetic analysis of Phytosauria from the Late Triassic of the western United States. In *Dawn of the Age of Dinosaurs in the American Southwest,* ed. S. G. Lucas and A. P. Hunt, pp. 309–39. Albuquerque: New Mexico Museum of Natural History.

Bazard, D. R., and Butler, R. F. 1991. Paleomagnetism of the Chinle and Kayenta formations, New Mexico and Arizona. *Journal of Geophysical Research* 96:9837–46.

Belousova, Z. D. 1961. Ostrakody nizhny Triasovye [Ostracods of the Lower Triassic]. *Byuletin Moskovskova Ispytatyuschickh Prirody Otdel Geologii* 36:127–47.

Blakey, R. C., and Gubitosa, R. 1983. Late Triassic paleogeography and depositional history of the Chinle Formation, southern Utah and northern Arizona. In *Mesozoic Paleogeography of the West-Central United States,* ed. M. W. Reynolds and E. D. Dolly, pp. 57–76. Denver: RMS-SEPM.

Burke, D. B., and Silberling, N. J. 1973. *The Auld Lang Syne Group, of Late Triassic and Jurassic(?) Age, North-Central Nevada.* Bulletin 1394-E. U.S. Geological Survey.

Cadigan, R. A. 1963. *Tuffaceous Sandstones in the Triassic Chinle Formation, Colorado Plateau.* Professional paper 475-B. U.S. Geological Survey.

Camp, C. L. 1930. A study of the phytosaurs with description of new material from western North America. *Memoirs, University of California* 10:1–170.

Case, E. C. 1922. *New Reptiles and Stegocephalians from the Upper Triassic of Western Texas.* Publication 321. Carnegie Institution of Washington.

Caster, K. E. 1938. A restudy of the tracks of *Paramphibius. Journal of Paleontology* 12:3–60.

Caster, K. E. 1944. Limuloid trails from the Upper Triassic (Chinle) of the Petrified Forest National Monument, Arizona. *American Journal of Science* 242:74–84.

Colbert, E. H., and Gregory, J. T. 1957. Correlation of continental Triassic sediments by vertebrate fossils. *Bulletin of the Geological Society of America* 68:1456–67.

Cornet, B. 1986. The leaf venation and reproductive structures of a Late Triassic angiosperm, *Sanmiguelia lewisii. Evolutionary Theory* 7:231–309.

Cornet, B. 1993. Applications and limitations of palynology in age, climatic and paleoenvironmental analyses of Triassic sequences in North America. In *The Nonmarine Triassic,* ed. S. G. Lucas and M. Morales, pp. 75–93. Bulletin 3. Albuquerque: New Mexico Museum of Natural History and Science.

Cox, C. B. 1965. New Triassic dicynodonts from South America, their origins and relationships. *Philosophical Transactions of the Royal Society of London B* 248:457–516.

Daugherty, L. H. 1941. *The Upper Triassic Flora of Arizona.* Publication 526. Carnegie Institution of Washington.

Dodge, C. N. 1973. Pebbles from the Chinle and Morrison Formations. *New Mexico Geological Society Guidebook* 24:114–21.

Dubiel, R. F. 1989. Depositional and climatic setting of the Upper Triassic Chinle Formation, Colorado Plateau. In *Dawn of the Age of Dinosaurs in the American Southwest,* ed. S. G. Lucas and A. P. Hunt, pp. 171–87. Albuquerque: New Mexico Museum of Natural History.

Dunay, R. E., and Fisher, M. J. 1974. Late Triassic palynofloras of North America and their European correlatives. *Review of Palaeobotany and Palynology* 17:179–86.

Dunay, R. E., and Fisher, M. J. 1979. Palynology of the Upper Triassic Dockum Group (Upper Triassic), Texas, U.S.A. *Review of Palaeobotany and Palynology* 28:61–92.

Ellenberger, P. 1970. Les niveaux paleontologiques de premiere apparition des Mammiferes primordiaux en Afrique du Sud et leur ichnologie: establissement de zones stratigraphiques detaillees dans le Stormberg du Lesotho, (Afrique du Sud) (Trias Superieur a Jurassique). In *IUCS 2nd Symposium on Gondwana Stratigraphy and Paleontology*, ed. S. H. Haughton, pp. 343–70. Pretoria: Council for Scientific and Industrial Research.

Forster, S. C., and Warrington, G. 1985. Geochronology of the Carboniferous, Permian and Triassic. In *The Chronology of the Geological Record*, ed. N. J. Snelling, pp. 99–113. Oxford: Blackwell Scientific Publishers.

Fraser, N. C. 1993. A new sphenodontian from the early Mesozoic of England and North America: implications for correlating early Mesozoic continental deposits. In *The Nonmarine Triassic*, ed. S. G. Lucas and M. Morales, pp. 135–9. Bulletin 3. Albuquerque: New Mexico Museum of Natural History and Science.

Furrer, H. 1993. Entdeckung und Untersuchung der Dinosaurierfahrten im Nationalpark. *Parc Naziunal Svizzer, Cratschla Ediziuns Specialis* 1:1–19.

Gallet, Y., Besse, J., Krystyn, L., Marcoux, J., and Theveniaut, H. 1992. Magnetostratigraphy of the Late Triassic Bolucektasi Tepe section (southwestern Turkey): implications for changes in magnetic reversal frequency. *Physics of the Earth and Planetary Interiors* 73:85–108.

Good, S. C. 1989. Nonmarine Mollusca in the Upper Triassic Chinle Formation and related strata of the Western Interior: systematics and distribution. In *Dawn of the Age of Dinosaurs in the American Southwest*, ed. S. G. Lucas and A. P. Hunt, pp. 233–48. Albuquerque: New Mexico Museum of Natural History.

Good, S. C. 1993a. Molluscan paleobiology of the Upper Triassic Chinle Formation, Arizona and Utah. Unpublished Ph.D. dissertation, University of Colorado, Boulder.

Good, S. C. 1993b. Stratigraphic distribution of the mollusc fauna of the Chinle Formation and molluscan biostratigraphic zonation. In *The Nonmarine Triassic*, ed. S. G. Lucas and M. Morales, pp. 155–9. Bulletin 3. Albuquerque: New Mexico Museum of Natural History and Science.

Gregory, J. T. 1957. *Significance of Fossil Vertebrates for Correlation of Late Triassic Continental Deposits of North America*. Report of the 20th session, International Geological Congress, Section II, pp. 7–25.

Hahn, G., LePage, J. C., and Wouters, G. 1984. Cynodontier-Zahne aus der Ober-Trias von Medernach, Grossherzogtum, Luxemburg. *Bulletin de la Societé Belge de Geologie* 93:357–73.

Harland, W. B., Armstrong, R. L., Cox, A. V., Craig, L. E., Smith, A. G., and Smith, D. G. 1990. *A Geologic Time Scale, 1989*. Cambridge University Press.

Harland, W. B., Cox, A. V., Llewellyn, P. G., Pickton, C. A. G., Smith, A. G., and Walters, R. 1982. *A Geologic Time Scale*. Cambridge University Press.

Hasiotis, S. T., and Dubiel, R. F. 1993a. Continental trace fossils of the Upper Triassic Chinle Formation, Petrified Forest National Park, Arizona. In *The Nonmarine Triassic*, ed. S. G. Lucas and M. Morales, pp. 175–8. Bulletin 3. Albuquerque: New Mexico Museum of Natural History and Science.

Hasiotis, S. T., and Dubiel, R. F. 1993b. Crayfish burrows and their paleohydrologic significance – Upper Triassic Chinle Formation, Fort Wingate, New Mexico. In *The Nonmarine Triassic*, ed. S. G. Lucas and M. Morales, pp. 624–6. Bulletin 3. Albuquerque: New Mexico Museum of Natural History and Science.

Hasiotis, S. T., and Mitchell, C. E. 1989. Lungfish burrows in the Upper Triassic Chinle and Dolores Formations, Colorado Plateau – discussion: new evidence suggests origin by a burrowing decapod crustacean. *Journal of Sedimentary Petrology* 59:871–5.

Hasiotis, S. T., and Mitchell, C. E. 1993. A comparison of crayfish burrow morphologies: Triassic and Holocene fossil, paleo-, and neo-ichnological evidence, and the identification of their burrowing signatures. *Ichnos* 2:291–314.

Haubold, H. 1984. *Saurierfahrten*. Wittenberg Lutherstadt: A. Ziemsen Verlag.

Hester, P. M. 1988. Depositional environments in an Upper Triassic lake, east-central New Mexico. M.S. thesis, University of New Mexico, Albuquerque.

Huber, P., Lucas, S. G., and Hunt, A. P. 1993a. Revised age and correlation of the Upper Triassic Chatham Group (Deep River basin, Newark Supergroup), North Carolina. *Southeastern Geology* 33:171–93.

Huber, P., Lucas, S. G., and Hunt, A. P. 1993b. Vertebrate biochronology of the Newark Supergroup, Triassic, eastern North America. In *The Nonmarine Triassic*, ed. S. G. Lucas and M. Morales, pp. 179–86. Bulletin 3. Albuquerque: New Mexico Museum of Natural History and Science.

Huber, P., Lucas, S. G., and Hunt, A. P. 1993c. Late Triassic fish assemblages of the North American Western Interior and their biochronologic significance. *Bulletin of the Museum of Northern Arizona* 59:51–66.

Hunt, A. P. 1991. The early diversification pattern of dinosaurs in the late Triassic. *Modern Geology* 16:43–60.

Hunt, A. P. 1992. Late Pennsylvanian coprolites from the Kinney Brick Quarry, central New Mexico, with notes on the classification and utility of coprolites. *Bulletin of the New Mexico Bureau of Mines and Mineral Resources* 138:221–9.

Hunt, A. P. 1993a. Vertebrate paleontology and biostratigraphy of the Bull Canyon Formation, east-central New Mexico, with revisions of the families Metoposauridae (Amphibia: Temnospondyli) and Parasuchidae (Reptilia: Archosauria). Ph.D. dissertation, University of New Mexico, Albuquerque.

Hunt, A. P. 1993b. A revision of the Metoposauridae (Amphibia: Temnospondyli) of the Late Triassic with description of a new genus from the western United States. *Bulletin of the Museum of Northern Arizona* 59:67–97.

Hunt, A. P., Lockley, M. G., and Lucas, S. G. 1993a. Vertebrate and invertebrate tracks and trackways from Upper Triassic strata of the Tucumcari basin, east-central New Mexico, USA. In *The Nonmarine Triassic*, ed. S. G. Lucas and M. Morales, pp. 199–201. Bulletin 3. Albuquerque: New Mexico Museum of Natural History and Science.

Hunt, A. P., and Lucas, S. G. 1990. Re-evaluation of "*Typothorax*" *meadei*, a Late Triassic aetosaur from the United States. *Paläontologische Zeitschrift* 64:317–28.

Hunt, A. P., and Lucas, S. G. 1991a. The *Paleorhinus* biochron and the correlation of the nonmarine Upper Triassic of Pangaea. *Palaeontology* 34:487–501.

Hunt, A. P., and Lucas, S. G. 1991b. A new rhynchosaur from West Texas (USA) and the biochronology of Late Triassic rhynchosaurs. *Palaeontology* 34:487–501.

Hunt, A. P., and Lucas, S. G. 1991c. A new aetosaur from the Upper Triassic of eastern New Mexico. *Neues Jahrbuch für Geologie und Paläontologie, Abhandlungen* 1991:728–36.

Hunt, A. P., and Lucas, S. G. 1992a. The first occurrence of the aetosaur *Paratypothorax andressi* (Reptilia: Aetosauria) in the western United States and its biochronological significance. *Paläontologische Zeitschrift* 66:147–57.

Hunt, A. P., and Lucas, S. G. 1992b. Stratigraphic distribution and age of vertebrate tracks in the Chinle Group (Upper Triassic), western North America. *Geological Society of America, Abstracts with Programs* 24:19.

Hunt, A. P., and Lucas, S. G. 1993a. Taxonomy and stratigraphic distribution of Late Triassic metoposaurid amphibians from Petrified Forest National Park, Arizona. *Journal of the Arizona-Nevada Academy of Science* 27:89–95.

Hunt, A. P., and Lucas, S. G. 1993b. A new phytosaur (Reptilia: Archosauria) genus from the uppermost Triassic of the western United States and its biochronological significance. In *The Nonmarine Triassic,* ed. S. G. Lucas and M. Morales, pp. 193–6. Bulletin 3. Albuquerque: New Mexico Museum of Natural History and Science.

Hunt, A. P., Lucas, S. G., and Kietzke, K. K. 1989. Dinosaur footprints from the Redonda Member of the Chinle Formation (Upper Triassic), east-central New Mexico. In *Dinosaur Tracks and Traces,* ed. D. D. Gillette and M. G. Lockley, pp. 277–80. Cambridge University Press.

Hunt, A. P., Lucas, S. G., and Lockley, M. G. 1993b. Fossil limuloid trackways from Petrified Forest National Park, Arizona, USA. In *The Nonmarine Triassic,* ed. S. G. Lucas and M. Morales, pp. 205–7. Bulletin 3. Albuquerque: New Mexico Museum of Natural History and Science.

Jain, S. L. 1983. Spirally coiled "coprolites" from the Upper Triassic Maleri Formation, India. *Palaeontology* 26:813–29.

Kent, D. V., Witte, W. K., and Olsen, P. E. 1993. A complete late Triassic magnetostratigraphy from the Newark basin. In *The Nonmarine Triassic,* ed. S. G. Lucas and M. Morales, p. 100. Bulletin 3. Albuquerque: New Mexico Museum of Natural History and Science.

Kietzke, K. K. 1987. Calcareous microfossils from the Upper Triassic of northeastern New Mexico. *New Mexico Geological Society Guidebook* 38:119–26.

Kietzke, K. K. 1989. Calcareous microfossils from the Triassic of the southwestern United States. In *Dawn of the Age of Dinosaurs in the American Southwest,* ed. S. G. Lucas and A. P. Hunt, pp. 223–32. Albuquerque: New Mexico Museum of Natural History.

Kietzke, K. K., and Lucas, S. G. 1991a. Ostracoda from the Upper Triassic (Carnian) Tecovas Formation near Kalgary, Crosby County, Texas. *Texas Journal of Science* 43:191–7.

Kietzke, K. K., and Lucas, S. G. 1991b. Triassic nonmarine "*Spirorbis*": gastropods not worms. *New Mexico Geology* 13:93.

Kozur, H. 1993. Annotated correlation tables of the Germanic Buntsandstein and Keuper. In *The Nonmarine Triassic,* ed. S. G. Lucas and M. Morales, pp. 243–8. Bulletin 3. Albuquerque: New Mexico Museum of Natural History and Science.

Litwin, R. J., Traverse, A., and Ash, S. R. 1991. Preliminary palynological zonation of the Chinle Formation, southwestern U. S. A., and its correlation to the Newark Supergroup (eastern U. S. A.). *Review of Palaeobotany and Palynology* 68:269–87.

Lockley, M. G., Conrad, K., Paquette, M., and Farlow, J. O. 1992. Distribution and significance of Mesozoic vertebrate trace fossils in Dinosaur National Monument. *University of Wyoming, National Park Service Research Report* 16:74–85.

Long, R. A., and Ballew, K. L. 1985. Aetosaur dermal armor from the late Triassic of southwestern North America with special reference to material from the Chinle Formation of Petrified Forest National Park. *Bulletin of the Museum of Northern Arizona* 54:35–68.

Lozovsky, V. R. 1993. The most complete and fossiliferous Lower Triassic section of the Moscow synclise: the best candidate for a nonmarine global scale. In *The Nonmarine Triassic,* ed. S. G. Lucas and M. Morales, pp. 293–9. Bulletin 3. Albuquerque: New Mexico Museum of Natural History and Science.

Lucas, S. G. 1991a. *Revised Upper Triassic Stratigraphy in the San Rafael Swell, Utah.* Publication 19. Utah Geological Association.

Lucas, S. G. 1991b. Sequence stratigraphic correlation of nonmarine and marine late Triassic biochronologies, western United States. *Albertiana* 9:11–18.

Lucas, S. G. 1992. Nonmarine standards for Triassic time. *Albertiana* 10:35–40.

Lucas, S. G. 1993. The Chinle Group: revised stratigraphy and chronology of Upper Triassic nonmarine strata in the western United States. *Bulletin of the Museum of Northern Arizona* 59:27–50.

Lucas, S. G., and Anderson, O. J. 1993. Lithostratigraphy, sedimentation and sequence stratigraphy of Upper Triassic Dockum Formation, West Texas. In *1993 Southwest Section Geological Convention, AAPG Transactions and Abstracts,* ed. R. E. Crick, pp. 55–65. Arlington: University of Texas at Arlington.

Lucas, S. G., and Huber, P. 1993. Revised internal correlation of the Newark Supergroup, Triassic, eastern United States and Canada. In *The Nonmarine Triassic,* ed. S. G. Lucas and M. Morales, pp. 311–19. Bulletin 3. Albuquerque: New Mexico Museum of Natural History and Science.

Lucas, S. G., and Hunt, A. P. (eds). 1989. *Dawn of the Age of Dinosaurs in the American Southwest.* Albuquerque: New Mexico Museum of Natural History.

Lucas, S. G., and Hunt, A. P. 1992. Triassic stratigraphy and paleontology, Chama basin and adjacent areas, north-central New Mexico. *New Mexico Geological Society Guidebook* 43:151–72.

Lucas, S. G., and Hunt, A. P. 1993a. A dicynodont from the Upper Triassic of New Mexico and its biochronological significance. In *The Nonmarine Triassic,* ed. S. G. Lucas and M. Morales, pp. 321–5. Bulletin 3. Albuquerque: New Mexico Museum of Natural History and Science.

Lucas, S. G., and Hunt, A. P. 1993b. Tetrapod biochronology of the Chinle Group (Upper Triassic), western United States. In *The Nonmarine Triassic,* ed. S. G. Lucas and M. Morales, pp. 327–9. Bulletin 3. Albuquerque: New Mexico Museum of Natural History and Science.

Lucas, S. G., Hunt, A. P., and Hayden, S. N. 1987. The Triassic system in the Dry Cimarron Valley, New Mexico, Colorado and Oklahoma. *New Mexico Geological Society Guidebook* 38:97–117.

Lucas, S. G., and Kietzke, K. K. 1993. Calcareous microfossils from the Upper Triassic of Petrified Forest National Park, Arizona. *Journal of the Arizona-Nevada Academy of Science* 27:55–68.

Lucas, S. G., and Luo, Z. 1993. *Adelobasileus* from the Upper Triassic of West Texas: the oldest mammal. *Journal of Vertebrate Paleontology* 13:309–34.

Lucas, S. G., and Marzolf, J. E. 1993. Stratigraphy and sequence stratigraphic interpretation of Upper Triassic strata in Nevada. In *Mesozoic Paleogeography of the Western United States,* vol. 2, ed. G. Dunne and K. McDougall, pp. 375–8. Pacific Section SEPM, Book 71.

Lucas, S. G., and Morales, M. (eds). *The Nonmarine Triassic.* Bulletin 3. Albuquerque: New Mexico Museum of Natural History and Science.

Lucas, S. G., Oakes, W., and Froehlich, J. W. 1985. Triassic microvertebrate locality, Chinle Formation, east-central New Mexico. *New Mexico Geological Society Guidebook* 36:205–12.

Lupe, R. D., and Silberling, J. N. 1985. Genetic relationship between Lower Mesozoic continental strata of the Colorado Plateau and marine strata of the western Great Basin: significance for accretionary history of Cordilleran lithotectonic terranes. In *Tectonostratigraphic Terranes of the Circum-Pacific Region,* ed. D. G. Howell, pp. 263–71. Houston: Circum-Pacific Council for Energy and Mineral Resources.

McIntosh, W. C., Hargraves, R. B., and West, C. L. 1985. Paleomagnetism and oxide mineralogy of Upper Triassic–Lower Jurassic red beds and basalts in the Newark Basin. *Geological Society of America Bulletin* 96:463–80.

Marzolf, J. E. 1993. Palinspastic reconstruction of early Mesozoic sedimentary basins near the latitude of Las Vegas: implications for the early Mesozoic Cordilleran cratonal margin. In *Mesozoic Paleogeography of the Western United States,* vol. 2, ed. G. Dunne and K. McDougall, pp. 433–62. Pacific Section SEPM, Book 71.

Miller, D. S., and Kulp, J. L. 1963. Isotopic evidence on the origin of the Colorado Plateau uranium ores. *Bulletin of the Geological Society of America* 74: 609–30.

Miller, G. L., and Ash, S. R. 1988. The oldest freshwater decapod crustacean, from the Triassic of Arizona. *Palaeontology* 31:273–9.

Molina-Garza, R. S., Geissman, J. W., and Lucas, S. G. 1993. Late Carnian–early Norian magnetostratigraphy from nonmarine strata, Chinle Group, New Mexico. Contributions to the Triassic magnetic polarity time scale and the correlation of nonmarine and marine Triassic faunas. In *The Nonmarine Triassic,* ed. S. G. Lucas and M. Morales, pp. 345–52. Bulletin 3. Albuquerque: New Mexico Museum of Natural History and Science.

Molina-Garza, R. S., Geissman, J. W., Van der Voo, R., Lucas, S. G., and Hayden, S. N. 1991. Paleomagnetism of the Moenkopi and Chinle Formations in central New Mexico: implications for the North American polar wander path and Triassic magnetostratigraphy. *Journal of Geophysical Research* 96:14239–62.

Nichols, K. M. 1972. Triassic depositional history of China Mountain and vicinity, north-central Nevada. Ph.D. dissertation, Stanford University.

Nichols, K. M., and Silberling, N. J. 1977. *Stratigraphy and Depositional History of the Star Peak Group (Triassic), Northwestern Nevada.* Special paper 178. Boulder: Geological Society of America.

Ochev, V. G. 1974. Nekotorye zamechaniya o koprolitakh triasovyx pozvonbochnykh [Some observations on Late Triassic coprolites]. *Paleontologicheskii Zhurnal* 2:146–8.

Odin, G. S. 1982. The Phanerozoic time scale revisited. *Episodes* 5:3–9.

Oldow, J. S. 1984. Evolution of a late Mesozoic back-arc fold and thrust belt, northwestern Great Basin, U.S.A. *Tectonophysics* 102:245–74.

Oldow, J. S., Bartel, R. L., and Gelber, A. W. 1990. Depositional setting and regional relationships of basinal assemblages: Pershing Ridge Group and Fencemaker Canyon sequence in northwestern Nevada. *Bulletin of the Geological Society of America* 102:193–222.

Olsen, P. E., and Galton, P. M. 1984. A review of the reptile and amphibian assemblages from the Stormberg of South Africa, with special emphasis on the footprints and the age of the Stormberg. *Palaeontologia Africana* 25:87–110.

Palmer, A. R. 1983. Decade of North American geology (DNAG) geologic time scale. *Geology* 11:503–4.

Pang, Q. 1993. The nonmarine Triassic and Ostracoda in northern China. In *The Nonmarine Triassic,* ed. S. G. Lucas and M. Morales, pp. 383–92. Bulletin 3. Albuquerque: New Mexico Museum of Natural History and Science.

Reeside, J. B., Jr. 1927. The new unionid pelecypods from the Upper Triassic. *Journal of the Washington Academy of Sciences* 17:476–8.

Reeve, S. C., and Helsley, C. E. 1972. Magnetic reversal sequence in the upper part of the Chinle Formation, Montoya, New Mexico. *Bulletin of the Geological Society of America* 83:3795–812.

Rogers, R. R., Swisher, C. C., III, Sereno, P. C., Monetta, A. M., Forster, C. A., and Martinez, R. C. 1993. The Ischigualasto tetrapod assemblage (late Triassic, Argentina) and $^{40}Ar/^{39}Ar$ dating of dinosaur origins. *Science* 260:794–7.

Rusconi, C. 1949. Coprolitos triasicos de Mendoza. *Revista Museo Historia Natural Mendoza* 3:241–51.

Salvador, A. 1985. Chronostratigraphic and geochronometric scales in COSUNA stratigraphic correlation charts of the United States. *Bulletin of the American Association of Petroleum Geologists* 69:181–9,

Schultz, L. G. 1963. *Clay Minerals in Triassic Rocks of the Colorado Plateau.* Bulletin 1147-C. U.S. Geological Survey.

Shields, O. 1990. Terrestrial replacements in the Lower Liassic. *Modern Geology* 14:265–6.

Silberling, N. J. 1961. Upper Triassic marine molluscs from the Natchez Pass Formation in northwestern Nevada. *Journal of Paleontology* 35:535–42.

Silberling, N. J., and Tozer, E. T. 1968. *Biostratigraphic Classification of the Marine Triassic in North America.* Special paper 110. Geological Society of America.

Silberling, N. J., and Wallace, R. E. 1969. *Stratigraphy of the Star Peak Group (Triassic) and Overlying Rocks, Humboldt Range, Nevada.* Professional paper 592. U.S. Geological Survey.

Simpson, J. H. 1850. Journal of a military reconnaissance from Santa Fe, New Mexico to the Navajo country made in 1849. U.S. Congress, 31st Congress, 1st Session, Senate Executive Document 64:56–131, 148–9.

Smith, D. G. 1982. Stratigraphic significance of a palynoflora from ammonoid-bearing early Norian strata in Svalbard. *Newsletters in Stratigraphy* 11:154–61.

Speed, R. C. 1978a. Basinal terrane of the early Mesozoic marine province of the western Great Basin. In *Mesozoic Paleogeography of the Western United States,* ed. D. G. Howell and K. McDougall, pp. 237–52. Pacific Section SEPM.

Speed, R. C. 1978b. Paleogeography and plate tectonic evolution of the early Mesozoic marine province of the western Great Basin. In *Mesozoic Paleogeography of the Western United States,* ed. D. G. Howell and K. McDougall, pp. 253–70. Pacific Section SEPM.

Stewart, J. H., Anderson, T. H., Haxel, G. B., Silver, L. T., and Wright, J. E. 1986. Late Triassic paleogeography of the southern Cordillera: the problem of a source for the voluminous volcanic detritus in the Chinle Formation of the Colorado Plateau region. *Geology* 14:567–70.

Stewart, J. H., Poole, F. G., and Wilson, R. F. 1972. Stratigraphy and origin of the Chinle Formation and related Upper Triassic strata in the Colorado Plateau region. Professional paper 690. U.S. Geological Survey.

Symons, D. T. A., Bormann, R. E., and Jans, R. P. 1989. Paleomagnetism of the Triassic beds of the lower Fundy Group and Mesozoic tectonism of the Novia Scotia platform, Canada. *Tectonophysics* 164:13–24.

Tasch, P. 1978. Clam shrimps. In *Geology, Paleontology, and Paleoecology of a Late Triassic Lake, Western New Mexico*, ed. S. R. Ash, pp. 61–5. Geology studies 25. Brigham Young University.

Tozer, E. T. 1984. *The Trias and Its Ammonoids: The Evolution of a Time Scale*. Miscellaneous report 35. Geological Survey of Canada.

Visscher, H. 1992. The new STS stage nomenclature. *Albertiana* 10:1.

Visscher, H., and Brugman, W. A. 1981. Ranges of selected palynomorphs in the Alpine Triassic of Europe. *Review of Palaeobotany and Palynology* 34:115–28.

Waters, A. C., and Granger, H. C. 1953. *Volcanic Debris in Uraniferous Sandstones, and Its Possible Bearing on the Origin and Precipitation of Uranium*. Circular 224. U.S. Geological Survey.

Webb, J. A. 1981. A radiometric time scale of the Triassic. *Journal of the Geological Society of Australia* 28:107–21.

Weedon, M. J. 1990. Shell structure and affinity of vermiform 'gastropods'. *Lethaia* 23:297–309.

Witte, W. K., Kent, D. V., and Olsen, P. E. 1991. Magnetostratigraphy and paleomagnetic poles from Late Triassic–earliest Jurassic strata of the Newark basin. *Bulletin of the Geological Society of America* 103:1648–62.

Wyman, R. V., Stewart, J. H., Anderson, L. T., Haxel, G. B., Silver, L. T., and Wright, J. E. 1987. Comment and reply on "Late Triassic paleogeography of the southern Cordillera: the problem of a source for voluminous volcanic detritus in the Chinle Formation of the Colorado Plateau region." *Geology* 15:578–9.

Xu, M. Y. 1988. Ostracods from the Mesozoic coal-bearing strata of northern Shaanxi, China. In *Evolutionary Biology of Ostracoda*, ed. Y. Hani et al., pp. 1283–91. Beijing: Science Press.

Young, R. G. 1964. Distribution of uranium deposits in the White Canyon–Monument Valley districts, Utah–Arizona. *Economic Geology* 59:850–73.

24 Otapirian Stage: its fauna and microflora

J. D. CAMPBELL

Classification of Lower Mesozoic strata in New Zealand began with the work of Ferdinand von Hochstetter. A geologist with European experience, he recognized *Monotis* in the field in 1859, and it was his material, collected on a South Island expedition near Nelson, that provided the basis for the variety *richmondiana* of *Monotis salinaria* that was described as new by K. A. von Zittel in 1864 – the *M. (Entomonotis) richmondiana* of later writing.

In an 1864 publication, Hochstetter compared and effectively correlated New Zealand *Monotis*-bearing strata with the Hallstätt Beds of the Austrian Alps, which in turn were considered to be the lateral equivalents of the Upper Triassic of the Germanic succession (Hochstetter, 1959).

In an early attempt to arrange New Zealand fossiliferous strata in order of age, James Hector (1870) introduced the term "Otapiri Series" for Triassic strata and by 1878 was citing an "Otapiri Series, Upper Trias (Rhaetic)" (Hector, 1878, p. vii). Cox (1878) published a description of Triassic and Jurassic strata in the Hokonui Hills, Southland, that year, noting for the Otapiri Series that "the series [is] either Lower Lias or Rhaetic" (p. 44). The term "Rhaetic" had been used in European writing since 1861. Correlation of the Otapiri Series with the Rhaetic was important. There was considerable agreement in the Old World as to what constituted the Rhaetic, even though the boundary between the Triassic and Jurassic systems was drawn variously at its upper and lower limits.

Cox (1878) used the term "ammonite beds" for the lower part of the "Bastion Series," the unit that succeeded the Otapiri Series stratigraphically in the early classification. His and Hector's suspicion that the often conspicuous ammonites in lower Bastion strata were of early Jurassic age was to be confirmed later by Spath's identification of the genus *Psiloceras* (Spath, 1923). Spath considered the ammonite fauna to represent a closely constrained horizon within the Lower Lias (Spath, 1923; Arkell, 1956).

The 1878 survey of the Hokonui Hills (S. H. Cox, geologist; A. McKay, fossil collector) demonstrated the existence of a body of clastic strata at least 500 m thick for which a post-*Monotis* Triassic age seemed to be imperative. Subsequent work in the Hokonui Hills and elsewhere has upheld that assertion.

Otapirian local stage and the Otapiri Valley section

Marwick (1951, 1953) proposed the time–stratigraphic unit Otapirian Stage as part of a local, but New Zealand–wide scheme. He nominated a section about 2 km long in Otapiri Valley, Hokonui Hills, Southland, as the type locality. Otapiri Stream flows south through the western flanks of the Hokonui Hills to intersect northwest-striking strata. The steeply dipping beds are part of the north limb of the Southland Regional Syncline, and their overall age range spans at least mid-Triassic to mid-Jurassic. The axial zone of the syncline can be traced from the southeast coast of South Island to the Hokonui Hills and beyond (Figure 24.1).

Following Marwick's proposal, Otapirian successions have been recognized widely in New Zealand, such as at the Kawhia Regional Syncline (Martin, 1975) and at Nelson (Campbell and Johnston, 1984), as well as in New Caledonia, about 1,400 km northwest of New Zealand (Campbell, Grant-Mackie, and Paris, 1985).

The type locality was described in some detail by Campbell and McKellar (1956), who chose the first appearance of *Rastelligera diomedea* (Trechmann) in the type section as the datum for the base of the stage. They recorded Otapirian faunas through a thickness of about 1,200 m in the Otapiri Valley section. The base of the overlying Aratauran local stage was drawn at the first appearance of psiloceratid ammonites.

Work during recent decades has led to refinements in the map and to a better understanding of the Otapiri Valley section in relation to the Hokonui Hills succession (Figures 24.2 and 24.3).

Refinements in correlation of pre- and post-Otapirian strata

Monotis, the key fossil of the Warepan local stage, has worldwide distribution. It appears and disappears within the Norian Stage in an interval spanning two ammonite zones in the North American scheme of Tozer (1984). Within any region of occurrence, the genus can be present as a succession of taxonomic associations (intrageneric faunas). New Zealand sequences (within the Warepan local stage) have four subgeneric groupings. The youngest *Monotis*-bearing strata there correlate with the *cordilleranus* Zone (Grant-Mackie, 1985; Grant-Mackie and

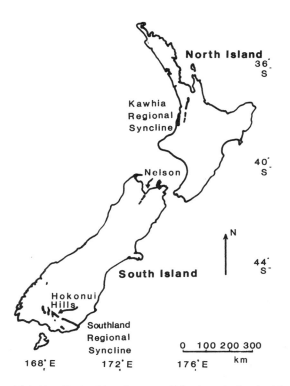

Figure 24.1. New Zealand location map. Triassic rocks involved in the Southland and Kawhia regional synclines are shown in black.

Silberling, 1990) and with the top of the Lower Sevatian (L. Krystyn, personal communication, 1989), representing the lower part of the Upper Norian (Figure 24.4).

Post-Otapirian rocks are characterized by psiloceratid ammonites, whose first appearance is used to define the lower boundary of the Aratauran local stage in South Island and New Caledonia sections. Although the ammonites were placed in a hemera within the *planorbis* Zone, Hettangian Stage, by Spath (1923), on the basis of three taxa that he described and figured – *Psiloceras (Euphyllites?)* sp. nov.(?) indet., *Psiloceras (Euphyllites)* sp. indet., and *Psiloceras* sp. cf. *calcimontanum* (Waehner) – there remains some doubt as to the presence there of that earliest Jurassic zone. Stevens (1968) accepted a Hettangian correlation, citing *Saxoceras* sp., among other taxa. In the scheme of Guex (1987), that would indicate the *liasicus* Zone, middle Hettangian. There are strong correlations with the succeeding *angulata* Zone (late Hettangian) and with the Sinemurian Stage.

Otapirian correlation

The time that the Otapirian local stage represents falls within the post-*columbianus* Zone interval at the end of the Triassic. In aggregate, the *stuerzenbaumi* Zone and the *marshi* Zone of Krystyn (1987) are broadly equivalent, and the *Tosapecten efimovae* Zone of Siberia (Dagys, 1988; Kazakov and Kurushin,

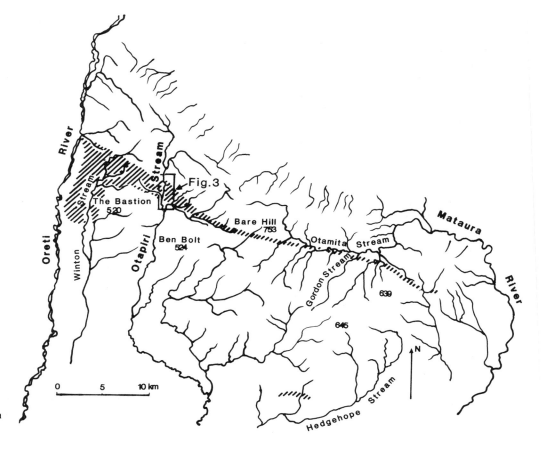

Figure 24.2. Outline map of the Hokonui Hills, Southland. Otapirian rocks are shown by hachuring.

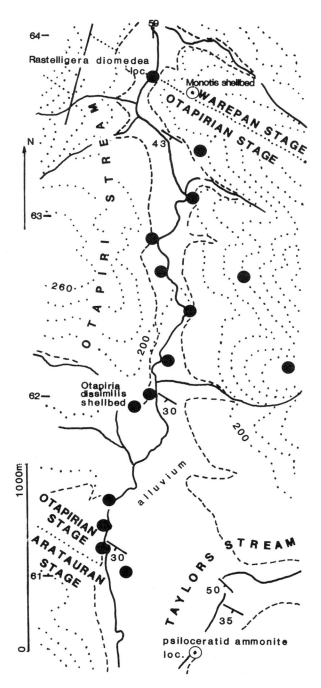

Figure 24.3. Geologic map of part of the Otapiri Valley, Hokonui Hills. Otapirian fossil localities are shown by solid circles. Topographic contours are dotted (contour interval, 20 m).

1992) must be closely analogous. All are Rhaetian in the sense of Krystyn (1987) (Campbell, 1991a).

Otapirian palaeontology

Brachiopods are important components in Otapirian faunas. Campbell (1991a) reviewed the occurrences of spiralium-bearing

groups, noting their pre-Otapirian local derivation, their moderate degree of taxonomic diversification (especially *Rastelligera* and *Clavigera*) within the stage, and the fate of the groups at the end of Otapirian time: *Rastelligera* and *Clavigera* did not survive; *Psioidiella* disappeared from New Zealand, but was still extant in New Caledonia in the Hettangian. The *Mentzelia* lineage continued on to diversify in a modest way in New Zealand and in Argentina (Damborenea and Mancenido, 1992). In a later paper, Campbell (1991b) described the Otapirian occurrence of *Zugmayerella* in New Zealand and New Caledonia.

In a monograph on Triassic and Jurassic rhynchonellids, MacFarlan (1992) recognized early and late Otapirian faunas. The generalized *Sakawairhynchia* lineage was present in both, and the long-ranging and often abundant *S. marokopana* MacFarlan was the most characteristic Otapirian rhynchonellid. The early fauna included *Fissirhynchia pacifica* MacFarlan, which had a pre-Otapirian range, and *Herangirhynchia otapiriensis* MacFarlan, a precursor of an important early Jurassic line. The late Otapirian fauna was dominated by *Sakawairhynchia*, and *Vincentirhynchia*, which may have been a derivative of *Sakawairhynchia*, also appeared. Both generic groups continued into the Hettangian, but became extinct at the end of the early Jurassic. Rhynchonellids were little affected by events at the Otapirian–Aratauran boundary.

Otapirian terebratulids are poorly known. The form described as *Coenothyris* sp. by Trechmann (1918) from *Monotis* beds ranged into the early Otapirian. *Zeilleria* has been tentatively identified from the highest levels of the stage (MacFarlan, 1992). It also occurs in Aratauran strata.

Otapirian rocks were referred to as "Trigonia beds" by workers in the nineteenth century. Fleming's monographic treatment of New Zealand Mesozoic trigoniaceans confirmed their Otapirian importance and included descriptions and illustrations of four species from the stage (Fleming, 1987). All are placed in Minetrigoniinae, within the family Trigoniidae, three in *Maoritrigonia* along with *Minetrigonia otapiriense* Fleming. All became extinct in the New Zealand region at the end of Otapirian time, but the lineage has been shown to have continued into the early Jurassic in the southern Andes (Damborenea and Mancenido, 1992).

Otapiria (family Monotidae) (Begg and Campbell, 1985) was proposed by Marwick (1935) for early Jurassic pteriomorphs. He placed the wholly Otapirian species *dissimilis* in *Otapiria* in 1953, noting that it was distinguished from the type species *O. marshalli* (Trechmann) by "its almost smooth right valve and its stronger, more irregular sculpture on the left valve." *O. dissimilis* (Cox) ranges through the upper half of the Otapirian succession in New Zealand. *O. marshalli* succeeded it in the Aratauran Stage.

Antiquilima n. sp. (family Limidae) (Braithwaite, 1983) first appeared in the earliest Otapirian in the Taringatura Hills (Coombs, 1950; Campbell, 1956) and formed a shell bed at the base of the Otapirian sequence. Less abundant *Pseudolimea* spanned the Otapirian within a Triassic–Jurassic range.

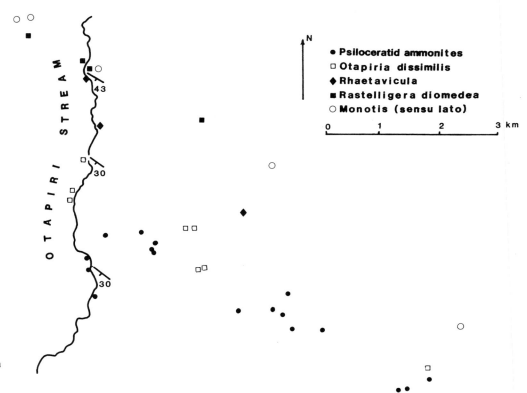

Figure 24.4. Schematic map of an area within the catchment areas of Otapiri Stream and Taylor's Stream showing the occurrences of selected fossils.

Torastarte (Family Cardiniidae) (Fleming, 1957) has been described from Otapirian collections. The sole species, *T. bensoni* Marwick, did not survive beyond the stage. *Kalentera marwicki* Grant-Mackie (Family Cypricardiacea) has been described from an Otapirian locality. It belonged to a lineage that was present in New Zealand from at least the beginning of the late Triassic to the end of the early Jurassic. Its Jurassic geographic range included western Argentina (Damborenea and Mancenido, 1992). *Rhaetavicula* (Family Pteriidae) first appears in the lowest levels of the Otapirian succession (Campbell, 1956; Fleming, 1979) and is an uncommon fossil in the higher parts of the stage. The oyster *Lopha* cf. *haidingeriana* Emmrich was figured and described from an Otapirian locality by Fleming (1953).

Gastropods are only sparsely present in Otapirian rocks. *Pleurotomaria hectori* Trechmann ranges upward from the Lower Norian, and there are at least two undescribed species within the *P. hectori* group (J. G. Begg, personal communication, 1993). *Talantodiscus trechmanni* Marwick has been described from an Upper Otapirian locality – not Lower Jurassic as Marwick (1953) suggested. The status of *Atlantobellerophon zealandicus* Trechmann remains obscure.

Two distinctive specimens of Otapirian coleoids have been described and figured: *Aulacoceras otapiriense* (Hector) (Marwick, 1953) from a basal horizon, and *Prographularia* sp. (Marwick, 1953; Jeletzky and Zapfe, 1967) from Otapirian strata in Nelson. Nautiloids are represented by orthocones, thus far undescribed.

The only ammonite described and figured from Otapirian rocks is *Arcestes* cf. *rhaeticus* Clark (Stevens, 1970), but poorly preserved arcestids have been collected from a number of localities, including one at the top of the stage (Figure 24.5), and from New Caledonia.

An early comatulid crinoid (calyx) from the New Caledonia Otapirian has recently been described (Hagdorn and Campbell, 1993). Crinoid columnals are often abundant in Otapirian beds.

The conulariid, *Paraconularia matauraensis* Waterhouse, has been recorded from an Otapirian locality in the catchment area of Otapiri Stream (Waterhouse, 1979) and from New Caledonia (Hagdorn and Campbell, 1993). The nodosariid foraminiferan *Astacolus* cf. *pediacus* Tappan has been described and figured from a basal Otapirian horizon (Strong, 1984).

Microfloral study of New Zealand Triassic rocks has begun only recently. Dickson (1972) recorded two palynomorph species from North Island Otapirian strata. Wilson and Helby (1986) noted the occurrence of dinoflagellates in Otapirian rocks, as did de Jersey and Raine (1990). Of 52 genus-level taxa of miospores described and figured by de Jersey and Raine from New Zealand Triassic and earliest Jurassic strata, some 38 occur in collections from Otapirian localities, with *Rogalskiasporites fenestratus* de Jersey and Raine and *Rugaletes awakinoensis* Raine having Otapirian type localities. The miospores allow recognition of seven zonal subdivisions, of which the *Foveosporites moretonensis* Zone is restricted to the Otapirian Stage. There are zonal links with eastern Australia, and the New Zealand flora is placed in the Ipswich province of southern high latitudes. Otapirian

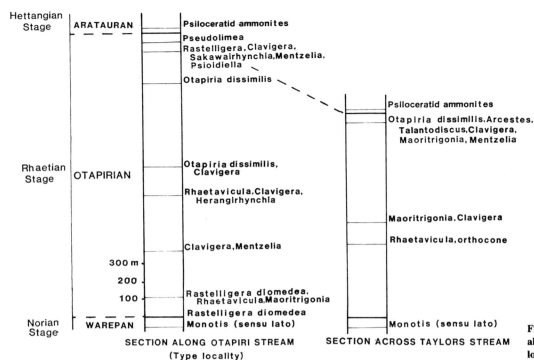

Figure 24.5. Stratigraphic columns along Otapiri Stream and across Taylor's Stream.

acritarchs, including the fresh-water species *Circulisporites parvus* de Jersey, have also been recorded.

Acknowledgments

I am grateful to many colleagues for their helpful interest, their comments, and their field companionship, especially John Begg, Hamish Campbell, Andrew Grebneff, Alan Orpin, and Stuart Owen. This work has benefited from discussions with Douglas Coombs, Jack Grant-Mackie, Daphne Lee, and Donald Mac-Farlan. Bill Wadworth (The Cornwalls) and Henry Wadworth (Warwick Downs) are thanked for allowing access to their properties and for their continued interest. Adrien Dever helped with typing and layout. Financial support from the New Zealand Foundation for Science and Technology is gratefully acknowledged.

References

Arkell, W. J. 1956. *Jurassic Geology of the World*. London: Oliver & Boyd.

Begg, J. G., and Campbell, H. J. 1985. *Etalia*, a new Middle Triassic (Anisian) bivalve from New Zealand, and its relationship with other pteriomorphs. *New Zealand Journal of Geology and Geophysics* 28:725–41.

Braithwaite, L. R. S. 1983. *Some Triassic and Jurassic Limidae from New Zealand* (abstract). Miscellaneous publication 30A. Geological Society of New Zealand.

Campbell, H. J., Grant-Mackie, J. A., and Paris, J.-P. 1985. Geology of the Moindou–Téremba area, New Caledonia. Stratigraphy and structure of Téremba Group (Permian–Lower Triassic) and Baie de St. Vincent Group (Upper Triassic–Lower Jurassic). *Geologie de la France* 1:19–36.

Campbell, H. J., and Johnston, M. R. 1984. *Fossil Localities of the Richmond Group Triassic of Nelson*. Report 106. New Zealand Geological Survey.

Campbell, J. D. 1956. The Otapirian Stage of the Triassic System of New Zealand. Part 2. *Transactions of the Royal Society of New Zealand* 84:45–50.

Campbell, J. D. 1991a. Latest Triassic (Rhaetian) brachiopods of New Zealand and New Caledonia. In *Brachiopods through Time*, ed. D. I. MacKinnon, D. E. Lee, and J. D. Campbell, pp. 389–92. Rotterdam: Balkema.

Campbell, J. D. 1991b. A Late Triassic spiriferinacean brachiopod (family Laballidae) from the Taringatura Hills, Southland, New Zealand. *New Zealand Journal of Geology and Geophysics* 34:359–63.

Campbell, J. D., and McKellar, I. C. 1956. The Otapirian Stage of the Triassic system of New Zealand. Part 1. *Transactions of the Royal Society of New Zealand* 83:695–704.

Coombs, D. S. 1950. The geology of Northern Taringatura Hills, Southland. *Transactions of the Royal Society of New Zealand* 78:426–48.

Cox, S. H. 1878. Report on the geology of the Hokonui Ranges, Southland. In *Reports of Geological Exploration 1877–78*, pp. 25–48. Wellington: New Zealand Geological Survey.

Dagys, A. S. 1988. An alternative interpretation of the Rhaetian. *Albertiana* 7:4–6.

Damborenea, S., and Mancenido, M. O. 1992. A comparison of Jurassic marine benthonic faunas from South America and New Zealand. *Journal of the Royal Society of New Zealand* 22:131–52.

de Jersey, N. J., and Raine, J. I. 1990. *Triassic and Earliest Jurassic Miospores from the Murihiku Supergroup, New Zealand*. Palaeontological bulletin 62. New Zealand Geological Survey.

Dickson, M. 1972. First records of *Annulispora folliculosa* (Rogalska) de Jersey and *Polycingulatisporites mooniensis* de Jersey and Paten from the Upper Triassic of New

Zealand. *New Zealand Journal of Geology and Geophysics* 15:169–70.

Fleming, C. A. 1953. A Triassic oyster from New Zealand. *New Zealand Journal of Science and Technology* B35:276–8.

Fleming, C. A. 1957. The Triassic lamellibranch *Torastarte bensoni* Marwick. *Proceedings of the Malacological Society of London* 32:173–5.

Fleming, C. A. 1979. *The Geological History of New Zealand and Its Life.* Auckland University Press.

Fleming, C. A. 1987. *New Zealand Mesozoic Bivalves of the Superfamily Trigoniacea.* Paleontological bulletin 53. New Zealand Geological Survey.

Grant-Mackie, J. A. 1960. On a new *Kalentera* (Pelecypoda, Cypricardiacea) from the Upper Triassic of New Zealand. *New Zealand Journal of Geology and Geophysics* 3:74–80.

Grant-Mackie, J. A. 1985. The Warepan Stage of the Upper Triassic: redefinition and subdivision. *New Zealand Journal of Geology and Geophysics* 28:701–24.

Grant-Mackie, J. A., and Silberling, N. J. 1990. New data on the Upper Triassic bivalve *Monotis* in North America, and the new Subgenus *Pacimonotis. Journal of Paleontology* 64: 240–54.

Guex, J. 1987. *Sur la phylogenèse des ammonites du Lias inférieur.* Bulletin de Geologie Lausanne no. 292.

Hagdorn, H., and Campbell, H. J. 1993. *Paracomatula triadica* sp. nov. – an early comatulid crinoid from the Otapirian (Late Triassic) of New Caledonia. *Alcheringa* 17:1–17.

Hector, J. 1870. *Catalogue of the Colonial Museum.* Wellington: Colonial Museum.

Hector, J. 1878. Progress report. In *Reports of Geological Exploration 1877–78,* pp. i–xv. Wellington: New Zealand Geological Survey.

Hochstetter, F. von. 1959. *Geology of New Zealand. Contributions to the Geology of the Provinces of Auckland and Nelson,* trans. C. A. Fleming. Wellington: Government Printer. (First published 1864.)

Jeletzky, J. A., and Zapfe, H. 1967. Coleoid and orthocerid cephalopods of the Rhaetian Zlambach Marl from the Fischerwiese near Aussee, Styria (Austria). *Annalen Naturhistorische Museum Wien* 71:69–106.

Kazakov, A. M., and Kurushin, N. I. 1992. *Stratigraphy of Norian and Rhaetian Deposits in Northern Middle Siberia* (in Russian). Publication 378. Novosibirsk: Russian Academy of Sciences, Siberian Branch, Geology and Geophysics.

Krystyn, L. 1987. Zur Rhät-stratigraphie in den Zlambach-Schichten (vozläufiger Bericht). *Sitzungsberichten der Oesterreiche Akademie der Wissenschaften Mathematische-naturwissenschaft Klasse, Abt 1* 196:21–36.

MacFarlan, D. A. B. 1992. *Triassic and Jurassic Rhynchonellacea (Brachiopoda) from New Zealand and New Caledonia.* Bulletin 31. Royal Society of New Zealand.

Martin, K. R. 1975. Upper Triassic to Middle Jurassic stratigraphy of south-west Kawhia, New Zealand. *New Zealand Journal of Geology and Geophysics* 18:909–38.

Marwick, J. 1935. Some new genera of Myalinidae and Pteriidae of New Zealand. *Transactions of the Royal Society of New Zealand* 65:295–303.

Marwick, J. 1951. Series and stage divisions of New Zealand Triassic and Jurassic rocks. *New Zealand Journal of Science and Technology* B32:8–10.

Marwick, J. 1953. *Divisions and Faunas of the Hokonui System (Triassic and Jurassic).* Paleontological bulletin 21. New Zealand Geological Survey.

Spath, L. F. 1923. On ammonites from New Zealand. *Quarterly Journal of the Geological Society of London* 73:286–312.

Stevens, G. R. 1968. *The Jurassic System in New Zealand.* Report 35. New Zealand Geological Survey.

Stevens, G. R. 1970. Comments on New Zealand Triassic and Cretaceous correlations. *New Zealand Journal of Geology and Geophysics* 13:718–20.

Strong, C. P. 1984. *Triassic Foraminifera from Southland Syncline, New Zealand.* Paleontological bulletin 52. New Zealand Geological Survey.

Tozer, E. T. 1984. *The Trias and Its Ammonoids: The Evolution of a Time Scale.* Miscellaneous report 35. Geological Survey of Canada.

Trechmann, C. T. 1918. The Trias of New Zealand. *Quarterly Journal of the Geological Society of London* 73:165–241.

Waterhouse, J. B. 1979. Permian and Triassic conulariid species from New Zealand. *Journal of the Royal Society of New Zealand* 9:475–89.

Wilson, G. J., and Helby, R. 1986. New Zealand Triassic dinoflagellates – a preliminary survey. *Newsletter, Geological Society of New Zealand* 71:48–9.

25 Upper Palaeozoic glaciation and Carboniferous and Permian faunal changes in Argentina

C. R. GONZÁLEZ

As the western portion of Gondwana in the late Palaeozoic, South America extended over a great range of palaeolatitudes, from southern polar to tropical. At the time of the global lowering of temperature that triggered the "ice age," strong climatic differences became established between those extreme palaeolatitudes. The portion that was close to the southern palaeopole had low temperatures and was covered by ice sheets; the rest of the continent, which was at lower palaeolatitudes, had higher temperatures (Figure 25.1).

As was the case in the Cenozoic, late Palaeozoic glaciation in South America encompassed several glacial and interglacial stages. In western and southern Argentina, evidence of that glaciation is widespread in sediments whose dates range from as old as Visean–Namurian to as young as late Asselian. Discrimination of discrete glacial stages has been based on striated pavements covered with diamictites and associated varved sediments, dropstone laminites, and so forth (Frakes, Amos, and Crowell, 1969; López Gamundi and Amos, 1983; González, 1983). Sections that afforded doubtful or questionable evidence of glaciation were neglected. Interglacial stages can be recognized on the basis of intercalations of well-sorted sediments, without glacigenic features (Dickins, 1985). Fossiliferous beds, or a transgression with abundant marine fauna between two discrete glacial members, may suggest important defrosting (González, 1990). Elsewhere I have commented on the chronology and durations of those glaciations (González, 1990). Ice sheets could spread over large areas, and occasionally glaciers may have behaved as temporary dams, actually isolating inland seas. Some late Palaeozoic glacial sediments of western and southern Argentina were deposited in near-shore environments, where striated pavements (González, 1981b) and other erosion surfaces (González Bonorino, 1992) provide evidence of grounded ice. In places, dropstone laminites and other mixtites suggest that bodies of ice were active as floating ice shelves or "icebergs."

The evidences of the origins and extinctions of the major Carboniferous and early Permian faunal groups are closely associated with sedimentological evidence of climatic alterations due to temperature changes. A summary of the compositions of those major faunal groups is provided elsewhere (González, 1993).

Faunas and Palaeogeography

During the late Palaeozoic, the Argentinian Precordilleran region consisted of two structurally linked basins: the Rio Blanco Basin in the north and the Calingasta-Uspallata Basin in the south (Figure 25.2). "Pacific" transgressions did not cover them simultaneously, but alternately, forming two separated embayments. The area between those two basins was emergent throughout all of that period, and no marine sediments, but only continental sediments, were deposited at the Tocota-Tucunuco latitude (Furque, 1962; Cuerda and Furque, 1981; Bossi and Andreis, 1983). That palaeogeographic disconnection is reflected in the striking differences between the faunas of the northern and southern portions of the Precordillera (González, 1985a). The early Carboniferous Malimänian fauna and the late Carboniferous *Buxtonia-Heteralosia* assemblage of the Rio Blanco Basin had "northern" affinities, showing links with the faunas in tropical latitudes of central South America. They lived in relatively warm waters (González, 1993) and had no affinities with the "cold" faunas of the Calingasta-Uspallata Basin and Patagonia. On the other hand, the Calingasta-Uspallata Basin was freely connected with the Languiñeo-Genoa Basin in Patagonia (Figure 25.3). Those regions shared the same faunal assemblages from the Namurian until the earliest Permian.

Eastern Argentina experienced a different tectonic development, with deposition beginning at the early Permian in the Sauce Grande Basin (Figure 25.4). There, a poorly diversified *Eurydesma* fauna (Harrington, 1955) indicates a close connection with South Africa, but suggests an indirect link with Australia and a less probable link with the Paraná Basin (González, 1989).

Two major sea-level rises were related to significant ice waning in the late Palaeozoic in Argentina. The first was at the latest Carboniferous, the outcome of a long-lasting interglacial stage coeval with the *Buxtonia-Heteralosia* fauna (González, 1993). The second gave rise to the "sea of *Eurydesma*," related to the postglacial transgression at the early Sakmarian (Dickins, 1985). Minor floodings during that period are regarded as a reflection of interglacials, and their strata contain marine fossils and plant remains.

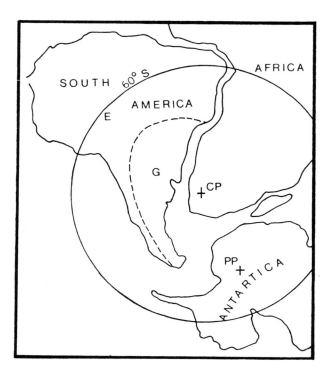

Figure 25.1. Relative position of South America during the Late Palaeozoic. G: Gondwana area; E: extra-Gondwanic realm. CP; Mid-Carboniferous South Palaeopole and parallel of 60° S latitude; PP: Permian South palaeopole. Palaeomagnetic data from Creer (1972) and Valencio (1973, 1981).

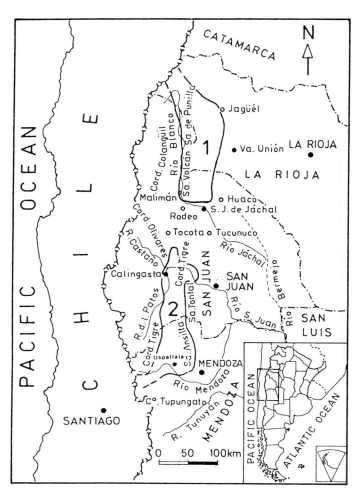

Figure 25.2. Precordilleran region of western Argentina, showing localities mentioned in the text: 1, Rio Blanco Basin; 2, Calingasta-Uspallata Basin. (Adapted from González, 1985a.)

Factors controlling faunal changes

The biofacies and lithofacies associations strongly suggest climatic controls that affected the behaviours of Carboniferous and early Permian faunas. Benthic populations underwent more or less severe changes, concomitantly with variations in temperatures, that certainly influenced evolution.

The difference in temperatures between the present era and the previous ice age has been estimated at around 5°C (Schneider, 1993). Extrapolation of that figure to the late Palaeozoic may not be justifiable; however, we can guess that glacial–interglacial variations in temperatures were not much different during that period. Variations in temperatures determined the extent of glacial activity, which in turn initiated flooding or emergence in littoral areas or shallow inland seas. That may have caused fluctuations in sea level, causing destruction of ecosystems and breaks in faunal continuity. The consequences of shifting shorelines (Roberts, 1981) probably would have had more catastrophic effects on benthic life than would variations in temperatures of around 5°C. Hence, climatic instability during that period indirectly brought about periodic massive killings and extinctions. Furthermore, that would have favoured the origin and development of new faunal assemblages. Also, significant global thermal variations would have provoked shifts in climatic zonations and rearrangements in the latitudinal distributions of biomes (Waterhouse and Bonham-Carter, 1975). In this regard, faunas from the "Pacific" and central parts of South America arrived at the Rio Blanco Basin during times of warm climatic conditions (i.e., at the early Carboniferous preglacial epoch and at the late Carboniferous interglacial); those faunas disappeared when the temperature fell. It is possible that the poor diversification and endemism of late Palaeozoic faunas in Argentina were related to such unstable environmental conditions.

Late Palaeozoic glaciations and bioevents

The periodic variations in temperatures in the South American Gondwana area were reflected in sequences of glacial–interglacial phases in the early Permian (Keidel, 1921, 1940; Du Toit, 1927; Frakes and de Figueiredo, 1967; Gravenor and Rocha-Campos, 1983) and in the Carboniferous (González Bonorino, Vega, and Guerin, 1988; González, 1990). At least three significant faunal episodes can be attributed to major

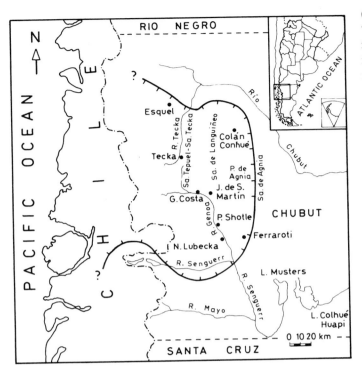

Figure 25.3. Extent of the late Palaeozoic transgressions in the Languiñeo-Genoa Basin in central Patagonia. (Adapted from González, 1984.)

Figure 25.4. Sauce Grande Basin (Southern Hills of Buenos Aires).

climatic changes. Two of them occurred in the Carboniferous, and the other in the early Permian. Less important faunal alterations due to seasonal advances and retreats of glaciers probably took place, but an understanding of those changes will require further study.

The occurrences of some invertebrates appear to have been related to temperature variations. The remains of rhynchonellids such as the early Carboniferous *Rossirhynchus chavelensis*

(Amos) and the late Carboniferous *Nudirostra cuyana* Amos and "*Camarotoechia*" sp. (González and Bossi, 1986) are found only in nonglacial deposits. It is assumed that those brachiopods lived in warm waters exclusively. Desmodont bivalves are relatively rare in western Argentina; *Vacunella*(?) is present in the Lower Carboniferous (Tournaisian) Malimăn Formation, which probably was deposited during a period of warm climate. Those shells become more common in Middle Carboniferous strata, associated with the "cold" *Levipustula* fauna. *Pyramus* and *Megadesmus*(?), which preferred near-shore shallow sands, lived during glacial and cold interglacial stages. Elements of the Barrealian fauna, such as *Myofossa antiqua* González, *Leptodesma variabilis* González, some atomodesmids, and other taxa that were dwellers in fine sediments on the seafloor, may have lived on the shelf under floating ice or a waxing glacier. They are frequently associated with dropstones and mudstones between diamictites.

The early Permian *Eurydesma* fauna was ubiquitous in Gondwana, closely associated with glacial deposits; *Eurydesma* has been well identified as a conspicuous constituent of the "cold" faunas of that age (Dickins, 1957). The mid-Carboniferous Barrealian fauna, which invariably is found in glacigenic sediments, is also regarded as a "cold" fauna (Campbell and McKellar, 1969; Roberts, 1981). In a broad sense, those "cold" faunas featured low diversities and in some cases low numbers of individuals. That situation cannot be attributed solely to low temperatures; it may also have been related to unstable palaeoecologic conditions and to the palaeogeographic restrictions of the peri-Gondwanan inland seas (González, 1989).

Carboniferous

Although dispersed in separate basins, the records of the Carboniferous system and faunas in Argentina are almost continuous since the Tournaisian. During that period, the climate in South America must have been quite homogeneous and probably warm or mild, much like that during the middle–late Devonian (González, 1990). Neither the fossils nor the sediments of Malimănian age in western and southern Argentina suggest low temperature, and Campbell and McKellar (1969) have pointed out that the Tournaisian climate in eastern Australia was warm. Although the faunas of that age in Argentina and Australia are not closely related, we assume that the climates may not have been that different in those two regions of Gondwana (González, 1981a, 1990).

The oldest Carboniferous bioevent is reflected in strata containing the Malimănian fauna, which is based on the *Protocanites scalabrinii–Rossirhynchus chavelensis* assemblage. In Argentina, that fauna occurs only in the Rio Blanco Basin, and it has some species in common with the Sierra de Cuevitas fauna of northern Chile. The Malimănian fauna includes brachiopods, bivalves, gastropods, corals, trilobites, and cephalopods; herein it is regarded as rather moderately diversified, probably because of environmental restrictions (González, 1990; cf. Dutro and Isaacson, 1991). *Posidoniella malimanensis* González is a conspicuous element of the Malimănian fauna; those bivalves had thin shells that became thickened at the umbonal region, a characteristic perhaps due to its habit of living in groups or clusters attached to the sea bottom by its byssus, like modern *Mytilus*. *Posidoniella* is a common genus in the Carboniferous limestones of Ireland, in the equatorial belt. Also, the goniatitids were common floating mollusks in the equatorial belt that seem to have preferred warm waters. At present, only one species of the genus *Protocanites* is known in the Rio Blanco Basin, a circumstance perhaps related to basinal restrictions, rather than to cold waters. There is no lithologic nor palaeontologic evidence of glaciation during the time of the Malimănian fauna (González, 1990).

The late Palaeozoic "ice age" was triggered at around the time of the early–late Carboniferous boundary (Visean–Namurian), and the sharp lithologic "warm–cold boundary" reflected by the Tepuel Group suggests relatively rapid climatic changes. At that time, the former Malimănian fauna was completely extinguished, being replaced by the Barrealian fauna. Although there is a stratigraphic gap between those two assemblages, their extinction and origin, respectively, are linked to the climatic changes.

The Barrealian fauna encompassed the *Rugosochonetes-Bulahdelia* assemblage and the *Levipustula levis* Zone and lived in cold waters (Campbell and McKellar, 1969; Roberts, 1981; González, 1981a, 1990). In western Argentina and in Patagonia, that fauna is everywhere associated with sediments of glacial origin. In this regard, the range of the whole Barrealian fauna comprises all of the Carboniferous glacial period, which lasted from the Visean–Namurian until probably the early Westphalian. It has "southern affinities" because of its strong similarities to eastern Australian faunas (Amos and Sabattini, 1969; Campbell and McKellar, 1969; Taboada, 1991). That suggests similar environmental conditions along the peri-Gondwanan belt between eastern Australia and southwestern South America, allowing faunal connections via the Antarctic coast.

At around the middle Westphalian, a sensible rise of temperature brought about a climatic amelioration that probably lasted until the end of the Carboniferous Period. That seems to have been the longest interglacial stage of the late Palaeozoic "ice age" in Argentina (González, 1981a, 1990). That climatic amelioration allowed a "Pacific" transgression to bring into the Rio Blanco Basin extra-Gondwanan elements from lower latitudes in Bolivia, Perú, and the Amazonas Basin (González, 1989, 1993). That was the conspicuous *Buxtonia-Heteralosia* assemblage, which is regarded as a "warm" fauna linked to a tropical biome (Waterhouse and Bonham-Carter, 1975). The *Buxtonia-Heteralosia* assemblage is not found in marine beds of equivalent age in Patagonia, where the temperature had to be cooler, because that region was closest to the south palaeopole (Valencio, 1973). *Levipustula levis* and many other conspicuous species of the former Barrealian fauna were extinguished at the beginning of that interglacial stage, but some few taxa, informally grouped in the "intermediate fauna," remained longer, mostly confined to the colder waters of the Languiñeo-Genoa Basin.

The *Buxtonia-Heteralosia* fauna occurs in the upper section of the Rio del Peñón Formation. The lower section of the formation is nonmarine or continental, and immediately below the upper marine member there are beds bearing species of the genus *Carbonicola* and other limnic bivalves (Leanza, 1948; Frenguelli, 1945). Associated with that horizon are carbonaceous beds with *Nothorhacopteris* and other elements of the NBG flora (González and Bossi, 1986). Those sediments probably were deposited in quiet lakes or coastal swamps, suggesting temperatures much like those of the equatorial belt, where the genus *Carbonicola* and its allies are commonly associated with coal seams (Calver, 1969).

Early Permian

At the beginning of the Permian, the late Carboniferous interglacial ended; temperatures fell, and the South American Gondwana area was again under a glacial regime. Evidence of early Permian glaciation is sparse and scattered in Argentina (Keidel, 1921, 1940; Du Toit, 1927), southern Bolivia, and eastern Paraguay (Helwig, 1972), but it is better displayed in the Brazilian portion of the Paraná Basin. In the peri-Gondwanan belt of Argentina, glacigenic deposits of that age occur in the lower Agua del Jagüel Formation and in the upper Tepuel Group. They seem to represent the earliest Permian glacial episodes, perhaps older than the Sauce Grande glaciation of eastern Argentina.

Important biologic events occurred at the Carboniferous–Permian boundary, associated with the initiation of the early Permian glacial period: The *Buxtonia-Heteralosia* assemblage was extinguished and replaced by the *Cancrinella* fauna. Those two faunas were completely different, and no palaeontologic or stratigraphic relationships between them are known.

Two major faunal groups are recognizable from the early Permian: the Uspallatian, ascribed to the Asselian, and the Bonetian, of early Sakmarian age (González, 1993). The Uspallatian fauna, with *Cancrinella* cf. *farleyensis* (Etheridge & Dun)* as guide fossil, marks the oldest Permian deposits in southern and western Argentina. The stratigraphic range of that species is nearly 800 m in the Agua del Jagüel Formation, where it has intercalated diamictites, probably representing a glacial episode (Taboada, 1987) (Figure, 25.5). In Patagonia, beds bearing *Glossopteris* leaves are interbedded with the *Cancrinella* fauna (González, 1981a).

In the Sauce Grande Basin of eastern Argentina, the Bonetian (*Eurydesma*) fauna is associated with evidence of a transgression that followed what was probably the last glacial episode in that sector of Gondwana (Dickins, 1985). The Bonete Formation and Piedra Azul Formation (Figures 25.4 and 25.5) overlap thick massive diamictites of glacial origin (Coates, 1969; Massabie and Rossello, 1984). The palaeogeographic separation of the late Palaeozoic basins of eastern and western Argentina makes it difficult to determine the stratigraphic relationships between the

---

*The brachiopods that are referred to *Cancrinella* aff. *farleyensis* (Eth. & Dun.), the guide fossil of the *Cancrinella* Zone in Argentina, need revision (*fide* Taboada, in conversation, 1993).

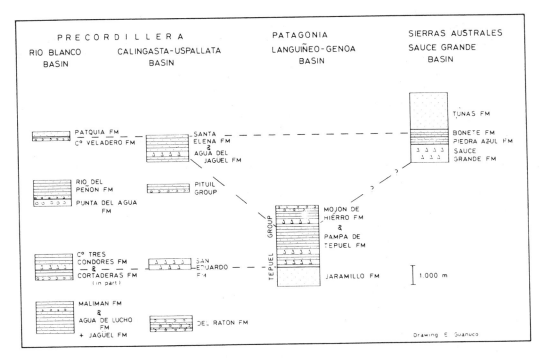

Figure 25.5. Selected late Palaeozoic sequences of Argentina, showing positions of glacigenic beds and their probable correlation.

Uspallatian and Bonetian faunas. For practical purposes, the upper biostratigraphic limit of the Uspallatian is assumed to be at the base of the Bonetian. However, a fauna younger than the Uspallatian occurs in the upper section of the Santa Elena Formation near Uspallata (Figures 25.2 and 25.5). It was in those beds that the late Dr. P. Aparicio discovered the existence of "eurydesmatids," later reported by Rocha-Campos (1970). Although those bivalves were not confirmed with new findings, it is probable that specimens of *Eurydesma* may occur in western Argentina. Material from that locality, lodged in the Instituto de Paleontologia of the Miguel Lillo Foundation, has close similarities to the fauna of the Callytharra Formation of western Australia (J. M. Dickins, personal communication, 1991). That assemblage is regarded as Bonetian, and it has some affinities with (but is perhaps older than) that of eastern Argentina.

The Bonetian was followed by a generalized regression; post-Bonetian strata do not bear glacigenic sediments, but show fairly clear nonglacial facies. The late Palaeozoic sedimentary cycle in Argentina ended with deposition of continental redbeds. There was progressive warming and aridity that climaxed during the Permian (Harrington, 1962; Limarino and Spalletti, 1984; Limarino and Sessarego, 1987). Those facies strongly recall Permo-Triassic sequences of the equatorial belt, suggesting the return of the South American Gondwana area to a global climatic uniformity.

The typical Bonetian genera *Eurydesma, Merismopteria, Megadesmus,* and *Deltopecten* share a preference for sandy bottoms, moderately swept by currents in cold, shallow waters near the coast (Dickins, 1957). The origin and extinction of the *Eurydesma* fauna coincided with the onset and end of the early Permian glacial period. The latest glacigenic sediments in the

Calingasta-Uspallata Basin, as well as in the Languiñeo-Genoa Basin, are associated with the Uspallatian fauna. It seems that in eastern Argentina, cold conditions persisted longer than in the western sector. The Paraná Basin and Sauce Grande Basin were fed glacigenic detritus from one or more large ice sheets, whereas in southern Argentina, and perhaps also in western Argentina, glaciation probably was relict. That may have been the cause of the higher faunal diversity in Patagonia and in the Precordillera than in the Sauce Grande and Paraná basins during that period (cf. Rocha-Campos, 1967; Harrington, 1955; González, 1985b).

Surrounding the western border of the South American Gondwana there extended an extra-Gondwanan realm, where both the biofacies and lithofacies of Permo-Carboniferous age suggest warm waters and palaeoequatorial affinities. The existence of those contrasting realms has been explained (González, 1989) on the basis of palaeogeographic conditions, and their latitudinal differences were increased by deviation of the south palaeopole to Antarctica during the Permian (Creer, 1972; Valencio, 1981). In that scheme (Figure 25.1), the "Pacific" coast of South America was farthest "north" from the south palaeopole, and was then in a warmer zone than the Gondwana area.

After the Sakmarian, progressively rising temperatures provoked climatic changes, and the early Permian glacial period ended, being followed by continental deposition and aridity. The Permo-Triassic was characterized by accumulations of huge volcaniclastic deposits over the rising Andean belt.

Conclusions

The late Palaeozoic "ice age" in Argentina encompassed several glacial episodes, separated by interglacials. Those phases were concomitant with changes in sea level that affected benthic communities to greater or lesser extents. Three main faunal changes were closely related to strong thermal variations during that period. The first occurred at the onset of the Carboniferous glaciation (Visean–Namurian). That climatic event marked the extinction of the "warm" Malimănian fauna and the debut of the "cold" Barrealian fauna. The second faunal change occurred during the late Carboniferous, probably at the middle Westphalian, related to a climatic amelioration and the beginning of a long-lasting interglacial. The Barrealian fauna was almost extinguished at that time, though some few elements endured, mainly retracted at higher latitudes. During that climatic episode, the *Buxtonia-Heteralosia* assemblage became installed in the northern Precordillera; it is regarded as a "warm" fauna that was linked to faunas in Chile, Perú–Bolivia, and the Amazonas Basin, but was disconnected from the southern Precordillera and Patagonia. The third faunal change was associated with the lowered temperatures of the Asselian, which saw the first appearance of the Uspallatian fauna. At the early Sakmarian, the Uspallatian fauna was replaced by the Bonetian fauna, which was also associated with the early Permian glaciation. The Bonetian fauna was extinguished as a consequence of global increases in temperatures, which finished the glacial regime and brought progressive aridity.

Acknowledgments

I am indebted to the National Geographic Society and the National Research Council of Argentina for supporting the field work for this study. My thanks to Dr. A. C. Taboada for assistance in the field, and to Pamela Díaz Saravia for typing the manuscript.

References

Amos, A. J., and Sabattini, N. 1969. Upper Palaeozoic faunal similitude between Argentina and Australia. In *IUGS Symposium on Gondwanan Stratigraphy, Buenos Aires, 1967, UNESCO. Earth Sciences* 2:235–48.

Bossi, G., and Andreis, R. R. 1983. Secuencias deltaicas y lacustres del Carbónico del centro-oeste argentino. In *Anales X Congreso Internacional de Estratigrafia y Geología del Carbonifero, Madrid*, pp. 285–309.

Calver, M. A. 1969. Westphalian of Britain. In *Compte Rendu 6e Congrès International Stratigrafie et Géologie Carbonifère, Sheffield, 1967*, vol. 1, pp. 233–54.

Campbell, K. S. W., and McKellar, R. G. 1969. Eastern Australian Carboniferous invertebrates: sequence and affinities. In *Stratigraphy and Palaeontology. Essays in Honor of D. Hill*, ed. K. S. W. Campbell, pp. 77–119. Canberra: Australian National University Press.

Coates, D. A. 1969. Stratigraphy and sedimentation of the Sauce Grande Formation, Sierra de la Ventana, southern Buenos Aires. In *IUGS Symposium on Gondwanan Stratigraphy, Buenos Aires, 1967, UNESCO. Earth Sciences* 2:799–814.

Creer, K. M. 1972. Paleomagnetism of Permocarboniferous rocks with special reference to South American forma-tions. *Anales Academia brasileira de Ciencias (Suplemento) (Rio de Janeiro)* 44:99–112.

Cuerda, A., and Furque, G. 1981. Depósitos carbónicos de la Precordillera de San Juan. Parte I. Comarca del cerro La Chilca (Rio Francia). *Revista de la Asociación Geológica Argentina (Buenos Aires)* 36:187–92.

Dickins, J. M. 1957. Lower Permian pelecypods and gastropods from the Carnarvon Basin, Western Australia. *Bulletin B.M.R., Geology & Geophysics (Canberra)* 41:1–75.

Dickins, J. M. 1985. Late Paleozoic glaciation. *Journal of Australian Geology and Geophysics* 9:163–9.

Du Toit, A. L. 1927. *A Geological Comparison of South America with South Africa.* Carnegie Institution of Washington.

Dutro, J. T., Jr., and Isaacson, P. E. 1991. Lower Carboniferous brachiopods from Sierra de Almeida, northern Chile. In *Brachiopods through Time,* ed. D. I. MacKinnon, D. E. Lee, and J. D. Campbell, pp. 327–32. Proceedings of the 2nd International Brachiopod Congress. Rotterdam: Balkema.

Frakes, L. A., Amos, A. J., and Crowell, J. C. 1969. Origin and stratigraphy of Late Paleozoic Diamictites in Argentina and Bolivia. In *IUGS Symposium on Gondwanan Stratigraphy, Buenos Aires, 1967, UNESCO. Earth Sciences* 2:821–43.

Frakes, L. A., and de Figueiredo, P. M. 1967. Glacial rocks of the Paraná Basin exposed along the Sorocaba–Itapetininga road. In *Problems in Brazilian Gondwana Geology,* ed. J. J. Bigarella, R. D. Becker, and I. D. Pinto, pp. 103–6. Curitiba:

Frenguelli, J. 1945. Moluscos continentales en el Paleozoico Superior y en el Triásico de la Argentina. *Notas Museo de La Plata, 10, Paleontologia* 83:181–204.

Furque, G. 1962. Perfil geológico de la Cordillera de Olivares, Iglesia, San Juan. *Anales de las I Jornadas Geológicas Argentinas (Buenos Aires)* 2:79–88.

González Bonorino, G. 1992. Carboniferous glaciation in Gondwana. Evidence for grounded marine ice and continental glaciation in south western Argentina. *Palaeogeography, Palaeoclimatology, Palaeoecology* 91:363–75.

González Bonorino, G., Vega, V., and Guerin, D. 1988. Ambientes de plataforma neritica dominada por tormentas en la sección glacigénica del Grupo Tepuel (Paleozoico superior), en las sierras de Tepuel y Tecka, Chubut noroccidental, Argentina. *Revista de la Asociación Geológica Argentina (Buenos Aires)* 43:239–52.

González, C. R. 1981a. El Paleozoico superior marino de la República Argentina. Bioestratigrafia y Paleoclimatología. *Ameghiniana (Buenos Aires)* 18:51–65.

González, C. R. 1981b. Pavimento glaciario en el Carbónico de la Precordillera. *Revista de la Asociación Geológica Argentina (Buenos Aires)* 36:262–6.

González, C. R. 1983. Evidences for the neopaleozoic glaciation in Argentina. In *Tills and Related Deposits,* ed. E. B. Evenson, C. Schlüchter, and J. Rabassa, pp. 271–7. Rotterdam: Balkema.

González, C. R. 1984. Rasgos paleogeográficos del Paleozoico superior de Patagonia. In *Actas IX Congreso Geológico Argentino, San Carlos de Bariloche,* vol. 1, pp. 191–205.

González, C. R. 1985a. Esquema bioestratigráfico del Paleozoico superior marino de la Cuenca Uspallata–Iglesia, República Argentina. *Acta Geológica Lilloana (Tucumán)* 16:231–44.

González, C. R. 1985b. El Paleozoico superior marino de la Patagonia extraandina. *Ameghiniana (Buenos Aires)* 21:125–42.

González, C. R. 1989. Relaciones bioestratifgráficas y paleogeográficas del Paleozoico superior marino en el Gondwana sudamericano. *Acta Geológica Lilloana* (*Tucumán*) 17:5–20.

González, C. R. 1990. Development of the Late Paleozoic glaciations of the South American Gondwana in western Argentina. *Palaeogeography, Palaeoclimatology, Palaeoecology* 79:275–87.

González, C. R. 1993. Late Paleozoic faunal succession in Argentina. *In Compte Rendu XII International Congress on Carboniferous and Permian Stratigraphy and Paleontology, Buenos Aires, 1991*, ed. S. Archangelsky.

González, C. R., and Bossi, G. E. 1986. Los depósitos carbónicos al oeste du Jagüel, La Rioja. In *Actas IV Congreso Argentino de Paleontologia y Bioestratigrafia, Mendoza*, vol. 1, pp. 231–6.

Gravenor, C. P., and Rocha-Campos, A. C. 1983. Patterns of late Paleozoic glacial sedimentation on the southeast side of the Paraná Basin, Brazil. *Palaeogeography, Palaeoclimatology, Palaeoecology* 43:1–39.

Harrington, H. J. 1955. The Permian *Eurydesma* fauna of eastern Argentina. *Journal of Paleontology* 29:112–28.

Harrington, H. J. 1962. Paleogeographic development of South America. *Bulletin of the American Association of Petroleum Geologists* 46:1773–814.

Helwig, J. 1972. Stratigraphy, sedimentation, paleogeography and paleoclimates of Carboniferous ("Gondwana") and Permian of Bolivia. *Bulletin of the American Association of Petroleum Geologists* 56:1008–33.

Keidel, J. 1921. Sobre la distribución de los depósitos glaciares del Pérmico conocidos en la Argentina y su significación para la estratigrafia de la serie de Gondwana y la paleogeografia del hemisferio austral. *Boletín Academia Nacional de Ciencias de Córdoba* 25:239–368.

Keidel, J. 1940. Paleozoic glaciation in South America. In *Proceedings of the 8th American Scientific Congress, Geological Sciences*, vol. 4, pp. 89–108. Washington.

Leanza, A. F. 1948. Braquiópodos y pelecipodos carboniferos en la provincia de La Rioja (Argentina). *Revista del Museo de La Plata (n.s.), Paleontologia* 3:237–64.

Limarino, C. O., and Sessarego, H. L. 1987. Algunos depósitos lacustres de las formaciones Ojo del Agua y De la Cuesta (Pérmico). Un ejemplo de sedimentación para regiones áridas o semiáridas. *Revista AGA (Buenos Aires)* 42:267–79.

Limarino, C. O., and Spalletti, L. A. 1984. Areniscas eólicas en unidades pérmicas del oeste y noroeste de la Rep.

Argentina. In *Abstracts, Annual Meeting, W.G. Project 211, San Carlos de Bariloche*, p. 49.

López Gamundi, O. R., and Amos, A. J. 1983. Criteria for identifying old glacigenic deposits. In *Tills and Related Deposits*, ed. E. B. Evenson et al., pp. 279–85. Rotterdam: Balkema.

Massabie, A. C., and Rossello, E. A. 1984. La discordancia pre-Formación Sauce Grande y su entorno estratigráfico, Sierras Australes de Buenos Aires. In *Actas IX Congreso Geológico Argentino, San Carlos de Bariloche*, vol. 1, pp. 337–52. Buenos Aires:

Roberts, J. 1981. Control mechanisms of Carboniferous brachiopod zones in eastern Australia. *Lethaia* 14:123–34.

Rocha-Campos, A. C. 1967. The Tubarao Group in the Brazilian portion of the Paraná Basin. In *Problems in Brazilian Gondwanan Geology*, ed. J. J. Bigarella, R. D. Becker, and I. D. Pinto, pp. 27–102. Curitiba.

Rocha-Campos, A. C. 1970 Upper Paleozoic bivalves and gastropods of Brazil and Argentina: a review. In *Proceedings and Papers, 2nd Gondwana Symposium, South Africa, 1969*, pp. 605–12. Pretoria: CSIR.

Schneider, S. H. 1993. Degrees of certainty. *Research & Exploration* (*Washington*) 9:173–90.

Taboada, A. C. 1987. Estratigrafia y contenido paleontológico de la Formación Agua del Jagüel, Pérmico inferior de la Precordillera mendocina. In *Actas 1as, Jornadas Geológicas de la Precordillera, San Juan, 1985*, vol. 1, pp. 181–6.

Taboada, A. C. 1991. Bioestratigrafia y facies del Paleozoico superior marino de la Subcuenca Calingasta-Uspallatta, provincias de San Juan y Mendoza. Inedit these, Universidad Nacional de Tucumán.

Valencio, D. A. 1973. El significado estratigráfico y paleogeográfico de los estudios paleomagnéticos de formaciones del Paleozoico superior y del Mesozoico inferior de América del Sur. In *Actas V Congreso Geológico Argentino, (Buenos Aires)*, vol. 5, pp. 71–9.

Valencio, D. A. 1981. Magnetic correlation of sequences of sediments and igneous rocks assigned to the Late Paleozoic and Triassic from northwestern Argentina. *Anales Academia brasileira de Ciências* 53:393–7.

Waterhouse, J. B., and Bonham-Carter, G. F. 1975. Global distribution and character of Permian biomes based on brachiopod assemblages. *Canadian Journal of Earth Sciences* 12:1085–146.

Index